International Association of Fire Chiefs

National Fire Protection Association

O9-BUD-196

Fire Officer
Principles and Practice

JONES AND BARTLETT PUBLISHERS

Sudbury, Massachusetts

BOSTON   TORONTO   LONDON   SINGAPORE

Jones and Bartlett Publishers

World Headquarters
40 Tall Pine Drive
Sudbury, MA 01776
978-443-5000
www.jbpub.com

Jones and Bartlett Publishers Canada

2406 Nikanna Road
Mississauga, ON L5C 2W6
Canada

Jones and Bartlett Publishers International

Barb House, Barb Mews
London W6 7PA
United Kingdom

National Fire Protection Association

1 Batterymarch Park
Quincy, MA 02169-7471
www.NFPA.org

**International Association
of Fire Chiefs**

4025 Fair Ridge Drive
Fairfax, VA 22033
www.IAFC.org

Jones and Bartlett's books and products are available through most bookstores and on-line booksellers. To contact Jones and Bartlett Publishers directly, call 800-832-0034, fax 978-443-8000, or visit our website www.jbpub.com.

Substantial discounts on bulk quantities of Jones and Bartlett's publications are available to corporations, professional associations, and other qualified organizations. For details and specific discount information, contact the special sales department at Jones and Bartlett via the above contact information or send an email to specialsales@jbpub.com.

Production Credits

Chief Executive Officer: Clayton E. Jones
Chief Operating Officer: Donald W. Jones, Jr.
*President, Jones and Bartlett Higher Education
  and Professional Publishing:* Robert W. Holland, Jr.
V.P. of Sales and Marketing: William J. Kane
V.P. of Production and Design: Anne Spencer
V.P. of Manufacturing and Inventory Control: Therese Bräuer
Publisher, Public Safety Group: Kimberly Brophy
Editor: Jennifer L. Reed

Associate Production Editor: Jenny McIsaac
Senior Photo Researcher: Kimberly Potvin
Director of Marketing: Alisha Weisman
Cover Design: Kristin Ohlin
Interior Design Studio Montage
Composition: Graphic World
Cover Photograph: © Keith Cullom
Text Printing and Binding: The Courier Company
Cover Printing: The Courier Company

Library of Congress Cataloging-in-Publication Data

Ward, Michael.
  Fire officer : principles and practice / Michael Ward.
    p. cm.
  Includes bibliographical references and index.
  ISBN 0-7637-2247-2 (pbk.)
  1. Fire departments—Management—Vocational guidance. 2. Fire extinction—Vocational guidance. 3. Fire chiefs—Certificatic I.
Title.
  TH9119.W37 2005
  363.37'068—dc22

                                            2004023562

Printed in the United States of America
09 08 07 06 05   10 9 8 7 6 5 4 3 2 1

Brief Contents

Contents

Resource Preview

The National Fire Protection Association (NFPA) and the International Association of Fire Chiefs (IAFC) are pleased to bring you *Fire Officer: Principles and Practice,* a modern integrated teaching and learning system for the Fire Officer I and II levels.

Fire officers need to know how to make the transition from fire fighter to leader. *Fire Officer: Principles and Practice* is designed to help fire fighters make a smooth transition to the fire officer level.

Fire Officer: Principles and Practice combines current content with dynamic features and interactive technology to better support instructors and help prepare tomorrow's fire officers.

Chapter Resources

Fire Officer: Principles and Practice thoroughly supports instructors and prepares tomorrow's fire officers for the job. The text covers NFPA 1021, *Standard for Fire Officer Professional Qualifications, 2003 Edition,* for the Fire Officer I and II levels, from fire officer communications to managing fire incidents. The text is the core of the teaching and learning system with features that will reinforce and expand on the essential information and make information retrieval a snap. These features include:

Chapter Objectives
NFPA 1021, Additional NFPA Standards, Knowledge Objectives, and Skills Objectives are listed at the beginning of each chapter.
- Portions of NFPA 1021 that are highlighted in red are applicable to the chapter.
- Additional NFPA Standards that apply to the chapter are listed for reference.
- Knowledge Objectives outline the most important topics covered in the chapter.
- Skills Objectives map skills provided in the chapter.

Navigation Toolbar
Found at the beginning of each chapter, the navigational toolbar will guide you through the technology resources and text features available for that chapter.

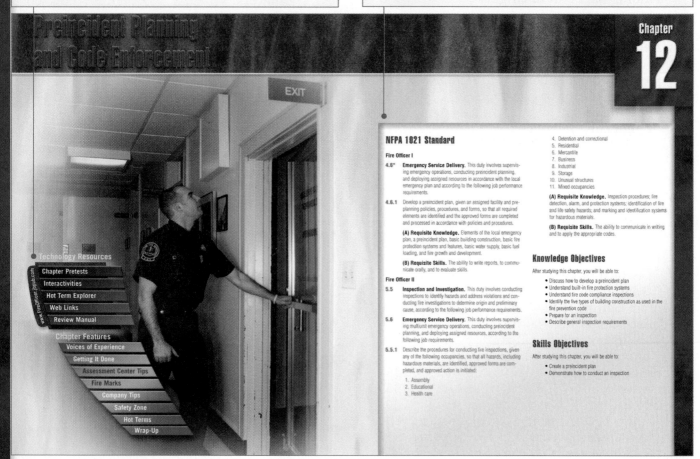

You are the Fire Officer

Each chapter opens with a case study that will stimulate classroom discussion, capture student's attention, and provide an overview of the chapter.

You Are the Fire Officer

A promotional announcement for fire lieutenant has been posted on the bulletin board in your fire station. The announcement lists 15 reference books and a description of the test that will be held in 6 months. There will be a 100-question multiple-choice test, followed by an in-basket exercise, an emergency incident simulation, and an interpersonal interaction.

At the fire station kitchen table, the "A" shifters are discussing the test. You are a 4-year fire fighter and have recently become eligible to take your first promotional examination. You ask the veterans for advice.

Captain Keene says you should plan to study for 6 hours every day, on duty and off, to get ready for the examination. On the last promotional list, the difference between the #1 ranked candidate and the #13 ranked candidate was only 0.215 of a point. You need to be prepared with the right answer for every possible question.

Lieutenant Cropper says it is a waste of time studying for the test. You can get more benefit from using those 42 hours a week to work overtime, go to college, or set up a business. Cropper tells you to go ahead and take the examination, but not to spend a lot of time preparing for it.

Fire Fighter Meade points out that some fire fighters started studying for this examination 18 months ago. You should study and take the examination for practice, but do not expect to score very high against the candidates who have been working so hard to prepare. Maybe next time you will be ready to compete seriously.

After consulting with your coworkers you feel confused and frustrated.

1. Is it worth the effort to study and prepare, if the chances of being successful are so low?
2. If you work hard and pass the test, are you really ready to be an officer?
3. There will be plenty of chances for promotions down the road. What is wrong with just concentrating on being a good fire fighter for now?

Fire Marks

Fire Marks offer history, lore, legends, and other interesting information.

Introduction to Preparing for Promotion

The purpose of this chapter is to provide a general description of the civil service promotional examination process that is used by most fire departments today. There are many variations of the testing procedures and promotional processes. It is vital that you understand the specific process that is used in your organization. The structure of promotional examinations in your organization may vary, depending on the rank. Many organizations also change elements or assessment tools from examination to examination.

The Origin of Promotional Examinations

Prior to the Civil War, most government jobs were awarded according to the patronage or **spoils system**; those in power could appoint people to public office based on a personal relationship or political affiliation, rather than on merit. The best jobs went to political supporters and often required a payment to the individual who had the power to make appointments. The system was ripe for corruption.

Fire Marks

When New York City created the paid Metropolitan Fire Department in 1865, Tammany Hall was the seat of political power. William March "Boss" Tweed was the chairman of the Democratic Party and the Grand Sachem (leader) of the Society of St. Tammany, which had been founded in 1789 as a club for patriotic and fraternal purposes. Between 1865 and 1871, Boss Tweed and Tammany Hall exploited the spoils process in New York City, swindling an estimated $75 to $200 million from the city. The public began to demand political reform.

Congress enacted the Pendleton Civil Service Reform Act in 1883 in response to the abuses at Tammany Hall in New York City and many other publicized abuses of the patronage system. The Pendleton Act established the civil service system in the federal government and provided a model for the civil service systems that were developed by many states and cities in the subsequent years. The spoils system was gradually replaced

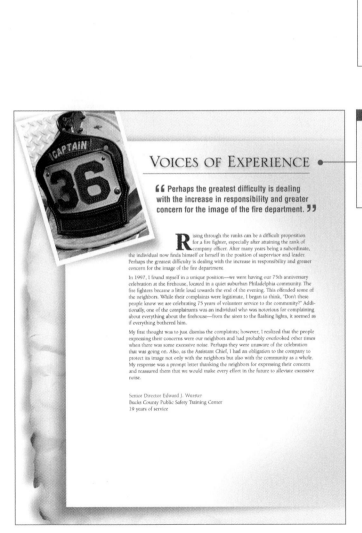

VOICES OF EXPERIENCE

❝ Perhaps the greatest difficulty is dealing with the increase in responsibility and greater concern for the image of the fire department. ❞

Rising through the ranks can be a difficult proposition for a fire fighter, especially after attaining the rank of company officer. After many years being a subordinate, the individual now finds himself or herself in the position of supervisor and leader. Perhaps the greatest difficulty is dealing with the increase in responsibility and greater concern for the image of the fire department.

In 1997, I found myself in a unique position—we were having our 75th anniversary celebration at the firehouse, located in a quiet suburban Philadelphia community. The fire fighters became a little loud towards the end of the evening. This offended some of the neighbors. While their complaints were legitimate, I began to think, "Don't these people know we are celebrating 75 years of volunteer service to the community?" Additionally, one of the complainants was an individual who was notorious for complaining about everything about the firehouse—from the siren to the flashing lights, it seemed as if everything bothered him.

My first thought was to just dismiss the complaints; however, I realized that the people expressing their concerns were our neighbors and had probably overlooked other times when there was some excessive noise. Perhaps they were unaware of the celebration that was going on. Also, as the Assistant Chief, I had an obligation to the company to protect its image not only with the neighbors but also with the community as a whole. My response was a prompt letter thanking the neighbors for expressing their concern and reassured them that we would make every effort in the future to alleviate excessive noise.

Senior Director Edward J. Wurster
Bucks County Public Safety Training Center
19 years of service

Voices of Experience

In the Voices of Experience essays, veteran fire officers share accounts of memorable incidents while offering advice and encouragement. These essays highlight what it is truly like to be a fire officer.

Assessment Center Tips

Assessment Center Tips offer hints on how to successfully complete Assessment Center activities in the promotional exam process.

Getting It Done

Getting It Done provides suggestions and tips for applying theories, knowledge, and skills to complete a task.

Hot Terms

Hot Terms are easily identifiable within the chapter and define key terms the student must know. A comprehensive glossary of Hot Terms is in the end-of-chapter Wrap-Up. The Hot Term Explorer on **www.FireOfficer.jbpub.com** provides interactivities for students.

Safety Zone

Safety Zone reinforces safety concerns for fire officers and their crew.

Sample page 54:

54   FIRE OFFICER: PRINCIPLES AND PRACTICE

where it is difficult to maintain confidentiality. The fire officer must insist on and enforce confidentiality during the investigation.

- **Write it all down.** Take notes during all interviews. Before the interview is over, go back through your notes with the interviewee to ensure accuracy. Keep a journal of the investigation. Write down the steps you have taken to get at the truth, including dates and places of interviews. Keep a list of all documents that are reviewed. Document any action taken against the accused or the reasons for deciding not to take action. Anticipate that all of this written record will be used in any subsequent civil service or court actions.
- **Cooperate with government agencies.** If the fire fighter files a complaint with another government agency (either the federal EEOC or an equivalent state agency), that agency may investigate. Notify your supervisor as soon as you receive a call or visit from an EEOC investigator. You will probably be asked to provide certain documents, to give your side of the story, and to explain any efforts you made to deal with the complaint yourself. Be cautious, but cooperative.

Regardless of where the complaint is filed, the fire officer is required to take corrective action if the investigation confirms that the complaint has merit. If the department concludes that some form of discrimination or harassment occurred, formal corrective action could include mandatory training, work location transfer, or demotion. Termination may be proposed for the more egregious kinds of discrimination and harassment, such as threats, stalking, or repeated and unwanted physical contact.

The Fire Station as a "Business Work Location"

The fire officer needs to consider the fire station or other fire department facility as a business work location. This is a drastic change from the concept of a fire station as a home away from home for a group of fire fighters, but it is necessary to ensure that the fire station maintains a professional work environment. In the eyes of the law and probably in the opinion of administrators and elected officials, the same rules of behavior apply to an office in city hall at 2 P.M. or a fire station at 2 A.M.

The fire officer can help maintain an appropriate work environment by encouraging and enforcing acceptable behavior whenever fire fighters are on duty. The fire officer accomplishes this by:

- Educating the employees on the workplace rules and regulations that define expected behavior. Start with the local government's "Code of Conduct" or other documents that outline the chief administrative officer's expectations of all municipal employees. This can usually be found in the municipality's mission statement, core values, or personnel regulations.
- Promote the use of "on duty speech." The goal is not to change the thoughts or feelings of individual fire fighters, but to establish a workplace environment where certain behaviors and words are not used. Fire fighters can think what they want. However, while they are on duty, in the fire station, or in uniform, they cannot use certain words or phrases or act out certain behaviors.
- Be the designated adult. This requires the fire officer to model appropriate behavior as well as encourage and enforce the same behavior by the fire fighters. The fire officer must identify and correct unacceptable workplace behavior whenever it is observed.

Assessment Center Tips

Know Your Organization's Procedure

Even if the normal practice for your department is to have a specialized or designated person to conduct an investigation of a harassment or hostile workplace complaint, the promotional candidate must know both local and federal procedures that must be followed when an employee files a complaint. Harassment and hostile workplace complaints require a specific and detailed response by the first-line supervisor.

An assessment center scenario may involve a fire fighter coming to the candidate for advice. During the interview process, the fire fighter reveals that he or she may be a victim of harassment or hostile workplace. The candidate would be expected to explain the options available to the fire fighter in filing either a federal or a local complaint.

Getting It Done

The Fire Company Defines Its Nonhostile Environment

A fire officer who unilaterally imposes new and severe requirements that affect the language fire fighters use in the workplace would encounter much resistance. One method of fostering the concept of "on duty speech" or appropriate workplace behavior is to have the fire fighters develop their own set of rules or behaviors. Provide the fire fighters with the source documents (federal and municipal codes and regulations), and ask them to consider what words or behaviors are inappropriate while at work.

Sample page 108:

108   FIRE OFFICER: PRINCIPLES AND PRACTICE

occurred. Future fire fighters may well look back at the practice of underutilizing thermal imaging devices in the same manner. The fire officer must provide frequent training to integrate the thermal imager into interior firefighting operations.

Situation Awareness

One of the primary responsibilities of a fire officer is to maintain a continual connection between the functions being performed by the company and the overall situation. This is one of the most important reasons for establishing and maintaining an effective incident command structure at every incident. The fire officer maintains situation awareness by staying oriented, making observations, providing and receiving regular updates within the incident command system, listening to fireground radio communications, and continually assessing the risk/benefit model.

It is very easy to become distracted or preoccupied with a particular task or function and to lose track of a larger situation that is occurring and changing simultaneously. This can be a very serious problem at a dynamic emergency incident, particularly for an officer who is supervising a crew that is performing a complex task deep inside a smoke-filled building. Conditions can change very quickly, and crews might have no way of observing what is going on around them. A fire officer must always maintain the link between the immediate situation and the overall incident situation.

Risk/Benefit Analysis

Risk/benefit analysis is based on a hazard and situation assessment that weighs the risks involved in a particular course of action against the benefits to be gained for taking those risks. Fire fighters sometimes use the simple axiom, "Risk a little to save a little. Risk a lot to save a lot." However, this is a simplified description of the risk/benefit process. Risk/benefit analysis must always be approached in a structured and measured manner.

Life safety is a paramount goal, including the lives of fire fighters. If the situation requires placing fire fighters in extreme danger with little chance of success, the operation should not be undertaken. There is no justification for risking the lives of fire fighters to save property that is already lost or has no real value. Conducting an interior attack on a fire in an unoccupied abandoned building needlessly places fire fighters at risk. Similarly, entering a burning building to search for missing occupants who could not possibly still be alive cannot be justified.

Every interior fire attack operation and many other types of situations and circumstances expose fire fighters to a set of unavoidable inherent risks. These risks are managed within a measured and controlled system that is based on training, coordination, and the use of protective clothing and equipment. The nature of the mission requires fire fighters to be able to work safely in situations that are inherently dangerous. Fire officers are responsible for keeping this system in balance.

The only situation that truly justifies exposing fire fighters to a high level of risk is one where there is a realistic chance that a life can be saved. Even in this circumstance, fire fighters must use all of the resources at their disposal to limit the risks. All of the fire department's training, equipment, and systems are designed to enable fire fighters to be effective in the face of such challenging and dangerous situations.

The fire officer starts the risk/benefit analysis by preparing a preincident plan, which is a written document that provides information that can be used by responding personnel to determine the appropriate actions in the event of an emergency at a specific facility. Building construction, occupancy, building use, building contents, and condition of the structure are all factors that should be used to develop the risk/benefit analysis.

When operating at an emergency incident, the fire officer reviews the preincident plan and makes observations about current conditions. These two inputs are combined to produce an incident action plan, which is developed by the incident commander and incorporates the overall incident strategy, tactics, risk management evaluation, and organization structure for that particular situation. Incident action plans are updated throughout the incident, based on updates from operating crews and observations by the incident management team.

Incident Safety Officer

An incident safety officer is a designated individual at the emergency scene who performs a set of duties and responsibilities that are specified in NFPA 1521, Standard for Fire Department Safety Officer. The incident safety officer functions as a member of the incident command staff, reporting directly to the incident commander. The incident commander is personally responsible for performing the functions of the incident safety officer if this assignment has not been assigned or delegated to another individual.

Many fire departments assign a designated officer to respond to emergency scenes to fill this position. In the absence of a predesignated safety officer, this position may be assigned to a qualified individual by the incident commander. NFPA 1521 also specifies that the fire department must have a stan-

Safety Zone

The NFFF summarizes situational awareness with the following items that support the interior firefighting plan:

- Work as a team
- Stay together
- Stay oriented
- Manage your air supply
- Get off the apparatus with tools and a thermal imager for EVERY interior operating team
- Provide a radio for EVERY member
- Provide regular updates
- Constantly assess the risk/benefit model

CHAPTER 7 Training and Coaching 127

Company Tips

Demonstrating Skills

In the late 1990s, the American Heart Association (AHA) conducted research to determine the most effective way to teach cardiopulmonary resuscitation (CPR). Most fire fighters have received training. The program has to prepare the individual to perform CPR correctly under high-stress conditions.

The old AHA psychomotor skill training called for the instructor to deliver an initial perfect demonstration in real time, followed by a slower step-by-step version. The important issue was that the first demonstration must be perfect. When the AHA observed hundreds of CPR classes, it discovered very little consistency or accuracy in the demonstrations performed by lead instructors. As a result, the AHA changed the psychomotor skill training practice in 2000 and now provides a videotape that includes a perfect performance of every basic life support evolution. The lead instructor is no longer required to provide the perfect demonstration.

Figure 7-5 Fire officers should deliver a perfect demonstration of how skills are performed.

a training session to address this behavior. If, after the training session, Fire Fighter Jones always comes to a complete stop at red traffic lights and stop signs, there has been an observable change of behavior. If the behavior does not change, the training objective has not been accomplished.

Ensure Proficiency of Existing Skill Sets

The fire officer is responsible for ensuring that every fire fighter is proficient in performing a series of skill sets. This is both an individual and a company-level requirement. Some departments have a standard set of evolutions that have to be performed proficiently, either by a company or by an individual fire fighter, in which both skill and time requirements must be met. The fire officer must invest some of the available training time practicing and reviewing these standard evolutions.

It is important that some of these practice sessions be performed while the fire fighters are wearing full personal protective clothing and simulating realistic fireground situations. This may require construction of some training props, such as an assembly that allows the fire fighters to practice opening a roof using a power saw. The fire officer should always be on the alert for opportunities to acquire abandoned structures where realistic fireground skills can be practiced.

Provide New or Revised Skill Sets

On occasion, the fire officer is required to provide the initial training for a new or revised skill set. This is often related to a new device that has been acquired, such as a new type of

breathing apparatus or a thermal image camera. The fire fighters need to become familiar with the new equipment and proficient in its use, especially during emergency procedures. In the case of a thermal imaging camera, the fire fighters have to learn how it works and how to maintain it, as well as how to incorporate its use into fireground operations. This type of training could also be required in order to introduce a change in standard operating procedures. Sometimes procedures are changed or additional training is mandated in response to a near miss that caused a problem to be recognized.

Teaching new skills takes more time than maintaining proficiency of existing skills. The fire officer should obtain as much information as possible about the device or procedure, especially identifying any fire fighter safety issues. The emphasis of fire station–based training should be on the safe and effective use of the device or procedure. The fire officer should plan to spend a couple of training periods developing competency and showing how it relates to existing procedures. Avoid spending more than 15–20 minutes on any lecture or video presentation; the adult learner drifts away if the presentation takes more time.

Ensure Competence and Confidence

The fire officer works as a <u>coach</u> when providing training to an individual or a team. After team members have learned the basic skills and can appropriately demonstrate them, the coach has to work with them to build competence and confidence. The coach has to provide the guidance that advances them from being capable of performing the basic required skills to being able to perform them effectively, efficiently, and consistently.

Company Tips
Company Tips offer practical advice and information on teamwork and communication.

Wrap-Up
End-of-chapter activities reinforce important concepts and improve students' comprehension. Additional instructor support and answers to all activities are contained in the Instructor's Resource Manual.

- Ready for Review summarizes chapter content to help students prepare for exams.
- Hot Terms provide key terms and definitions from the chapter.
- Fire Officer in Action promotes critical thinking through the use of case studies and provides you with discussion points for your classroom presentation.
- Technology Resources guide exploration of topics online at **www.FireOfficer.jbpub.com** where additional activities reinforce and expand on important information from the chapter.

Wrap-Up

Ready for Review

- NFPA 1620 provides a six-step method of developing a preincident plan.
  Step 1: Evaluate physical elements and site considerations.
  Step 2: Evaluate occupant considerations.
  Step 3: Evaluate fire protection systems and water supply.
  Step 4: Evaluate special hazards.
  Step 5: Evaluate emergency operation considerations.
  Step 6: Evaluate special or unusual characteristics of common occupancy.
- The state or commonwealth determines the range and scope of local community fire code enforcement. The local community adopts an ordinance or regulation to establish a fire code.
- Automatic sprinkler systems come in wet pipe and dry pipe versions.
- There are three classes of standpipes, based on the size of the hose coupling outlets and the intended use of the standpipe.
- There are five building construction classifications.
- Every inspection is completed when the fire officer conducts an exit interview with the owner or representative and leaves a copy of a fire inspection report.

Hot Terms

Adoption by reference Method of code adoption in which the specific edition of a model code is referred to within the adopting ordinance or regulation.

Adoption by transcription Method of code adoption in which the entire text of the code is published within the adopting ordinance or regulation.

Authority having jurisdiction An organization, office, or individual responsible for enforcing the requirements of a code or standard, or for approving equipment, materials, an installation, or a procedure.

Automatic sprinkler system A system of pipes with water under pressure that allows water to be discharged immediately when a sprinkler head operates.

Catastrophic theory of reform When fire prevention codes or firefighting procedures are changed in reaction to a fire disaster.

Construction type The combination of materials used in the construction of a building or structure, based on the varying degrees of fire resistance and combustibility.

Fire prevention division or hazardous use permit A local government permit renewed annually after the fire prevention division performs a code compliance inspection. A permit is required if the process, storage, or occupancy activity creates a life-safety hazard. Restaurants with more than 50 seats, flammable liquid storage, and printing shops that use ammonia are three examples of occupancies that may require a permit.

Masonry wall May consist of brick, stone, concrete block, terra cotta, tile, adobe, precast, or cast-in-place concrete.

Mini/max codes Codes developed and adopted at the state level for either mandatory or optional enforcement by local governments; these codes cannot be amended by local governments.

Model codes Codes generally developed through the consensus process with the use of technical committees developed by a code-making organization.

Occupancy type The purpose for which a building or a portion thereof is used or intended to be used.

Ordinance A law of an authorized subdivision of a state, such as a city, county, or town.

Ongoing compliance inspection Inspection of an existing occupancy to observe the housekeeping and confirm that the built-in fire protection features, such as fire exit doors and sprinkler systems, are in working order.

Preincident plan A written document resulting from the gathering of general and detailed data to be used by responding personnel for determining the resources and actions necessary to mitigate anticipated emergencies at a specific facility.

Regulations Governmental orders written by a governmental agency in accordance with the statute or ordinance authorizing the agency to create the regulation. Regulations are not laws but have the force of law.

Standpipe system An arrangement of piping, valves, hose connections, and allied equipment installed in a building or structure, with the hose connections located in such a manner that water can be discharged in streams or spray patterns through attached hose and nozzles, for the purpose of extinguishing a fire, thereby protecting a building or structure and its contents in addition to protecting the occupants. This is accomplished by means of connections to water supply systems or by means of pumps, tanks, and other equipment necessary to provide an adequate supply of water to the hose connections.

Use group Found in the building code classification system whereby buildings and structures are grouped together by their use and by the characteristics of their occupants.

Fire Officer in Action

241

Your administration has just announced that they will be instituting a new company-level fire inspection program in a month. There is a lot of dissent in the room from some of the other officers, but you look forward to the opportunity to get out and do this level of inspection. The fire marshal's office has been too overwhelmed with the amount of new construction and cannot keep up with the inspection schedule, prompting the new program. In the next month, the fire marshal will be conducting training with each of the shifts to ensure that the program is carried out properly.

1. What is type II construction type?
   - **A.** Fire resistive
   - **B.** Limited combustible
   - **C.** Heavy timber
   - **D.** Noncombustible

2. What is occupancy type?
   - **A.** Who lives there
   - **B.** What a building is used or intended to be used for
   - **C.** What processes occur at a building
   - **D.** Development

3. What is one of the major use classifications?
   - **A.** Retail stores
   - **B.** Hospitals
   - **C.** Residential
   - **D.** Arena

4. What tools should you have with you to complete the inspection?
   - **A.** Hose
   - **B.** Digital camera with flash attachment
   - **C.** NFPA 1901
   - **D.** PPE

- Chapter Pretests
- Interactivities
- Hot Term Explorer
- Web Links
- Review Manual

www.FireOfficer.jbpub.com

Resources

Instructor Resources

A complete teaching and learning system developed by educators with an intimate knowledge of the obstacles you face each day supports *Fire Officer: Principles and Practice*.

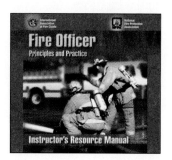

Instructor's ToolKit CD-ROM

ISBN: 0-7637-3202-8

Preparing for class is easy with the resources found on this CD-ROM, including:

PowerPoint® Presentations, providing you with a powerful way to make presentations that are both educational and engaging to your students. The slides can be edited and modified to meet your needs.

Lecture Outlines, providing you with complete, ready-to-use lesson plans that outline all of the topics covered in the text. The lesson plans can be modified and customized to fit your course.

Electronic Test Bank, containing multiple-choice and scenario-based questions, allows you to originate tailor-made classroom tests and quizzes quickly and easily by selecting, editing, organizing, and printing a test along with an answer key, that includes references to the text.

Image and Table Bank, providing you with many of the images and tables found in the text. You can use them to incorporate more images into the Power-Point® Presentations, make handouts, or enlarge a specific image for discussion.

The resources found on the Instructor's ToolKit CD-ROM have been formatted so that you can seamlessly integrate them into the most popular course administration tools. Please contact Jones and Bartlett technical support at any time with questions.

The supplements provide practical, hands-on, time-saving tools like PowerPoint® Presentations and customizable test banks to better support you and your students. This system includes the text plus the following resources:

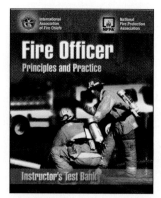

Instructor's Resource Manual

ISBN: 0-7637-3203-6

The Instructor's Resource Manual is your guide to the entire teaching and learning system. This indispensable manual contains:

- Detailed lesson plans that are keyed to the PowerPoint® Presentations with sample lectures, lesson quizzes, and teaching strategies.
- Teaching tips and ideas to enhance your classroom presentation.
- Answers to all end-of-chapter student questions found in the text.

Instructor's Test Bank (print)

ISBN: 0-7637-3205-2

This is the printed version of the Test Bank available on the Instructor's ToolKit CD-ROM. It contains multiple-choice and scenario-based questions with page references to the text.

Student Resources

To help students retain the most important information and to assist them in preparing for exams, we have developed the following resources:

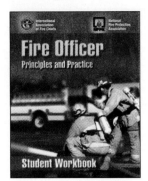

Student Workbook

ISBN: 0-7637-3207-9

This resource is designed to encourage critical thinking and aid comprehension of the course material through:

- Case studies and corresponding questions
- In-basket activities
- Matching, fill-in-the-blank, short answer, and multiple-choice activities

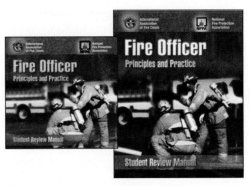

Student Review Manual

Print – ISBN: 0-7637-3208-7
CD-ROM – ISBN: 0-7637-3209-5
Online – ISBN: 0-7637-3210-9

This Review Manual has been designed to prepare students for exams by including the same type of questions that they are likely to see on classroom and certification examinations. The manual contains multiple-choice question exams with an answer key and page references. It is available in print, on CD-ROM, and online.

Technology Resources

A key component to the teaching and learning system are interactivities and simulations to help students become great fire officers.

www.FireOfficer.jbpub.com

Make full use of today's teaching and learning technology with **www.FireOfficer.jbpub.com.** This site has been specifically designed to compliment *Fire Officer: Principles and Practice* and is regularly updated. Some of the resources available include:

Chapter Pretests to prepare students for training. Each chapter has a pretest and provides instant results, feedback on incorrect answers, and page references for further study.

Hot Term Explorer is your virtual dictionary. Here, students can review key terms, test their knowledge of key terms through quizzes and flashcards, and complete crossword puzzles.

Web Links provide up-to-date information for instructors and students including current trends in the fire service community.

Review Manual allows students to prepare for classroom and certification examinations providing instant results, feedback, and page references.

Acknowledgments

Editorial Board

David Hall
Springfield Missouri Fire
    Department
Rogersville, Missouri

Brian Johnson
International Association of
    Fire Chiefs
Fairfax, Virginia

Pam Powell
National Fire Protection
    Association
Quincy, Massachusetts

J. Gordon Routley
Fire Protection Engineer
Montreal, Quebec
Canada

Gary Tokle
National Fire Protection
    Association
Quincy, Massachusetts

Michael Ward
Northern Virginia
    Community College
Annadale, Virginia

Charles Werner
Charlottesville Fire
    Department
Charlottesville, Virginia

Contributors

Mike DeVirgillio
City of Tempe Fire
    Department
Tempe, Arizona

Jerrod Vanlandingham
Longmont Fire Department
Longmont, Colorado

Gerard P. Forte
City of Palm Coast Fire
    Department
City of Palm Coast, Florida

Gregg Lord
Cherokee County Fire-
    Emergency Services
Cherokee, Georgia

Bart Ridings
Air National Guard Fire
    Department
182nd Airlift Wing
Peoria, Illinois

David Hall
Springfield Missouri Fire
    Department
Rogersville, Missouri

John House
Springfield Fire Department
Springfield, Missouri

William Shouldis
Philadelphia Fire
    Department
Philadelphia, Pennsylvania

Edward J. Wurster
Bucks County Public Safety
    Training Center
Doylestown, Pennsylvania

Ken Farmer
Fuquay Varina Fire
    Department
Fuquay Varina, North
    Carolina

Matthew B. Thorpe
City of King Fire Department
King, North Carolina

Timothy A. Hawthorne
Cranston Fire Department
Cranston, Rhode Island

Cecil V. "Buddy" Martinette Jr.
Lynchburg Fire and EMS
    Department
Lynchburg, Virginia

Reviewers

Julie Coffman
Alabama Fire College
Tuscaloosa, Alabama

Paul Sanchez
Alaska Fire Service
Fort Richardson, Alaska

Cliff W. Hadsell
Rio Hondo Community
    College
Whittier, California

Bob O'Brien
City of Fremont Fire
    Department
Fremont, California

Todd R. Gilgren
Arvada Fire Protection
    Agency
Arvada, Colorado

Jerrod Vanlandingham
Longmont Fire Department
Longmont, Colorado

Dan Nolan
Hartford Fire Department
Hartford, Connecticut

Robert Foraker
Delaware Technical and
    Community College
Stanton, Delaware

David Casey
Division of State Fire
    Marshal
Bureau of Standards
    and Training
Ocala, Florida

Julie Downey
Town of Davie Fire Rescue
Davie, Florida

Randy E. Novak
Iowa Fire Service Training
    Bureau
Ames, Iowa

Ron Hopkins
Eastern Kentucky University
Richmond, Kentucky

Sandy Davis
Shreveport Fire Department
Shreveport, Louisiana

Dean Melanson
Hyannis Fire Department
Hyannis, Massachusetts

Guy Hubbard
City of Perry Fire
    Department
Perry, Michigan

Reviewers, continued

Roger J. Land
Lake Superior State
    University
Sault Ste. Marie, Michigan

Todd Seitz
Anoka Technical College
Anoka, Minnesota

Attila J. Hertelendy
University of Mississippi
    Medical Center
Brandon, Mississippi

Michael Arnhart
High Ridge Fire Protection
    District
High Ridge, Missouri

David Hall
Springfield Missouri Fire
    Department
Rogersville, Missouri

John C. Huff
City of Lincoln Fire and
    Rescue
Lincoln, Nebraska

Bruce Evans
Community College of
    Southern Nevada
Henderson, Nevada

David Yows
San Juan County Fire
    Department
Aztec, New Mexico

Mark Gregory
Suffolk County Fire
    Academy
New York, New York

Gerard F. Sikorski
Nassau County Fire, Police,
    and EMS Academy
East Meadow, New York

Ken Farmer
Fuquay Varina Fire
    Department
Fuquay Varina, North
    Carolina

Matthew B. Thorpe
City of King Fire Department
King, North Carolina

Leon Schlafmann
Fargo Fire Department
Fargo, North Dakota

Glen Jirka
Miami Township Fire
    Department
Miamisburg, Ohio

William Shouldis
Philadelphia Fire Academy
Philadelphia, Pennsylvania

Edward Wurster
Bucks County Community
    College
Doylestown, Pennsylvania

Robert Christofaro
Navy Fire Department
Newport, Rhode Island

Timothy A. Hawthorne
Cranston Fire Department
Cranston, Rhode Island

J. Curtis Varone
Providence Fire Department
Providence, Rhode Island

Byrn Crandell
Louis F. Garland Department
    of Defense Fire Academy
Goodfellow Air Force Base
San Angelo, Texas

James M. Gaston
Fire Training Online
Texas Fire Chiefs Association
Austin, Texas

Neil Fulton
Norwich Fire Department
Norwich, Vermont

Pat Cornell
Albemarle Fire and Rescue
    (Retired)
Charlottesville, Virginia

Thomas L. Harper
Jefferson College of Health
    Sciences
Roanoke, Virginia

John Lemley
West Virginia University
Morgantown, West Virginia

Tim Olson
Madison Area Technical
    College
Madison Fire Department
Madison, Wisconsin

Phillip Oakes
Wyoming State Fire
    Marshal's Office
Cheyenne, Wyoming

Special thanks to the Portland, Maine Fire Department for hosting the January photoshoot. Thank you to Kimberly Potvin for taking the photographs.

Introduction to Fire Officer

Technology Resources

www.FireOfficer.jbpub.com

- Chapter Pretests
- Interactivities
- Hot Term Explorer
- Web Links
- Review Manual

Chapter Features

- Voices of Experience
- Getting It Done
- Assessment Center Tips
- Fire Marks
- Company Tips
- Safety Zone
- Hot Terms
- Wrap-Up

NFPA 1021 Standard

Fire Officer I

4.1* **General.** For certification at Fire Officer I, the candidate shall meet the requirements of Fire Fighter II as defined in NFPA 1001, Fire Instructor I as defined in NFPA 1041, and the job performance requirements defined in Sections 4.2 through 4.7 of this standard.

4.1.1 **General Prerequisite Knowledge.** The organizational structure of the department; geographical configuration and characteristics of response districts; departmental operating procedures for administration, emergency operations, incident management systems, and safety; departmental budget process; information management and record keeping; the fire prevention and building safety codes and ordinances applicable to the jurisdiction; current trends, technologies, and socioeconomic and political factors that affect the fire service; cultural diversity; methods used by supervisors to obtain cooperation within a group of subordinates; the rights of management and members; agreements in force between the organization and members; generally accepted ethical practices, including a professional code of ethics; and policies and procedures regarding the operation of the department as they involve supervisors and members.

Fire Officer II

5.1 **General.** For certification at Level II, the Fire Officer I shall meet the requirements of Fire Instructor I as defined in NFPA 1041 and the job performance requirements defined in Sections 5.2 through 5.7 of this standard.

5.1.1 **General Prerequisite Knowledge.** The organization of local government; the enabling and regulatory legislation and the law-making process at the local, state/provincial, and federal levels; and the functions of other bureaus, divisions, agencies, and organizations and their roles and responsibilities that relate to the fire service.

Additional NFPA Standards

NFPA 1001, *Standard for Fire Fighter Professional Qualifications*

NFPA 1041, *Standard for Fire Service Instructor Professional Qualifications (IH)*

Knowledge Objectives

After studying this chapter, you will be able to:

- Describe the roles and responsibilities of the Fire Officer I.
- Describe the roles and responsibilities of the Fire Officer II.
- Describe the fire service in the United States.
- Describe fire department organization.
- Describe the functions of management.
- Describe rules and regulations, policies, and standard operating procedures.
- Describe working with other organizations.
- Describe the opportunities for the 21st century fire officer.

Skills Objectives

There are no skills objectives for this chapter.

You Are the Fire Officer

Congratulations! Today is your first day as a fire officer assigned to Ladder Company 11. Sitting in the fire station parking lot, your stomach has butterflies. You studied long and hard for your promotion to lieutenant. Now that the testing is over, the real work begins. There is a little trepidation as you look at that shiny aerial platform. Your head is a jumble with concepts, facts, figures, and "what if" scenarios. You settle down as you walk into the fire station.

1. How does your role as the lieutenant assigned to the ladder company fit with the station commander and the battalion chief?
2. How do all of these organizational management concepts work in real life?
3. What are your responsibilities as the fire officer?

Introduction

This book provides information to meet the standards of the National Fire Protection Association (NFPA) 1021, *Standard for Fire Officer Professional Qualifications*, at the Fire Officer I and Fire Officer II levels. The professional qualifications standards for fire officers are documented in NFPA 1021.

The NFPA 1021 standard defines four levels of fire officer. The technical committee that developed the current edition of NFPA 1021 performed a task analysis to validate the existence of four distinct fire officer levels and the specific requirements that should apply at each level.

The Fire Officer I level is the first step in a progressive sequence and is generally associated with an officer supervising a single fire company or apparatus. A Fire Officer I could also be assigned to supervise a small administrative or technical group. The next step, Fire Officer II, generally refers to the senior non–chief officer level in a larger fire department. An officer at this level could be the overall supervisor of a multiple-unit fire station. A Fire Officer II could also be in charge of a larger group performing a specialized service or a significant administrative section within the fire department.

Fire Officer III and IV generally refer to chief officer positions. An individual who is qualified at the Fire Officer III level might be qualified to work as a battalion or district chief in a large department and possibly as a deputy or assistant chief in a smaller organization. Fire Officer IVs tend to be fire chiefs or hold senior positions in charge of a major component of the fire department.

A successful fire officer uses a diversity of skills and knowledge. An officer is responsible for being a leader and supervisor to a crew of fire fighters, managing a budget for the station, understanding the response district, knowing departmental operational procedures, and being able to manage an incident. The officer must also understand fire prevention methods, fire and building codes and applicable ordinances (laws enacted by a local government or municipality, e.g., a city or town), and the department's records management system. At each higher level, there are increasing requirements for knowledge, technical skills, and management skills.

Fire Officer I

The Fire Officer I classification is generally bestowed upon an individual who supervises a single fire suppression unit or a small administrative group within a fire department. At this level, the emphasis is placed on accomplishing the department's goals and objectives by working through subordinates to achieve desired results. It is essential for the Fire Officer I to be able to prioritize multiple demands on the time of the company or work group members and to delegate tasks to subordinates. These demands may be related to emergency operations, nonemergency tasks, or administrative functions.

The Fire Officer I performs many administrative duties and supervisory functions that are related to a small group of fire department members. Typical administrative duties include record keeping, managing projects, preparing budget requests, initiating and completing station maintenance requisitions, and conducting preliminary accident investigations. Supervisory duties include making work assignments, conducting performance appraisals, and ensuring that health and safety procedures are followed. Nonemergency duties could include developing preincident plans, providing company-level training, delivering public education programs, and responding to community inquiries.

Emergency duties include the ability to supervise a group of fire fighters performing company-level tasks, functioning as the initial arriving officer at an emergency scene, performing a size-up, establishing the incident management system, developing and implementing an incident action plan, deploying resources, and maintaining personnel accountability. Once the emergency incident has been mitigated, the Fire Officer I is expected to conduct a preliminary investigation to determine the cause, secure the scene to preserve evidence, and conduct a postincident analysis. Fire Officer I candidates are also required to meet all of the requirements of Fire Fighter II as defined in NFPA 1001, *Standard for Fire Fighter Professional Qualifications* and Fire Instructor I as defined in NFPA 1041, *Standard for Fire Service Instructor Professional Qualifications (IH)*.

Fire Officer II

The requirements for Fire Officer II begin with meeting all of the requirements of Fire Officer I as defined in NFPA 1021. Like with Fire Officer I, the duties of Fire Officer II can be divided into administrative, nonemergency, and emergency activities.

Administrative duties include evaluating a subordinate's job performance, correcting unacceptable performance, and completing formal performance appraisals on each member. Other duties include developing a project or divisional budget, including the related activities of purchasing, soliciting, and rewarding bids, and preparing news releases and other reports to supervisors.

Nonemergency duties include conducting inspections to identify hazards and address fire code violations; reviewing accident, injury, and exposure reports to identify unsafe work environments or behaviors; and taking approved action to prevent reoccurrence of an accident, injury, or exposure. Other duties could include developing a preincident plan for a large complex or property; developing policies and procedures appropriate for this level of supervision; analyzing reports and data to identify problems, trends, or conditions that require corrective action; and then developing and implementing the required actions.

Emergency duties include supervising a multiunit emergency operation using an **incident management system (IMS)** and developing an operational plan to safely deploy resources to mitigate the incident. IMS is a system that defines the roles and responsibilities to be assumed by personnel and the operating procedures to be used in the management and direction of emergency operations. The Fire Officer II is also expected to determine the area of origin and preliminary cause of a fire and to develop and perform a postincident analysis of a multicompany operation.

Roles and Responsibilities of Fire Officer

The roles and responsibilities of the fire officer differ greatly from those of a fire fighter. Understanding the new roles is essential for the new fire officer to succeed.

Roles and Responsibilities for Fire Officer I

- Supervises and directs the activities of a single unit
- Instructs members of the company regarding operating procedures, including duty assignments and giving special instructions when fighting fires

- Responds to alarms for fires, vehicle extrications, hazardous materials incidents, emergency medical incidents, and other emergencies as required
- Assumes command of emergency scenes, per the Incident Command System, analyzes situations, and determines proper procedures until being relieved by a higher ranking officer
- Administers emergency medical first aid, cardiopulmonary resuscitation, and attends to victims until primary medical personnel arrive
- Oversees routine and preventive maintenance and makes periodic inspections of their assigned apparatus
- Receives direction and instruction from the fire captain and battalion chief regarding station operations, grounds and building maintenance, and overall fire scene action
- Provides training to crew members regarding the apparatus operations, including leading practical training exercises; participates in departmental in-service training and drills
- Reads, studies, interprets, and applies departmental procedures, technical manuals, building plans, and so on
- Completes and maintains manual or computer records and prepares necessary reports on incidents, accidents, and personnel training
- Performs prefire planning activities, including touring and studying businesses for physical layout, possible hazards, location of water sources, exposure problems, potential life loss, and so on
- Participates, prepares, and delivers various public education programs regarding fire prevention and safety and conducts tours of the fire station as required
- Assists in fire safety inspections of public and private buildings or property
- Oversees and participates in the care and cleaning of the apparatus
- Participates in and oversees the periodic inspection and testing of equipment, such as hoses, ladders, and engines
- Works directly in firefighting activities; utilizes tools, equipment, portable extinguishers, hoses, ladders, and so on

Additional Roles and Responsibilities for Fire Officer II

- Supervises and directs activities of a multi-unit station
- Completes employee performance appraisals
- Conducts occupancy inspections
- Determines cause and a preliminary origin of the fire
- Leads water rescue, hazardous materials, or other special teams as assigned
- Ensures the safe and proper use of equipment, clothing and protective gear and enforces departmental policies

- Participates in the formulation or evaluation of departmental or agency policies as assigned, implements new or revised policies, and encourages team efforts of fire personnel
- Participates in the formulation of the departmental budget and makes purchases within it
- Takes appropriate action on the maintenance needs of equipment, buildings, and grounds
- Supervises and performs maintenance and cleaning work on fire equipment, buildings, and grounds

This book only covers the roles and responsibilities of Fire Officers I and II according to NFPA 1021. Fire Officers III and IV have more training and responsibilities.

The transition to fire officer is a big step in one's career. Not only does it involve increased responsibility and usually increased pay, but also the role is very different from that of a fire fighter. The officer is a part of management and is responsible for the conduct of others. The officer has to apply policies, procedures, and rules to subordinates and to different situations. This means being consistent and fair and not playing favorites. These changes often require difficult adjustments for the new officer. As an officer, you will be required to take actions that might not make you happy or popular, but they are your responsibility. It is very much like the role of a parent—often difficult, but ultimately rewarding.

Fire Service in the United States

Historically, the American fire service originated as companies of volunteers who responded to calls when a fire broke out. Fighting fires was considered a civic duty, and no compensation was provided. These men volunteered their time to answer the call of public service, and each fire fighter had a regular occupation that provided a living. Over time, the fire service has evolved and now has many different methods of providing personnel to answer when the alarm sounds.

Today, there are still many fully volunteer departments composed of members who are notified to respond when there is an alarm. These men and women drop whatever they are doing to respond to the emergency. Some volunteer departments have a sufficient number of personnel and volume of calls that members are scheduled to be on standby or present at the fire station for specific shifts, according to a duty roster. In both forms of fully volunteer departments, the personnel are still unpaid for their services.

Other departments abandoned the purely volunteer method of staffing, often in response to an increasing number of alarms and a decreasing availability of volunteers. Some departments provide an incentive for fire fighters by paying them for each response to an alarm, as well as for training. These departments are termed "paid on call" or use part-time paid personnel.

Recruitment and retention have become problematic in volunteer fire departments that respond to several calls each day, particularly in areas where few individuals are available or willing to serve as volunteers. The demands often exceed the amount of time that volunteers are able to commit to the fire department, even if compensation is provided. A combination department uses full-time career personnel along with volunteer or paid-on-call personnel. This system usually provides faster response times because some personnel are on duty at the stations, ready to respond immediately. Frequently, the full-time staff consists of a minimum number of fire fighters, allowing for apparatus to respond and handle routine emergencies, such as medical assists, motor vehicle collisions, and incipient fires. The volunteer or paid-on-call staff are dispatched as a back-up force when an incident exceeds the capabilities of the full-time personnel, such as a working structure fire.

A fully career department is staffed by full-time, paid personnel whose regular job is working for the fire department. These fire departments are typically found in urban and suburban areas, where the level of risk and call volumes require personnel to be on duty at the station at all times. In some jurisdictions, the need to establish a fully career fire department was driven by the lack of people in the area who were able to volunteer.

Although there are four common forms of staffing fire department organizations, most discussions divide fire fighters into two categories: career and volunteer. According to the NFPA, there are approximately 1.1 million fire fighters in the United States. Of this total, approximately 27% are full-time, career fire fighters and 73% are volunteers, which includes part-time and paid-on-call fire fighters. Three out of four career fire fighters work in communities with populations of 25,000 or more. More than half of the volunteer fire fighters work in fire departments that protect small, rural communities with populations of 2,500 or less.

Safety Zone

Fire Statistics for the United States

- There were 402,000 residential fires in the United States in 2003.
- A total of 3,145 people died in residential fires in 2003.
- The total number of people killed in fires in 2003 was 3,925.
- Every 20 seconds, a fire department responds to a fire somewhere in the United States.
- An estimated 37,500 intentionally set structure fires occurred in 2003.
- There were 753,000 outside fires in 2003.

There are approximately 30,000 fire departments in the United States. Some of these departments are fully career organizations and some are operated entirely by volunteers. Combination departments have varying proportions of career and volunteer members and can be mostly career or mostly volunteer.

History of the Fire Service

Since prehistoric times, controlled fire has been a source of comfort and warmth, but uncontrolled fire has brought death and destruction. Historic accounts from the ancient Roman Empire describe community efforts to suppress uncontrolled fires. In 24 BC, the Roman emperor Augustus Caesar created what was probably the first fire department. Called the *Familia Publica*, it was composed of about 600 slaves who were stationed around the city to watch for and fight fires. But because the *Familia Publica* were slaves, they had little interest in preserving the homes of their masters and little desire to take risks, so fires continued to be a problem. By about 60 AD, under the emperor Nero, the *Corps of Vigiles* had been established as the fire protectors. This group of 7,000 free men was responsible for firefighting, fire prevention, and building inspections. The *Corps* adopted the formal rank structure of the Roman military, which is still used by most fire departments.

The first documented fire in North America was in Jamestown, Virginia, in 1607. The fire started in the community blockhouse and almost burned down the entire settlement. At that time, most structures were built entirely of combustible materials, such as straw and wood. Local ordinances soon required the use of less flammable building materials and mandated that fires be banked, or covered over, throughout the night. In 1630, Boston, Massachusetts, established the first fire regulations in North America when it banned wood chimneys and thatched roofs. In the Dutch colony of New York in 1647, Governor Peter Stuyvesant not only banned wood chimneys and thatched roofs but also required that chimneys be swept out regularly. Fire wardens imposed fines on those who did not obey the regulations; the money collected was used to pay for firefighting equipment.

Colonial fire fighters had only buckets, ladders, and fire hooks (tools used to pull down burning structures). Homeowners were required to keep buckets filled with water outside their doors and to bring them to the scene of the fire. Some towns also required that ladders be available so that fire fighters could access the roof to extinguish small fires. If all else failed, the fire hook was used to pull down a burning building and prevent the fire from spreading to nearby structures. The "hook-and-ladder truck" evolved from this early equipment.

The first organized volunteer fire company was established in Philadelphia in 1735, under the leadership of Benjamin Franklin. Franklin recognized the many dangers of fire and continually sought ways to prevent it. For example, he developed the lightning rod to help draw lightning strikes away from homes, a common cause of fires. Another early volunteer fire fighter, George Washington, imported one of the first fire engines from England, which he donated to the Alexandria Fire Department in 1765.

In 1871, two major fires significantly affected the development of both the fire service and fire codes. At the time, the city of Chicago was a boom town with 60,000 buildings—40,000 of which were constructed wholly of wood, with roofs of tar and felt or wooden shingles. With the lax construction regulations and no rain for 3 weeks, the city was extremely dry. On October 8, 1871, a fire, which was said to have been caused by a cow kicking over a lantern, started in a barn on the west side of the city. The fire department was already exhausted from fighting a four-block fire earlier in the day. Errors in judging the location of the fire and in signaling the alarm resulted in a delayed response time. The Great Chicago Fire burned through the city for 3 days (▼ **Figure 1-1**). When it was over, more than 2,000 acres and 17,000 homes had been destroyed, the city had suffered over $200 million in damage, 300 people were dead, and 90,000 were homeless.

At the same time, another major fire was raging just 262 miles north of Chicago in Peshtigo, Wisconsin. Although not as highly publicized as the Chicago Fire, the Peshtigo Fire would become the deadliest fire in United States history. Throughout the summer, the north woods of Wisconsin had experienced drought-like conditions. Logging operations had left pine branches carpeting the forest floor. A flash forest fire created a "tornado of fire" more than 1,000 feet high and 5 miles wide. More than 2,400 square miles of forest

Figure 1-1 The Great Chicago Fire of 1871 caused the deaths of 300 people and led to changes in firefighting operations in America.

Canadian Perspectives

In Canada, fire loss statistics are compiled in each province by Human Resources Development Canada. The results are forwarded to the Canadian Council of Fire Marshals and Fire Commissioners, which produces an annual fire loss report for Canada. In 1999, the latest year for published results, the statistics revealed:

- There were 22,150 residential fires, resulting in the loss of 284 lives.
- The total number of people killed in fires was 388.
- Every 2.5 minutes, a fire department responds to a fire somewhere in Canada.
- An estimated 12,738 intentionally set fires occurred.

Fire Marks

Benjamin Franklin also organized the first fire insurance company in the United States and coined the phrase, "an ounce of prevention is worth a pound of cure," an apt motto for fire safety. Early insurance companies marked the homes of their policy holders with a plaque, or __fire mark__, that showed the name or the logo of the insurance company. The insurance company paid fire departments to respond to those buildings displaying their fire mark. Sadly, others were left to burn. There are still some communities in the United States that fund their fire departments through annual subscription fees.

land burned, several small communities were destroyed, and more than 2,200 people lost their lives. The Peshtigo firestorm even jumped the 60-mile-wide Green Bay to destroy several hundred square miles of land and additional settlements on Wisconsin's northeast peninsula.

These events would forever change the American fire service and the role of the fire officer. Particularly as a result of the Great Chicago Fire, but also in response to the Peshtigo Fire, communities began to enact strict building and fire codes. The development of water pumping systems, advances in firefighting equipment, and improvements in communications and alarm systems all helped ensure that such tragedies would not recur.

Fire Equipment

Today's equipment and apparatus developed over the years, as new inventions were adapted to the needs of the fire service. Buckets gave way to hand-powered pumpers in 1720, when Richard Newsham developed the first such pumper in London, England. Several strong men powered the pump, making it possible to propel a steady stream of water from a safe distance. By 1829, more powerful, steam-powered pumpers had been developed and began to replace the hand-powered pumpers. Many volunteer fire fighters felt threatened by the steam engines and fought against their use. Steam engines were heavy machines that were pulled to the fire by a trained team of horses. They required constant attention, which limited their use to larger cities that could meet the costs of maintaining the horses and the steamers.

The advent of the internal combustion engine in the early 1900s greatly changed the fire service and enabled even small towns to have machine-powered pumpers. Today, both staffed and unstaffed firehouses have fire engines ready to respond at any hour of the day or night. Although they require regular maintenance, current equipment does not require the constant

attention that was demanded by horses and steam engines. Modern fire apparatus carry water, a pumping mechanism, hoses, equipment, and personnel. A single modern apparatus can outperform several older vehicles.

The progress in fire protection equipment extends beyond trucks. Without an adequate water supply, modern day apparatus would be helpless. The advent of municipal water systems provided large quantities of water to extinguish major fires.

Romans developed the first municipal water systems, just as they had developed the first fire companies. It wasn't until the 1800s that water distribution systems were developed to support fire suppression efforts. George Smith, a fire fighter in New York City, developed the first fire hydrants in 1817. He realized that using a valve to control access to the water in the pipes would enable fire fighters to tap into the system when there was a fire. These valves, or fireplugs, were used with both aboveground and belowground piping systems.

Communications

Because small fires are more easily controlled, the sooner a fire department is notified, the more likely it will be able to extinguish the fire and minimize losses. During the Colonial period, a fire warden or night watchman patrolled neighborhoods and sounded the alarm if a fire was discovered. Some towns, including Charleston, South Carolina, built a series of fire towers where wardens would watch for fires. In many towns, ringing the community fire bell or church bells alerted citizens to a fire.

The introduction of public call boxes in Washington, D.C. during the 1850s and 1860s was a major advance. The call boxes, located around the city, enabled citizens to send a coded telegraph signal to the fire department in a series of bells. The fire department would know the location of the fire alarm box by the number of bells in the signal. When fire fighters arrived at the alarm box, the caller could direct them to the exact location of the fire. Similar systems are still used in some areas, but most have being replaced by more immediate and effective communications systems, particularly with the proliferation of telephone systems. Cellular telephones enable citizens to report an emergency from almost anywhere, anytime. The introduction of computer-aided dispatch facilities has improved response times because the closest available fire units can be quickly sent to the emergency.

Communications are also vital for a fire officer to effectively coordinate the firefighting efforts. During the firefight, officers must be able to communicate with fire fighters or summon additional resources. Improvements in communication systems are tied to the history of the fire service. Today's two-way radios enable fire units and individual fire fighters to remain in contact with each other at all times. Before electronic amplification and two-way radios became available, the chief officer would shout commands through his trumpet. The **chief's trumpet**, or bugles, eventually became a symbol of authority. Although chief officers no longer use trumpets for communicating, the use of trumpets to symbolize the rank of chief also signifies the chief's need to communicate clearly.

Building Codes

Throughout history, fires have served as an impetus for communities to establish building codes. Although the first building codes, developed in ancient Egypt, focused on preventing building collapse, building codes were quickly recognized as an effective means of preventing, limiting, and containing fires.

Colonial communities had few building codes. The first settlers had a difficult time erecting even primitive shelters, which often were constructed of wood with straw-thatched roofs. The fireplaces used for cooking and heating may have had chimneys constructed of smaller logs. The all-wood construction and open fires meant that fires were a constant threat. As communities developed, they enacted codes restricting the hours during which open fires were permitted and the materials that could be used for roofs and chimneys. In 1678, Boston required that "tyle," or slate be used for all roofs. After the British burned Washington, D.C. in 1814, codes prohibiting the building of wooden houses were adopted. Building codes also began to require the construc-

tion of a fire-resistive wall, or firewall, of brick or mortar between two buildings.

Today's building codes not only govern construction materials, they also frequently require built-in fire prevention and safety measures. Required fire detection equipment notifies both building occupants and the fire department. Built-in fire suppression or sprinkler systems help to contain a fire to a small area and prevent small fires from becoming major fires. Fire escapes, stairways, doors that unlock when the alarms sounds, and doors that open outward all enable occupants to escape a burning building safely. Without modern building code requirements, high-rise buildings and large shopping centers could not be built safely.

Fire officers are often on the front line of ensuring that the fire codes are obeyed. They must also understand built-in fire protection systems and how they affect firefighting operations.

Code Development

Building codes and fire codes have evolved over many years. The first codes were locally developed and were often influenced by local preferences and customs. The insurance industry played a major role in the development of the first model codes, which were offered to local jurisdictions as a proposed minimum standard. Local communities could make the code stricter as needed.

Today, model codes and standards are written by national organizations, such as the NFPA. Volunteer committees of citizens and representatives of businesses, insurance companies, and government agencies research and develop proposals that are debated and reviewed by various groups. The final document, or the **consensus document**, is then presented to the public. Many states and municipalities adopt selected NFPA codes and standards into their laws.

Paying for Fire Service

The question of who should pay for firefighting equipment and fire fighters has also evolved over many years. Many volunteer fire departments were funded by donations or subscriptions, and many still rely on this source of revenue to purchase equipment and pay operating expenses. The first fire wardens were employed by communities and paid from community funds.

Fire insurance companies were established in England soon after the Great Fire of London in 1666, to help cope with the financial loss from fires. The companies would collect fees (premiums) from homeowners and businesses and pledged to repay the owner for any losses due to fire. The first fire insurance company in America was established in 1736 in Charleston, South Carolina, but it folded shortly after a major fire in 1740.

Because the insurance companies could save money if the fire was put out before much damage was done, they often

Fire Marks

The historic symbol of the chief officer's trumpet is still used on a chief's badge. This series of crossed trumpets is one of the cherished traditions of the fire service.

VOICES OF EXPERIENCE

❝ As a fire officer, you need to mentor your fire fighters, coach them, and provide them with the tools they need to reach the journey's final destination. ❞

I was once asked to take command of a difficult division in my previous department, a job that no one envied. However, as with many things in life, the most agonizing and least anticipated period in my career turned out to be the most productive and challenging time. While I was contemplating how to start motivating a less-than-enthusiastic division forward, I discovered that many of the most difficult employees were boat enthusiasts. So I decided to use that information to describe how I would lead the division forward. I told my team that I envisioned the division as two groups of rowers, each rowing their oars on opposite sides of the boat to keep it balanced. While the supervisors would coordinate the work effort to make sure that equal rowing was occurring on both sides, and I would look toward the horizon and guide the boat's direction.

Expanding on this idea, I added:

- If I am not looking out at the horizon for obstacles while guiding the direction of the boat, then the boat may run into something.

- If I do not coordinate the rowing, the boat will steer more to one side and end up traveling in a circle.

- If one group quits or becomes tired, the boat will veer off course and never reach its destination.

I also mentioned that anybody who did not row the boat would find himself or herself swimming. There were a couple of people who went swimming, but they later decided that rowing was less trouble and got back in the boat.

The challenge for the fire officer is to create an environment in which the people rowing are proud to do so. You also need to mentor them, coach them, and provide them with the tools they need to reach the journey's final destination. You will find that most people will row the boat and keep it on course. In the end, all of you will arrive at the destination together, happy, safe, and full of energy for the next voyage.

My talk that day proved to effective: the division's performance immediately began to improve. When I left that division, they presented me with a personally signed and expertly finished oar. It remains one of my most prized possessions.

Cecil V. "Buddy" Martinette Jr.
Lynchburg, Virginia Fire and EMS Department
30 years of service

agreed to pay a fire company for trying to extinguish a fire. Houses that had insurance were designated with a fire mark (▼ **Figure 1-2**). Most fire companies were loosely governed and organized, and more than one company might show up to fight a fire. If two fire companies arrived at a fire, however, a dispute might arise over which company would collect the money. This hastened many municipalities to begin assuming the responsibility for providing fire protection. Local tax revenues pay for most career fire departments today and support many volunteer organizations. Regardless of the form of the organization, good fire officers are vital to efficient and effective operations.

Training and Education

Fire service training and education has also come a long way over the years. The first fire fighters required simply muscular strength and endurance to pass buckets or operate a hand pumper. As equipment became more complex, the importance of formalized training and good judgment increased.

Today's fire fighters operate high-tech, costly equipment, including apparatus, radios, thermal imaging cameras, and self-contained breathing apparatus. These tools, as well as

(**Figure 1-2**) A fire mark indicated the homeowner had insurance that would pay the fire company for extinguishing the blaze.

better fire-detection devices, increase the safety and effectiveness of modern fire fighters. However, the most important machines on the fire scene remain the knowledgeable, well-trained, physically capable fire fighters who have the ability and the determination to attack a fire. These changes have increased the importance of the technical, management, and leadership skills required of fire officers to coordinate such a highly skilled work force.

Fire Department Organization

Source of Authority

Governments—whether municipal, county, state, provincial, or national—are charged with protecting the welfare of the public against common threats. Fire is one such peril; an uncontrolled fire threatens everyone in the community. Citizens accept certain restrictions on their behavior and pay taxes to protect themselves and support the common good. People charged with protecting the public are given certain authority to enable them to perform effectively. For example, fire departments can legally enter a locked home without permission to extinguish a fire and protect the public. Extinguishing a fire is considered to be an important measure to protect the community.

In most areas, the fire service draws its authority from the governing entity responsible for protecting the public from fire—whether it is a town, a city, a county, a township, or a special fire district. The head of the fire department, the fire chief, is accountable to the leader of the governing body, such as the city council, the county commission, the mayor, or the city manager. Because of the relationship between a fire department and a local government, fire fighters should consider themselves civil servants, working for the taxpaying citizens who fund the fire department.

Federal and state governments also grant authority to fire departments and operate their own fire departments and fire protection agencies to protect federal or state properties, particularly for military installations and wildland areas. Some private corporations have government contracts to provide fire protection services or offer subscription services to private property owners.

Most urban and suburban fire departments are organized by a municipal or county government. Typically, these agencies fall under the organizational umbrella as a department, just like the police department, the public works department, or a human resources department.

Another form of organization that is usually similar to that of a municipal or county department is a fire protection district. A fire protection district is a special political subdivision that can be established by a state or a county, with the single purpose of providing fire protection within a defined geographic area. It has oversight by a fire district board that is usually elected by the voters in the district. A fire district operates very much like a school district and has the ability to set a tax rate, collect taxes, and issue bonds.

In some states, a volunteer fire department can be established by charter and is independent of any local government body. The fire department may be a private association rather than a governmental entity. This type of department is not funded directly through taxes, although it may receive a grant or contract with the local government to provide services. Funding can also be obtained from fund-raising events, donations, subscriptions, and fees for service.

Chain of Command

The organizational structure of a fire department consists of a **chain of command**. The ranks may vary in different departments, but the basic concept is generally the same. The chain of command creates a structure for managing the department as well as for directing fire ground operations. Fire fighters usually report to an officer, typically **lieutenant** or **captain**, who is responsible for a single fire company (e.g., an engine company) on a single shift. Some fire departments have only one officer rank (lieutenants *or* captains) and some have two officer ranks (lieutenants *and* captains).

When there are two levels of officers in a fire department, a captain has more authority than a lieutenant. Either a captain or a lieutenant would be in charge of each company on each shift. A captain could be directly responsible for supervising a fire company on one shift and also responsible for coordinating all of the company's activities with other shifts. A captain could also be in charge of all of the companies on one shift in a multi-unit fire station.

Captains report directly to chief officers. In a large organization, there are often several levels of chiefs. **Battalion chiefs**, or district chiefs, are responsible for managing the activities of several fire companies within a defined geographic area, usually in more than one fire station. A battalion chief is usually the officer in charge of a single-alarm working fire.

Above battalion chiefs are **division chiefs**, deputy chiefs, and/or **assistant chiefs**. Officers at these levels are usually in charge of major functional areas, such as training, emergency operations, support services, and fire prevention within the department. They can also have responsibility for relatively large geographic areas, including several battalions or districts. These officers report directly to the chief of the department.

The **fire chief** (or chief of the department) has overall responsibility for the administration and operations of the department. The chief can delegate responsibilities to other members of the department but is still responsible for ensuring that these activities are properly carried out.

The chain of command is used to implement department rules, policies, and procedures. This organizational structure enables a fire department to determine the most efficient and effective way to fulfill its mission and to communicate this information to all members of the department (▼ **Figure 1-3**). Using the chain of command ensures that a given task is carried out in a uniform manner.

Basic Principles of Organization

The fire department uses a paramilitary style of leadership. Most fire departments are structured on the basis of four management principles:
1. Unity of command
2. Span of control
3. Division of labor
4. Discipline

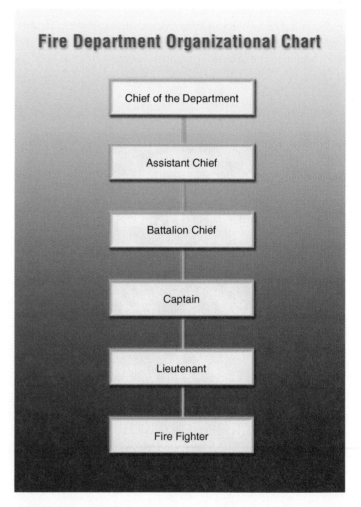

Fire Department Organizational Chart

- Chief of the Department
- Assistant Chief
- Battalion Chief
- Captain
- Lieutenant
- Fire Fighter

Figure 1-3 The chain of command ensures that the department's mission is carried out efficiently and effectively.

Unity of Command

Unity of command is the theory that each fire fighter answers to only one supervisor, each supervisor answers to only one boss, and so on (▼ **Figure 1-4**). In this way, the chain of command ensures that everyone is answerable to the fire chief and establishes a direct route of responsibility from the chief to the fire fighter.

At a fire ground, all functions are assigned according to incident priorities. A fire fighter with more than one supervising officer during an emergency could be overwhelmed with various conflicting assignments. The incident priorities may not get accomplished in a timely and efficient manner.

Span of Control

Span of control refers to the maximum number of personnel or activities that can be effectively controlled by one individual (usually three to seven). Most experts believe that span of control should extend to no more than five people, but this number can change, depending on the assignment or task to be completed. A fire officer must recognize his or her own span of control in order to be effective.

Division of Labor

Division of labor is a way of organizing an incident by breaking down the overall strategy into smaller tasks. Some fire departments are divided into units based on function. For example, the functions of engine companies are to establish water supplies and flow water; truck companies perform forcible entry and rescue functions. Each of these functions can be divided into multiple assignments, which can then be assigned to individual fire fighters. With division of labor, the specific assignment of a task to an individual makes that person responsible for completing the task and prevents duplication of job assignments.

Discipline

Discipline is the set of guidelines that a department establishes for fire fighters. Discipline encompasses behavioral requirements, such as always following orders from superior officers and performing up to expectations. Standard operating procedures, suggested operating guidelines, policies, and procedures are all forms of discipline because they outline how things are to be done, and usually how far a person can go without requesting further guidance. Firefighting

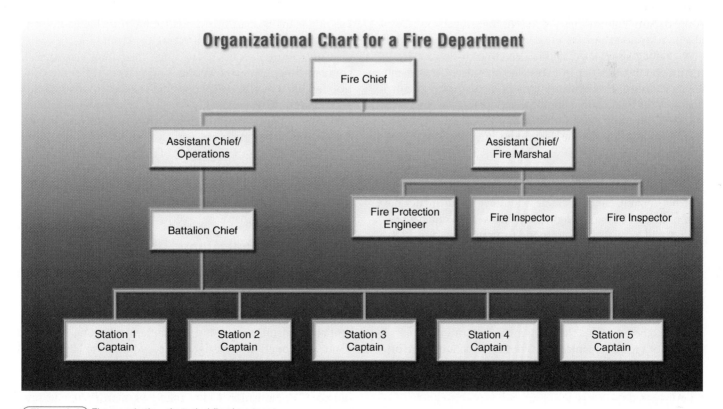

Figure 1-4 The organization of a typical fire department.

demands strong discipline to operate safely and effectively. Discipline can also be positive, when it defines appropriate action, or it can be corrective, when it responds to inappropriate actions or behaviors.

Other Views of Organization

There are several different ways to look at the organization of a fire department.

Function

Fire departments can be organized along functional lines. For example, the training division is responsible for leading and coordinating department-wide training activities. Engine companies and truck companies have certain defined functional responsibilities at a fire. Hazardous material squads have different functional responsibilities that support the overall mission of the fire department.

Geography

Each fire department is responsible for a specific geographic area. Fire stations are located throughout a community to ensure a rapid response time to every area, and each station is responsible for a specific geographic area. This also enables the fire department to distribute and use specialized equipment efficiently throughout the community.

Staffing

A fire department must have sufficient trained personnel available to respond to a fire at any hour of the day, every day of the year. Staffing issues affect all fire departments—career departments, combination departments, and volunteer departments.

In volunteer departments, it is particularly important to ensure that there are enough responders available at all times, particularly during the day. In the past, when many people worked in or near the community where they lived, volunteer response time was not an issue. Today, people often have longer commutes to work and work longer hours, so the number of people available to respond during the day can be severely limited. Some volunteer departments have had to hire full-time fire fighters to ensure that sufficient personnel will be available to respond to an incident during daytime hours.

The Functions of Management

Fire officers are managers, and like all other managers, they have specific functions that they must perform, regardless of the size or type of organization. The four functions were originally identified by Henri Fayol and include:

- Planning
- Organizing
- Leading
- Controlling

Planning means developing a scheme, program, or method that is worked out beforehand to accomplish an objective. The fire officer is responsible for developing a work plan for a fire company or an administrative work group. The fire officer develops plans to achieve departmental, work unit, and individual objectives. Short-range planning covers developing a plan that extends up to a year. Medium-range planning covers planning that is 1–3 years in advance. Long-range planning covers events longer than 3 years in advance.

Planning includes establishing goals and objectives and then developing a plan to meet and evaluate those goals and objectives. This could be as simple as planning the daily activities for the fire company or as complex as developing an annual budget. Planning can also include emergency activities, such as developing strategies and tactics and an incident action plan.

Organizing means putting together into an orderly, functional, structured whole. The fire officer takes the available people, equipment, structure, and time and develops them into an orderly, functional, and structural unit to implement the plan and deliver the expected services. Fire officers decide which companies will perform certain duties when they arrive on an emergency scene. They also decide what station duties each member of the crew will perform.

Leading means guiding or directing in a course of action. The act of **leadership** is a complex process of influencing others to accomplish a task. When most people think of managing, they think of the leading function of management. Leading is the human side of managing. It includes motivating, training, guiding, and directing the employees.

Controlling means restraining, regulating, governing, counteracting, or overpowering. Fire officers are in the controlling function when they consider the impact on the budget before making purchases, when they conduct employee performance appraisals, or when they ensure compliance with departmental policies.

Fire officers use the functions of management to get work accomplished by and through others. The four functions are a continuous cycle. Although fire officers at all levels use all four functions, each level may use them to different degrees.

Rules and Regulations, Policies, and Standard Operating Procedures

Fire officers must thoroughly know the department's regulations, policies, and standard operating procedures. This is essential to ensure a safe and harmonious working environment. Although a fire fighter is required to follow all regulations, policies, and procedures, the fire officer must not only follow them but also ensure compliance by subordinates. The transition to fire officer is often a difficult one because of the new role the officer is now required to fill. For officers to enforce the organizational rules, they must understand the critical difference between rules and regulations, policies, and standard operating procedures.

__Rules and regulations__ are developed by various government or government-authorized organizations to implement a law that has been passed by a government body. Rules may also be established by the local jurisdiction that sets the conditions of employment or internally within a fire department. For example, an organization may have a rule that states that employees with more than 15 years of service will receive 10 shifts of vacation. A fire department may have a rule that requires all members to wear their seat belts when riding in vehicles. Rules and regulations do not leave latitude or discretion.

__Policies__ are developed to provide definite guidelines for present and future actions. Fire department policies outline what is expected in stated conditions. Policies often require personnel to make judgments and to determine the best course of action within the stated policy. Policies governing parts of a fire department's operations may be enacted by other government agencies, such as personnel policies that cover all employees of a city or county. An example of a policy is one that states that the fire officer shall ensure that station sidewalks are maintained to provide safety from slips and falls during snow and ice accumulation. Because it gives the officer latitude in determining how to ensure the safety of pedestrians, this is a policy.

__Standard operating procedures (SOPs)__ are written organizational directives that establish or prescribe specific operational or administrative methods to be followed routinely for the performance of designated operations or actions. SOPs are developed within the fire department, are approved by the chief of the department, and ensure that all members of the department approach a situation or perform a given task in the same manner. SOPs provide a uniform way to deal with emergency situations, enabling different stations or companies to work together smoothly, even if they have never worked together before. They are vital because they enable everyone in the department to function properly and know what is expected for each task. Fire officers must learn and frequently review departmental SOPs. An example of an SOP is a statement of the step-by-step process (procedure) whenever vertical ventilation is required.

Some fire departments prefer the term suggested operating guidelines (SOGs) instead of SOPs because conditions often require the fire officer to use personal judgment in determining the most appropriate action for a given situation. The term SOG suggests that there is a specific step-by-step procedure that should be used, but it allows the officer to deviate from the step-by-step procedure if the conditions warrant. The distinction between an SOP and an SOG is very subjective.

Working with Other Organizations

Fire departments are part of the structure of the community. In order to fulfill its mission, the fire department must often interact with other organizations. A motor vehicle crash provides a good example of this need. An area with a centralized 9-1-1 call center could dispatch the fire department, a separate emergency medical service provider, law enforcement officials, and tow truck operators to the same incident. At the scene, they all must work together to solve the problem. Fire officers frequently have to request and then interact with other agencies.

21st Century Fire Officer Opportunities

The start of the 21st century includes some of the most dramatic changes in U.S. fire departments since World War II. The basic organizational structure of fire departments has remained relatively unchanged for the past 300 years. However, in recent years, there has been a gradual tempering of the traditional paramilitary structure. It is still recognized that a rigid command and control process is required at emergency scenes; however, for a department to be successful today, employee empowerment, decentralized __decision making__, and delegation in nonemergency activities are required. These changes are driven in part by a younger generation of fire fighters; the greater demand for increased services, such as emergency medical service; and a greater concern about terrorism.

Aging Fire Fighter Workforce

The massive baby boomer demographic is reaching retirement age, resulting in higher-than-average retirements from municipal fire departments. Some departments noted that in 2004, up to half of their workforce is eligible to retire. The other half of the workforce is young, most with less than 5 years' experience.

Suburban volunteer fire departments are also undergoing major changes. Many of them were established in the years immediately after World War II, when young soldiers were creating middle-class communities out of farmland. Driven by a local fire disaster, these veterans jumped right in to help their neighbors by establishing community fire companies. Now, the founding fire department members are fading away, and many fire companies are struggling to keep enough active members to respond to calls for help.

The 1990s witnessed an explosion of combination career-volunteer fire departments in the suburbs, with local government supplementing the volunteers with partial or full weekday staffing by career fire fighters. This creates a different work environment for the fire officer, regardless of his or her pay status. Changes in the work force are discussed in more detail in Chapter 2, Preparing for Promotion.

Changing Emergency Service Workload

The image of a fire fighter includes battling a structure fire. Structural firefighting is a focus of the Fire Fighter I and II training and much of the public perception of the fire fighter.

In actual fact, in the 21st century, structural firefighting is one of the least frequent activities performed by fire companies.

Protecting the Homeland

Since the September 11, 2001 terrorist attacks, fire officers have become part of the first line of homeland protectors. This responsibility continues to evolve. Fire officers have received additional training in responding to chemical, biological, and nuclear weapons of mass destruction. The federal government reorganized emergency services resources into the U.S. Department of Homeland Security. Local fire department incident management is affected by the National Incident Management System, which must be implemented if the jurisdiction plans to receive federal assistance after a major natural or man-made disaster.

These new challenges are changing the federal and, in some cases, regional or local, organizational structures. The fire officer is also observing changes in the community's expectations from the fire department. For example:

- Hundreds of thousands of responses to investigate "white powder" incidents after the anthrax letters were discovered in late 2001
- An equal number of responses to handle "suspicious packages"
- Any hint of a fire problem results in the complete evacuation of a high-rise office building, whether or not the fire department orders an evacuation. The vacating occupants use all of the stairwells and elevators.

In addition to the citizens, other public safety agencies are getting into areas that used to be the primary domain of the fire department. Hundreds of police departments have created new hazardous materials response teams, training police officers to the hazardous materials technician level and procuring equipment that may exceed the fire department resources.

The combining of the roles covers all agencies. The Washington, D.C. area had a massive response of multiple public safety agencies in response to the 2002 sniper attacks. Every agency had a role to play after a shooting, many performing nontraditional activities. Every activity included minute-by-minute media coverage from the ground and the air. These events disrupt the normal fire department activities and emergency responses. The public expectation is that the public safety services function as a seamless operation, with total cooperation and coordination among the police, fire, and emergency medical service agencies and at the local, state, and federal levels.

To manage this effectively, a fire officer must understand the other agencies and how they interact with the department. The Department of Homeland Security has brought many agencies that respond to an emergency under one umbrella. Others are at the state or local level. Federal, state, and local response plans identify which organizations are responsible to what areas of the incident.

For a local incident, the local fire service is often given primary responsibility for search, rescue, and fire extinguishment. Law enforcement is given responsibility for criminal investigation and scene security. Emergency management is responsible for evacuation notification. The American Red Cross may be responsible for establishing evacuation shelters.

As incidents grow, the Federal Bureau of Investigation may take a lead role in an investigation. The Federal Emergency Management Agency may become the primary agency responsible for coordination of the incident, which could include the use of Federal Emergency Management Agency's urban search and rescue teams. Fire officers must understand their role within the local, state, and federal response plans.

Fire officers must also be aware that there are individuals and organizations that wish to create chaos and harm emergency service workers. They research fire department activities to exploit weaknesses and identify opportunities. The fire officer must be vigilant for threats to fire fighters and the department.

Cultural Diversity

Like the world in general, the fire service is experiencing a change in the diversity of the organization. In many areas and until fairly recently, the fire service was composed of virtually all white males. However, this began to change in the 1960s and 1970s. At first, the integration of women and minorities was one of assimilation. The fire service desired to make those who were different to fit into the mold of the traditional fire service. Although the fire service began including a few women and minorities, they either assimilated or they left.

This view is beginning to change. It is becoming more commonplace to have a blend of men, women, Caucasian, Hispanic, African American, Native American, and others, each bringing their strengths and unique perspectives to the fire service. This integration is far from complete, and many departments struggle with bridging these relationships. The fire officer of the future must look beyond the physical attributes of individuals and match the individual's strengths with the organization's needs in order for the organization, the individual, and the officer to be successful.

Ethics

Ethics has been pushed to the forefront of nearly every organization. Virtually every day, one can read about unethical behavior. Unfortunately, the fire service is not exempt. Most of the time, fire officers make ethical choices, choosing to do the "right" thing; however, some officers make unethical choices. When they do, these choices often appear in the newspaper and have very negative consequences for the individual and the organization.

Ethical choices are based on a value system. The officer has to consider each situation, often subconsciously, and make a decision based on his or her values. If the organization has clear values that are part of a strong organizational culture, the officer uses the organization's value system. If the values are not clear, the individual substitutes his or her own value system.

The key to improving ethical choices is to have clear organizational values. This can be accomplished by:

- Having a code of ethics that is well known throughout the organization
- Selecting employees who share the values of the organization
- Ensuring that top management exhibit ethical behavior
- Having clear job goals

- Having performance appraisals that reward ethical behavior
- Implementing an ethics training program

Even at the company level, these values can be implemented to help prevent undesirable ethical choices. One way to help judge a decision is to ask yourself three questions:

- What would my parents and friends say if they knew?
- Would I mind if the paper ran it as a headline story?
- How does it make me feel about myself?

Asking these questions can help prevent an event that could devastate the department's and the fire officer's reputation for years to come.

You Are the Fire Officer: Conclusion

You are not alone. Every day, there is a newly promoted lieutenant walking into a fire station for a first shift in a new role. It is critical for the new officer to understand the relationship between levels within the organization.

As a lieutenant, you will be responsible for ensuring that your piece of apparatus is well maintained and ready to go. You will also need to ensure your crew is properly trained and physically and mentally prepared. This preparation allows the station commander and the battalion chief to focus on the broader picture of the preparation and training of multiple units and stations.

The station commander is responsible for the station as a whole, including your apparatus. You are each directly responsible for one piece of apparatus and one crew. A clear understanding of the roles of each position is essential.

There are a myriad of managerial concepts that can be learned and applied in the fire station setting. The key is to learn the various methods and find the proper technique based on the situation, the individuals involved, and your personal preference. In reality, it takes years of practice and experience to intuitively apply concepts. As a new lieutenant, with years of supervision ahead of you, it is time to get your feet wet.

Realize that it is perfectly normal to have butterflies on your first day in a position that has a vastly different degree of responsibility. Your superiors have confidence in your abilities; otherwise, you would not have been promoted. Now you have a job to do, and your concerns will go away as you learn it.

Ready for Review

- Major Fire Officer I responsibilities:
  - Supervising a single unit
  - Conducting crew training
  - Conducting prefire plans
  - Presenting public education presentations
- Major Fire Officer II responsibilities:
  - Supervising multiple units
  - Conducting occupancy inspections
  - Conducting employee performance appraisals
- There are more than 1.1 million fire fighters in the United States, of which 27% are career and 73% are volunteer.
- Benjamin Franklin established the first organized volunteer fire company in 1735 in Philadelphia.
- Speaking trumpets were originally acquired for company officers to shout orders at the scene of a fire.
- Historical events of the past have shaped the fire service of today.
- A chain of command is a continuous line of authority that exists from the top of the organizational structure to the lowest level.
- Military organizational concepts that still serve the fire officer are:
  - Unity of command
  - Division of labor
  - Span of control
  - Discipline
- Fire officers use leadership and delegation to accomplish departmental objectives and goals.
- The four-step management process includes:
  - Planning
  - Organizing
  - Leading
  - Controlling
- Rules are statements that do not allow for latitude.

- Policies give officers decision-making abilities within specified parameters.
- Standard operating procedures establish standardized step-by-step methods for handling a given situation.
- The 21st century fire officer opportunities include:
  - Aging fire fighter work force
  - Changing emergency service workload
  - Protecting the homeland
  - Cultural diversity
  - Ethics

Hot Terms

Assistant or division chiefs Midlevel chief who often has a functional area of responsibility, such as training, and answers directly to the fire chief.

Battalion chiefs Usually the first level of fire chief; also called district chief. These chiefs are often in charge of running calls and supervising multiple stations or districts within a city. A battalion chief is usually the officer in charge of a single-alarm working fire.

Captain The second rank of promotion, between the lieutenant and battalion chief. Captains are responsible for a fire company and for coordinating the activities of that company among the other shifts.

Chain of command The superior-subordinate authority relationship that starts at the top of the organization hierarchy and extends to the lowest levels.

Chief's trumpet An obsolete amplification device that enabled a chief officer to give orders to fire fighters during an emergency; precursor to a bullhorn and portable radios.

Consensus document A code or standard developed through agreement between people representing different organizations and interests. NFPA codes and standards are consensus documents.

Controlling Restraining, regulating, governing, counteracting, or overpowering.

Hot Terms (continued)

Decision making The process of identifying problems and opportunities and resolving them.

Discipline A moral, mental, and physical state in which all ranks respond to the will of the leader. Also, the guidelines that a department sets for fire fighters to work within.

Division of labor The production process in which each worker repeats one step over and over, achieving greater efficiencies in the use of time and knowledge; also, the formal assignment of authority and responsibility to job holders.

Fire chief The highest ranking officer in charge of a fire department. The individual assigned the responsibility for management and control of all matters and concerns pertaining to the fire service organization.

Fire mark Historically, an identifying symbol on a building to let fire fighters know that the building was insured by a company that would pay them for extinguishing the fire.

Incident management system (IMS) A system that defines the roles and responsibilities to be assumed by personnel and the operating procedures to be used in the management and direction of emergency operations; the system is also referred to as an incident command system (ICS).

Leadership A complex process by which a person influences others to accomplish a mission, task, or objective and directs the organization in a way that makes it more cohesive and coherent.

Leading Guiding or directing in a course of action.

Lieutenant A fire officer who is usually responsible for a single fire company on a single shift; the first in line of officers.

Organizing Putting together into an orderly, functional, structured whole.

Planning Developing a scheme, program, or method that is worked out beforehand to accomplish an objective.

Policies Formal statements that provide guidelines for present and future actions; policies often require personnel to make judgments.

Rules and regulations Developed by various government or government-authorized organizations to implement a law that has been passed by a government body.

Span of control The maximum number of personnel or activities that can be effectively controlled by one individual (usually three to seven).

Standard operating procedures (SOPs) A written organizational directive that establishes or prescribes specific operational or administrative methods to be followed routinely for the performance of designated operations or actions.

Unity of command The management concept that a subordinate should have only one direct supervisor, and a decision can be traced back through subordinates to the manager who originated it.

As a new lieutenant, you realize that there is still a large amount of information that you need to learn to become comfortable as a fire officer. You feel confident in your new role with regards to technical information and the ability to command and lead, but some of the more detailed aspects of administration are overwhelming. In order to be a better lieutenant, you request approval to attend some supervision and administration courses at the local community college.

1. What is the span of control?
   - **A.** Fire fighters only answer to one supervisor
   - **B.** Number of people one supervisor can supervise effectively
   - **C.** Way of organizing an incident
   - **D.** A set of guidelines established for fire fighters to follow

2. Most fire departments use a paramilitary type of leadership. What is one of the four management principles used under this type of leadership?
   - **A.** Discipline
   - **B.** Chain of command
   - **C.** Planning
   - **D.** Incident management

3. What is one of the four manager functions that a fire officer must perform as identified by Henri Fayol?
   - **A.** Function
   - **B.** Incident management
   - **C.** Unity of command
   - **D.** Organizing

4. What is a standard operating procedure?
   - **A.** Formal statement providing definite guidelines for actions
   - **B.** Rules and regulations for an agency
   - **C.** Form of incident management
   - **D.** Specific information on actions that should be taken to complete a task

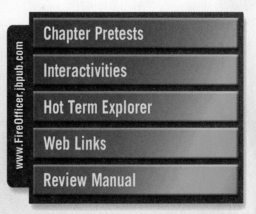

www.FireOfficer.jbpub.com

- Chapter Pretests
- Interactivities
- Hot Term Explorer
- Web Links
- Review Manual

Preparing for Promotion

Technology Resources

www.FireOfficer.jbpub.com

- Chapter Pretests
- Interactivities
- Hot Term Explorer
- Web Links
- Review Manual

Chapter Features

- Voices of Experience
- Getting It Done
- Assessment Center Tips
- Fire Marks
- Company Tips
- Safety Zone
- Hot Terms
- Wrap-Up

NFPA 1021 Standard

Fire Officer I

NFPA 1021 contains no Fire Officer I Job Performance Requirements for this chapter.

Fire Officer II

NFPA 1021 contains no Fire Officer II Job Performance Requirements for this chapter.

Knowledge Objectives

After studying this chapter, you will be able to:

- Discuss the origin of civil service promotional examinations.
- Describe how a promotional examination is prepared.
- Identify the elements of a promotional examination.
- Identify the components of an assessment center.
- List techniques for studying for a promotional examination.

Skills Objectives

There are no skills objectives for this chapter.

A promotional announcement for fire lieutenant has been posted on the bulletin board in your fire station. The announcement lists 15 reference books and a description of the test that will be held in 6 months. There will be a 100-question multiple-choice test, followed by an in-basket exercise, an emergency incident simulation, and an interpersonal interaction.

At the fire station kitchen table, the "A" shifters are discussing the test. You are a 4-year fire fighter and have recently become eligible to take your first promotional examination. You ask the veterans for advice.

Captain Keene says you should plan to study for 6 hours every day, on duty and off, to get ready for the examination. On the last promotional list, the difference between the #1 ranked candidate and the #13 ranked candidate was only 0.215 of a point. You need to be prepared with the right answer for every possible question.

Lieutenant Cropper says it is a waste of time studying for the test. You can get more benefit from using those 42 hours a week to work overtime, go to college, or set up a business. Cropper tells you to go ahead and take the examination, but not to spend a lot of time preparing for it.

Fire Fighter Meade points out that some fire fighters started studying for this examination 18 months ago. You should study and take the examination for practice, but do not expect to score very high against the candidates who have been working so hard to prepare. Maybe next time you will be ready to compete seriously.

After consulting with your coworkers you feel confused and frustrated.

1. Is it worth the effort to study and prepare, if the chances of being successful are so low?
2. If you work hard and pass the test, are you really ready to be an officer?
3. There will be plenty of chances for promotions down the road. What is wrong with just concentrating on being a good fire fighter for now?

Introduction to Preparing for Promotion

The purpose of this chapter is to provide a general description of the civil service promotional examination process that is used by most fire departments today. There are many variations of the testing procedures and promotional processes. It is vital that you understand the specific process that is used in your organization. The structure of promotional examinations in your organization may vary, depending on the rank. Many organizations also change elements or assessment tools from examination to examination.

The Origin of Promotional Examinations

Prior to the Civil War, most government jobs were awarded according to the patronage or **spoils system**; those in power could appoint people to public office based on a personal relationship or political affiliation, rather than on merit. The best jobs went to political supporters and often required a payment to the individual who had the power to make appointments. The system was ripe for corruption.

Fire Marks

When New York City created the paid Metropolitan Fire Department in 1865, Tammany Hall was the seat of political power. William March "Boss" Tweed was the chairman of the Democratic Party and the Grand Sachem (leader) of the Society of St. Tammany, which had been founded in 1789 as a club for patriotic and fraternal purposes. Between 1865 and 1871, Boss Tweed and Tammany Hall exploited the spoils process in New York City, swindling an estimated $75 to $200 million from the city. The public began to demand political reform.

Congress enacted the Pendleton Civil Service Reform Act in 1883 in response to the abuses at Tammany Hall in New York City and many other publicized abuses of the patronage system. The Pendleton Act established the civil service system in the federal government and provided a model for the civil service systems that were developed by many states and cities in the subsequent years. The spoils system was gradually replaced by merit selection and promotion for most government employees. The process of developing promotional examinations for fire officer positions, based on testing for specific knowledge and skills, was derived directly from this concept.

Sizing Up Promotion Opportunities

The opportunities for promotion that could be available to you depend on many factors, particularly the size and organizational structure of your fire department. In a very small department, there could be only a few officer positions, and promotional openings may not be available unless someone retires. If you are working in a **metropolitan (metro-sized) fire department**, an agency with more than 400 fully paid fire fighters, your opportunities for promotion are higher than average simply because of the number of openings that can be anticipated.

A combination of growing communities and retiring baby boomers is creating the need for more fire fighters and more fire officers. The fire fighters and fire officers who began their careers in many cities and urban counties in the 1970s are reaching retirement age. Some departments will see more than half of their existing work force retire by 2015.

Many suburban and rural communities are also adding more career fire fighters to cover increasing workloads, often to supplement volunteer coverage for weekday emergencies. Loudoin County, Virginia, located 35 miles northwest of Washington, D.C., is a rural agricultural and outer suburb bedroom community. In the 1980s, only a dozen career fire fighters worked for the county, and most of the fire stations were operated entirely by volunteers. During the 1990s, Loudoin County became the third fastest growing county in the nation, fueled by affordable housing close to Washington, D.C. and a booming information technology industry. As residential communities replaced farms, the number of career fire fighters increased from 14 in 1991 to 148 in 2003.

The age of the work force and different retirement systems also have a major influence on promotional opportunities. In many fire departments, the fire fighters who entered the work force in the 1970s are now becoming eligible for retirement. The elimination of an upper-limit entry-age restriction in some fire departments brought hundreds of older recruits into the ranks during the 1990s. Due to their age, some of them have to work for only 5 years to become vested in the retirement systems, so they are likely to be retiring years before their younger classmates. The fire fighters who become eligible for promotion should have many opportunities to advance in those departments.

Of course, not all fire departments are expanding, and fire officers do not automatically retire at the first opportunity. Firefighting is a lifestyle decision, like the military, medicine, and other public safety careers. Rotating shift work, the unique team-based work force, and hours of tedium mixed with the moments of intense life-saving activity create a special work environment. The rich tradition of service and sacrifice extends beyond the duty hours and becomes a major component of an individual's life. Some fire fighters represent the third or fourth generation of a firefighting family. Many fire fighters and officers enjoy their work so much that it is common to find them working a decade beyond their retirement eligibility dates or fighting to eliminate the mandatory retirement age **▼ Figure 2-1**.

Figure 2-1 Many fire fighters and officers enjoy their work so much that it is common to find them working beyond their retirement eligibility.

Many other factors can influence the number of promotional opportunities that are available in a fire department. Sometimes, the numbers decrease, particularly when the organization is downsized or restructured.

During the Clinton administration, one of the first tasks was to have Vice President Al Gore oversee an effort to reinvent the federal government. Similar efforts were launched within many state and local governments.

One of the prominent actions in local government reinvention was to reduce the number of middle managers. For fire departments, this often resulted in the elimination or civilianization of headquarters and staff positions that previously had been held by fire officers.

In most fire department systems, completion of a promotional examination process creates an eligibility list that lasts 2 to 4 years. Depending on local practice, the list may be either rank ordered or banded. On a rank-ordered list, the highest-scoring candidate is ranked #1, the second highest-scoring candidate is ranked #2, and so forth. Past practice is usually to promote individuals in the order in which they placed on the eligibility list.

Other departments use a banded list, in which the candidates are placed into bands, or groups, of promotional candidates. The bands are usually identified as "Highly Qualified," "Qualified," and "Not Qualified." In this case, all of the candidates on the "Highly Qualified" list are considered equally qualified for promotional consideration, followed by all of the "Qualified" candidates.

Post-Examination Promotional Considerations

Regardless of the eligibility-ranking scheme, the jurisdiction has the ability to make promotions to meet departmental and community needs. A basic requirement is that the candidate must be medically qualified to assume the new job. Most promotional job announcements require successful candidates to have an appropriate medical or physical per-

formance rating. A candidate who is medically restricted or cannot complete the required physical ability assessment may not be considered for promotion until these issues are resolved (▼ Figure 2-2).

In addition, the candidate must be free of active formal discipline. Departments often require a set time to pass between a formal disciplinary action and consideration for an officer promotion. The period varies from a month to a full year. That information is found in the jurisdiction's personnel regulations, the department's administrative procedures, or the current labor contract. For example, a candidate who receives a 2-day suspension without pay for repeated tardiness in reporting for duty may not be considered for promotion for 365 days, even after ranking #1 on a lieutenant eligibility list.

The increasing complexity of the fire department mission adds variables to the decision of who should be promoted. Some specialty supervisory positions require additional certifications or work experience. For example, to be promoted into the position of arson squad captain, the candidate also needs to have certification as an NFPA 1033 Fire Investigator II and 2 years of experience as a fire investigator. Other assignments may include preferred qualifications. For example, the department may prefer promoting bilingual candidates in fire companies that serve ethnic communities.

The promotional process is constantly changing to meet community needs and respond to legal and political challenges. Each jurisdiction has a promotion process that evolved as a result of community needs, consent decrees, lawsuit settlements, arbitration decisions, grievance settlements, labor contracts, and memoranda of understanding. It is important to learn what rules, goals, or practices affect the department's promotional practice.

Company Tips

In addition to the love of the lifestyle, some fire fighters continue to work several years after they become eligible to retire in order to earn the maximum pension benefits. For example, a captain who retires at the first opportunity, which could be after 20 or 25 years of service, might get a monthly annuity payment that is equivalent to 50% of a captain's pay rate. If the captain continues to work, the monthly payment increases for every additional year of service. In some retirement systems, a captain who retires with 38 years of service receives an annuity payment that is almost equal to a working captain's pay rate. Working those extra years can be a very worthwhile investment for the future.

Figure 2-2 Most promotional job announcements require that the candidate have an appropriate physical performance rating.

Preparing a Promotional Examination

The preparation of a promotional examination usually involves a combined effort between the fire department and the municipality's human resources section. Public safety promotional examinations often are the most complex and detailed examinations conducted by a municipality.

The fire department usually establishes a test preparation committee consisting of three or more officers, usually chaired by a chief officer. These individuals are the subject matter experts and are responsible for validating the technical content of the promotional examination. If the examination is developed within the agency, the test committee members write the questions and establish the content of the promotional examination (▶ **Figure 2-3**).

Charting the Knowledge, Skills, and Abilities

The municipality's personnel or human resources department uses two documents to define the **knowledge, skills, and abilities (KSAs)** that are required for every classified

position within the municipality: a narrative job description and a technical class specification. Fire officers need to be familiar with both of these documents to prepare for a promotional examination.

A narrative **job description** summarizes the scope of the job and provides examples of the typical tasks that a person holding that position would be expected to perform. The job description also lists the KSAs needed at the time of appointment and any special requirements that apply. Examples of special requirements could include possession of a valid driver's license and an EMT-Basic certificate. Other requirements might include a Class "A" physical rating, passing a physical Work Performance Test, or possession of technical or professional certifications such as NFPA Fire Officer I. When the human resources department posts a promotional announcement, a full or partial narrative job description is usually included.

Human resources departments also prepare a technical **class specification** worksheet to quantify the KSA components of every classified municipal job (▶ **Figure 2-4**). The classification system is a core component of the civil service system and is used to determine the compensation level for a position. The classification details would show why the base salary for a battalion chief is set 35% higher than for a lieutenant, based on the KSAs.

The worksheet also provides a map for the promotional examination in order to identify the factors that need to be evaluated. A promotional examination should focus on the unique, high-importance tasks that distinguish one position from another.

Periodically, a classification specialist from the human resources department surveys the individuals within a classification to rank the frequency and importance of a wide range of job tasks. This survey is performed to validate and update the job description and the KSA technical worksheet.

Company Tips

Uniformed officers staffed most of the 1970s-era fire department administrative supervisory positions. In a typical fire department, the personnel director was often a deputy chief, a battalion chief ran the fire academy, the boss of the apparatus shop was a captain, and a lieutenant was in charge of the warehouse. In addition to reducing the number of middle-management and administrative positions in local government during the 1990s, many of the remaining middle manager jobs were converted to civilian positions. Here is an example of what happened to those four uniformed supervisory positions in one fire department.

The city consolidated the fire and police department personnel directors, replacing two uniformed positions with a single public safety personnel director. The public safety personnel director is a 27-year-old with a master's degree in human resources. The department privatized the fire academy, and the battalion chief position at the academy was eliminated. A community college now runs the recruit school under a service contract. A city management analyst in charge of contract compliance is responsible for ensuring that the college delivers the required training services. The city also civilianized the fire apparatus shop supervisor, replacing the fire captain with a civilian employee from the city's fleet management division. The new boss of the apparatus shop is a 50-year-old master mechanic with decades of experience running a fleet of heavy equipment rigs.

The fire department warehouse is gone, absorbed into the citywide resource management system. A logistics specialist and an accountant coordinate the fire department supply requests.

The combination of civilianization and reduction of middle-management positions has reduced the number of uniformed fire officer positions in many fire departments, thereby reducing the number of potential officer promotions.

(**Figure 2-3**) If the examination is developed within the agency, test committee members establish the content of the promotional examination.

NEW HANOVER COUNTY, NC
CLASS SPECIFICATION

CLASS TITLE: Fire Lieutenant

CLASS CODE:		
DEPARTMENT: Fire Services	**ACCOUNTABLE TO:** Fire Captain	**FLSA STATUS:** Non-exempt

CLASS SUMMARY:

Incumbents are responsible for shift supervision of County firefighters. Duties include: supervising and evaluating staff; overseeing responses to reports of fires; supervising rescue operations; preparing work schedules; writing reports of firefighting and rescue activities; training personnel; coordinating equipment and facility maintenance; presenting fire prevention programs; and, representing the department at special events.

DISTINGUISHING CHARACTERISTICS:

The Fire Lieutenant is the second level in a five level firefighter series. The Fire Lieutenant is distinguished from the Firefighter/Apparatus Operator in that it has shift supervisory responsibilities. The Fire Lieutenant is distinguished from the Fire Captain which has full supervisory authority.

DUTY NO.	TYPICAL CLASS ESSENTIAL DUTIES: (These duties are a representative sample; position assignments may vary.)	FRE-QUENCY	
1.	Supervises two or more full-time staff to include: prioritizing and assigning work; conducting performance evaluations; ensuring staff are trained; and, making hiring, termination, and disciplinary recommendations.	Daily 25%	
2.	Oversees response to reports of fires which includes: preparing pre-incident surveys; securing the scene; extinguishing the fire; salvaging structures and their contents; and providing emergency medical services to injured parties.	Daily 15%	
3.	Supervises rescue operations by overseeing extrication activities and providing emergency medical services to injured parties.	Daily 15%	
4.	Prepares daily and weekly work schedules for firefighters.	Daily 10%	
5.	Prepares written reports of fire and rescue activities.	Daily 10%	
6.	Provides training to shift personnel and students at the County fire academy on firefighting, rescue, and emergency medical topics.	Daily 10%	
7.	Coordinates firefighter maintenance of vehicles, fire and rescue equipment, and facilities.	Daily 5%	
8.	Presents fire prevention and education programs to businesses, schools, and community groups.	Weekly 5%	
9.	Represents the department at special events including parades and open houses.	Occasion-ally 5%	
10.	Performs other duties of a similar nature or level.	As Required	
11.	Performs work during emergency/disaster situations.	As Required	

Figure 2-4 Sample technical class specification worksheet.

NEW HANOVER COUNTY, NC
CLASS SPECIFICATION

CLASS TITLE: Fire Lieutenant

POSITION SPECIFIC RESPONSIBILITIES MIGHT INCLUDE:
• Does not apply.

Knowledge (position requirements at entry):
Knowledge of:
• General principles of fire science;
• Emergency management techniques;
• Basic principles of rescue;
• Hazardous materials management techniques;
• Emergency medical practices;
• Departmental policies and practices;
• Local and state fire ordinances.

Skills (position requirements at entry):
Skill in:
• Performing fire suppression and rescue operations;
• Driving a vehicle;
• Preparing and making presentations;
• Preparing written incident reports;
• Using a computer and related software applications;
• Supervising and evaluating employees;
• Communication, interpersonal skills as applied to interaction with coworkers, supervisor, the general public, etc. sufficient to exchange or convey information and to receive work direction.

Training and Experience (positions in this class typically require):
High School Diploma or General Equivalency Diploma (GED) and five years of related firefighting experience, including two years of progressively responsible supervisory experience; or, an equivalent combination of education and experience sufficient to successfully perform the essential duties of the job such as those listed above; Associate's Degree in Fire Science preferred.

Licensing/Certification Requirements (positions in this class typically require):
• Class B Driver's License;
• Firefighter II Certification;
• Must be able to obtain EMT and Fire Instructor Level II certifications within one year and Level I Fire Inspector and ERT Certifications within two years.

Physical Requirements/Working Conditions:
Positions in this class typically require: climbing, balancing, stooping, kneeling, crouching, crawling, reaching, standing, walking, pushing, pulling, lifting, fingering, grasping, feeling, talking, hearing, seeing, and repetitive motions.

Heavy Work: Exerting up to 100 pounds of force occasionally, and/or up to 50 pounds of force frequently, and/or up to 20 pounds of forces constantly to move objects.

Incumbents may be subjected to moving mechanical parts, electrical currents, vibrations, fumes, odors, dusts, gases, poor ventilation, chemicals, oils, extreme temperatures, inadequate lighting, work space restrictions, intense noises, and travel.

NOTE:
The above job description is intended to represent only the key areas of responsibilities; specific position assignments will vary depending on the business needs of the department.

Classification History:
Draft prepared by Fox Lawson and Associates LLC (CC).
Date: 10/99

Figure 2-4 cont'd, Sample technical class specification worksheet.

Promotional Examination Components

Multiple-Choice Written Examination

Multiple-choice written examinations are widely used in promotional processes because they can be structured to focus on very specific subjects and factual information. There is no element of style or creativity in answering a multiple-choice question; the candidate only has to select the appropriate answer. The scoring is equally straightforward because an answer is simply right or wrong.

The multiple-choice written examination concentrates on facts that can be found within the materials on a reading list. The reading list generally includes textbooks, reference books, standard operating procedures manuals, rules and regulations, and other locally developed reference materials.

Using the KSA technical worksheet, the test committee determines the number of questions to include from each knowledge area.

The first-level supervisory examination usually includes many technical questions covering Emergency Medical Service, engine, truck, and rescue company operations. These questions are directed toward a candidate's ability to demonstrate the basic knowledge that would be important for a fire officer. In general, the first-level supervisory examination has the longest and most diverse reading list. Up to 70% of this test focuses on the technical aspects of a first-line company commander.

Typical technical questions cover building construction, incident management, hydraulics, and firefighting tactics. Many examinations also test the candidate's knowledge in specialized company tasks, such as technical/heavy rescue, truck company operations, and rapid intervention teams.

Supervisory questions tend to focus on the immediate requirements affecting fire company preparedness, such as a subordinate unfit for duty or injured on the job. A first-level supervisor examination might include a question about how to handle a fire fighter reporting late for duty. The hierarchical nature of the fire department requires that the first-level supervisor identify the problem, stop the behavior, and report to or consult with a senior officer or chief before taking any significant supervisory action.

A second-level supervisory examination usually includes fewer technical questions and more management and administration questions because the higher-level position involves more management responsibilities. The process would assume that the candidate for a higher-level position has already met the qualifications for the lower-level position. The second-level supervisor might also be expected to develop a fire fighter's work improvement plan, prepare a budget proposal, evaluate fire company performance, or develop a department-wide training plan. For example, a second-level supervisor could be asked questions that relate to evaluating a fire fighter's reporting-time performance over a year. The supervisor should be able to identify patterns and determine the underlying causes for tardy reporting.

There are three options for constructing a multiple-choice written examination. Sometimes, the local test committee develops the test and the committee members write the questions. There are also private companies that develop generic examinations used by many fire departments. The department has the third option of hiring a consultant to write a more specific examination that is directed toward local priorities. Larger fire departments are more likely to develop an examination internally because they have the necessary resources. For example, the fire chief in a large city can probably assign a command officer and five to seven lieutenants and captains to write the questions for a written lieutenant examination.

The committee first determines how many questions are needed to assess the candidate's knowledge within each particular area. For example, the committee might decide that nine building construction questions should be included in an examination. Two of the committee members could be assigned to work together and develop 15 multiple-choice questions on building construction. The whole committee would then meet to select the nine best questions for the examination.

A smaller department would be more likely to go to an external source for the examination questions, simply because of the time and effort that would be required to write and validate them internally. Writing good examination questions is difficult work, particularly for individuals who do not have experience in this field. Each question has to be fully researched and carefully worded. A question can be challenged and invalidated if the wording is not clear, if there is no correct answer, or if more than one of the answers that are provided could be correct.

In many fire departments, multiple-choice written examinations were the only assessment tools used for promotional examinations from the 1950s to the 1980s. The simplicity of a multiple-choice examination can also be viewed as a weakness. A candidate who knows all of the facts can do very well on this type of examination, even if the individual is unable to apply the information in a real-life situation. The description "Book smart, street dumb" was often used in complaints about examination processes that relied solely on written multiple-choice questions to promote officers.

In the late 1970s, some public safety agencies began to use **assessment centers** in promotional examinations. An assessment center is a series of simulation exercises that are used to evaluate a candidate's competence in performing the actual tasks associated with a job. The assessment center process was developed in the officer corps of the German army in the 1920s. The goal of the program was to select future officers based on their predictive performance in a 2- to 3-day assessment procedure. The same type of process can be developed for fire officers as well as military officers.

In-Basket

One of the most common assessment center events is the **in-basket** exercise (▼ **Figure 2-5**). The candidate has to deal with a stack of correspondence and related items that have accumulated in a fire officer's in-basket. This timed exercise measures the candidate's ability to organize, prioritize, delegate, and follow up on administrative tasks. Studies have shown that those who do well on in-basket exercises tend to demonstrate successful managerial performance in real life.

A typical in-basket contains the following items:
- Instructions for the exercise
- A calendar
- An organizational chart or list of personnel
- Ten to 30 exercise items

An in-basket assessment generally begins by positioning the candidate as a newly promoted officer reporting to the fire station for the first time. The former officer suddenly retired or left the fire department, leaving a stack of issues that require attention in the in-basket. The candidate has a specified time to go through the officer's in-basket and decide what to do with each item.

The situation is contrived so that the candidate is unable to directly call or contact anyone else for advice or assistance. The candidate might be provided with copies of the department's standard operating procedures and reference manuals. Hidden within the in-basket are surprises and time-critical issues, such as an important report that is overdue or an activity that has to be performed immediately. Frequently, the

candidate encounters a scheduling conflict, such as an order from the operations chief to have the engine company report to a multiple-company drill at the same time the apparatus is scheduled for an annual pump test at the shop. Many in-baskets also include a writing exercise that could require the candidate to prepare a letter for the chief's signature or complete an incident report using correct formatting, spelling, and grammar.

From the test administrator's viewpoint, in-baskets are generally easy to organize and grade. This type of test can be difficult to develop for the fire officer level because many of the typical job tasks in a fire station do not readily lend themselves to an in-basket assessment.

A suggested method for handling promotional in-baskets:
- **Review**—Look at every item and determine which items are important (critical) and urgent, (time constraints), which are important but not urgent, and which are unimportant.
- **Prioritize**—Handle the important and urgent items first, followed by the important and non-urgent items. When these are finished, you will have handled all items or will have only unimportant items left.
- **Identify resources/options/alternatives**—Determine who can handle the items, what options you may have in how they can be handled, and possible alternative courses of action.
- **Follow-up**—Provide some form of follow-up on all delegated items.
- **Make notifications**—Ensure that appropriate persons are notified of necessary and critical information.

Candidates can prepare for an in-basket exercise by practicing on sample exercises that can be obtained from study materials publishers. Both lieutenant and captain in-basket exercises are available; these include answers with explanations and descriptions of the dimensions assessed.

Emergency Incident Simulations

Emergency incident simulations are often included in a promotional examination to test the candidate's ability to perform in the role of an officer at a fire or some other type of situation. Emergency incident management usually occupies a small amount of an officer's actual time, yet the required skill sets can be the most critical for evaluation in a promotional examination. An officer must be able to lead, supervise, and perform a set of essential skills at an emergency incident scene. Emergency incident simulations fall into one of four formats.

In the first format, the candidate is provided information concerning an emergency situation, usually in a written format that includes pictures or preincident plan information. The candidate has to explain the actions that would be taken in the situation that is described and what factors would

Figure 2-5 The in-basket exercise is a common assessment center event.

Voices of Experience

❝ **Preparation for promotion begins on your first day on the job.** ❞

During the initial years of my career, I observed others during the promotion preparation process. At various times before the actual test, fire fighters would begin the studying ritual—reading the manuals, talking to peers, and practicing test-taking tactics. Some fire fighters would begin to prepare a few months before the test, while some fire fighters would begin almost a year in advance.

As my eligibility time for promotional testing approached, I came to a realization: preparation for promotion begins on your first day on the job. I had closely observed and listened to each of my fire officers, especially as a probationary fire fighter. Information, guidance, instruction—my fire officers were my primary learning resource. Each of the fire officers modeled what they considered a fire officer should be. By paying close attention to their actions, I built in my mind the model fire officer. Then I tried to bring that model fire officer to life during the testing process and after promotion. Even if you do not realize it, you begin the preparation process on your first day—I know I did. After you become a fire officer, you need to realize that you are modeling what a fire officer should be to current fire fighters and future fire officers.

Mike DeVirgilio
Tempe Fire Department, Tempe, Arizona
24 years of service

be considered in the making of those decisions. This is known as a "<u>data dump" question</u> because it provides an opportunity for the candidate to demonstrate a depth of knowledge about a particular subject. For example, the evaluators might find out how much expertise the candidate can demonstrate regarding basement fires in mercantile properties.

In the second format, the candidate is provided with a set of basic information concerning an emergency incident to begin the exercise. As the simulation progresses, the candidate is provided with additional updates of the situation, such as "Immediately after you arrive on Engine 1, the store manager reports an acrid, unknown odor coming from a leaking package in the rear of the store. It has made a dozen people sick. What actions will you take? What will you report to dispatch?" The candidate has to react to the unfolding situation, which can change based on the answers to the previous questions.

In the third format, the candidate participates in an interactive emergency scene simulation in a classroom or incident simulation trainer. The candidate is presented with a dynamic emergency incident through a multimedia format that typically includes full-color pictures depicting the incident. The graphics often include simulated smoke and flames superimposed on photos of an actual building in the community. Using a hand-held radio, the candidate is expected to command the event using the local jurisdiction's standard operating procedures and incident command tools (accountability boards, command post clipboards, Incident Management System worksheets) that are appropriate to what a person at that rank would use on the street (▼ **Figure 2-6**).

This type of exercise is complex and can be expensive because the test-site administrator might need to arrange for role players and technical support staff to run the simulation event for every candidate.

The fourth simulation format attempts to make the conditions as realistic as possible by actually having the candidate don protective clothing, climb into the officer's seat on an apparatus, and respond to a realistic scenario with a crew of fire fighters. This type of exercise is usually performed at a large urban or regional training academy, where there are buildings and props that allow for a consistent simulation for all of the candidates. The real-life format allows a candidate to fully function in the role of an officer; however, it requires extensive preparations and a large supporting cast.

Interpersonal Interaction

An interpersonal interaction exercise is designed to test a candidate's ability to perform effectively as a supervisor. In a typical interpersonal interaction scenario, the candidate has to deal with a role player who has a problem, complaint, or dilemma. The candidate is usually provided with background information to prepare for the situation, then has 10 to 20 minutes of face-to-face interaction with the role player (▼ **Figure 2-7**).

Many of these assessments require the candidate to deal with a poorly performing or troubled employee. The background information could include the employee's last evaluation report, a sick leave use report, or some recent disciplinary action that the employee received. In many cases, the role player has important information related to the performance issue that is

Figure 2-6 The candidate is expected to command the simulated event.

Figure 2-7 An interpersonal interaction exercise tests a candidate's ability in performing effectively as a supervisor.

revealed only if the candidate asks the right questions during the interview phase. After the interview, the candidate may have to write up an employee improvement plan or a discipline letter (▼ **Figure 2-8**).

The key points of an interpersonal interaction assessment are to:

1. Maintain control of the interview.
2. Tell the employee the exact behavior you expect (show up on time, wear your seat belt, etc.).
3. Give the employee a deadline to demonstrate a consistent behavioral change.
4. Specifically arrange for follow-up meetings ("We will meet once a week to review your progress in meeting this goal.").
5. Attempt to get the employee to buy into or take personal responsibility for the improvement plan.
6. Be an empathetic listener, but remain focused on the reason for the interview.
7. Clearly explain the consequences if the employee's behavior does not change or improve.
8. Try to finish the session on a positive note ("You are a valuable member of this company and I am counting on you . . .").

A frequently successful approach for an interpersonal interaction exercise is for the candidate to demonstrate the qualities of an "extreme supervisor." That means that any issue is handled strictly "by the book" according to the rules and regulations of the organization.

An alternative scenario for this type of assessment is to have the role player confront the candidate as an angry or frustrated citizen. This exercise would assess how well the candidate can handle a customer service problem or a citizen complaint. This type of exercise is usually based on some situation that has actually occurred, either locally or in some other fire department.

Writing or Speaking Exercise

A candidate should expect to deliver a short oral presentation or write a memo or report as part of the assessment center process. The oral presentation assesses oral communication, planning and organizing skills, and persuasiveness. Sometimes, the candidate is required to write a report or a letter for a superior officer's signature. The written report assesses the candidate's written communication, problem analysis, and decision-making skills.

The oral presentation could be incorporated into one of the other exercises. For example, a candidate might be required to make a brief oral presentation to explain the size-up considerations of the emergency incident simulation.

Many assessment centers include an oral interview session in which the candidate is asked questions by an interview board (▼ **Figure 2-9**). One of the most popular questions for this type of interview is to ask the candidate why he or she should be selected for promotion. This provides an opportunity for the candidate to demonstrate skills or knowledge in a particular area or even to correct an error in a previous exercise. In some cases, a candidate's ability to recognize an error and correct a response could be viewed as equal to having the right answer to every question.

Videos are available from emergency service publishers and individual subject matter experts to assist candidates in preparing to make a great presentation in an oral interview. Chapter 14, Fire Officer Communications, discusses fire officer communication skills in more detail.

Technical Skill Demonstration

Fire officers are expected to be working supervisors and to be skilled in both task and tactical level activities. For some first-line supervisor examinations, the candidates might also

(**Figure 2-8**) Employee work improvement plan.

(**Figure 2-9**) Many assessment centers include an oral interview session.

be required to demonstrate very specific technical skills. This type of exercise is often included in promotional tests for officers who would command highly specialized teams, such as urban search and rescue, hazardous materials, fire investigations, and paramedic teams.

Preparing for a Promotional Examination

To prepare for an examination, the candidate must master both the content of the examination (technical and reference information) and the process (answering multiple choice questions or role playing as an officer). Most of the content comes from the materials on the reference reading list; however, you should also be prepared for questions that relate to current issues and recent local history. For example, if your department recently had a problem with fire fighters using fire station computers to surf to non–work-related web sites, do not be surprised to find that issue somewhere in the examination. You should keep your knowledge current by reading fire service publications and periodicals to keep up with trends and issues. Special interest bulletins provide excellent additional study material.

Mastering the content for an examination requires a personal study plan. Many tests require the candidates to absorb a tremendous quantity of factual information, which can require a major investment of time and effort. In some fire departments, it is not unusual for candidates to spend more than a year preparing for an examination, setting aside a period of time every day for reading and answering practice questions.

In many cases, fire fighters who participate in structured programs and study groups perform better than fire fighters who study on their own. Many fire departments have officially organized programs to help candidates prepare for examinations. In other cases, more informal study groups or preparatory programs are available for candidates who want to be well prepared. If your fire department does not have a promotional preparation program, you can set up your own study group. The individuals support each other, share materials, and even develop their own multiple-choice questions from the local standard operating procedures (▼ Figure 2-10).

Building a Personal Study Journal

A valuable tool in studying for a promotional examination is to create a **personal study journal**. This journal can be used to set up your personal study schedule, keep track of your progress, and make notes about confusing, interesting, or important facts that you will discover.

For the journal, you could use a three-ring binder that would be divided into sections. The first section could include all of the information about the promotional examination— a copy of the official announcement and copies of all of the

Assessment Center Tips

NIOSH Publication 99-146, *Preventing Injuries and Deaths of Firefighters Due to Structural Collapse,* is a good example of a special interest bulletin that could provide valuable information for a promotional examination. This particular publication makes 10 recommendations that every fire officer should know:

1. Ensure that the incident commander conducts an initial size-up and risk assessment at the incident scene before beginning interior firefighting.
2. Ensure that the incident commander always maintains close accountability for all personnel at the fire scene—both by location and by function.
3. Ensure that at least four fire fighters are on the scene before entering a structure and beginning interior firefighting at a structural fire (two fire fighters inside the structure and two fire fighters outside).
4. Establish rapid intervention crews and make sure they are positioned to immediately respond to emergencies.
5. Equip fire fighters who enter hazardous areas (e.g., burning structures or structures suspected of being unsafe) to maintain two-way communications with the incident commander.
6. Ensure that standard operating procedures and equipment are adequate and efficient to support radio traffic at multiple-responder fires.
7. Provide all fire fighters with Personal Alert Safety System devices and make sure that they wear and activate them when they are involved in firefighting, rescue, or other hazardous duties.
8. Conduct prefire planning and inspections that cover all building materials and components of a structure.
9. Transmit an audible tone or alert immediately when conditions become unsafe for fire fighters.
10. Establish a collapse zone around buildings with parapet walls.

(**Figure 2-10**) Fire fighters who establish study groups often perform better than fire fighters who study on their own.

Getting It Done

Textbook Study Guides

A small industry has developed to assist promotional candidates in preparing for written examinations. There are publishers that produce study guides with hundreds of multiple-choice questions taken from the most popular textbooks and references. The answer to each question is keyed to a page number in the textbook or source material. Some book publishers release their own study guides, whereas others specialize in producing promotional study guides that are appropriate for specific examination levels.

In addition to printed study guides, some publishers produce computer-based programs, accessible either on compact disk or on-line via the Internet.

documents that the candidate was required to submit. Any test-related documents from the department can also be placed in this section.

The second section contains a calendar. Usually, in the one-month-per-page format, the calendar covers the period from the start of the preparation process through the last promotional test activity. Important deadlines are highlighted, and the candidate can create the study plan, working backward from the examination date.

The third section contains the written reference materials. Subdivided by reference, these sections would cover the items used to master the written material. Some candidates outline each chapter; others make up practice tests. You should make a note of any questions that you miss in prac-

tice tests to be sure that you go back and obtain the correct answer. This information can be extremely valuable during the final review before the examination date.

The fourth section covers information about the announced or anticipated components of the promotional examination.

Preparing for Role Playing

The most effective candidates in role-playing exercises are the ones who act naturally during the examination, while making supervisory decisions that are strictly based on policies and regulations. This is sometimes described as performing in the "extreme supervisor" role. Many successful candidates model their performance in the examination, as well as afterward, on the behavior of successful and respected officers. It is a good idea to pay attention to the supervisors and officers who are known for being efficient and effective.

One of the best ways to prepare for role playing is to experience working in one of the busier or larger fire stations within your department. More supervisory issues come up in a fire station with 14 fire fighters and three officers on each shift than in a station with three fire fighters and one officer. Similarly, a fire company that runs 15 calls a day encounters more problem-solving opportunities than a company that runs three calls a day. Working under one of the widely respected officers is also a good way to develop supervisory skills. Candidates who have worked for a supervisor who demonstrates leadership and provides a positive role model have a significant advantage over candidates who have never had that experience.

You Are the Fire Officer: Conclusion

Throughout the day, you have been calling your friends about the lieutenant examination. One of your recruit school partners points out that many fire officers will be retiring during the life of this lieutenant eligibility list. The competition will be fierce, but there will be more promotional opportunities than usual. She recommends that you give it your best shot in the next 6 months and see what happens. You cannot be promoted if you do not take the examination.

While cleaning the dinner pots and pans, you ask Captain Keene for suggestions for a study plan. Keene brings out his personal study journal and points out that he ranked #6 on his first lieutenant examination, with just 5 years on the job.

Wrap-Up

Ready for Review

- Promotional examinations were a product of the 1883 civil service reform act.
- Every civil service job has a narrative classified job description and a knowledge, skills, and abilities (KSA) worksheet.
- Promotional examinations assess the important KSAs and measure performance dimensions.
- Multiple-choice examinations concentrate on facts from the reading list.
- A local test committee, comprised of existing officers with technical subject matter expertise, validates promotional examinations.
- Assessment centers provide a variety of role-playing exercises.
- During an in-basket assessment, you should remember to review; prioritize; identify resources, options, and alternatives; follow-up; and notify.
- Emergency incident simulations fall into four formats: the "data dump" question, a mock emergency incident, an interactive emergency scene simulation, and a full-scale incident simulation.
- During an interpersonal assessment, remember to remain in control, exactly state the desired behavior, give the employee a deadline, arrange follow-up meetings, get the employee to buy into or take personal responsibility for the improvement plan, be empathetic but focused, explain consequences, and finish the session on a positive note.
- Candidates may be required to deliver a short presentation or write a memo or report as part of the assessment center process.
- Technical skills may be evaluated during promotion tests for highly specialized positions.
- The candidate needs to develop a personal study plan to master the content for a promotional examination. Study techniques include keeping a study journal and participating in role-playing activities.

Hot Terms

Assessment centers A series of simulation exercises to identify a candidate's competency to perform the job that is offered in the promotion examination.

Class specification A technical worksheet that quantifies the knowledge, skills, and abilities (KSAs) by frequency and importance for every classified job within the local civil service agency.

"Data dump" question A promotional question that asks the candidate to write or describe all of the factors or issues covering a technical issue, such as suppression of a basement fire in a commercial property.

Dimensions Attribute or quality that can be described and measured during a promotional examination. There are between five and 15 dimensions that would be measured on a typical promotional examination. The six most common include Oral Communication, Written Communication, Problem Analysis, Judgment, Organizational Sensitivity, and Planning/Organizing.

In-basket A promotional examination component in which the candidate deals with correspondence and related items accumulated in a fire officer's in-basket.

Job description A narrative summary of the scope of a job. Provides examples of the typical tasks.

Knowledge, skills, and abilities (KSAs) The traits required for every classified position within the municipality. Defined by a narrative job description and a technical class specification.

Metropolitan (metro-sized) fire department A department with more than 400 fully paid fire fighters. The Metropolitan Fire Chiefs is a special interest group organized by the International Association of Fire Chiefs in 1965 to address the needs of large fire departments. Metro Chiefs are also a section of the National Fire Protection Association.

Personal study journal A personal notebook to aid in scheduling and tracking a candidate's promotional preparation progress.

Spoils system Also known as the patronage system. The practice of making appointments to public office based on a personal relationship or affiliation rather than because of merit. A problem since 1850, the spoils system scandals of the New York City "Tweed Ring" and the Tammany Hall political machine (1865–1871) resulted in Congress passing the Pendleton Civil Service Reform Act of 1883.

You are taking the upcoming fire officer examination and know that there are at least ten other fire fighters who will be testing as well. You have been studying for two weeks and you have read every book and have completed every practice exam that you can find. However, you still do not feel very comfortable about the exam and know the other candidates have been doing the same activities to prepare. You ponder what steps you can take to make yourself a more competitive candidate.

1. What is the first step you can take to make yourself a stronger candidate when hearing about a promotional exam opportunity?

   A. Read more books on promotional exams

   B. Read the test announcement again, including the job description for details

   C. Meet with some current officers for helpful information that could make you a stronger candidate

   D. Ask to study with some of the other candidates

The testing announcement states that there will be a written exam, assessment center, and oral board as part of the test process. You are comfortable with both the written exam and oral board; however, the assessment center is something new. The announcement states that there will be an emergency incident simulation and an in-box exercise.

2. What is normally involved in an emergency incident simulation?

   A. Evaluating leadership skills

   B. Your ability to command and control

   C. Your ability to extinguish a fire

   D. Hands-on fire fighter skills evaluation

3. What is one document required for every classified position within a municipality and should be familiar to fire officers?

   A. EMT-Basic certificate

   B. Driver's license

   C. Special requirements for position

   D. Narrative job description

4. What is the role of the test preparation committee?

   A. Validate the technical content of the exam

   B. Review candidates who may apply

   C. Perform the test

   D. Conduct candidate interviews

www.FireOfficer.jbpub.com

Chapter Pretests

Interactivities

Hot Term Explorer

Web Links

Review Manual

www.FireOfficer.jbpub.com

Fire Fighters and the Fire Officer

Technology Resources

www.FireOfficer.jbpub.com

- Chapter Pretests
- Interactivities
- Hot Term Explorer
- Web Links
- Review Manual

Chapter Features

- Voices of Experience
- Getting It Done
- Assessment Center Tips
- Fire Marks
- Company Tips
- Safety Zone
- Hot Terms
- Wrap-Up

NFPA 1021 Standard

Fire Officer I

4.1.1 **General Prerequisite Knowledge.** The organizational structure of the department; geographical configuration and characteristics of response districts; departmental operating procedures for administration, emergency operations, incident management systems, and safety; departmental budget process; information management and recordkeeping; the fire prevention and building safety codes and ordinances applicable to the jurisdiction; current trends, technologies, and socioeconomic and political factors that impact the fire service; cultural diversity; methods used by supervisors to obtain cooperation within a group of subordinates; the rights of management and members; agreements in force between the organization and members; generally accepted ethical practices, including a professional code of ethics; and policies and procedures regarding the operation of the department as they involve supervisors and members.

4.2.5* Apply human resource policies and procedures, given an administrative situation requiring action, so that policies and procedures are followed.

> **(A)* Requisite Knowledge.** Human resource policies and procedures.

> **(B) Requisite Skills.** The ability to communicate orally and in writing and to relate interpersonally.

4.4 **Administration.** This duty involves general administrative functions and the implementation of departmental policies and procedures at the unit level, according to the following job performance requirements.

4.4.2 Execute routine unit-level administrative functions, given forms and record-management systems, so that the reports and logs are complete and files are maintained in accordance with policies and procedures.

> **(A) Requisite Knowledge.** Administrative policies and procedures and records management.

> **(B) Requisite Skills.** The ability to communicate orally and in writing.

Fire Officer II

NFPA 1021 contains no Fire Officer II Job Performance Requirements for this chapter.

Knowledge Objectives

After studying this chapter, you will be able to:

- Describe the fire officer's vital tasks.
- Describe a typical fire station workday.
- Describe the transition from a fire fighter to a fire officer.
- Describe the activities that a fire officer performs to maintain an effective working relation with his or her supervisor.
- Describe integrity and ethical behavior.
- Describe how to maintain workplace diversity.
- Describe the concept of the fire station as a business work location.

Skills Objectives

After studying this chapter, you will be able to:

- Function as a newly assigned fire officer, with a description of a fire station, work group, and schedule, prepare a morning report or activity plan.
- Function as a fire officer, demonstrate the effective issuing of an unpopular order to a fire company.
- Demonstrate making a decision consistent with the department's core values, mission statement, and value statements given an ethical dilemma.
- Conduct an initial interview and notifications consistent with the department's policy, rules, and regulations given a harassment or hostile workplace complaint.

You Are the Fire Officer

It is 8:01 A.M. on your first day as the officer-in-charge of Engine 46. You receive a fax from the fire chief's office with the banner message "IMMEDIATELY READ AND IMPLEMENT." The fire chief has issued an order suspending all "round ball" sports activities by on-duty fire companies. The established routine at Engine 46 includes an hour every morning playing basketball as part of physical fitness training. This makes for an uncomfortable morning meeting with the crew. It is clear that two of the fire fighters are very angry and plan to play basketball anyway. One is going to call the shop steward. You enjoy playing basketball and think that this order is unreasonable.

The fax indicates that the reason for the round ball ban is that the department has been incurring excessive medical expenses and disability retirements due to the "friendly" games of volleyball and basketball. The chief administrative officer of the municipality has told the fire chief to immediately halt all round ball sports.

1. As the fire officer, what will you do?

Introduction to the Fire Officer's Job

When we think about the duties of a fire officer, the most prominent functions usually relate to leading a team of fire fighters into a challenging emergency situation. Emergency duty is certainly an interesting and challenging part of the job; however, much of what a fire officer does during a normal workday is often routine administrative activities related to the work group and the physical facility. The fire officer is responsible for managing a work unit within the fire department, and many basic functions have to be performed to ensure that the work unit will be prepared to function effectively and efficiently when it is needed. A large part of the officer's time has to be spent managing personnel, resources, and programs.

Fire officers usually report to higher-ranking chief officers. In a large fire department, the company officer's direct supervisor is usually a battalion chief or an individual at a rank level equivalent to that of battalion chief. In a smaller organization, a fire officer might report directly to the fire chief or to a deputy or assistant chief. In most cases, this supervisor is an individual who has spent time coming up through the ranks and has a good understanding of what to expect from an effective fire officer.

The Fire Officer's Vital Tasks

A group of battalion chiefs in Fairfax County, Virginia, were asked to list their expectations from a good fire officer. They produced a list of four basic tasks that they consider vital: morning report, notifications, decision making, and problem solving. They believed that a fire officer who meets these expectations is on the right track.

The Morning Report

The battalion chiefs emphasized the importance for a new fire officer to provide a prompt and accurate morning report (**▼ Figure 3-1**). The morning report is a form that is faxed from every fire station to the battalion chief within 15 minutes of the 0700 reporting time. The battalion chiefs rely on this information to make staffing adjustments at the beginning of the shift. An accurate report is needed to ensure that adequate staffing and equipment are in place and ready for the next 24 hours (**▶ Figure 3-2**).

The first part of the report provides the on-duty staffing information and sick leave list and identifies any positions that need to be filled for that day. The positions are a priority because someone who worked the previous 24-hour shift

Figure 3-1 The morning report is a vital company officer task.

Today's Date is: May 29

	Total	Paramedics	Fire Officer	EMS Officer	Prearranged Callback	Annual Leave	Vacancies	Detail Out-of-Operations	Injury Lv or Light Duty	LWOP	Fire OIC	EMS OIC
Minimums	10	3	2	1							**Fire OIC**	**EMS OIC**
Today's staffing	9	3	1	1	1	1	0	1	0	0	Smyth	Willow
Next day staffing	8	2	1	0							Smyth	????

Today's shortage	Why?	PM Surplus	Next Day's Shortage	Why?
Engine Officer	O/R	none	Engine Officer	Off Rep
			Medic Officer	Leave

Sick Leave	Detailed Out of Ops	Next Day APPROVED Leave
None	Capt. Johnson	FF Tolliver
		Lt. Willow

Injury Leave/Light Duty
None

	Vehicle Status	
	Engine 7746	Eng 46
Messages for the Chief:	Rescue 7099	Res 46
Vehicle 7234 overdue for preventative maintenance	Medic 6322	Med 46
Rescue 46 thermal imager broken	Reserve Engine 7234	Eng 35
Furnace malfunctioning	Reserve Medic 4276	Med 11
Fire Chief at 46 for dinner @ 1830	Battalion 9 5040	shop
	Reserve Suburban 5107	BC 09

Figure 3-2 Example of a municipal morning report.

has to remain on duty until a relief person shows up to fill any position that is not covered by the on-coming shift. The battalion chief moves available staff to cover the vacancies noted on the morning reports, and then any vacancies that remain have to be covered by fire fighters working overtime. It takes time to reassign on-duty personnel, call the overtime personnel, and get everyone to their work locations so that the holdover personnel can be released.

The goal of the six on-duty battalion chiefs is to calculate their staffing ratios so that all of the personnel moves are in place by 0745. It is maddening when a new fire officer sheepishly calls the battalion chief at 0830 to report that a fire fighter from the previous shift is still on involuntary holdover because someone failed to note the problem on the morning report. By the time the position is covered, the fire fighter on involuntary holdover may have remained on duty for several more hours.

The morning report also projects the anticipated staffing for the next day. This allows the battalion chief to make assignments in anticipation of the known absences for the next workday, instead of trying to solve all of the problems during the first 45 minutes of the shift.

In addition, the morning report notes the location and condition of all of the apparatus or rolling stock, such as a reserve pumper that has been loaned to another station or an ambulance at the shop. Finally, the report provides the battalion chief with any "must know" information that will require the chief's immediate attention.

Notifications

The second most important issue noted by the battalion chiefs was that the new fire officer must make prompt notifications. Some types of information have to be passed up the chain of command quickly. For example, all injury and infectious disease exposure reports have to be processed without delay. If a fire fighter is exposed to a possible blood-borne pathogen early Saturday morning, the exposure report cannot sit on the fire officer's desk until Monday morning before it gets to the battalion chief. The "designated officer" has to be informed immediately to get the patient information while it can still be easily obtained. The same priority applies to any information that the battalion chief needs to know about when it is current, particularly before someone at a higher level calls to ask about it.

Decision Making and Problem Solving

The battalion chiefs rated the third and fourth issues as equal in importance. Some chiefs complained that new fire officers were hesitant to make decisions. They seemed to want the battalion chief to make the hard decisions and to enforce unpopular rules. The chiefs wanted the new officers to run their companies and make the decisions that are within their scope of responsibility. The chiefs were available for consultation, but they expected their officers to run the fire stations.

Some of the chiefs also noted that they spent a lot of time listening to fire officers complain about problems without proposing any solutions. The chiefs preferred the officers who would think through the problem and propose a solution. The most valuable proposals would consider the larger picture of how a possible solution would affect the other fire companies in the battalion ▶ Figure 3-3 .

Example of a Typical Fire Station Workday

A fire officer is responsible for accomplishing the fire department mission through the efforts of the fire fighters under his or her command. At the company level, this requires a balance of management and leadership skills. The officer has to organize the work and provide the leadership to ensure that it gets done safely and effectively.

The fire department has an agency-wide mission that is translated into annual goals. These goals are used to develop annual, quarterly, and monthly objectives for each fire company. The battalion chief and the officer meet regularly to set the objectives and to review progress. The monthly goals show up as the planned activities on the fire company daily planner.

The following is an example of a typical workday schedule for a 24-hour shift in a fire station. The starting point is a schedule that ensures that all of the required tasks are completed in a logical order. The fire officer has to anticipate that emergency incidents will alter the workday and will require adjustments in the schedule.

0700	Line-up and equipment check. Send morning report to battalion chief. Clean quarters, empty trash, clean dishes.
0800	Dust and vacuum all carpeted areas. Sweep all tile floors.
0830	Physical training and outside skill drill.
1100	Heavy cleaning (while still in physical training clothes).
	• Monday—Air out bunkroom and rotate mattresses, clean all windows.
	• Tuesday—Clean utility rooms and shop area.
	• Wednesday—Clean and inventory EMS, SCBA, and decontamination areas.
	• Thursday—Move recyclables outside for pickup, then clean weight room and lockers.
	• Friday—Scrub kitchen and clean out refrigerators.
1130	Scrub bathrooms after fire fighters clean up from physical training.
Noon	Lunch
1330	Scheduled productivity activity (e.g., fire safety inspections, school visits, inside or outside training).
1800	Dinner, followed by kitchen clean-up. Run dishwasher.
1930	Individual study time, occasional fire safety inspections (nightclubs) or drills.
2130	Remove all trash, tidy up day room, and make final pass through the kitchen.

Figure 3-3 Discuss possible solutions to a problem with the chief.

Special station activity
- First and third Thursdays of the month—Scrub apparatus bay floor.
- Second and fourth Fridays of the month—Wax kitchen floor.
- February—Safety officer inspection of facility, apparatus and PPE.
- March and September—Steam clean carpeted areas.
- April or May—Chief's annual inspection of the fire station.
- October—Fire Prevention Open House.

Typical Volunteer Duty Night

The following example of a volunteer duty night is less detailed but has the same essential tasks as a 24-hour shift in a municipal fire department, including equipment main-tenance, training, and station maintenance. Many departments use a day book to plan the duty crew activities.

1700	Evening duty crew starts. Equipment check.
1800	Dinner, followed by kitchen clean-up. Run dishwasher.
1930	Classroom session, skill drill, or community outreach activity.
2130	Remove all trash, tidy up, and make final pass through the kitchen. Take clean dishes and cups out of dishwasher and put away.

General station and apparatus cleaning are conducted on weekends. One section of the fire station gets a major cleaning every 2 weeks (e.g., wax floors, scrub apparatus floors). Specialized heavy cleaning, such as steam cleaning carpets and waxing apparatus, is scheduled throughout the year (▼ **Figure 3-4**).

Duty Crew for: Tuesday, Feb xx

| Duty Chief | Alban |
| Duty Officer | Denton |

Staffing:

Engine 712

Officer	O'Sullivan
Driver	Cropper
Nozzle	Jenkins
Hook	Carol Smith
Right bucket 2	J. Thompson (rookie)
Left bucket 2	Harrold (rookie)

Equipment status:

Engine 712	at shop
Engine 712-2	as E712
Reserve engine	in quarters
Truck 712	in quarters
Brush 712	in quarters
Utility 712	in quarters
Asst Chief 712	at dealership
Chief 712	with chief

Truck 712

Officer	Denton
Driver	Stevens
Forcible Entry	Brize
Can	K. Anderson
Outside 1	E. Walters (rookie)
Outside 2	

Planned Activities:

Training
Ground ladders
CPR practice

Inspection

Community outreach

Upcoming activities:
15 Feb CPR - AED recert
08 Mar SCBA proficiency

Station maintenance
Small bunkroom
Truck compartment 4
Follow-up
Rookie Test #4

Figure 3-4 Example of a volunteer duty day book entry.

The Transition From Fire Fighter to Fire Officer

There are four times in a fire fighter's career when a major change occurs in how the individual relates to the formal fire department organization. The first change occurs when the fire fighter completes the probationary training period. Completing the initial training and probationary period is a major milestone. It is marked in many departments by a change in helmet color or shield. The second change takes place when the fire fighter successfully completes a promotional process and starts working as a fire company commander. The third event is when the fire officer completes another level of training, advances through the promotional process, and starts working as a chief officer. The fourth event is when a fire fighter retires from active duty.

In all four situations, a significant change occurs in the individual's relationship to the organization and to the other members of the fire department. Part of this change is the individual's sphere of responsibility within the formal organization. As a fire fighter, there is a mutual responsibility among the crew members on the rig. They work together to accomplish fireground tasks and look out for each other.

A promotion from fire fighter or driver/operator to company-level officer is a large step. The company-level officer is directly responsible for the supervision, performance, and safety of a crew of fire fighters. There is a sacred duty to ensure that all of the fire company team members remain safe when operating in a hostile or hazardous work environment. The company-level officer functions as a working supervisor, sharing the hazards and work conditions with the fire company team (▼ **Figure 3-5**).

This change in the sphere of responsibility often requires the new officer to change some on-duty behaviors or practices. The formal fire department organization considers a fire officer to be the fire chief's representative at the work location. This creates an expectation that the fire officer will behave in a way that is appropriate for a first-line supervisor. Behavior that was acceptable as a fire fighter may be unacceptable for a fire officer. For example, consider a fire fighter who is known for developing elaborate practical jokes within the fire station environment. As a new fire officer, this individual would have to consider the impact that being seen as a practical joker would have on his or her ability to function as an effective company commander.

Conversely, the new fire officer needs to consider how to respond to pranks and verbal jabs from fire fighters. What may have been an appropriate response from another fire fighter may no longer be the best response from an officer. Consider that wearing the fire officer badge enhances the effect and consequences of any action or response.

Promotion to a chief officer rank changes the individual's relationship to the organization and the members to an even greater degree. A command officer has less of a hands-on role than an officer of a company. The command officer is typically directly responsible for several fire companies and must depend on company officers to provide direct supervision over crews performing fireground tasks, often in hazardous conditions. In many cases, the command officer works outside the hazardous area but is still responsible for whatever happens inside.

Fire Officer as Supervisor-Commander-Trainer

In his book, *Effective Company Command,* James O. Page divided the company officer's duties into three distinct roles: supervisor, commander, and trainer.

Supervisor

In this role, the fire officer functions as the official representative of the fire chief. That means that the fire chief expects that every fire officer will issue orders and directives and

Figure 3-5 The company-level officer is responsible for the supervision, performance, and safety of a crew of fire fighters.

conduct business in a way that meets the chief's objectives. The fire officer will supervise the fire company in a manner that is consistent with the rules and regulations of the fire department. For example, if all fire companies are expected to spend Tuesday afternoon conducting fire safety inspections in commercial properties, the fire officers are responsible for ensuring that their individual fire stations complete this task.

Unpopular Orders and Directives

On occasion, a fire officer may be required to issue and enforce unpopular orders (▼ **Figure 3-6**). Even if the officer disagrees with a particular directive, the formal organization requires and expects the officer to carry out that directive to the best of his or her ability. A fire officer can improve his or her effectiveness in handling an unpopular order by determining the story or history behind the order, which would enable the fire officer to put the directive in perspective.

When faced with an unpopular order, the fire officer should express any concerns and objections with his or her supervisor in private. This is the time for the fire officer to discuss suggestions for modifying or reversing the order. Occasionally, special circumstances may make the order difficult to implement, and the supervisor might be able to authorize adjustments. Once the meeting with the officer's supervisor is over, the formal organization expects the officer to enforce the order as issued or amended.

Telling the fire fighters that their officer does not agree with an order undermines the officer's authority and supervisory ability. The fire chief expects an officer to perform the required supervisory tasks, and the fire fighters have to understand that the fire officer does not make all the rules or have a choice about which ones to enforce. Enforcing unpopular orders is part of the job.

Commander

When operating at the scene of an emergency incident, the fire officer is expected to demonstrate a special type of supervisory technique. A fire officer has to function as a commander and exercise strong direct supervision over the company members. In some cases, the fire officer could be responsible for directing the actions of additional resources, or the fire officer might be functioning as the initial incident commander (▼ **Figure 3-7**).

Functioning as the initial incident commander on a major emergency is one of the higher-profile roles of a fire officer. The ability to bring order out of the chaos of an emergency incident is an art that requires a well-developed skill set. The fire officer needs to be clear, calm, and concise in the initial radio transmissions. The communication of incident size-ups must be consistent with the organization's requirements and incident management system.

Developing a command presence is a key part of mastering the art of incident command. Command presence is the ability of an officer to project an image of being in control of the situation. In order to be a successful leader, the officer must convince others to follow by demonstrating the ability to take charge and make the right things happen. A fire officer who is going to establish command upon arriving at an emergency incident should have a detailed knowledge of the responding companies, a mastery of the local procedures, and the ability to issue clear direct orders. Fire fighters are aggressive, action-oriented people who also can be compulsive. It is important that a new fire officer develop a command presence in order to focus the efforts of this action-oriented team (▶ **Figure 3-8**).

Figure 3-6) An officer sometimes has to issue and enforce unpopular orders.

Figure 3-7) At the scene of an emergency incident, the fire officer may function as the initial incident commander.

Figure 3-8 It is important for the fire officer to develop a command presence.

Figure 3-9 The officer must ensure that fire fighters are confident and competent in their skills.

Trainer

The fire officer has the responsibility of making sure that the fire fighters under his or her command are confident and competent in their skills. The company-level officer is responsible for the performance level of the fire company and has to establish a set of expectations that the company will perform at the highest level possible. ▶ **Figure 3-9**. In a large department, one fire company may need to have a higher level of specialized skill in one area than in another. For example, a fire company may have a specialty assignment or a co-located specialized unit that the fire company staffs when it is requested. Technical rescue rigs, decontamination units, foam pumpers, and mobile command posts are examples of co-located specialized units.

In addition, the fire company's response district may require a higher level of fire fighter skill or knowledge. Consider hose handling skills. A fire company that is in a high-rise district would require different hose handling skills and competencies than those required of a company in a mountainous wildland interface area. Although both companies need to demonstrate the basics of operating a hoseline from a standpipe, the high-rise fire company should have a much higher level of expertise and competence in this skill. Conversely, the mountain fire company usually has a higher level of expertise and competence in rope rescue evolutions.

The company-level officer is the key to developing these competencies within the company. Page makes three specific recommendations to assist fire officers in this task: develop a personal training library, know the neighborhood, and use problem-solving scenarios.

Developing a Personal Training Library

James O. Page's personal training library starts with a three-ring notebook with subject matter tabs. The subject tabs could be from the topic headings in NFPA 1001, from the local recruit school curricula, or from a personal list of important topics. Every time the fire officer attends a training event, the notes from that session are placed into the three-ring binder ▶ **Figure 3-10**. Every related handout, product information sheet, and other related items are also placed

Fire Marks

When retired FDNY Deputy Chief Vincent Dunn was a chief's aide in the 1960s, he drove for two different deputy chiefs on alternating days. Deputy chiefs were dispatched on multiple-alarm fires to function as the senior fireground commander. They directed the incident with the assistance of four to six battalion chiefs. Dunn compared the command presence of the two chiefs.

One of the deputy chiefs would arrive at the scene of a fire and not announce his arrival over the radio. He would don his helmet and fire coat but stay in his office shoes while he walked around the incident scene. He would stand next to the battalion chief who was running the incident and observe. This deputy chief would have little impact on emergency operations.

The other deputy chief always maintained a high command presence when he responded to an incident. He would immediately announce his arrival on the fireground over the radio. He would put on his fire boots, fire coat, and helmet and immediately report to the battalion chief who was running the incident. While the battalion chief provided a face-to-face status report to the deputy chief, Dunn would go to the rear of the fire building to observe conditions and report back to the deputy. Even if the battalion chief continued to function as the incident commander, everyone knew that the deputy chief was on the scene and providing direction. This deputy chief's command presence made him a much more effective leader of his team.

Figure 3-10 Notes from training events should be placed in a three-ring binder.

Figure 3-11 Highlight, tab, and write in your personal copies of textbooks used for training.

into the three-ring binder. When the officer is preparing to present a class that covers that topic, the personal training library is the first stop.

Today, a fire officer can make an electronic version of Page's three-ring binder with the use of a laptop computer and a digital scanner. The fire officer can collect PowerPoint® presentations and video clips, download files from the Internet, and convert various paper media into Adobe® portable document format (.pdf) files. A projector allows all these materials to be used in classroom sessions.

In addition to the three-ring notebook or laptop computer, Page recommends that the fire officer obtain personal copies of the textbooks used in fire fighter training and promotional examinations. You can highlight, tab, and write in this copy of the book until it becomes your personal reference. You can write notes in the margins and even note disagreements with the author's statement or point of view to personalize the learning experience. This is especially valuable when you are reading a book in preparation for a promotional exam (▶ **Figure 3-11**).

Know the Neighborhood

Fire fighters should have a detailed knowledge of the environment they protect. That requires going out to walk through each nonresidential structure, from the roof through the subbasement; these walk-throughs reinforce the written preincident plan or diagrams (▶ **Figure 3-12**). During the walk-through, the fire officer should take pictures to capture significant details. A detailed overhead view of many communities can be obtained from aerial photographs and digital maps. This aerial view is especially valuable for apartment complexes, business parks, and retail areas. The maps and photos can be used to locate access routes, plan the placement of first-alarm apparatus, and identify exposure problems.

Knowing the neighborhood may include working with the building owners and occupants to practice incident action plans and procedures. After the September 11, 2001 terrorist attacks, many commercial and office buildings revised their internal emergency plans, upgraded their

Figure 3-12 Walk-throughs help fire fighters get to know the properties in the neighborhood.

security systems, and introduced access restrictions. Fire officers should strive to establish a working relationship with every building manager or emergency program manager in their response district.

Use Problem-Solving Scenarios

The fire officer can help the company members become more skilled and knowledgeable by providing opportunities to use their problem-solving skills. Instead of reading the code regulations to provide training for his company members, James O. Page would present fact-based situations and require them to use the code to solve the problem. This technique forced the fire fighters to identify the occupancy use group, identify the issues, look up the applicable regulations, and make decisions. This is an excellent way for adults to learn concepts, regulations, and decision-making skills. The same technique can be used to reinforce fire hydraulics principles and other applications of principles or concepts.

The same problem-solving approach can be used in reviews of preincident action plans. The fire officer can construct various emergency incident scenarios for training sessions, based on actual buildings and potential situations. There are computer-based fire incident simulators that can accept digital pictures taken during walk-through visits to create customized emergency incident scenarios.

The Fire Officer's Supervisor

Every fire officer has a supervisor. In municipal fire departments, a fire officer's supervisor is usually a command-level officer (a battalion chief, a district chief, or a battalion commander) who supervises numerous fire companies within a geographical area. The battalion chief's supervisor could be a deputy chief, who reports directly to the fire chief. In the same manner, the fire chief would report to the mayor, the city manager, the commissioner of public safety, or an individual at an equivalent level. Under the chain of command, the fire chief's orders and directives are passed down through the deputy chief, to the battalion or district chief, who then ensures that the fire officers enact the orders or directives.

Regardless of the organization structure, every fire officer has an obligation to work effectively with a supervisor. Three activities are necessary for this:

- Keep your supervisor informed
- Make appropriate decisions at your level of responsibility
- Consult with your supervisor before making major disciplinary actions or policy changes

No supervisor likes surprises. Part of the supervisor-subordinate relationship is to make sure that the supervisor is not surprised or blindsided (▶ **Figure 3-13**). For example, if a fire company has a vehicle crash at 10:00 P.M., the battalion chief needs to know about it within the hour. It would

be poor form for the battalion chief to find out about the accident when the fire chief calls to ask about a story that was on the 11 o'clock news.

Officers are expected to make decisions. Fire officers should not hesitate to make decisions when the decision should be made at their level of responsibility. This means that problems should be addressed and situations resolved where and when they occur. If a fire officer has the authority to solve a problem, he or she should not be waiting for the fire officer's supervisor to arrive to solve it.

A company-level officer is expected to make appropriate company-level decisions. This usually covers fire station level activities such as maintenance, training, and community outreach. At fire stations with rotating career work shifts, the expectation is that most issues between the shifts are resolved at the company officer level. A fire officer who is working in a staff position should make the decisions that are appropriate for that position. Volunteer officers should make decisions that are appropriate for their level within the organization.

However, some issues require the fire officer to consult with a supervisor before making a decision or taking action. Policy changes cannot be performed in an administrative vacuum. If a decision is going to have an impact that goes

Figure 3-13) Always keep your supervisor informed.

beyond the fire officer's scope of authority, it is time to talk to the supervisor.

This policy also applies before major disciplinary actions are taken. For most fire officers, consultation with a supervisor is required by the municipal personnel regulations to ensure that all major discipline is delivered in a consistent and impartial manner. This is also a recommended practice in most volunteer fire departments.

The Vital Importance of Integrity and Ethical Behavior

The formal organization provides the new fire officer with the symbols of power and authority that are associated with the badge, insignia, and distinctive markings on the officer's helmet. The individual needs to provide the core values of integrity and ethical behavior that, combined with the formal symbols, create an effective fire officer. An unethical fire officer is ineffective and damages the department's reputation. Left uncorrected, unethical and corrupt fire officers corrode the department's ability to deliver services and maintain the public trust.

Integrity

Integrity refers to the complex system of inherent attributes that determine a person's moral and ethical actions and reactions, including the quality of being honest.

The fire officer should "walk the talk" and demonstrate the behaviors that he or she says are important (▶ Figure 3-14). If the company officer says that physical fitness is important, then the fire fighters should see the their officer performing physical fitness training during the workday.

Integrity can be demonstrated by a steadfast adherence to a moral code. This is a combination of a fire officer's internal value system and the fire department's official organizational value system. Formal organizations publish their expectations as a code of ethics, a code of conduct, or a list of value statements.

Ethical Behavior

The fire officer position provides a wide range of opportunities to demonstrate ethical behavior. The fire officer demonstrating **ethical behavior** makes decisions and models behavior that are consistent with the department's core values, mission statement, and value statements.

Paragraph A.1.3 in the "Annex A: Explanatory Material" section of NFPA 1021, *Standard for Fire Officer Professional Qualifications*, 2003 edition, makes the following statement:

Fire officers are expected to be ethical in their conduct. Ethical conduct includes being honest, doing "what's right," and performing to the best of one's abil-

ity. For public safety personnel, ethical responsibility extends beyond one's individual performance. In serving the citizens, public safety personnel are charged with the responsibility of ensuring the provision of the best possible safety and service.

Ethical conduct requires honesty on the part of all public safety personnel. Choices must be made on the basis of maximum benefit to the citizens and the community. The process of making these decisions must also be open to the public. The means of providing service, as well as the quality of the service provided, must be above question and must maximize the principles of fairness and equity as well as those of efficiency and effectiveness.

The International Association of Fire Chiefs Code of Ethics provides an example of a code of ethics. This code was updated in 2003:

The purpose of the International Association of Fire Chiefs is to actively support the advancement of the fire service, dedicated to the protection and preservation of life and property against fire, provision of emergency medical services and other emergencies. Towards this endeavor, every member of the International Association of Fire Chiefs shall represent those ethical principles consistent with professional conduct as members of the IAFC:

- Recognize that we serve in a position of public trust that imposes responsibility to use publicly owned resources effectively and judiciously
- Not use a public position to obtain advantages or favors for friends, family, personal business ventures, or ourselves

Figure 3-14 The fire officer should demonstrate the behaviors that he or she says are important.

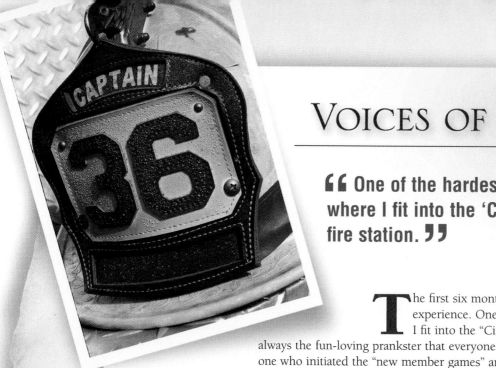

VOICES OF EXPERIENCE

❝ One of the hardest things to learn was where I fit into the 'Circle of Life' within the fire station. ❞

The first six months as a fire officer was a constant learning experience. One of the hardest things to learn was where I fit into the "Circle of Life" within the fire station. I was always the fun-loving prankster that everyone got along with. As a fire fighter, I was the one who initiated the "new member games" and I would rely on my fire officer to tell me if the games were going too far. Now that I was a fire officer, I was the one who had to ruin all the fun. In my mind, I had become the enemy. Fire fighters would see me coming and change the conversation. My own company would also avoid unnecessary conversation with me.

This weighed heavy on my heart. So I did something stupid—I put myself back in the role of "one of the guys." I started joking around and playing pranks on my company. Eventually, this got me into trouble when one of the senior members saw this. After a discussion with that senior member, I came away with a better understanding of what my duty as a fire officer was. You do not have to completely exclude yourself from your personnel, but you must know when to draw the professional line. You need to spend quality time with your personnel so that you can learn the different strengths of the men and women that you lead. You can have open and friendly communications with your company without crossing that professional line. You must let your people know what your role is. Becoming a fire officer **does** mean that you are no longer "just one of the guys."

Matthew Thorpe
City of King, North Carolina Fire Department
12 years of service

- Use information gained from our positions only for the benefit of those we are entrusted to serve
- Conduct our personal affairs in such a manner that we cannot be improperly influenced in the performance of our duties
- Avoid situations whereby our decisions or influence may have an impact on personal financial interests
- Seek no favor and accept no form of personal reward for influence or official action
- Engage in no outside employment or professional activities that may impair or appear to impair our primary responsibilities as fire officials
- Comply with local laws and campaign rules when supporting political candidates and engaging in political activities
- Handle all personnel matters on the basis of merit
- Carry out policies established by elected officials and policymakers to the best of our ability
- Refrain from financial investments or business that conflicts with, or is enhanced by, our official position
- Refrain from endorsing commercial products through quotations, use of photographs, testimonials, for personal gain
- Develop job descriptions and guidelines at the local level to produce behaviors in accordance with the code of ethics
- Conduct training at the local level to inform and educate local personnel about ethical conduct and policies and procedures
- Have systems in place at the local level to resolve ethical issues
- Orient new employees to the organization's ethics program during new employee orientation
- Review the ethics management program in management training experiences
- Deliver accurate and timely information to the public and to elected policymakers to use when deciding critical issues

Fire department activity tends to be high profile, regardless of the task. A simple trip to the grocery store to pick up dinner creates public attention. The fire officer should act as if someone is always documenting his or her actions when out of the fire station.

Workplace Diversity

The civil rights of Americans are established by federal laws, which are enforced by the EEOC. Title VII of the Civil Rights Act of 1964 covers state and local governments, schools, colleges, and unions. This law states that it is illegal for an employer to:

"(1) fail or refuse to hire or discharge any individual, or otherwise discriminate against any individual with re-

spect to compensation, terms, conditions, or privileges of employment because of such individual's race, color, religion, sex, or national origin, or (2) to limit, segregate, or classify employees or applicants for employment in any way that would deprive any individual of employment opportunities or otherwise adversely affect status as an employee because of such individual's race, color, religion, sex or national origin."

The Equal Employment Opportunity Act of 1972 amended the Civil Rights Act of 1964 and expanded its coverage to include almost all public and private employers with 15 or more employees. In general, the 1972 act covers volunteer fire departments and other nonprofit emergency service organizations.

The Civil Rights Act of 1991 provides additional compensatory and punitive damages in cases of intentional discrimination under Title VII and the Americans with Disabilities Act of 1990. The changes that were introduced in the 1991 act increased the number of discrimination lawsuits filed against organizations.

Many fire departments have made changes to their recruitment, hiring, and promotion practices in order to comply with the civil rights laws. Some departments took action to diversify their work forces without the prompting of the courts. Where this did not occur, the legal remedies have ranged from consent decrees, in which the fire department agrees to accomplish specific diversity goals within a specific time, to court orders outlining specific hiring practices.

Diversity as applied to fire departments means that the work force should reflect the community it serves. In all cases, the overarching goal is for the fire department to reflect the diversity of the community. Consider the example of a fire department that is 90% Caucasian in a community where the population is 40% African-American, 20% Latino, and 20% Asian. The fire department population is not reflecting the community.

If the department has agreed to an EEOC consent decree, it has promised the court that it will work to hire qualified individuals who reflect the community. A consent decree can require a variety of activities, including community outreach, job fairs, pre-employment preparation, and peer group coaching. The court formally meets with the fire department representative periodically to see how well the department is progressing. Some fire departments that worked under Justice Department consent decrees in the 1970s have successfully met the court's goals and are no longer under court supervision.

Some departments operate under a specific court-mandated hiring process. Using our earlier example, the court could require the department to hire two African-Americans, one Latino, and one Asian before it can hire one additional Caucasian. This requirement would remain in effect until the hiring diversity goal is accomplished.

The Fire Officer's Role in Workplace Diversity

Jack W. Gravely is a lawyer, subject matter expert, and trainer on workplace diversity. He has been a frequent speaker at public safety agencies on issues of racial and cultural diversity. From 1985 to 1988, he was the Special Assistant to the County Manager for Equal Employment Opportunity/ Affirmative Action (EEO/AA) in Arlington County, Virginia. Gravely developed the policies and staffing changes necessary to address the sudden influx of Central Americans, Asians, and Ethiopians who moved into the county. Gravely was named Director of Workplace Diversity for the Federal Communications Commission in 1995.

Gravely points out that a fire officer today has the benefits of four decades of EEO/AA court decisions to guide decision making. When he started diversity training classes, the emphasis was on the language of the regulations and their potential impact. The EEOC files about 400 lawsuits every year, so now there are hundreds of court decisions and a large body of case law that present a clear and generally consistent policy on how a supervisor should behave in the workplace. Gravely recommends that a fire officer focus on actionable items and hostile workplace.

Actionable Items

Actionable items are employee behaviors that require an immediate corrective action by the supervisor. Dozens of lawsuits have shown that failing to act when these situations occur is likely to create a liability and a loss for the department. The best example is the use of certain words in the workplace. An employee's use of derogatory or racist terms about people from other ethnicities or genders requires immediate corrective action by the supervisor. Regardless of the conditions, context, or situation, such words are inappropriate and represent a potential million-dollar liability to the organization.

The fire officer must act immediately in these situations. That means speaking with the offending fire fighter, in private, and counseling the fire fighter that the use of such words is unacceptable in the fire station or in any situation while the individual is representing the department (either on duty or off duty but in uniform.) The fire officer should provide the fire fighter with the fire department's or municipality's EEO/AA policy statement and, if applicable, the code of conduct. The fire officer should also maintain a formal or informal record of the counseling session ► **Figure 3-15**). If there is any doubt that the message was understood, the fire officer should ensure that a higher-level supervisor is informed of the action that has been taken.

The same policies should apply to fire fighters regularly assigned to the fire company, fire fighters detailed in or visiting the fire station, and other uniformed or civilian members

Figure 3-15 Privately discuss inappropriate behaviors with fire fighters.

of the fire department. Case law has shown that use of unacceptable language requires an immediate response. An officer's failure to act has been interpreted as the department's official condoning or encouragement of such behavior.

Although there is enough case law to provide general guidelines and some specific examples, the subject of what constitutes harassment remains a dynamic aspect of the work environment. For example, in one city, a judge ruled that use of the term "boy" contributed to a hostile workplace environment in the fire department. A new recruit complained that the term was used to harass him. In this instance, the recruit was Caucasian and the two senior fire fighters accused of harassment were African-American.

The fire officer needs to stay informed about the organization's EEO/AA and diversity policies. Most large organizations have a diversity or EEOO/AA office that can provide up-to-date information and answer questions.

Hostile Workplace and Sexual Harassment

In 1993, the EEOC amended the guidelines on sexual harassment. The amended regulations broadened the types of harassment that are considered illegal and included a requirement that employers have a duty to maintain a harassment-free work environment. The standard for evaluating sexual harassment is what a "reasonable person" in that same or similar circumstances would find intimidating, hostile, or abusive. The 1993 guidelines clearly state that employers are liable for the acts of those who work for them if the organization knew or should have known about the conduct and took no immediate, appropriate, corrective action. That is why a fire officer must immediately respond to any utterances of offensive or derogatory language in the work environment.

Sexual harassment is unwanted, uninvited, and unwelcome attention and intimacy in a nonreciprocal relationship.

The abuse of power is an essential component of sexual harassment. The 1993 EEOC guidelines state that verbal and physical conduct of a sexual nature is harassment when the following conditions are present:

- The employee is made to feel that he or she has to endure such treatment in order to remain employed
- Whether or not the employee submits or rejects such treatment is used when making employment decisions
- The employee's work performance is affected
- An intimidating, hostile, or offensive work environment is present

The term "hostile work environment" can be used to describe a broad range of situations in which an employee is subject to discrimination in the workplace. The generalized hostile workplace definition can apply to a variety of circumstances that do not necessarily include a specific abuse of supervisory power.

Gravely believes that hostile workplace complaints will shape workplace diversity in the 21st century. There are few quid pro quo sexual harassment cases—cases in which a supervisor specifically promises a work-related benefit in return for a sexual favor. The trend is toward more complaints about a hostile workplace, which can involve sexual issues.

Handling a Harassment or Hostile Workplace Complaint

Fire fighters who want to initiate a harassment complaint have a choice of three methods. They can start with the federal government, they can start with the local government, or they can start within the fire department—it is their choice. If the process starts within the fire department, a fire officer may be the first formal point of contact. In this case, the fire officer would function as the first step of a multistep procedure.

The fire officer should know the department's procedure for handling a harassment or hostile workplace complaint. The fire officer's designated role in conducting an investigation of an EEO complaint depends on the procedures adopted by the jurisdiction or the fire department. In some cases, an officer's role is limited to starting the process by making appropriate notifications. Many local governments have specially trained EEO staff that conduct an investigation. In other cases, the fire officer might be required to perform the initial investigation and submit a report. Below are some general guidelines:

- **Keep an open mind.** Many fire officers have a hard time believing that discrimination or harassment could be happening right under their noses. Failure to investigate a complaint is the most common reason that a local government is found liable. Every complaint must be investigated and documented. Do not come to any conclusions until your investigation is complete.

Follow your organization's procedures and inform your supervisor.

- **Treat the person who files the complaint with respect and compassion.** Employees often find it extremely difficult to complain about discrimination or harassment. When an employee comes to you with concerns about discrimination or harassment, be professional, but also be understanding.
- **Do not blame the person filing the complaint.** Case law and current practice dictate that it is the complainant who determines whether the situation is hostile or is harassment. Blaming the person bringing the complaint to you is the second most common way that local governments lose harassment or hostile workplace complaints.
- **Do not retaliate against the person filing the complaint.** It is against the law to punish someone for complaining about discrimination or harassment. The most obvious forms of retaliation are termination, discipline, demotion, or threats to do any of these things. More subtle forms of retaliation could include changing the shift hours or work location of the accuser, even if the intent is to remove the alleged victim from the problem.
- **Follow established procedures.** Local government personnel regulations generally provide a detailed procedure on how to handle harassment or hostile workplace complaints. Follow the procedures.
- **Interview the people involved.** An initial investigation usually involves conducting interviews of the people who are involved in the situation. This usually starts with the person who made the complaint. The interviewer needs to find out exactly what the employee is concerned about. Get details: what was said or done, when and where, and the names of those present? Then talk to any employees who are being accused of discrimination or harassment. Get details from them as well. Be sure to interview any witnesses who may have seen or heard any problematic conduct. Take notes about your interviews and gather any relevant documents.
- **Look for corroboration or contradiction.** Discrimination and harassment complaints often involve "he said/she said" situations. The accuser and the accused offer different versions of an incident, leaving you with no way of knowing who is telling the truth. You may have to turn to other sources for clues. Witnesses may have seen part of an incident. In some cases, documents, such as e-mails and posted notes, prove one side right.
- **Keep it confidential.** A discrimination complaint can polarize a workplace. The unique team nature of 24-hour fire service work creates a close environment

where it is difficult to maintain confidentiality. The fire officer must insist on and enforce confidentiality during the investigation.

- **Write it all down.** Take notes during all interviews. Before the interview is over, go back through your notes with the interviewee to ensure accuracy. Keep a journal of the investigation. Write down the steps you have taken to get at the truth, including dates and places of interviews. Keep a list of all documents that are reviewed. Document any action taken against the accused or the reasons for deciding not to take action. Anticipate that all of this written record will be used in any subsequent civil service or court actions.
- **Cooperate with government agencies.** If the fire fighter files a complaint with another government agency (either the federal EEOC or an equivalent state agency), that agency may investigate. Notify your supervisor as soon as you receive a call or visit from an EEOC investigator. You will probably be asked to provide certain documents, to give your side of the story, and to explain any efforts you made to deal with the complaint yourself. Be cautious, but cooperative.

Regardless of where the complaint is filed, the fire chief is required to take corrective action if the investigation confirms that the complaint has merit. If the department concludes that some form of discrimination or harassment occurred, formal corrective action could include mandatory training, work location transfer, or demotion. Termination may be proposed for the more egregious kinds of discrimination and harassment, such as threats, stalking, or repeated and unwanted physical contact.

The Fire Station as a "Business Work Location"

The fire officer needs to consider the fire station or other fire department facility as a business work location. This is a drastic change from the concept of a fire station as a home away from home for a group of fire fighters, but it is necessary to ensure that the fire station maintains a professional work environment. In the eyes of the law and probably in the opinion of administrators and elected officials, the same rules of behavior apply to an office in city hall at 2 P.M. or a fire station at 2 A.M.

The fire officer can help maintain an appropriate work environment by encouraging and enforcing acceptable behavior whenever fire fighters are on duty. The fire officer accomplishes this by:

- Educating the employees on the workplace rules and regulations that define expected behavior. Start with the local government's "Code of Conduct" or other documents that outline the chief administrative officer's expectations of all municipal employees. This can usually be found in the municipality's mission statement, core values, or personnel regulations.
- Promote the use of "on duty speech." The goal is not to change the thoughts or feelings of individual fire fighters, but to establish a workplace environment where certain behaviors and words are not used. Fire fighters can think what they want. However, while they are on duty, in the fire station, or in uniform, they cannot use certain words or phrases or act out certain behaviors.
- Be the designated adult. This requires the fire officer to model appropriate behavior as well as encourage and enforce the same behavior by the fire fighters. The fire officer must identify and correct unacceptable workplace behavior whenever it is observed.

Assessment Center Tips

Know Your Organization's Procedure

Even if the normal practice for your department is to have a specialized or designated person to conduct an investigation of a harassment or hostile workplace complaint, the promotional candidate must know both local and federal procedures that must be followed when an employee files a complaint. Harassment and hostile workplace complaints require a specific and detailed response by the first-line supervisor.

An assessment center scenario may involve a fire fighter coming to the candidate for advice. During the interview process, the fire fighter reveals that he or she may be a victim of harassment or hostile workplace. The candidate would be expected to explain the options available to the fire fighter in filing either a federal or a local complaint.

Getting It Done

The Fire Company Defines Its Nonhostile Environment

A fire officer who unilaterally imposes new and severe requirements that affect the language fire fighters use in the workplace would encounter much resistance. One method of fostering the concept of "on duty speech" or appropriate workplace behavior is to have the fire fighters develop their own set of rules or behaviors. Provide the fire fighters with the source documents (federal and municipal codes and regulations), and ask them to consider what words or behaviors are inappropriate while at work.

Ignoring a problem is, in reality, permitting it to continue. The fire officer who is a candidate for promotion is tested and is expected to identify, explain, and enforce the limits of unprofessional behavior (▼ **Figure 3-16**).

Figure 3-16 The company officer must identify and correct unacceptable behavior.

A company-level officer should make it a practice to walk around the fire station at various times during the workday to observe what is going on. This walk around is more important when the officer is in a big station with multiple companies and in combination career-volunteer departments, where there is a constant flow of people coming in and out of the fire station. This practice is not designed to catch someone doing something wrong; rather, it is to make sure that everything is functioning properly. The officer should routinely determine what the crew members are doing and check on the safety of the facility and equipment. At the same time, the officer should look out for unexpected situations and surprises. Knowing that the officer knows what is going on in the station and will react to inappropriate situations goes a long way toward encouraging appropriate workplace behaviors (▼ **Figure 3-17**).

Figure 3-17 The officer should routinely check in with crew members.

You Are the Fire Officer: Conclusion

As the new fire officer, you have three choices: enforce the ban as outlined by the chief's fax, modify the ban after consulting with your supervisor, or ignore the ban and permit basketball during today's physical fitness period. Ignoring the ban may make you popular with the fire fighters, but it could cost you your promotion. Ignoring a directive from the fire chief is not a career-enhancing activity.

The safest and most appropriate response to a new directive is to enforce the rule as outlined by the fire chief. Many sudden directives are the direct result of situations that occur several levels above the fire officer. The formal organization expects that the fire fighters will immediately comply.

If there is time, consult with your supervisor to get the background on the ban. In this case, the chief informs you that the round ball ban came from the city manager to the fire chief. The city manager read an evaluation of the injuries and discovered that disability costs showed profoundly excessive medical costs related to round ball injuries.

Discuss your concerns about the round ball ban with your supervisor in private. If your points are reasonable, the command officer might be able to authorize a local adjustment to the ban. In this particular case, you learn that the subject is not open for negotiations at the company level, and you will have to enforce the rule it until the situation changes.

Wrap-Up

Ready for Review

- A fire officer is responsible for accounting for the people and resources at a fire station and work location. This may require a report at the beginning of duty, outlining staffing and equipment status.
- Transitioning from fire fighter to fire officer changes how the individual relates to the formal fire department organization and the role the fire officer plays with fellow fire fighters.
- A fire officer has a larger sphere of responsibility when supervising a work group than he or she had as a fire fighter.
- A fire officer should "walk the talk" and demonstrate integrity by behaving ethically.
- Fire officers have a supervisor. Keep your supervisor informed, make appropriate company-level decisions, and consult with your supervisor before making major disciplinary or policy changes.
- The Equal Employment Opportunity Commission is the federal agency empowered to enforce compliance with the Civil Rights Act of 1964, the Equal Employment Opportunity Act of 1972, and the Civil Rights Act of 1991. Current fire department recruitment, hiring, and promotion practices are guided by the EEOC to achieve workplace diversity.
- An employee can file an EEO complaint in one of three ways: with the federal EEOC office, with the municipality's EEOC/diversity office, or with the fire department.
- The 1993 amendment of the 1980 EEOC sexual harassment guidelines clarified the issue of hostile workplace complaints. The complainant determines whether conditions are creating hostile workplace or sexual harassment.
- Follow the local procedures when encountering a harassment or hostile workplace complaint. Take notes, make appropriate notifications, and do not take sides.
- The trend in local government is to expect the fire station to comply with the same behavior rules as are applied to other offices. The fire station is a business work location.
- Fire officers should educate and encourage the concept of "on duty speech" to maintain a nonhostile workplace.

Hot Terms

Actionable items Employee behavior that requires an immediate corrective action by the supervisor because dozens of lawsuits have shown that failing to act will create a liability and a loss for the department.

Diversity A fire work force that reflects the community it serves.

Ethical behavior The fire officer makes decisions and demonstrates behavior that is consistent with the department's core values, mission statement, and value statements.

Fire Officer in Action

You arrive at the station early in the morning for your duty shift and prepare yourself for the day. After getting your morning coffee you arrive in your office and contemplate the schedule for today. You know by looking at your calendar that there are several tasks that need to be completed today. You have a public education presentation at the local elementary school, volunteer drill this evening, daily duties, two prefire plans to complete, and maybe some calls thrown into the mix. There has been some miscommunication between two of your fire fighters the last couple shifts and you are anticipating having to deal with the situation today as well.

1. What is included in a morning report to your battalion chief or supervisor?

    A. Physical training for the day

    B. Status of apparatus

    C. How people are feeling today

    D. What training is scheduled

2. As a commander, what is one of the responsibilities of the company-level officer?

    A. Provide hands-on training

    B. Discipline personnel

    C. Provide good communications

    D. Develop competencies

3. You are meeting with your supervisor as you do once a week to keep him up-to-date. What is your obligation as a company-level officer to your supervisor?

    A. Wait until your weekly meeting to tell him about a new problem

    B. Apprise him of a suspension

    C. Keep him informed

    D. Advise him of a needle-stick incident that happened a week ago

Technology Resources

www.FireOfficer.jbpub.com

- **Chapter Pretests**
- **Interactivities**
- **Hot Term Explorer**
- **Web Links**
- **Review Manual**

Chapter Features

- **Voices of Experience**
- **Getting It Done**
- **Assessment Center Tips**
- **Fire Marks**
- **Company Tips**
- **Safety Zone**
- **Hot Terms**
- **Wrap-Up**

NFPA 1021 Standard

Fire Officer I

4.2   Human Resources Management. This duty involves utilizing human resources to accomplish assignments in accordance with safety plans and in an effective manner. This duty also involves evaluating member performance and supervising personnel during emergency and nonemergency work periods, according to the following job performance requirements.

4.2.5   Apply human resource policies and procedures, given an administrative situation requiring action, so that policies and procedures are followed.

   (A) Requisite Knowledge. Human resource policies and procedures.

   (B) Requisite Skills. The ability to communicate orally and in writing and to relate interpersonally.

4.2.6   Coordinate the completion of assigned tasks and projects by members, given a list of projects and tasks and the job requirements of subordinates, so that the assignments are prioritized, a plan for the completion of each assignment is developed, and members are assigned to specific tasks and supervised during completion of the assignments.

   (A) Requisite Knowledge. Principles of supervision and basic human resource management.

   (B) Requisite Skills. The ability to plan and to set priorities.

Fire Officer II

5.2.1   Initiate actions to maximize member performance and/or correct unacceptable performance, given human resource policies and procedures, so that member and/or unit performance is improves or the issue is referred to the next level of supervision.

   (A) Requisite Knowledge. Human resource policies and procedures, problem identification, organizational behavior, group dynamics, leadership styles, types of power, and interpersonal dynamics.

   (B) Requisite Skills. The ability to communicate orally and in writing, to solve problems, to increase team work, and to counsel members.

Knowledge Objectives

After studying this chapter, you will be able to:

- Understand principles of supervision and basic human resource management.
- Coordinate the completion of assigned tasks and projects.

Skills Objectives

There are no skills objectives for this chapter.

You Are the Fire Officer

You are frustrated. As the rookie officer of Metro City Engine 5, you are determined to do everything by the book, but it seems that nothing you do works. This morning's line-up was particularly painful. You were reviewing the brand-new Metro City Worldview that had just been issued as a replacement for last year's MCFD Mission Statement. You were reading aloud each of the fifteen "affirmations" that described the Worldview, and you had only gotten to Affirmation 5, *We value each other's time*, when the engine driver suddenly said, "And it is MY time to check the pumper!" The line-up dissolved as the rest of the crew agreed it was time to start station duties.

A few minutes later the battalion chief's truck pulls onto the fire station ramp. Captain Mandel, your old skipper at Rescue 1, is filling in for the chief. He takes one look at the expression on your face and says, "I thought you wanted to be a lieutenant. What's wrong?"

You start to tell Mandel about the problems you are having with the fire fighters not taking you seriously. When you describe the difficulty in reviewing the Metro City Worldview, Mandel's jaw drops. "Don't tell me that you were reading that to the crew at line-up!"

1. Why are there so many management concepts with contradictory information?
2. How do these theories assist you to execute supervisory tasks?
3. Is there a way for you to evaluate a new management concept or procedure before using it at the fire station?

Introduction to Management Concepts

Management is the science of using available resources to achieve desired results. A formal definition of management would probably refer to the systematic pursuit of practical results, using available human and knowledge resources in a concerted and reinforcing way.

A fire officer is a manager who has been given the responsibility to direct and supervise a group of fire fighters, as well as apparatus, equipment, facilities and other resources, in order to achieve certain outcomes. The desirable outcomes begin with protecting people and property from a variety of undesirable situations. Additional desirable outcomes include ensuring that the work is performed safely, efficiently, promptly, and in accordance with a long list of rules, regulations, procedures, and additional concerns.

Most fire officers will find that their greatest challenge has to do with managing people (▶ Figure 4-1). In fire departments, like most other organizations, it is the workers who ultimately get the job done. The manager is responsible for performing a set of functions that are essential to direct and coordinate their efforts, provide them with the necessary tools and resources, and ensure that the outcome meets the standards of desirability. Fire departments are particularly labor intensive, since most of the work has to be performed manually by skilled workers. In order to be effective as a manager, a fire officer has to develop an important set of skills that are directly related to managing human resources.

Management, as we know it today, is a product of the Industrial Revolution. In Europe, during the late 1700s, the advent of steam power led to the creation of large factories, which in turn created the need for management. Adam Smith noted that prior to the Industrial Revolution it took several hundred years for a country or region to develop a tradition of labor and the expertise in manual and managerial skills that are needed to produce and market a given topic. That was far too long to meet the needs of the rapidly growing textile factories, railroads, and other forces of the Industrial Revolution. Management was essential to produce results quickly and on a large scale.

Human resource management is built from two generalized schools of management thought: Scientific Management and Humanistic Management. Each school developed a set of theories on how supervisors can manage people to accomplish tasks in the work environment. Think of these theories and the related practices as tools in the fire officer's management toolbox. The fire officer will not necessarily use all of the tools or the same tools for every situation. The selection of management tools must consider the situation, the task, and the individuals or the team that will be involved in producing a desired outcome. Some of these theories are directly applicable for use by a fire officer in assigning tasks or responsibilities to crew members.

Figure 4-1 Managing people can be the fire officer's greatest challenge.

Scientific Management

Frederick Winslow Taylor

Frederick Winslow Taylor decided to forsake Harvard for a career in industry. In 1874, the skills that he needed generally were not taught in universities; they were learned on the shop floor. Taylor was hired as a laborer in the machine shop of Midvale Steel.

At Midvale, Taylor developed and put into place the basic elements of what later came to be known as "scientific management." **Scientific management** is based on the breaking down of work tasks into constituent elements; the timing of each element based on repeated stopwatch studies; the fixing of piece rate compensation based on those studies; the standardization of work tasks on detailed instruction cards; and generally, the systematic consolidation of the shop floor's brain work in a "planning department."

In 1911, Frederick Winslow Taylor published, *The Principles of Scientific Management*, in which he described how the application of the scientific method to the management of workers could greatly improve productivity. Scientific management methods called for optimizing the ways that tasks were performed and simplifying the jobs so that workers could be trained to perform a specialized sequence of motions in the one "best" way.

Prior to scientific management, work was performed by skilled craftsmen who served lengthy apprenticeships to learn their trades. Individual workers made their own decisions about how their job was to be performed. Scientific management took away much of this autonomy and converted skilled crafts into a series of simplified jobs that could be performed by unskilled workers. To determine the opti-

mal way to perform a job, Taylor performed experiments that he called time studies, (also known as time and motion studies). These studies featured use of a stopwatch to measure a worker's sequence of motions, with the goal of determining the most efficient way to perform each task. That worker and others could then be trained to perform those same specific tasks in the same efficient manner.

Taylor's Four Principles of Scientific Management

After years of various experiments to determine optimal work methods, Taylor proposed the following four principles of scientific management:

- Replace "rule-of-thumb" work methods with methods based on a scientific study of the tasks
- Scientifically select, train, and develop each worker, rather than passively leaving them to train themselves
- Cooperate with the workers to ensure that the scientifically developed methods are being followed
- Divide work nearly equally between managers and workers, so that the managers apply scientific management principles to planning the work and the workers actually perform the tasks

Many factories implemented these principles and achieved very impressive improvements in productivity. Henry Ford applied Taylor's principles in his automobile factories. Some families even began to perform their household tasks based on the results of time and motion studies. Scientific management changed the nature of work. Versions of scientific management can still be found in the 21st century workplace.

Many competitors in the Fire Fighter Combat Challenge use time and motion studies to improve their performance. When the first Combat Challenge was held at the Maryland Fire and Rescue Institute in May 1991, the team from Prince William County, Virginia, won with a time of 10:08 minutes. Within a decade, the winning time to complete the same course had dropped to less than two minutes. By closely analyzing every movement, the competitors learned how to perfect their techniques and shave seconds from every step.

Los Angeles City Fire Department "Measure of Effectiveness System"

When California voters passed Proposition 13 in 1978, local government revenues were significantly reduced due to restrictions on the property tax rate. This required immediate and serious reductions in local government staffing and services. The Los Angeles Fire Department (LAFD) was faced with the possibility of a reduction in staffing of their single-unit engine companies. At that time about half of the 103 LAFD fire stations operated a single-unit engine company.

The LAFD developed a task analysis of typical initial fireground scenarios that included a list of required on-scene performance objectives for an engine company. They broke each objective into fundamental and discreet tasks and performed extensive time and motion studies using different staffing levels to accomplish the tasks. The Measure of Effectiveness System identified the tasks performed by each fire fighter in chronological order. The analysis was performed with variations in crew size from three to six members and documented significant increases in the time that was required to accomplish the standard fireground objectives as the size of the crews decreased. The results justified retaining five fire fighters on single-unit engine companies.

Humanistic Management

One of the problems with the scientific management school was that people were considered as carbon-based cogs in a production line. Taylor considered the workers as cheap, stupid, and interchangeable. Each worker was trained to perform just one task, a small portion of a production process, and to repeat that task at a mind-numbing rate. The humanistic management school shifted the focus to pay more attention to the workers and to working conditions that would make them more productive.

The humanistic management school started with the Hawthorne experiments conducted by George Elton Mayo, a Harvard University industrial psychology professor. It continued through the work of Douglas McGregor and Abraham H. Maslow.

McGregor—Theory X and Theory Y

Douglas McGregor, a social psychologist and professor of management at the Massachusetts Institute of Technology, studied the work environment from the mid-1930s through the mid-1950s. The economic conditions during this time period were much different from the situation at the start of the Industrial Revolution. McGregor's study period covered the Depression, World War II, and the booming prosperity of the 1950s that resulted in the establishment of the middle class. McGregor's 1960 publication, *The Human Side of Enterprise*, summarized the results of his research and added the Theory X and Theory Y concepts to the manager's motivational toolbox.

While the new immigrants to America in the early 20th century worked to survive, most workers in the mid-20th century enjoyed much better retirement and financial resources. McGregor observed that in spite of these improvements, many mid-20th century workers were dissatisfied with their jobs. He concluded that worker motivation is directly related to autonomy and responsibility; workers with greater autonomy are more likely to be motivated in their jobs. McGregor also felt that modern employment often stifled human creativity and impaired motivation.

McGregor developed the Theory X and Theory Y concepts to define the problem in terms of the manager's or the organization's view of the workers.

A Theory X manager believes that people do not like to work, so they need to be closely watched and controlled. Theory X makes sense when you look at the Industrial Age working conditions of the late 1800s. The assembly line factories were horrible. They had scant light, poor ventilation, and unsafe and unsanitary working conditions. Some workers had to relieve themselves at their workstations. Many of the factory workers were recently arrived immigrants, with no industrial work skills and limited ability to read or write English. Many were children. No wonder they had to be coerced and threatened!

A Theory Y manager has an entirely different view of employee creativity and motivation. A Theory Y manager believes that people do like to work and that they need to be encouraged, not controlled.

These two distinctly different philosophical models of worker motivation and supervisory strategies can both be observed in many organizations today. The working world has changed significantly since 1960, especially with the increased participation of female managers, and Theory Y is definitely the prevailing trend. Nevertheless, academic and business writers point out that job satisfaction and employee loyalty are far lower in 2000 than they were in 1960 and Theory X remains a common managerial style.

Theory X Versus Theory Y for the Fire Officer

McGregor's Theory X and Theory Y models can be valuable tools for a fire officer. Most individuals who decide to become fire fighters, whether they select the fire service as a career or join a volunteer organization, are highly attracted to the nature of the work. Few people are forced by economic conditions to become fire fighters. The fire service requires a strong personal commitment, beginning with the initial investment in physical and technical training. Many fire fighter activities are difficult, physically demanding and unpleasant, yet most fire fighters love their work. They tend to operate from a different worldview, when compared to most workers in 21st century jobs, such as information technology or service industries.

The fire officer is supervising a self-selected work-force that is dedicated and enthusiastic about their responsibilities as fire fighters. The officer's challenge is often to steer their efforts in the right direction and to create an effective team (▶ **Figure 4-2**). The Theory Y concepts can be used effectively in many situations to encourage fire fighter creativity.

Theory Y does not work for every situation. There are three specific situations where the fire officer must behave as a Theory X manager, at least temporarily. The first situation is when operating at a fire or other high-risk activity—the fire officer must provide close and autocratic supervision. The second situation is when the fire officer must take control of a workplace conflict and issue specific directions to defuse the situation. The final situation is when a fire officer is near the end of a series of negative disciplinary measures;

Figure 4-2 The fire officer must steer the fire fighter's efforts in the right direction to create an effective team.

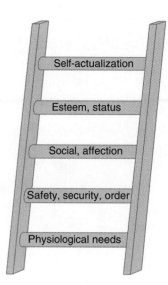

Figure 4-3 Maslow's Hierarchy of Needs.

the Theory X concept and techniques are applicable in an involuntary work performance meeting with a subordinate.

Maslow—Hierarchy of Need

Abraham H. Maslow was a psychologist that researched mental health and human potential. He is most often associated with the concepts of a hierarchy of needs, self-actualizing persons, and peak experiences. Maslow's concept saw human needs arranged like a ladder or a pyramid (▶ **Figure 4-3**).

Level One: Physiological Needs

The most basic human needs, at the bottom of the ladder, are physical—air, water, food, and shelter. It is hard to be creative or productive when you are exhausted and hungry. For the fire officer, this means to stay alert for conditions where the fire fighters are hungry, exhausted, dehydrated, too hot, or too cold. Sometimes fire fighters have to work at the extremes of discomfort and personal inconvenience due to the nature of emergency activities, but these situations should be the exceptions and not a normal way of getting

the job done. Making a fire company work without a rest and rehydration period will fail to meet the most basic needs (▶ **Figure 4-4**). Meals may be delayed, but food is needed to keep fire fighters working, just as fuel is needed to keep apparatus operating.

Level Two: Safety, Security, and Order

Safety, security, and order needs are next. Safety is obviously a primary concern in the fire service, although a fire fighter's perceptions of safety are probably quite different from a typical office or factory worker. If the fire fighters feel that their fire officer is leading them in an unsafe manner or a particular policy or practice in the fire department is exposing them to an avoidable risk, safety is likely to become a very significant issue.

Security is more closely associated with maintaining employment or status within the organization. It could be as subtle as a fear that a personality clash with a particular company officer will destroy or impede a fire fighter's career. An individual who feels threatened may do what is necessary to survive, but will not feel highly motivated to advance the fire officer's or the fire department's objectives.

Imagine the impact on your fire company if a notice from fire administration suddenly announced that all incumbent fire fighters would be required to achieve paramedic training, and that any fire fighter who does not pass the National Registry certification exam within one year will be terminated. While this would be a great example of Theory X coercion, it would also create tremendous fire fighter stress.

The need for security and order could be a factor when a fire department is undergoing a significant reorganization, such as the appointment of a new fire chief or the combining of two or more departments into a regional authority. Any type of change in the established order is likely to arouse

Figure 4-4 Fire fighters need periods of rest and rehydration to meet their most basic physiological needs.

Figure 4-5 Fire fighters often engage in social and recreational activities with their co-workers.

insecurities. The anticipation of significant changes in the organization or budget-related cutbacks can easily impede fire fighter creativity and motivation.

Level Three: Social and Affection

These are the psychological or social needs that are related to belonging to a group and feeling acceptance by the group. Group acceptance and belonging are particularly strong forces within the fire service. The majority of fire fighters have some type of identification on their personal vehicles that visibly declares their local fire department affiliation and membership in the fire service at large. Fire fighters wear their company patches proudly and often engage in social and recreational activities with their co-workers **Figure 4-5**. Transferring a fire fighter away from a "good" assignment can be a strong de-motivator, just as suspending a volunteer from riding can have a significant impact at the social and affection levels.

Level Four: Esteem and Status

Promotions, gold badges, special awards, and take-home fire department vehicles are all symbols that apply to the esteem and status level **Figure 4-6**. Membership in an elite fire department unit is often an indicator of special qualifications or achievements. Fire fighters will invest their money and off-duty time in order to compete for a high-esteem or high-status position. For example, fire fighters will spend months preparing for a fire officer promotional exam or taking courses to qualify for a rescue company. Most elite fire units have more applicants than available positions.

Level Five: Self-Actualization

Peak experiences are profound moments of love, understanding, happiness, or rapture, when a person feels more whole, alive, self-sufficient and yet a part of the world; more aware of truth, justice, harmony, and goodness. Self-actualizing people have many such peak experiences. Such people tend to focus on problems outside of themselves and have a clear sense of what is true.

Maslow felt that unfulfilled needs at the lower levels on the ladder would inhibit the person from climbing to the

Company Tips

Building Trust in the New Fire Officer

It is vital for a new supervisor to work to build trust with the workers. The failure to establish trust from the beginning of the relationship can quickly undermine a supervisor's career. This is particularly true for a new fire officer. Fire fighters make or break a new officer and they will make every effort to protect themselves from an officer who appears to be unsafe, unstable, or unprepared for the job.

How can the new fire officer build trust? Here are five suggestions:

1. Know the fire officer job, both administrative and tactical.
2. Be consistent. Strive to provide a measured response to any problem, emergency, or challenge.
3. Walk your talk. Actions speak much louder than words.
4. Support your fire fighters. Make sure that you help meet their physiological, safety, security, and order needs.
5. Make fire fighters feel strong. Help them to become competent and confident in their emergency service skills. Show how they can control their destiny.

Figure 4-6 Promotions and special awards are symbols that apply to the esteem and status level.

next step. As he pointed out, a person who is dying of thirst quickly forgets the thirst when deprived of oxygen.

Maslow's thinking was original. Most psychology research before him had been concerned with the abnormal and the ill, including Maslow's first textbook. He wanted to know what constituted positive mental health. Maslow observed that "Human nature is not nearly as bad as it has been thought to be."

Blake and Mouton's Managerial Grid

The Exxon Corporation hired behavioral scientists Robert Blake and Jane Mouton in the early 1960s to perform a series of experiments designed to increase leadership effectiveness. The Grid theory assumes that every decision made and every action taken in the workplace is driven by people's values, attitudes, and beliefs. At the individual level, these values are based on two fundamental concerns that influence behavior—a concern for people and a concern for results. Blake and Mouton developed a survey document with 35 questions that measures a person's level of concern in each area. The two values are plotted on an X-Y chart. Blake and Mouton describe five behavioral models based on a person's position on this Grid. In the 1990s the grid theory was applied to crew management of aircraft and aerospace teams—a concept later adopted by the fire service as "crew resource management" to improve incident safety.

Indifferent—Evade and Elude

The indifferent style represents the lowest level of concern for both results and people. The key word for this style is neutral. This is the least visible person in a team; he or she is a follower who maintains distance from active involvement whenever possible. An indifferent person carefully goes through the motions of work, doing enough to get by, but rarely making a deliberate effort to do more.

The stereotypical image of this personality is the bureaucratic government agency where everyone is treated like a number. This sort of workplace allows the person to blend in without attracting attention. In fact, he or she often seeks work that can be done in isolation in order to carry on without being disturbed or noticed. The indifferent manager relies heavily on instructions and process, depending on others to outline what needs to be done. Reliance on instructions avoids the need to take personal responsibility for results. If problems arise, the indifferent person is often content to ignore or overlook them, unless the instructions specify how to react to that particular problem. The indifferent person might point out the problem to someone else, but would be unlikely to offer a solution. With no instructions, he or she simply carries on with the attitude that "This is not my problem."

Controlling—Direct and Dominate

The controlling person demonstrates a high concern for results, along with a low concern for others. The high concern for results brings determination, focus, and drive for success. This person is usually highly trained, organized, experienced, and qualified to lead a team to success. The low concern for others prevents the controlling person from being aware of others involved in an activity, beyond what is expected of them in relation to results. The controlling person expects everyone else to "keep up" with his or her efforts, and so moves ahead, intensely focused on results, often leaving others lost in the wake of his or her forceful initiative.

Accommodating—Yield and Comply

The accommodating person demonstrates a low concern for results with a high concern for other people. This individual maintains a heightened awareness of the personal feelings, goals, and ambitions of others, and always considers how proposed actions will affect them. He or she is approachable, fun, friendly, and always ready to listen with sympathy and encouragement.

Some cornerstone phrases of the accommodating attitude are "Let's talk about it," "What can I do to help?" and "Let me know what you think." The main weakness in this behavior lies in the focus of the discussions. Discussions with an accommodating person tend to include an overwhelming emphasis on personal feelings and preferences, while avoiding concrete issues. The discussion itself becomes the goal, so conversations can meander in any direction instead of concentrating on solving the problem.

The controlling and accommodating styles are diametrically opposed in their perspectives. Each of these orientations leads in a narrow and singularly focused manner by ignoring the other primary concern in the workplace.

VOICES OF EXPERIENCE

❝ **Follow the rule of the three 'F's'.** ❞

After being elected as the volunteer chief of a combination fire department, I knew that I had many challenges ahead. There had been a request by paid fire fighters to merge with our local suburban town and end the non-profit corporation so that the paid fire fighters could enter the retirement system. The volunteer members felt very threatened and did not want to lose their status in our local fire department. At that time, we were also facing a huge demand in calls and a growing problem with volunteers not being able to respond to all of the calls.

The first business meeting of the fire department was held in July and I had a chance to set the tone for the upcoming year. I struggled for weeks over what to say and how to say it. I was in a difficult situation and needed both the paid fire fighters and the volunteers to support the organization and our new directions.

On the night of the meeting, I chose to use both symbolism and substance as the measure of our success. I told my fellow fire fighters that I would use the rule of the three "F's" during my term as the fire chief. These are:

1. **FAIR:** I would treat everyone fairly. Regardless of his or her perspective, status or past, everyone would be treated fairly. All persons would follow the same rules, and be praised or punished by the same procedures regardless if they were volunteer or paid personnel.
2. **FAMILY:** Family is an important element of this department. I will respect your family needs as a high priority. If, as a volunteer, you had to choose between committing to a family event instead of responding to a call; then you should be with your family and it would be acceptable to the department. As a paid member, don't put the needs of this department ahead of the needs of your family. If your family cannot support you as a paid or volunteer member of this department, you will not be able to succeed.
3. **FUN:** Having fun is an important part of life and this department. We need to laugh and enjoy our time together when ever possible. The world of emergency response is a serious business and we will approach all calls in a professional manner at all times. When we are training attending meetings and just hanging around, then we should always try to find ways to make these times fun and maintain a positive attitude.

Simple rules? Yes, but important and simple ones. Did it work? Yes, at the end of the year our department had merged with the town. We kept most of the volunteer members and the town hired its first paid fire chief. Eight years later, the agency is still a combination department, financially and organizationally strong. Both the volunteer and paid personnel have a positive relationship. Everything is not perfect but there remains a strong spirit of cooperation and mutual support.

Ken Farmer
Fuquay Varina, North Carolina Fire Department
35 years of service

Status Quo—Balance and Compromise

The status quo person believes there is an inherent contradiction between the two concerns, but does not value one concern over the other. Instead, the status quo person sees a high level of concern for either people or results as too extreme and tries to moderate both in the workplace.

The objective of the status quo person is to play it safe and work toward acceptable solutions that follow proven methods. This is a politically-motivated approach that seeks to avoid risk by maintaining the tried and true course, following popular opinion and norms without pushing too hard in any direction.

Another key aspect of the status quo approach is to maintain popular status within the team and organization. He or she must be intelligent and informed enough to persuade people and companies to settle for a compromise—often less than they want and less than they could achieve. This requires being well liked, keeping well informed, and effectively convincing people that the consequences are not worth the risk. On the surface, this might make the status quo appear unbiased and impartial, but more accurately, the status quo approach represents a narrow view that underestimates people, results, and the power of change.

Sound—Contribute and Commit

The sound person sees no contradiction in demonstrating a high concern for both people and results at the same time. He or she feels no need to restrain, control, or diminish the concerns for both people and results in a relationship. The consequence is a freedom to test the limits of success with enthusiasm and confidence. The sound attitude leads to more effective work relationships based on "what's right" rather than "who's right."

The sound behavioral model is preferred for a candidate to become a successful fire officer. The full integration of concerns for both people and results is in contrast with the other behavioral styles. Each of the other models represents a weakness in one or both critical areas. The controlling person feels that a high concern for results is more important than a high concern for people. The accommodating person feels the reverse—that a high concern for people is more important than results. The status quo feels that a high concern for either people or results is too risky, and prefers to maintain the status quo, holding the safe middle ground. The indifferent sees any major concern for people or results as unrealistic and too demanding.

Applying Human Resource Management Principles

Managing fire fighters requires utilizing physical, financial, human, and time resources. Human resource management focuses on the task of managing people. Since much of a fire

officer's time is spent dealing with subordinates, the fire officer must understand human resource management. Human resource management includes a wide variety of activities. Some of these are at the company level while others are typically addressed at the departmental or organizational level. Although some areas affect career departments more than volunteer departments, all functions are performed by all departments regardless of their structure. The typical human resource management functions include:

- Human resource planning
- Employee (labor) relations
- Staffing
- Human resource development
- Performance management
- Compensation and benefits
- Employee health, safety, and security

Human resource planning is the process of having the right number of people in the right place at the right time who can accomplish the task efficiently and effectively. Most often, this includes forecasting future staffing needs and determining how those needs can be met. When a fire department projects the number of retirements for the following year and determines that a new recruit class should begin prior to those vacancies, the fire department is fulfilling the human resource planning function. Typically, this is not an activity that is conducted at the company level other than at emergency scenes.

Employee relations include all activities designed to maintain a rapport with the employees. Typically, this is associated with working with the employee's union. Fire officers must know and understand all agreements between the union and the fire department. The fire officer should also be aware of the vast number of laws that regulate the relationship between employees and their employer. Ignorance of the law is no excuse and fire officers that violate the law may quickly find themselves and their department faced with a federal lawsuit. This human resource function is so important that Chapter 5, *Organized Labor and the Company Officer*, is devoted to it in its entirety.

Staffing is the process of attracting, selecting, and maintaining an adequate supply of labor. The fire service traditionally has been fortunate in its ability to attract applicants. However, with the increase in the services provided and the greater educational requirements, this is becoming more difficult. Although some fire departments have developed programs to actively seek qualified women and minorities, most fire departments have not.

This function also includes labor force reductions. There are many methods of shrinking the fire department's labor supply; most fire departments have opted for using attrition and layoffs rather than terminations, transfers, and reduced workweeks. The staffing function is typically accomplished at the organizational level.

Human resource development includes all activities to train and educate the employees. This function is heavily dependent on the fire officer at the company level. The development of the employee begins when he or she first arrives at the department. The fire department orients the recruits to the fire department's methods of operation through a recruit class. The process continues once they arrive at their fire station assignment. The fire company officer will usually sit down with the recruit and orient him or her to the job and the ways things are done at the fire station.

One method that is particularly effective is for the fire officer to select a fire fighter at the station who exemplifies good behavior. The fire officer discusses with the fire fighter why he or she believes the exemplary fire fighter would make a good role model for the recruit. The fire officer will then ask the fire fighter to become the "big brother" or "big sister" to the recruit. If the fire fighter accepts the responsibility, the recruit is then brought into the discussion so he or she knows that the "big brother" or "big sister" fire fighter is there to help him or her with any questions that he might have. This rewards the "big brother" or "big sister" fire fighter for his or her good performance and gives the recruit an experienced fire fighter as a resource for questions. It also allows the fire officer to focus on other higher level duties.

Most fire departments also require regular drilling and training to all personnel. This is also part of the human resource development function. The fire service is unlike most other private and public organizations in its ability to spend vast amounts of time solely to develop its employees while at work. Ensuring daily training is one of the most basic responsibilities the fire company officer has to do.

Performance management is the process of setting performance standards and evaluating performance against those standards. Generally, the standards are set at the fire organizational or fire departmental level; however, the evaluation of the fire fighter's performance is a primary function of the fire company officer. This issue is covered in great detail in Chapter 8, *Evaluation and Discipline*.

Human resource management also includes the setting of compensation and benefits. While the fire officer may not set the compensation policy, the fire officer must understand how their system is designed and what benefits the employees are entitled. Most fire departments operate on a step-and-grade pay system. The position of fire fighter is established on a particular pay grade level that is composed of a number of steps. If the fire fighter demonstrates satisfactory performance, he or she will progress from one pay step up to the next until he or she eventually reaches the top step of the pay grade. The fire fighter will remain at this pay step and grade until he or she is promoted. He or she will then move to a new pay grade and progress up through the steps until he or she again reaches the top step.

The other most common systems that are used include merit-based pay and skill-based pay. In merit-based pay systems, the fire fighter is typically paid a base amount and then receives additional compensation for a good performance. For example, a fire fighter that receives an outstanding evaluation might receive a bonus of 5% of his or her annual pay. Skill-based pay systems typically pay a base amount and then give additional compensation for any skills that the fire fighter can demonstrate. For example, a fire fighter who is also a paramedic and a hazardous materials technician would be paid an additional amount because of his or her additional skills.

Traditionally, fire departments have good benefits compared to the private sector. Benefits include defined benefit retirement systems where the employee contributes a set percent of his or her pay and receives a guaranteed benefit amount. It also includes defined contribution retirement plans where the employee pays in a percentage of pay and receives a benefit amount that is dependent upon the performance of the account. Typically, these accounts are in the form of a 457 deferred compensation plan. Benefits also include paid holidays, vacation leave, sick leave, health, dental, vision, and life insurance.

Health, safety, and security includes all activities to provide and promote a safe environment for fire fighters. The fire officer is responsible for many of these activities such as ensuring seat belts are utilized, floors are dry, and personal protective equipment is used correctly. The fire department or fire organization may provide other health activities such as an Employee Assistance Program, which is covered in detail in Chapter 8, *Evaluation and Discipline*. The fire department may also offer tobacco cessation classes and incentives or diet and exercise information.

Utilizing Human Resources

The fire officer's ability to utilize the human resources that are assigned is an essential function of the position. The fire officer must be able to accomplish the department's work with and through other people. The fire officer must ensure that duties are carried out in accordance with departmental policies. Normally, this is done through direct supervision.

Direct supervision requires that the fire officer directly observe the actions of the crew. If the crew makes an interior attack, the fire officer is present. When a hose is loaded, the fire officer oversees the reloading. Direct supervision allows the fire company officer to ensure that departmental safety policies are used. For activities that are more dangerous, the closer the direct supervision that is required. For nonemergency activities, less direct supervision is needed.

The more direct supervision that is needed, the less efficient the crew. The goal of the fire officer is to reduce the need for direct supervision and increase the utilization of the fire company. Frequently this is accomplished through a series of policies and procedures. Human resource policies and procedures guide the fire officer's decisions as they relate to personnel issues.

The fire officer must be familiar with the location and topical areas that are covered. The organization may have federal laws, a union contract, city regulations, and departmental policy that must all be followed. When presented with an issue, the fire officer must first determine what laws apply. The most common are the Fair Labor Standards Act (FLSA), the Civil Rights Act, the Age Discrimination Employment Act (ADEA), the Americans with Disability Act (ADA), the Family Medical Leave Act (FMLA), the Uniformed Service Employment and Reemployment Rights Act (USERRA), and the Health Insurance Portability and Accountability Act (HIPAA). The Department of Labor provides a great deal of information on the laws that affect fire fighters. The fire officer may also want to discuss issues that are affected by these laws with legal counsel.

Once the determination is made of any laws affecting a decision, the fire officer should review the fire organization's and fire department's policies to ensure compliance. If a labor contract exists, the fire officer must determine if the activity is addressed by the contract. If the fire organization has a human resource department, it may prove to be a valuable asset in determining the proper action by the fire officer.

Lastly, the fire officer must base his actions on fairness and equity to the parties involved. For example, if an employee desires to schedule leave time in short increments the fire officer must determine whether there are any laws, organizational policies, or departmental policies that address the issue. If not, the fire officer must consider the impact of the request on the achievement of the department's mission. Lastly, the fire officer must consider whether the action is fair to all employees.

Mission Statement

One basic principle is for the fire officer to know and understand the fire department's mission. Frequently, the fire department's mission is expressed through a written mission statement. The mission statement is a formal document that outlines the basic reason for the organization and how it sees itself. It is designed to guide the actions of all employees.

Getting Assignments Completed

One of the greatest demands on the fire officer is the effective use of time. Fire officers will find that they have a great number of demands on the company's time. This includes conducting public education, inspections, and other fire prevention efforts. It also includes training and education of the crew members. It includes routine duties such as cleaning the stations, doing paperwork, and maintaining the apparatus. And of course it includes responding to calls.

Some of these activities are known months in advance. Others may require the immediate response of the crew. The daily schedule of some crews is in a constant state of interruption due to the call volume. Other crews may only occasionally be interrupted. No matter which situation, the fire officer can ensure maximum efficiency by using good time management skills.

The first duty is to determine what activities are to be completed, when they must be completed, and how long it will take to complete. The fire officer can then categorize what needs to be done during the shift, the week, the month, and the year. Items that must be completed during the shift are a higher priority than those that must be completed next week. For example, completing the daily log is more important than completing a weekly inspection.

At times, there is not sufficient time to complete all the required tasks during the shift. When this occurs, the fire officer must determine the fire department's priorities. This allows the fire officer to determine which activity must be completed and which will have to wait. For example, meeting the deadline for the employee's timesheets may be a higher priority than getting the fire engine to the service center for an oil change.

Since many activities are not known until they occur, such as an emergency call, the fire officer must plan ahead and build in the expected interruptions. The fire officer must not wait until the last minute to have inspections completed because he may have calls that prohibit this from happening. The sooner the scheduled activity is completed, the more flexibility the fire officer has.

Once the activities have been prioritized and it is determined when they must be accomplished, the fire officer must develop a plan that lays out how the activities will be accomplished. Activities might require the entire company to be involved in the activity or it might require only a part of the crew to complete. Sending a crew member to the store while another cleans the fire station and yet another checks out the apparatus is an example.

Having many tasks to coordinate can easily become overwhelming. One method to assist in making sure that activities are accomplished is to place all scheduled events on a monthly calendar. Inspections, public education, and special training should be noted. Employee leave time should also be noted. The calendar provides a visual method of tracking upcoming events.

Another method is to create a "daily" file. Within this file, there is a page that describes each activity and when it is to be completed. The file is organized from the soonest activity to complete to the farthest. This allows the fire officer to quickly determine what needs to be done without having to look through a pile of papers.

One of the best tools to improve time efficiency is through delegation. As discussed previously, delegation allows tasks that subordinates are capable of performing to be completed by them. These duties should be ones that allow the subordinate to grow. An example is the fire officer delegating the responsibility for ordering supplies to a fire fighter. Delegation allows the fire officer to focus on duties that cannot be delegated, such as performance appraisals.

Once assigned, the fire officer must provide regular follow-up. There are many forms this may take. At the station level, this is most often a verbal progress report. On more formal projects, the progress report may be in writing to provide long-term documentation.

You Are the Fire Officer: Conclusion

Moving you away from the Engine 5 fire fighters, Captain Mandel says, "That's the stuff you put on the bulletin board." When you begin to quote an Employee Excellence Performance Principle, Mandel puts up his hand. "Let's get some coffee and go to the lieutenant's office. You need a reality check."

Mandel sighs and looks at you. "Remember how it feels when Rescue 1 first arrives at a challenging rescue? You want to do everything at once and use every tool we brought to the crash. However, there are not enough people and not enough time to dump all of the rescue tools on the pavement every time we have extrication. The smart rescue company fire fighter sizes up the situation and selects the best tool to accomplish the task."

Mandel continues, "Management concepts are like the tools on the rescue. You have collected a large set of management ideas, concepts, and practices that you carry in your management toolbox. The smart supervisor selects the best idea, concept, or practice for the situation. Reading the Metro City Worldview Affirmations during line-up is not a good management tool selection. You need to pick the idea, concept, or practice that accomplishes the objective without pummeling fire fighters or eroding your authority."

"You will see new management tools during your company officer tenure. Like new rescue tools, they will have high promise. Until you evaluate how the new management tool works, by understanding its history and development or using it in a training session, it remains an unknown resource. Just because it is new and comes from a best-selling book by a management guru does not mean it will work for you as a fire officer."

Wrap-Up

Ready for Review

- Scientific management breaks down work tasks into constituent elements, the timing of each element based on repeated stopwatch studies in order to standardize work tasks into simple, repeatable tasks.
- Frederick Winslow Taylor's four Principles of Scientific Management:
    1. Replace rule-of-thumb work methods with scientific study
    2. Scientifically select, train, and develop each worker
    3. Cooperate with workers to ensure methods are being followed
    4. Division of work: managers think, workers work
- Taylor considered workers cheap, stupid and interchangeable. They were trained to perform one small portion of a task that they would repeat at a mind-numbing rate.
- McGregor Theory X—People do not want to work.
- McGregor Theory Y—People do want to work.
- Maslow's Hierarchy of Needs is a ladder of five need levels. The supervisor's role is to meet the employee needs in order to proceed to the next level.
    - Level One: Physiological
    - Level Two: Safety, Security and Order
    - Level Three: Social and Affection
    - Level Four: Esteem and Status
    - Level Five: Self-actualization
- Managing fire fighters requires physical, financial, human, and time resources.
- Human resource planning is the process of having the right number of people in the right place at the right time.
- The fire officer's ability to utilize the human resources that are assigned is essential.
- Direct supervision requires the fire officer to directly observe the actions of the crew.
- One of the greatest demands on the fire officer is the effective use of time.

Hot Terms

Scientific management The breakdown of work tasks into constituent elements; the timing of each element based on repeated stopwatch studies; the fixing of piece rate compensation based on those studies; standardization of work tasks on detailed instruction cards; and generally, the systematic consolidation of the shop floor's brain work.

As a fire officer, two new rookies have recently been assigned to your crew. You are impressed with both fire fighters and find that they both are self-motivated and have a strong desire to be the best they can. They are constantly studying and completing tasks without ever being told. During incidents, while it is obvious that the two need experience, they appear to be very skilled and knowledgeable in basic firefighting practices. You are happy to have these individuals on your shift and praise them for their outstanding work. They should have no problem completing their probationary year.

1. What does Theory X say?

   A. Managers believe workers do like to work

   B. Employees should be allowed to be creative

   C. Employees have five level they try to achieve

   D. Give employees self-empowerment

2. Which theory can be used to encourage fire fighter creativity?

   A. Theory X

   B. Theory Y

   C. Maslow's Hierarchy of Needs

   D. MacGregor's theory

3. What is Level Four of Maslow's Hierarchy of Needs?

   A. Social and affection

   B. Self-actualization

   C. Safety and security

   D. Esteem and status

4. What does Maslow's Hierarchy of Needs say about Level Five: Self-Actualization?

   A. People need a sense of belonging to a group

   B. People have basic needs

   C. People expect some level of security

   D. People have a need to fulfill themselves

Chapter Pretests

Interactivities

Hot Term Explorer

Web Links

Review Manual

Organized Labor and the Fire Officer

Technology Resources

www.FireOfficer.jbpub.com

- Chapter Pretests
- Interactivities
- Hot Term Explorer
- Web Links
- Review Manual

Chapter Features

- Voices of Experience
- Getting It Done
- Assessment Center Tips
- Fire Marks
- Company Tips
- Safety Zone
- Hot Terms
- Wrap-Up

NFPA 1021 Standard

Fire Officer I

4.1.1   General Prerequisite Knowledge. The organizational structure of the department; geographical configuration and characteristics of response districts; departmental operating procedures for administration; emergency operations, incident management systems, and safety; departmental budget process; information management and recordkeeping; the fire prevention and building safety codes and ordinances applicable to the jurisdiction; current trends, technologies, and socioeconomic and political factors that affect the fire service; cultural diversity; methods used by supervisors to obtain cooperation within a group of subordinates; the rights of management and members; agreements in force between the organization and members; generally accepted ethical practices, including a professional code of ethics; and policies and procedures regarding the operation of the department as they involve supervisors and members.

4.2.5*   Apply human resource policies and procedures, given an administrative situation requiring action, so that policies and procedures are followed.

> **(A) Requisite Knowledge.** Human resource policies and procedures.

> **(B) Requisite Skills.** The ability to communicate orally and in writing and to relate interpersonally.

Fire Officer II

5.1.1   General Prerequisite Knowledge. The organization of local government; enabling and regulatory legislation and the law-making process at the local, state/provincial, and federal levels; and the functions of other bureaus, divisions, agencies, and organizations and their roles and responsibilities that relate to the fire service.

Knowledge Objectives

After studying this chapter, you will be able to:

- Discuss the lasting impact organized labor has had on fire fighter safety, working conditions, and procedures.
- Understand the diminished benefits of fire fighter strikes.
- Identify the increased benefits of political activism.
- Describe the grievance process steps.

Skills Objectives

After studying this chapter, you will be able to:

- Demonstrate the initial handling of an employee grievance.

You Are the Fire Officer

It is a cold Tuesday morning at Metro City Engine 6. "C" Platoon is starting a Tuesday-Thursday-Saturday schedule of 24-hour workdays. A few minutes before the start of the shift, you receive a message from the battalion chief's office that you should send a fire fighter to attend an EMT-recertification class at the Training Academy for the next 4 days, starting in 1 hour. After quickly checking to see who is due for recertification, you call Fire Fighter Trammel into your office and instruct him to report to the academy immediately.

"Sorry, but I can't go" states Trammel. "The EMT-refresher runs 9 to 4 Tuesday through Friday. I'm supposed to be off on Wednesday and Friday this week and I've made plans for tomorrow." You calmly repeat the instruction to report to the Training Academy 0900 hours today to attend the class.

"Are you really sure about this?" Trammel asks. "I believe that the new labor contract requires a 72-hour notice for any work schedule changes." You now advise him that this is not a request—it is a direct order!

"Okay," says Fire Fighter Trammel, "but I need to inform you that this is an unfair labor practice and a violation of the labor agreement between Local 9715 and Metro City Fire. This is your notice of my step 1 grievance. I am officially requesting that you comply with the 72-hour advance notice of a work hour change. I will be contacting the battalion representative."

1. Does a labor contract override personnel regulations?
2. How can this situation be resolved fairly and appropriately?

Introduction to Organized Labor and the Fire Officer

Within most municipal fire departments, wages, working conditions, and many other aspects of the work environment are directly influenced by labor-management relations. It is important for the fire officer, particularly in a career fire department, to understand some of the history of labor relations to better function as a first-line supervisor. The range, scope, and tasks of a fire officer's supervisory activities are defined by three primary components:

- The local labor contract
- The municipality's personnel regulations
- The fire department's rules, regulations, and procedures

In many cases, a fire officer is both a supervisor representing management and a member of the bargaining unit represented by the union. This is an unusual situation when compared with most work environments, in which there is a very clear distinction between labor and management. It is especially important for a fire officer to clearly understand how this applies to the specific organization and the position he or she is occupying.

Over the past century, the American fire service has been heavily influenced by organized labor activities and by several laws and regulations that have been shaped through the efforts of organized labor. Some of these activities have had a significant impact on all fire departments, from the largest all-career department to the smallest all-volunteer fire company. A career fire officer who is working in a municipal fire department is much more likely to be involved in an organized labor situation than a volunteer officer. However, many aspects of the relationship between the workers and the organization can be very similar.

At the fire company level, most career fire fighters work under a labor contract or some form of written agreement, such as a memorandum of understanding (MOU), between labor and management. The contract or MOU covers various working conditions, promotion/assignment practices, and problem-solving procedures. A labor contract is a negotiated legal agreement between the labor organization and the local jurisdiction. An MOU is a less powerful form of written agreement that is often used in jurisdictions where government employees do not have formal collective bargaining rights. **Collective bargaining** is a method whereby representatives of employees (unions) and employers determine the conditions of employment through direct negotiation, normally resulting in a written contract setting forth the wages, hours, and other conditions to be observed for a stipulated period (e.g., 3 years).

The nature of the relationship between the employer and the labor organization is determined by a wide variety of labor laws and regulations at the federal, state, and local levels. The balance of power between labor (representing the employees) and management (representing the fire chief/local government) swung back and forth like a pendulum during the 20th century. To a large extent, this balance of power depends on legislation and policies enacted by the federal government. In the early part of the century, the labor movement had a political advantage and was able to make considerable progress. Toward the close of the 20th century, management appeared to enjoy an advantage.

The Legislative Framework for Collective Bargaining

Collective bargaining is regulated by a complex system of federal and state legislation. As a general statement, the federal labor laws establish a basic framework that applies to all workers, and the states have discretionary powers to adopt labor laws and regulations that do not violate the federal requirements. The application of this basic model to governmental employees is much more complex. Some of the federal labor laws do not apply to federal government employees, state government employees, or employees of local government agencies within the states.

Since the New Deal era, four major pieces of federal legislation have established the groundwork for the rules and regulations of the present collective bargaining system. Before the adoption of these federal laws, each labor case was decided by a judge in a local court, who applied the broad concepts used in common-law decisions. The federal legislation provides a set of guidelines for how each state or commonwealth can regulate collective bargaining. Most fire fighters are employed by local government agencies and are subject to state law that can require, permit, or prohibit collective bargaining for local public employees. Each state has a unique set of laws that apply to particular situations.

The four federal laws that regulate the collective bargaining system are the Norris-LaGuardia Act of 1932, the Wagner-Connery Act of 1935, the Taft-Hartley Labor Act of 1947, and the Landrum-Griffin Act of 1959. These federal laws, together with the Railway Act of 1926 and some antitrust legislation, created the legal foundation for collective bargaining in the United States. Like most federal legislation, each act was designed to address a specific aspect of collective bargaining or to correct a problem. Each subsequent act built on the earlier legislation. Like a pendulum, each subsequent act may change the direction of the federal government in relation to collective bargaining issues.

Norris-LaGuardia Act of 1932

The Norris-LaGuardia Act of 1932 specified that an employee could not be forced into a contract by an employer in order to obtain and keep a job. During this time, employers required workers to sign a pledge that they would not join a union as long as the company employed them. These pledges were called **yellow dog contracts**.

The local courts generally sided with management in any labor dispute where a yellow dog contract was in effect. The judge would order an injunction that either prohibited striking or prohibited picketing during a strike. The police would enforce these injunctions.

In reaction to this practice, the Norris-LaGuardia Act of 1932 said that yellow dog contracts were not enforceable in any court of the United States. This act made it almost impossible for an employer to obtain an injunction to prevent a strike.

President Franklin Roosevelt took many steps to bolster economic growth during the Great Depression. One of his initiatives was the National Industrial Recovery Act (NIRA) of 1933. Section 7a of the NIRA guaranteed unions the right to collective bargaining in order to keep wages at a level that would maintain the purchasing power of the worker. After NIRA passed, workers flocked to join both the American Federation of Labor and the new Congress of Industrial Organizations. The Supreme Court struck down the NIRA as unconstitutional in 1935.

Wagner-Connery Act of 1935

When the NIRA was overturned, employers were free to use **unfair labor practices** in the management of their employees. This led to the Great Strike Wave of 1933–1934. During this period, labor organized against management and conducted citywide strikes and factory takeovers in numerous industrial sectors. In response, Senator Robert Wagner of New York introduced the Wagner-Connery Act, which was quickly passed in Congress in 1935 to mitigate the revolutionary labor climate and avert further economic disruption. A 1936 strike in the automobile industry quickly brought the Wagner-Connery Act before the Supreme Court, where it was upheld as constitutional. The Wagner-Connery Act established the procedures that are commonly called collective bargaining.

The Wagner-Connery Act forms the basis of formal labor relations in the United States. It grants workers the right to decide, by majority vote, which organization will represent them at the labor-management bargaining table. The act also requires management to bargain with duly elected union representatives and outlaws yellow dog contracts.

Provisions of the Wagner-Connery Act also established the National Labor Relations Board, which has the power to hold hearings, investigate labor practices, and issue orders and decisions concerning unfair labor practices. The act defined five types of unfair labor practices and declared them illegal:

1. Interfering with employees in a union
2. Stopping a union from forming and collecting money
3. Not hiring union members

4. Firing union members
5. Refusing to bargain with the union

The Wagner-Connery Act and favorable court decisions resulted in some unions becoming extremely powerful. Union membership swelled from 4 million members in 1935 to 16 million members in 1948. The next two acts were adopted to provide more balance of power between unions and management.

Taft-Hartley Labor Act of 1947

The Taft-Hartley Labor Act of 1947, which was passed over the veto of President Harry Truman, was formulated during the period that followed World War II. At that time, industry-wide strikes threatened to undermine a smooth return of the economy to civilian production. The Taft-Hartley Labor Act was designed to modify the Wagner-Connery Act and swing the pendulum back toward the middle by reducing the power of unions. It also spelled out specific penalties, including fines and imprisonment, for violation of the act.

Taft-Hartley gave workers the right to refrain from joining a union and applied the unfair labor practice provisions to unions as well as to employers. It specifically prohibited a union from forcing management to fire antiunion or non-union workers. Unions were required to engage in **good faith bargaining**, and a 60-day "cooling off" period was created, when a labor agreement ends without a new contract. The act also regulated much of the union's internal activities.

The most significant provision of the Taft-Hartley Labor Act that affects fire fighters is referred-to as "strikes during a national emergency." In the event that an imminent strike could affect a major part of an industry and imperil the health and safety of the nation, the President is granted certain powers to help settle the dispute. The President can order employees back to work, compel arbitration, or provide economic, judicial, or political pressure to achieve a resolution of the dispute.

Landrum-Griffin Act of 1959

After the American Federation of Labor and the Congress of Industrial Organizations merged in 1955, Senator John McClellan conducted hearings that revealed evidence of crime and corruption in some of the older local unions. The Labor-Management Reporting and Disclosure Act, otherwise known as the Landrum-Griffin Act, passed in 1959, at the height of the furor.

Landrum-Griffin established a bill of rights for members of labor organizations. It required that unions file an annual report with the government listing the assets of the organization as well as the names and assets of every officer and employee. Minimum election requirements were mandated, as were the duties and responsibilities of union officials and officers. This act also amended portions of the Taft-Hartley Labor Act.

Collective Bargaining for Federal Employees

Collective bargaining rights for public employees have traditionally lagged behind those in the private sector. Federal legislation passed in 1912 prohibits federal employees from striking. There were fewer than 1 million unionized government employees in 1956. Federal legislation during the 1950s and 1960s allowed unions to grow in the public sector. By 1970, the number of unionized government employees had grown to 4.5 million members, and relations between management and unions had matured.

President Kennedy issued Executive Order #10988 in 1963, which granted federal employees the right to bargain collectively under restricted rules. This was a significant milestone for public sector unions. President Nixon further

---

Fire Marks

Federal Labor-Management Conflicts

Two notable strikes involving federal government employees mark the beginning and end of a significant era in the evolution of labor-management relations: the postal service workers strike and the Professional Air Traffic Controllers Association (PATCO) air traffic controllers strike.

Postal Workers' Strike: 1970

US postal workers went on strike illegally in 1970. At that time, the Postmaster General was legislatively restricted from negotiating with the striking workers, but nevertheless, negotiations were conducted, and a settlement was reached. The striking postal workers were reinstated without penalty, and the negotiated wage increases were adopted. Subsequently, Congress recognized the postal workers union for the purposes of collective bargaining. This strike set the tone for future civil service strikes, including several that involved fire fighters.

PATCO Air Traffic Controllers' Strike: 1981

One reason that public employees are not striking frequently today could be in reaction to the 1981 strike of air traffic controllers. The PATCO went on strike after reaching an impasse in contract negotiations with the Federal Aviation Administration.

President Reagan used the example of the 1919 Boston, Massachusetts police strike when he described the actions he would take to end the walkout. When the PATCO workers failed to report to work as ordered, President Reagan fired every striking member and decertified the union. Military air traffic controllers were brought in as temporary replacements. Air traffic volume was reduced for 18 months while the Federal Aviation Administration hired and trained replacement workers.

Unlike the 1970 postal workers' strike, none of the PATCO strikers was hired back, and many of the union officials were subjected to years of aggressive litigation by the federal government. This action set the tone for employer/employee relations during the remainder of the Reagan administration. The frequency of public sector strikes in the United States decreased very significantly over the following two decades.

expanded the rights of employee unions within the federal government. Nixon also established a Federal Labor Relations Council that is similar to the National Labor Relations Board for private sector unions.

State Labor Laws

In addition to the federal laws that regulate collective bargaining and relationships between employers and employee organizations, each state exercises legislative control over several aspects of collective bargaining. Each state determines whether or not it will engage in collective bargaining with state government employees and whether local government jurisdictions within the state may engage in collective bargaining with their employees. Some states require municipal governments to bargain collectively with fire fighters' unions, some permit collective bargaining, and others limit or prohibit collective bargaining.

Strikes by state employees are illegal in all but 10 states, and many states also prohibit local government employees from striking. Forty percent of local governments have forbidden employee strikes; however, numerous municipal strikes occurred between 1967 and 1980. After this period, several states and municipalities adopted legislation to prohibit strikes or other adverse labor actions.

Fire fighters have the right to bargain collectively in half of the states. Those states authorize municipalities to recognize a local fire fighter labor organization as the bargaining unit and enter into a binding contract that covers fire fighter pay, benefits, working conditions, conflict resolution, promotions, and department practices. The remaining states place restrictions on bargaining with fire fighters (▶ **Table 5-1**).

Right to Work

The collective bargaining rights of public employees in different states are related to the right-to-work issue. Under the Taft-Hartley Labor Act, workers have the right to refrain from joining a union. In 2002, **right-to-work** laws were in effect in 22 states. In those states, a worker cannot be compelled, as a condition of employment, to join or not to join, or to pay dues to a labor union. The same states tend to limit or restrict collective bargaining for local government employees.

The purpose of the original legislation in 1947 was to prohibit the practice of closed shops, in which a worker must be a member of a particular union in order to work for the company. Thus, open shops provide the worker with the option of remaining outside the union. Proponents of the open shop statute believe that the practice eliminates union collusion and exclusionary practices, which are now deemed illegal, and protects an individual's right to refrain from joining an organization.

Opponents of open shops express the opinion that right-to-work statutes reduce a union's bargaining power and place an unfair burden on the union members. They state that, in essence, the union has to bargain for the entire work force, even for nonmembers. A worker who chooses not to join the union is afforded the same benefits as union members, without paying dues to support union activities, pay for attorney fees, or conduct research studies. Under the federal regulations, an individual who chooses not to join a union can still be compelled to pay a share of the cost of the union's representation in collective bargaining.

The right-to-work states remain a constant source of controversy among labor organizations. Although unions support overturning existing right-to-work statutes, advocacy groups defend the existing statutes and promote their adoption in additional states. Each side argues that income and cost-of-living statistics are either higher or lower in right-to-work states, depending on who is publishing the statistics. Workers entering the labor market should be aware of the prevailing labor laws that pertain to them.

The International Association of Fire Fighters

The largest fire service labor organization in the United States, the International Association of Fire Fighters (IAFF), represents 267,000 fire fighters and emergency medical service personnel in the United States and Canada (▶ **Figure 5-1**).

Table 5-1 States with Restrictions on Collective Bargaining	
Prohibit bargaining	• Virginia • North Carolina
Permit bargaining by local option but prohibit legally enforceable contracts	• Alabama • Arkansas • Missouri • South Carolina • Tennessee
Permit bargaining by local option	• Arizona • Colorado • Georgia • Indiana • Kansas • Kentucky • Louisiana • Maryland • Mississippi • New Mexico • Texas • Utah • West Virginia
Require bargaining for fire fighters but not for law enforcement	• Idaho • Wyoming

Provided by the International Association of Fire Fighters.

Figure 5-1 The International Association of Fire Fighters (IAFF) is the largest fire service labor organization.

The IAFF has provided almost a hundred years of support and advocacy for career fire fighters, and its accomplishments influence many aspects of a fire fighter's job. It is a very powerful organization, both politically and within the fire service at the national level. The IAFF also represents most career fire fighters in Canada.

The International Association of Fire Fighters was established on February 18, 1918. The three principle objectives of the IAFF at that time were to obtain pay raises, establish the two-platoon or 12-hour workday schedule, and ensure that appointments and promotions were based on individual merit, not political affiliation.

The IAFF is almost unique among labor organizations in its dominance in representing fire fighters. Most other public service professions, such as law enforcement and education, have at least two different national labor organizations competing to represent the employees. Although the International Brotherhood of Teamsters and the American Federation of State, County, and Municipal Employees also represent fire fighters; these organizations represent very few and have very little influence over the firefighting profession.

The influence of the IAFF has a major impact on the fire service as a whole. All fire fighters have benefitted from the efforts of the IAFF and its local and state affiliates, including volunteers. Their work has improved the quality of protective clothing, the safety of firefighting equipment, the content of training programs, and some of the advanced techniques of emergency incident operations. In addition, the conflict-resolution process developed by the IAFF to handle subordinate-supervisor issues is equally effective in volunteer organizations. The generalized process for handling a grievance, which is covered in this chapter, comes from the labor-management relationship.

Organizing Fire Fighters into Labor Unions

Regardless of time, statutory environment, or work culture, labor-management relationships have been present in every work environment that includes an employer and employees. The fire service is no exception. Historically, the first documented paid fire department in the United States was the Cincinnati Fire Department, established in 1853. Thus, one may extrapolate that the birth of labor-management relationships in the American fire service also occurred in 1853.

The career fire fighter's work environment at the start of the 20th century was grim in comparison to today. New York City fire fighters worked "continuous duty"—151 hours a week, with just 3 hours off each day to go home for meals. San Francisco fire fighters got 1 day off after five consecutive 24-hour duty periods. Career fire fighters spent most of their time in cramped, unhealthy, and unsafe fire stations, often sharing accommodations with the horses.

Most labor organizations in the fire service evolved from fraternal or benevolent support organizations. In many communities, special interest organizations began by organizing social activities to support volunteer fire fighters, such as the fire fighter's gala or ball. As paid fire departments were established, these organizations gradually became political advocates for the workers and began to define the labor-management relationship. The genesis of these organizations occurred entirely at the local level. Antitrust protection, legislated by the 1890 Sherman Act, was interpreted by the federal courts to prohibit union representation of multiple labor groups. Consequently, individual local unions could not organize to form regional or national labor organizations. This prohibition lasted until the Clayton Act was passed in 1914.

Labor-Management Conflicts in the Fire Service

The fire service has been not been immune to labor-management strife and has experienced many adverse labor actions in response to poor labor-management relations ▶ **Figure 5-2** .

A **strike**—the act of withholding labor for the purposes of effecting a change in wages, hours, or working conditions—is one of the most drastic labor actions. Although a strike is precipitated by a conflict between labor and management, the impact of a strike is often felt beyond the parties that are directly involved in the relationship. In the private sector, the consumer is often the ultimate victim of a strike, through a combination of inconvenience and economic impact. In the public sector, the safety and the welfare of the general public are often directly affected by major labor actions.

The potential impact on public safety is so severe that many states prohibit fire fighters from walking out. In many of the cases in which fire service strikes have occurred, the

Figure 5-2 Picketing is one of the most drastic labor actions.

public and media have questioned labor's right to strike as a matter of ethics. The IAFF started with the premise that fire fighter strikes are inadvisable and from 1930 to 1968; the IAFF charter included a no-strike clause. There were three periods during the 20th century when municipal fire fighters did go on strike in different US cities.

Striking for Better Working Conditions: 1918–1921

The 1918–1921 strikes occurred during a period of economic instability following World War I. During this period, price jolts, wage maladjustments, and numerous private industry strikes occurred. Most of the strikes by fire fighters were efforts to establish a two-platoon system, obtain more pay, or simply gain the right to form a labor organization that would be recognized by the municipality. Organized labor reached a turning point in 1919. First, there was an industry-wide strike in the steel industry. Then, the country witnessed the first "general strike" that shut down Seattle, Washington on February 6. A similar strike occurred in Winnipeg, Manitoba, Canada on May 15, 1919.

Fire fighters in Memphis, Tennessee threatened to strike in 1918 and succeeded in obtaining a two-platoon work schedule, allowing fire fighters more time to be at home. The old system in Memphis had allowed just 1–3 days off each month. The new system allowed fire fighters to spend every other day away from the fire station.

Many IAFF local unions were able to gain a two-platoon schedule and/or a pay raise by threatening to strike. Fire fighters on the two-platoon system could finally have a home away from the fire station, although they were still required to work 84 hours a week. They could marry and raise a family. Today, most municipal fire fighters have a workweek of

between 42 to 56 hours a week. The workweek for most federal and military fire fighters still exceeds 60 hours.

The progress of public safety organized labor screeched to a halt at 5:45 P.M. on Tuesday, September 9, 1919 in Boston, Massachusetts. At that time, the 1,117 Boston police officers scheduled to work the evening shift went on strike. They were striking for the right to form a union with the American Federation of Labor. By nightfall, looters were rioting in Boston and South Boston. On Wednesday evening, the Massachusetts State Guard was mobilized to establish order. During the 4-night effort to re-establish civil order, the State Guard and volunteer police officers killed eight civilians. Dozens were injured, and the city suffered hundreds of thousands of dollars of property damage.

Boston Police Commissioner Edwin U. Curtis fired every Boston police officer who had participated in the strike. American Federation of Labor President Samuel Gompers sent a telegram to Governor (later President) Calvin Coolidge asking the city to reinstate the men; their grievances could be negotiated later. Gompers' request was rebuffed on September 14 by Coolidge's immediately famous reply, "There is no right to strike against the public safety by anybody, anywhere, any time." The State Guard remained in the city until December. By that time, Curtis had created a replacement police force composed of World War I veterans.

The government's strong reaction to the Boston Police strike sent a chill through other public safety labor organizations. More than 20% of the brand-new IAFF local unions withdrew from the international organization in 1920. The local fire fighters felt that they would be unable to accomplish their goals if they were associated with a national labor organization. This national revulsion to public safety strikes did not stop local unions from working to improve conditions.

Despite the backlash from the Boston situation, the initial success of fire fighters striking for improved working conditions created the base of power for organized labor among paid fire fighters. The movement continued, although the rate of success slowed down. Memphis fire fighters seeking higher pay threatened to strike again in 1920. This time, they were unsuccessful.

Striking to Preserve Wages and Staffing: 1931–1933

The Great Depression began with the stock market crash of 1929. For the next several years, local governments, faced with greatly reduced tax revenues, reduced wages, initiated furloughs (unpaid leave), and reduced work-force size. Several fire fighter strikes between 1931 and 1933 occurred in reaction to wage reductions and elimination of fire fighter jobs. One local union went on strike because the fire fighters had not been paid by the city for 2 months.

VOICES OF EXPERIENCE

❝ It is as important for the fire officer to know every detail of the union contract as it is to know the fire department's SOPs. ❞

Having been a union officer, I know the importance of fire officers understanding the role of the union. It is as important for the fire officer to know every detail of the union contract as it is to know the fire department's SOPs. The union contract describes the limits that the fire officer has over certain situations. Without a clear understanding of the union contract, the fire officer may find himself or herself at the center of a grievance. If a fire officer violates the union contract and the grievance is ultimately decided in favor of the employee, the fire officer may feel undermined and may lose credibility with both subordinates and superiors. To avoid these situations, a clear understanding of the union contract is critical.

As a fire officer, you may find yourself having to discipline an employee who brings in the union. This should not be taken as the union taking the side of an employee, but rather as the union fulfilling its role to ensure that due process is followed. Just as a defense attorney works to ensure a client is provided due process according to the law, the union official seeks to ensure equity for the fire fighter.

The fire officer is also often a member of the union. This can create a dilemma for the new fire officer, who may find it intimidating to be both a loyal union member and a loyal supervisor. This requires a delicate balance. The fire officer must abide by the fire department's rules and work toward the goals of the fire department while at the same time ensuring fairness and open communication with the crew. Each of these goals are desired by both management and the union.

David Hall
Springfield, Missouri Fire Department
22 years of service

In Seattle, all fire fighters had to take two 25-day furloughs in 1933. In addition to the furloughs, the city closed seven fire stations and eliminated eight engine companies, two hose companies, two squad companies, two fireboats, and one truck company to reduce expenditures.

In general, organized labor and fire fighter strikes seemed to have accomplished the IAFF mission of improving working conditions. Davis Ziskind's 1940 book, *One Thousand Strikes of Government Employees*, examined 39 fire fighter strikes and lockouts from 1900 to 1937. About half of the strikes resulted in partial or complete victory by labor. Unlike the unfortunate result of the Boston police riot, Ziskind could only find one large-loss fire that occurred during a labor action. A large industrial fire occurred in Pittsburgh, Pennsylvania during the first IAFF organized strike in 1918.

Staffing, Wages, and Contracts: The 1973–1980 Fire Fighter Strikes

The most recent series of fire fighter strikes occurred between 1973 and 1980. The IAFF voted to eliminate the 50-year-old no-strike clause in their constitution at their 1968 convention, at a time when the United States was in political turmoil. Most of these strikes occurred after labor and management reached an impasse while negotiating a new contract. An **impasse** occurs when the parties have reached a deadlock in negotiations.

In the 1970s, local governments were facing another recession that hit hard. Several large cities were facing bankruptcy. In the depths of the recession, New York City laid off more than 40,000 city employees on July 2, 1975, including 1600 FDNY fire fighters. Although 700 of the fire fighters were hired back within 30 days, 900 others lost their permanent fire fighter jobs. It would take 2 years before the city could rehire the laid-off fire fighters.

1975 San Francisco Strike

In San Francisco in 1975, both the fire and the police labor contracts were due for renegotiation. The city balked at the union request for a 14% pay raise. The city's counteroffer was a 2.5% pay raise over 2 years. In addition, the city wanted the flexibility to reduce staffing of engine and truck companies.

On August 20, 1975 at 6:00 P.M., the fire fighters on duty in 44 San Francisco fire stations went on strike. The city police officers also went on strike at the same time. Within an hour, the city was down to operating just five fire companies, all staffed with chief officers. This skeleton crew handled two major fires that burned most of the night. By the following morning, three additional fire companies were placed in service.

The San Francisco fire fighter strike went on for 38 days. It ended when the California State Supreme Court compelled the parties to **binding arbitration** (the resolution of a dispute by a third and neutral party). As part of the arbitration, San Francisco fire fighters lost the contract-required distinction of being the highest paid fire fighters in the state.

The citizens of San Francisco reacted with anger after the strike ended on October 14, 1975. Two propositions passed on the November 1975 election ballot. Proposition P changed the city charter to provide for automatic dismissal of any public safety official who participates in a strike. Proposition Q changed the fire fighter work hours from 24-hour duty shifts to 10-hour days and 14-hour nights. The perception was that the 24-hour fire fighter workday allowed a fire fighter to maintain a second job. The citizens considered the 24-hour workday a luxury, and it was legislated out of existence.

The San Francisco strike was the largest of several fire fighter strikes in California during that time period. Fire fighters in San Diego and Santa Barbara County also walked out. California Division of Forestry crews and mutual aid fire fighters refused to fill in at the empty stations.

Other Fire Fighter Strikes

Several metro-sized fire departments experienced labor actions during the same time period. At 8:30 A.M. on November 6, 1973, the Uniformed Fire Fighters Association of New York City went on strike. They returned to work 5½ hours later, when Supreme Court Justice Sidney A. Fine ordered the city to negotiate with the union. Within a month, a new labor contract was ratified.

In 1978, there were strikes of police, fire fighters, sanitation workers, and teachers in Memphis, Tennessee, a right-to-work state where public-sector collective bargaining is not permitted. The situation in Memphis was particularly tense because business properties owned by local elected officials were destroyed by arson and fire equipment was sabotaged to fail if National Guard troops attempted to operate it. Organized labor initiated a recall petition to remove the mayor and the city council, with support from national labor organizations. Wharton Business School Professor Herbert Northrup examined the Memphis situation, particularly the impact of using violence as a coercive tool and labor's attempts to force a recall of elected officials.

During Mayor Daley's 20-year tenure in Chicago, Illinois, the union did not have a labor contract. His successor, Jane Byrne, promised the union a contract in return for their support in the election; however, after 10 months of negotiations, the contract talks broke down. On February 14, 1980, most of the Chicago fire fighters walked off the job, leaving 120 empty fire stations. The strike lasted for 23 days, and another 8 months passed before the bargaining unit was recognized by the city.

Negative Impacts of Strikes

The negative impacts of a strike can be severe and long lasting. The public perception of fire fighters can quickly change from positive to negative, particularly if lives and property are lost during a strike. The direct impact on lives and property is felt by ordinary citizens, who generally have only indirect control over the factors that can lead to a strike or a settlement. In Chicago, by the time the fire fighters returned to work, 20 people had died in structure fires, including two children in a fire that was just a few blocks away from an empty fire station. More than $3 million of property burned during fires on July 2 and 3 during the 1978 Memphis strike. The loss of public support and confidence had very negative consequences in both cities.

The negative impacts can also be specifically directed toward individuals who are involved in a strike, particularly in areas where public employees do not have the right to strike. During the Chicago strike, a federal judged jailed the Local 2 president and assessed significant fines against the union. Each fire fighter who went on strike received the discipline action of a 1-day suspension without pay. Battalion chiefs who went on strike received a 4-day suspension without pay. Local 2 managed to obtain a contract, and truck company staffing was increased from four to five; however, the negative impacts on individuals, as well as on morale and internal relationships, lasted for at least 2 decades.

Strikes often cause long-lasting internal divisions within an organization, particularly when the decision to call a strike was a highly emotional event. In some of the fire departments where strikes occurred, lists of who participated and who did not participate in the walkout were maintained for as long as those individuals were still employed.

Since the 1980s, there have been fewer fire fighter strikes in the United States or Canada. None of the strikes in the 1970s resulted in a net gain for organized labor. The extreme negative public reaction, the political response of changing legislation, and the lasting legacy of lost trust have caused fire fighter strikes to be viewed as counterproductive. Measures such as picketing and alternative pressure tactics that maintain emergency services have been used to influence public opinion in several fire departments. Fire service strikes have occurred more recently among other industrialized nations throughout the world, particularly in Europe, where military personnel have been used to fill in for fire fighters.

Positive Labor-Management Relations

The value of a positive and productive labor-management relationship has become widely recognized. A healthy labor-management relationship is essential to producing positive outcomes and avoiding the strife and consequences of a confrontational climate. Successful relationships are built on trust, respect, and open lines of communication. Each side must be willing and able to focus on the mutual benefits of a positive relationship or face the consequences of a negative outcome.

The root cause of almost every labor disturbance is a failure to properly manage the relationship between labor and management. The traditional way of thinking was based on the premise that either labor or management must score a victory over the other to settle every point. A philosophical shift in labor-management relationships is moving away from confrontational strategies to win-win relationships. As with many traditions in the fire service, the ability and the willingness to change were produced by necessity.

Tremendous amounts of time, energy, and money can be wasted in the process of two sides trying to overwhelm each other instead of working together. A poor labor-management relationship usually produces casualties on one or both sides. Both fire chiefs and union presidents can lose their positions of power and influence. Positive relationships, based on mutual respect and understanding, are much more likely to produce positive results.

Some fire departments exist merely to meet the public's basic requirements for public safety, whereas others are truly committed to excellence and continual progress. Most observers agree that the most successful and progressive fire departments put significant effort into managing their labor-management relationships instead of continual confrontations and power struggles (▼ Figure 5-3).

In some instances, management uses the moral ethos of public service as leverage against labor and vice versa. Public support is usually viewed as vital by both sides because elected officials, who represent the public, have ultimate control over the economic and policy issues.

(**Figure 5-3**) The most successful and progressive fire departments put significant effort into managing their labor-management relationships.

Fire Marks

The Fair Labor Standards Act and Fire-Based EMS

The <u>Fair Labor Standards Act (FLSA)</u>, which was originally passed in 1938 as part of Roosevelt's New Deal, is an example of federal legislation that significantly affected the fire service during the 1990s. Once passed, federal legislation remains in effect unless it is ruled unconstitutional (as with NIRA), amended (as with Taft-Hartley), or repealed. The FLSA gradually evolved with the adoption of new provisions and regulations that expanded its application to include fire departments.

The primary purpose of FLSA was to establish minimum standards for wages and to spell out administrative procedures covering work time and compensation, including overtime entitlement. For most workers, FLSA requires overtime to be compensated at time-and-a-half pay—150% of an employee's regular hourly wages. This legislation was adopted during a period of profound underemployment and was designed to encourage employers to hire more workers instead of paying current employees to work additional hours.

When it was originally adopted, the FLSA did not apply to government employees, but in 1974, additional provisions were adopted to make FLSA applicable to federal employees. In 1985, the Supreme Court ruled in *Garcia vs San Antonio MTA*, 469 US 528 (1985) that local union public employers must also abide by the FLSA regulations, unless Congress specifically legislates otherwise.

Amended regulations, adopted in 1986, established a special overtime rule for fire fighters. Most employees are entitled to be paid overtime starting at the 41st hour of a 7-day workweek. The average workweek for fire fighters across the United States at that time was determined to be 53 hours. Under the new federal regulations, public safety agencies do not have to pay fire fighters overtime until after they have worked 212 hours in a maximum 28-day cycle—which averages out to a 53-hour workweek. This is known as the 207(k) exemption and covers individuals whose work involves "the prevention, control, or extinguishment of fires" for 80% of their work time.

The inclusion of local union public safety employees within the FLSA regulations created tremendous complications in accounting for work hours because there is no simple duty rotation that results in a 53-hour workweek. The work schedules in most fire departments had to be adjusted, and thousands of local union personnel regulations had to be rewritten. It also resulted in hundreds of lawsuits to both change local union practices and obtain back pay. One lawsuit in 1990 revealed a significant oversight in the existing FLSA language.

Anne Arundel County, Maryland fire fighter/paramedics filed suit in 1990 for improperly calculated overtime payments. The FLSA description of fire fighter activities included "housekeeping, equipment maintenance, lecturing, attending community fire drills, and inspecting homes and schools for fire hazards" as incidental functions but did not include EMS activity. Some of the employees were assigned to paramedic ambulances and did not routinely participate in firefighting

activities. Other employees, who were assigned to fire suppression companies, were actually spending more time handling EMS first-responder calls than performing fire suppression or related activities.

The plaintiffs claimed that the 207(k) exemption did not apply to them because more than 20% of their time was spent on nonfirefighting activities. They argued that their overtime pay should start at the 41st hour and not the 54th hour of work in an average workweek. US District Court Judge Walter Black ruled in favor of the employees, creating a $4 million back payment obligation for the county.

Anne Arundel County appealed the decision to the fourth Circuit Court of Appeals. Seven years after the original lawsuit was filed in Maryland, the appeals court upheld the majority of Black's decision. Anne Arundel County then appealed the decision to the Supreme Court. In December 1998, the Supreme Court declined to hear the Anne Arundel case, letting the Black decision stand.

The Anne Arundel case was one of many similar lawsuits and union grievances that were going on throughout the nation during the same time period. Dozens of cities were involved in FLSA lawsuits and were becoming obligated for huge amounts of retroactive back pay. Some of the settlements went back to the 1986 adoption of the FLSA regulations. In Houston, Texas, which has the third largest fire department in the nation, with more than 3200 employees, the first FLSA settlement in 1997 resulted in the payment of $6.2 million of overtime pay to 173 dispatchers and arson investigators. A second settlement in 1999 provided $4.4 million in overtime pay for 2600 fire fighters. The fifth US Circuit Court of Appeals ruled on January 18, 2002 that Houston also owed overtime pay to 260 paramedics and EMTs who had been working more than 40 hours in a 7-day week.

The situation became even more complicated when the eighth US Court of Appeals made a ruling that was in conflict with the ruling of the fourth US Court of Appeals in the Anne Arundel case. When the Supreme Court declined to review the Anne Arundel case, the conflicting legal opinions were left to stand within the regions subject to each Circuit Court.

In response to this legal turmoil, US Representative Robert Ehlrich, a Maryland Republican, introduced House Resolution 1693, the Fire and Emergency Services Definition Act. This legislation expanded the FLSA definition of fire protection activities to include paramedics, emergency medical technicians, rescue workers, ambulance personnel, and hazardous materials workers. The Ehrlich resolution, which had the support of both the IAFF and the International Association of Fire Chiefs (IAFC), was passed on November 4, 1999. The Senate passed the same resolution a week later, and President Clinton signed the Fire and Emergency Services Definition Act into law on December 9, 1999. This 1999 law amended a portion of the original act that was passed in 1938 and expanded by the Supreme Court decision in 1985 to cover local government employees. It took almost a full decade to resolve the issue of FLSA coverage for fire fighters performing EMS duties.

Special Aspects of Fire Fighter Labor Contracts

A unique feature of labor contracts negotiated by local union IAFF leadership is the amount of control that the contract may have on staffing policy and field operations. Upheld in court as an employee safety issue, many labor contracts specify how many fire fighters are assigned to a fire company. In some cities, the contract specifies how many engine and ladder companies respond to a structure fire.

Growth of IAFF as a Political Powerhouse

The fire fighter strikes of 1973–1980 were much less successful than the first strikes 60 years earlier. The economic turmoil and wage reductions in the 1990s were as severe as they were in the 1930s, but organized labor took a different path. Instead of striking, labor worked to get political candidates who were supportive of labor into local and national political office. The attempted recall of Memphis elected officials in 1979 may have been the beginning of this new era of labor influence.

Fire fighters are particularly respected in political circles for the efforts that they can put behind a political candidate who supports the objectives of IAFF or a particular local union. In addition to providing off-duty fire fighters to assist with campaigning, organized labor found out that "money talks." The creation of FIREPAC, the IAFF's political action committee, in 1991 was a natural progression of the labor organization's activity as a major influence in improving the fire fighter's work environment.

FIREPAC is a **political action committee (PAC)**, a special interest group that can solicit funding and lobby local and national elected officials for their cause. Funded by donations from individuals, FIREPAC promotes the legislative and political interests of the IAFF. The money is used to educate members of Congress about issues important to fire fighters and emergency medical personnel and to elect candidates to office that support those issues. In the 2000 election cycle, FIREPAC raised more than $1.3 million in voluntary donations, which was contributed to 297 US Senate and Congressional candi-

dates. The election results showed that 89% of the candidates supported by FIREPAC won.

Using money and off-duty fire fighters to assist political candidates or causes may appear unseemly for public safety employees. Hal Bruno, former director of political reporting for ABC television and radio networks, has repeatedly pointed out in his monthly Firehouse® Magazine columns that the fire service must be fully engaged in the political process. That means being an active part of the local political process. Following are two examples.

Seattle/King County Medic One 2001 Tax Levy Referendum

It took an extra effort from the members of Seattle IAFF Local 27 to ensure passage of the Medic One tax levy at the November 2001 elections. Since 1979, a special tax levy in Metropolitan King County had paid for this high-performance paramedic ambulance service. A referendum is held every 6 years to reauthorize the special funding authorization.

In 1997, only 56% of the citizens voted for renewal of the tax levy, which failed to meet the number required to maintain funding for the Medic One program. A special referendum was passed in February 1998 to restore the funding and keep 22 paramedic ambulances in service throughout Metropolitan King County for 3 years. As the 2001 referendum vote approached, Seattle Local 27 worked hard to get out the vote to ensure another 6 years of Medic One funding, including four additional paramedic ambulances.

Corpus Christi Forces a Special Election to Settle Contract Impasse

On August 2, 2001, negotiations stalled between the City of Corpus Christi, Texas and IAFF Local 936. The union was asking for a 6% pay raise, and the city counteroffer was 3%. Negotiations ceased on November 15, 2001, after 105 days without an agreement. Both sides were in a fact-finding phase, with an outside mediator scheduled to resolve the impasse later in 2002. While waiting for the mediator, Local 936 formulated a ballot proposition calling for an 8.5% pay raise that would take effect on June 1, 2002. As the deadline approached, the city secretary confirmed that Local 936 had submitted more than enough verified signatures to compel the city to hold a special election in May 2002.

The city was opposed to having a special election for two reasons: they did not want city payroll decisions to become a ballot issue, and the special election would cost the city about $100,000. On the last day to call for a May special election, the city made a new pay offer. Fire fighters would get a 3% pay raise retroactive to August 1, 2001, a 2.4% pay raise retroactive to March 1, 2002, and a 3% raise effective in November 2002. In addition, the city promised to match any pay raise of greater than the 3% that is negotiated by the

Fire Marks

The Medic One paramedic ambulance service began operations in Seattle on March 7, 1970. The Seattle City Council declined to fund continuation of the Medic One project within the fire department budget in 1972. Members of IAFF Local 27 scrambled to fund the lifesaving project through a special countywide tax levy. CBS news magazine "60 Minutes" profiled Seattle as *The Best Place to Have a Heart Attack* in 1974.

police when their contract is renewed in 2003. Although the offer was less than the pay raise proposed for the special election, 81% of the 295 union members ratified the contract. Local 936 withdrew their special election petition.

Labor-Management Alliances

In addition to political action, the IAFF has developed strategic labor-management alliances to promote mutually agreeable goals. In some situations, both labor and management want the same goal. Instead of fighting with each other, they would work together to accomplish the goal. This chapter examines two national examples.

Fire Service Joint Labor Management Wellness-Fitness Task Force

The IAFF and the IAFC joined forces to develop a comprehensive and mutually beneficial fitness-wellness program. The program was released in 1997. It is designed to serve a fire fighter throughout a decades-long career. Ten fire departments and their IAFF local unions participated in a task force to develop a comprehensive and nonpunitive program:

- Austin, Texas and IAFF Local 975
- Calgary, Alberta and IAFF Local 255
- Charlotte, North Carolina and IAFF Local 660
- Fairfax County, Virginia and IAFF Local 2068
- Indianapolis, Indiana and IAFF Local 416
- Los Angeles County, California and IAFF Local 1014
- Miami-Dade County, Florida and IAFF Local 1403
- New York City and IAFF Locals 94 and 854
- Phoenix, Arizona and IAFF Local 493
- Seattle, Washington and IAFF Local 27

The completed physical fitness and wellness program package was released at both the IAFF Redmond Symposium and the IAFC Conference in August 1997. The task force is continuing to work on refining and supporting the original program and developing a Fire Service Peer Fitness Trainer certification program. The two organizations have also developed the Candidate Physical Ability Test, a standardized physical ability test that is validated as an assessment of a fire fighter candidate's existing physical ability. The physical tasks represent the typical tasks a fire fighter performs.

The Fire Service Leadership Partnership

During the mid- to late 1990s, the IAFC and the IAFF recognized the effect of human relations on labor-management relationships. Both organizations recognized the importance of developing good labor-management relationships to deal with emerging issues among fire departments across the United States. The IAFC/IAFF Fire Service Leadership Partnership was developed in 1999 to address the needs of today's fire chiefs and union presidents. The same fundamental

Company Tips

Phoenix Fire Department—A Case Study

The Phoenix Fire Department introduced the fire service to a new paradigm shift in labor-management relationships. In many industries during the 1980s, the nature of labor-management relations was closely examined for their existing culture, which were generally based on the traditional notion that labor and management should have an adversarial relationship. This doctrine of adversaries was prevalent in most industries for many decades.

In 1984, the Phoenix Fire Department embarked on a landmark initiative known as the relations by objectives (RBO) process. The RBO process had been used in other labor-management disputes in which the two sides had significantly different positions on values, goals, and trust for one another. The goal of RBO is to establish a positive relationship between labor and management, based on trust and mutual respect. Additionally, RBO provides the framework to fulfill the goals of the organization with participating work from both sides.

The Phoenix Fire Department and the United Phoenix Fire Fighters used RBO to develop a framework for decision making. The model outlines how each major issue is handled to include a means of analysis, decision, education, implementation, revision, and review. Each step has specific parameters in which labor and management cooperatively work on issues.

This cooperative style was new for labor-management relationships. With RBO, the Phoenix Fire Department has been able to address many issues without contention, such as incumbent drug testing, apparatus specifications and placement, safety, and employee wellness. Because both labor and management participated in the development of many programs and issue resolution, both sides of the employee/employer environment shared equal successes and failures and could pragmatically blame each other. This also led to ownership of the successes and failures; thus, the culture of the Phoenix Fire Department also changed from one of a traditional employee/employer-based culture to one in which each member (including their civilian staff) feels like a part owner of the department. Whether this cultural shift was intended or not, the result was the development of a work environment in which each member has the opportunity to make significant contributions. In most employee/employer environments, the work conditions are dictated from the top down from management to employee, often without input from the employees. Thus, the employees may develop a sense of apathy and place distance between themselves and the employer. RBO and the Phoenix Fire Department showed how the participative management can benefit a department and develop a culture in which each employee is treated and sees himself or herself as a member instead of a rank-in-file employee.

concepts that were involved in developing the Wellness Fitness Initiative provided the foundation for the Fire Service Leadership Partnership.

Today, scores of fire chiefs and union presidents have attended the Fire Service Leadership Partnership to learn how to enhance their labor-management relationships.

Many issues that face the fire service today and in the future will require hard problem solving from all facets of the fire service. Without the combined talent of all members of the fire service, fire fighters and fire chiefs will always be behind in their reaction to emerging issues, such as fire-based EMS, urban-wildland interface, occupational safety, and perhaps the greatest challenge to the fire service in the new millennium—the fire service response to terrorism. For tomorrow's leaders, the past provides perspective on where the fire service has historically been. It is clear in this new environment of new challenges, a good labor-management relationship is essential. Learning the humanistic skills required to handle these relationships should be a regular part of a fire officer's curriculum and not an ancillary elective.

The Fire Officer's Role as a Supervisor

One of the fundamental duties of a fire officer is to supervise the activities of subordinate fire fighters. The basic authority of a supervisor and the duties of subordinates are defined by the personnel rules of the city or governmental organization, as well as the specific rules, regulations, and procedures of the fire department. In addition, when a collective bargaining agreement is in effect, additional details of the relationship are spelled out in the contract or MOU. Supervisors are expected to follow all of the established rules and procedures in assigning duties and in all other aspects of the relationship with their subordinates. In many cases, it is a significant challenge for a newly promoted fire officer to learn which rules and regulations apply to which situation and how they are interpreted and applied.

In most organizational structures, there is a clear distinction between labor (the workers) and management (the managers and supervisors). The managers and supervisors represent the organization, and the union represents the workers. If there is any doubt or disagreement about the application or interpretation of the contract, a process should be in place for labor and management representatives to meet and resolve the problem.

This line between labor and management is more complicated in many fire departments because the first-level supervisors are often members of the same collective bargaining unit as the fire fighters they supervise. A fire officer's relationship to the organization is often covered by the same contract that the officer has to follow and enforce. The formal line between labor and management is often at a higher level, such as battalion chief. In some cases, the officers are members of a bargaining unit that is separate from that of the fire fighters.

As a supervisor, a fire officer is generally the first point of contact between the workers and the fire department organization (▶ **Figure 5-4**). If there is a disagreement relating to an interpretation or application of a work rule that is

Figure 5-4 A fire officer is generally the first point of contact between the workers and the organization.

covered by the contract, the first-level supervisor is the individual who should have the first awareness of the problem and the first opportunity to resolve it. An officer who is a member of the same bargaining unit as an individual who is dissatisfied must clearly understand the established problem-solving processes.

Application Example: A Grievance Procedure

A **grievance** is a dispute, claim, or complaint that any employee or a group of employees may have about the interpretation, application, and/or alleged violation of some provision of the labor agreement or personnel regulations. A **grievance procedure** is a formal structured process that is employed within an organization to resolve a grievance. In most cases, the grievance procedure is incorporated in the personnel rules or the labor agreement and specifies a series of steps that must be followed in order. If the problem cannot be resolved in a mutually acceptable manner at one level, it can be taken to the next level, up to some ultimate level, where an individual or body has the final authority to impose a binding decision.

Whether the grievance is related to the labor contract or to the municipal personnel regulations, the grievance procedure should specify a sequential process and a timeline to move through the steps. The grievance can be resolved at any point by management's accepting the complaint and the corrective action requested by the grievant or by both sides reaching a negotiated settlement that is acceptable to each. If management rejects the grievant's claim, the grievance can be taken to the next level. The timeline ensures that a grievance will not be stalled at any level for an excessive time period, awaiting a decision.

An employee can contact a union representative at any time to discuss a situation, including how the union interprets the rule in question and whether a grievance should be submitted. The employee's union representative usually becomes formally involved at either the first or the second step of the grievance process. The union representative acts as an advocate for the individual or group that submitted the grievance. The union becomes more involved as the process moves through the steps, particularly in cases in which the problem has broad impact within the organization.

The objective should always be to resolve the problem at the lowest possible level and in the shortest possible time. Grievances that have to be processed through multiple steps are disruptive, time consuming, and often costly to both sides. The ability to resolve problems at a low level is an indication of a healthy organization with a good labor-management relationship, whereas a steady stream of grievances moving up to the highest levels is a symptom of major problems in the relationship.

The grievance procedure outlined in the following steps is an example that comes from one particular fire department. The detailed procedures employed by different organizations usually include some variations from this model. A similar process can be established to resolve disputes in fire departments where there is no formal labor contract, even in volunteer organizations. The most important responsibility for a fire officer is to know and follow the procedures that apply in his or her organization.

Step One

The grievant presents his or her complaint verbally to a supervisor, shortly after the occurrence of the action that gave rise to the grievance. In some organizations, this nondocumented verbal notification is called an "informal grievance," or step zero. Even this informal step requires the grievant to provide three important pieces of information:

- The article and section of the labor agreement or personnel regulation alleged to have been violated
- A full statement of the grievance, giving facts, dates, and times of events, as well as specific violations
- A statement of the desired remedy or adjustment

Step Two

The second step initiates the formal part of a grievance procedure. If the problem is not resolved at step one, the employee may prepare and submit a written grievance. This is usually submitted on a specified grievance form document (often a carbonless copy form). The employee, the employee's supervisor, and the personnel office each receives a copy of the grievance.

The supervisor has 10 calendar days to reach a decision and provide a written reply to the grievant. Failure to respond to the grievance within 10 days means that the supervisor has denied the grievance and the grievant can immediately go to step three.

Step Three

A step-three grievance is written out on another specific grievance form and again specifies the article and section of the contract or personnel regulation alleged to have been violated; the dates, times, and specific violations that are alleged to have taken place; and the desired remedy or adjustment. Copies of the step-two grievance form and the supervisor's response are attached.

A step-three grievance is submitted to a second-level supervisor, typically a battalion chief, who has 10 calendar days to respond. If the grievance is denied or the battalion chief does not respond within the specified time, the grievant can move to step four and present his or her grievance to the fire chief.

Step Four

If the grievance remains unresolved, the grievant can present it to the fire chief or his or her designee as the fourth step. The same written information must be submitted, along with all of the documentation from the previous steps. Usually, the fire chief, as the agency head, has 14 to 30 days to respond to a step-four grievance.

If the fire chief does not respond within the time frame, or if the grievance remains unsettled, the process moves out of the fire department to a mediator, personnel board, or civil service board for resolution.

Summary

The fire officer is in a unique position regarding labor issues within the fire station. The fire officer is a working supervisor and is often both the designated representative of fire department administration and a dues-paying member of an IAFF local union. It is important that the fire officer knows the grievance procedures, labor contract provisions, and personnel regulations that create the limits to, and framework of, the scope of supervisory work.

One goal of this chapter was to demonstrate that the federal status of labor and management swings like a pendulum. Labor had an advantage with the Norris-LaGuardia and Wagner-Connery Acts and the 1970 postal service strike. Management had the advantage with the Taft-Hartley and Landrum-Griffin Acts and the 1981 air traffic controller strike.

Another goal was to show the changing methods of organized labor effectiveness. An old labor model would have had Corpus Christi Local 936 going out on strike when their contract negotiations stopped in November 2001. Instead, they mounted a signature drive to place their wage issue on a May 2002 special election. Placing municipal pay decisions on a ballot and incurring a $100,000 city expense for a special election seemed to be the motivation to get the city to complete contract negotiations.

You Are the Fire Officer: Conclusion

When the dust settles, you learn that the labor contract overrides the Metro City Personnel Regulations. Although the personnel regulations state that a supervisor may change an employee's work hours and do not make any mention of advance notice, the labor contract specifically requires a minimum of 72 hours notice. The labor contract also requires that employees be provided with the appropriate educational resources (books, practice tests, and so on) 2 weeks before they are scheduled to attend an employer-mandated recertification class.

The situation is resolved at the first step of the grievance process after you wisely call the battalion chief and ask for guidance. Following his advice, you apologize to Fire Fighter Trammel and admit that you had made a mistake. The battalion union representative calls Trammel later that day to confirm that the problem has been resolved.

The battalion chief clarifies the situation by explaining that he had just wanted you to see if any of the fire fighters would volunteer to attend the EMT refresher class this week because the academy had a last-minute vacancy and it could be filled by an employee who was willing to attend without the required notice. The contract did not allow a last-minute involuntary reassignment.

The labor contract was not one of the items you had to memorize for the promotional exam; however, you need to know what the contract says in order to succeed as a fire officer. Your ability to quickly correct the problem and apologize for having been wrong will create a positive impression with the members of your company.

Wrap-Up

Ready for Review

- Norris-LaGuardia Act of 1932 made yellow dog contracts unenforceable.
- Wagner-Connery Act of 1935 established the National Labor Relations Board and the procedures commonly called collective bargaining.
- Good faith bargaining and a 60-day "cooling off" period were created by the Taft-Hartley Labor Act of 1947.
- On occasion, striking public sector workers are replaced (Boston police in 1919 and air traffic controllers in 1981).
- Public sector bargaining rules vary by state.
- The grievance procedure is an incremental, multistep process that requires a timely response.
- Failure of management to provide a timely response immediately pushes the grievance to the next level of supervision.

Hot Terms

Binding arbitration The resolution of a dispute by a third and neutral party, one who is not personally involved in the dispute and may be expected to reach a fair and objective decision based on an informal hearing, at which the disputants may argue their cases and present all relevant evidence. It is usually agreed in advance that such a decision will be binding and not subject to appeal.

Collective bargaining Method whereby representatives of employees (unions) and employers determine the conditions of employment through direct negotiation, normally resulting in a written contract setting forth the wages, hours, and other conditions to be observed for a stipulated period (e.g., 3 years). Term also applies to union-management dealings during the terms of the agreement.

Fair Labor Standards Act (FLSA) 1938 act that provides the minimum standards for both wages and overtime entitlement and spells out administrative procedures by which covered work time must be compensated. Public safety workers were added to FLSA coverage in 1986.

Good faith bargaining A legal requirement arising out of Section 8(d) of the National Labor Relations Act on both the union and the employer. Enforced by the National Labor Relations Board, the parties are required: "To bargain collectively . . . to meet at reasonable times and confer in good faith with respect to wages, hours, and other conditions of employment, or the negotiation of an agreement or any question arising thereunder, and the execution of a written contract incorporating any agreement reached if requested by either party, but such obligation not to compel either party to agree to a proposal or require the making of a concession. . . ."

Grievance A dispute, claim, or complaint that any employee or group of employees may have in relation to the interpretation, application, and/or alleged violation of some provision of the labor agreement or personnel regulations.

Grievance procedure A formal structured process that is employed within an organization to resolve a grievance. In most cases, the grievance procedure is incorporated in the personnel rules or the labor agreement and specifies a series of steps that must be followed in order.

Impasse Occurs when the parties have reached a deadlock in negotiations. Also described as the demarcation line between bargaining and negotiation. A declaration of an impasse brings in a state or federal negotiator that will start a fact-finding process and will lead to a binding arbitration resolution.

Political action committee (PAC) Organizations formed by corporations, unions, and other interest groups who solicit campaign contributions from private individuals and who distribute these funds to political candidates.

Right-to-work A worker cannot be compelled, as a condition of employment, to join or not to join, or to pay dues to a labor union.

Strike A concerted act by a group of employees, withholding their labor for the purposes of effecting a change in wages, hours, or working conditions.

Unfair labor practices An employer or union practice forbidden by the National Labor Relations Board or state/local laws, subject to court appeal. It often involves the employer's efforts to avoid bargaining in good faith.

Yellow dog contracts Pledges that employers required workers to sign indicating that they would not join a union as long as the company employed them. Declared unenforceable by the Norris-LaGuardia Act of 1932.

Fire Officer in Action

The union and fire department are currently in negotiations for a new three-year contract. Negotiations have been underway for close to a year with no agreement in the areas of pay and benefits. As the fire officer, you have heard some of the rumblings on shift and have noticed a definite decline in the morale. Many of the personnel keep talking about wanting to strike against the department for what they perceive as a lack of any movement by the department.

1. If an employee was asked to sign a statement that they would not join the local, what would that be called?

    A. Collective bargaining

    B. Yellow dog contracts

    C. Unfair labor practices

    D. Right-to-work

2. What does right-to-work mean?

    A. Employees are required to be a member of a union.

    B. Employees have the right to bargain their positions.

    C. Workers cannot be compelled to join a union as a condition of employment.

    D. Employees are not permitted to strike.

3. If the contract dispute over pay and benefits continues, some states would require binding arbitration. What is this?

    A. Given a specific time period to agree on contract

    B. Dispute resolution by a third, neutral party

    C. Opens the ability to strike

    D. Mediation

4. What is included in the second step of the grievance procedure?

    A. Grievance presented to supervisor verbally

    B. A statement of the desired remedy or adjustment

    C. Grievance is written out and submitted to a second-level supervisor

    D. Initiates the formal part of the grievance procedure

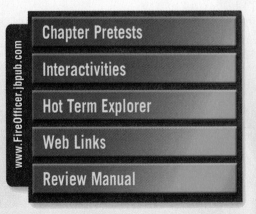

www.FireOfficer.jbpub.com

Chapter Pretests

Interactivities

Hot Term Explorer

Web Links

Review Manual

Technology Resources

NFPA 1021 Standard

Fire Officer I

4.2.1 Assign tasks or responsibilities to unit members, given an assignment at an emergency scene operation, so that the instructions are complete, clear, and concise; safety considerations are addressed; and the desired outcomes are conveyed.

> **(A) Requisite Knowledge.** Verbal communications during emergency situations, techniques used to make assignments under stressful situations, and methods of confirming understanding.

4.2.2 Assign tasks or responsibilities to unit members, given an assignment under nonemergency scene operation, so that the instructions are complete, clear, and concise; safety considerations are addressed; and the desired outcomes are conveyed.

4.2.3 Direct unit members during a training evolution, given a company training evolution and training policies and procedures, so that the evolution is performed in accordance with safety plans, efficiently, and as directed.

4.6.4 Develop and conduct a postincident analysis, given a single-unit incident and postincident analysis policies, procedures, and forms, so that all required critical elements are identified and communicated, and the approved forms are completed and processed in accordance with policies and procedures.

> **(A) Requisite Knowledge.** Elements of a postincident analysis, basic building construction, basic fire protection systems and features, basic water supply, basic fuel loading, fire growth and development, and departmental procedures relating to dispatch response tactics and operations and customer service.

> **(B) Requisite Skills.** The ability to write reports, to communicate orally, and to evaluate skills.

4.7* **Health and Safety.** This duty involves integrating safety plans, policies, and procedures into the daily activities as well as the emergency scene, including the donning of appropriate levels of personal protective equipment to ensure a work environment, in accordance with health and safety plans, for all assigned members, according to the following job performance requirements.

4.7.1 Apply safety regulations at the unit level, given safety policies and procedures, so that required reports are completed, in-service training is conducted, and member responsibilities are conveyed.

> **(A) Requisite Knowledge.** The most common causes of personal injury and accident to members, safety policies and procedures, basic workplace safety, and the components of an infectious disease control program.

> **(B) Requisite Skills.** The ability to identify safety hazards and to communicate orally and in writing.

4.7.2 Conduct an initial accident investigation, given an incident and investigation form, so that the incident is documented and reports are processed in accordance with policies and procedures.

> **(A) Requisite Knowledge.** Procedures for conducting an accident investigation and safety policies and procedures.

> **(B) Requisite Skills.** The ability to communicate orally and in writing and to conduct interviews.

NFPA 1021 Standard (continued)

Fire Officer II

5.6.1 Produce operational plans, given an emergency incident requiring multiunit operations, so that the required resources and their assignments are obtained and plans are carried out in compliance with approved safety procedures, resulting in the mitigation of the incident.

(A) Requisite Knowledge. Standard operating procedures; national, state/provincial, and local information resources available for the mitigation of emergency incidents; an incident management system; and a personnel accountability system.

(B) Requisite Skills. The ability to implement an incident management system, to communicate orally, to supervise and account for assigned personnel under emergency conditions, and to serve in command staff and unit supervision positions within the Incident Management System.

5.7 **Health and Safety.** This duty involves reviewing injury, accident, and health exposure reports; identifying unsafe work environments or behaviors; and taking approved action to prevent reoccurrence, according to the following job requirements.

5.7.1 Analyze a member's accident, injury, or health exposure history, given a case study, so that a report including action taken and recommendations made is prepared for a supervisor.

(A) Requisite Knowledge. The causes of unsafe acts, health exposures, or conditions that result in accidents, injuries, occupational illnesses, or deaths.

(B) Requisite Skills. The ability to communicate in writing and to interpret accidents, injuries, occupational illnesses, or death reports.

Additional NFPA Standards

NFPA 472, *Standard for Professional Competence of Responders to Hazardous Materials Incidents*

NFPA 1521, *Standard for Fire Department Safety Officer*

NFPA 1581, *Standard on Fire Department Infection Control Program*

NFPA 1583, *Standard on Health-Related Fitness Programs for Fire Fighters*

NFPA 1561, *Standard on Emergency Services Incident Management System*

NFPA 1975, *Standard on Station/Work Uniforms for Fire and Emergency Services*

Knowledge Objectives

After studying this chapter, you will be able to:

- Describe the most common causes of personal injury and accidents to fire fighters.
- Define incident safety officer.
- Describe safety policies and procedures and basic workplace safety.
- Describe the components of an infectious disease control program.
- Describe procedures for conducting an accident investigation and safety policies and procedures.

Skills Objectives

After studying this chapter, you will be able to:

- Identify safety hazards.
- Implement an incident management system and ensure the safety of personnel under emergency conditions.
- Interpret accident reports.

You are the officer in charge of a ladder company en route to a factory fire. The first-arriving fire company reported an explosion and fire in the building, with several injured civilians. They have called for additional ambulances and the hazardous materials team. Before you arrive, the incident commander designates your company the rapid intervention company and orders you to report to the incident safety officer.

1. What are the common causes of fire fighter deaths and injuries on the fireground?
2. What are the primary responsibilities of a rapid intervention crew?
3. What are the primary responsibilities of an incident safety officer?
4. What are the typical tasks expected of an incident safety officer?

Introduction to Safety

Fire department operations often include high-risk situations that can occur under any weather conditions at any time of the day or night. The fire officer is responsible for ensuring that every fire fighter survives every incident without injury. This is sometimes expressed as a fire officer's special obligation to ensure that "everyone goes home" at the end of a tour of duty.

It is a fundamental responsibility of every fire officer to always consider safety as a primary objective. The fact that firefighting is a dangerous occupation does not make death acceptable. Fire officers have a personal and professional obligation to prevent deaths and injuries to the fire fighters who are working under their supervision. This obligation also extends to the families and friends of those fire fighters, to the fire department, and to the community.

The responsibilities of a fire officer include identifying hazards and mitigating dangerous conditions in order to provide a safe work environment for fire fighters. The fire officer must also identify and correct behaviors that could lead to a fire fighter's injury or death. It is essential that the officer set a good example at all times because the crew members will follow their officer's lead. Safe practices must be the only acceptable behavior, and good safety habits should be incorporated into all activities, including training and nonemergency duties.

Fire Fighter Death and Injury Trends

There is no doubt that firefighting is a dangerous occupation, and the risk of injury or death is always present in the fire service work environment. It is almost inevitable that some line-of-duty deaths will occur and that some fire fighters will be injured on the job; however, death or injury must never be viewed as an acceptable or inevitable outcome of any situation.

If the work of fire fighters were not dangerous, there would be no need for highly trained and properly equipped fire fighters to protect our communities. We cannot eliminate all of the risks, but we do understand them and we can recognize them. It is our responsibility to plan for them, train for them, equip ourselves for them, and make allowances for them. In order to be successful and to survive the experience, fire fighters must be fully prepared to work safely in high-risk situations.

Understanding the causes of fire fighter deaths and injuries is an important first step in developing strategies to prevent them from recurring. Prevention depends on our ability to break the sequence of events that leads to injuries and deaths. The National Fire Protection Association (NFPA), the U.S. Fire Administration, and the International Association of Fire Fighters (IAFF) all publish annual reports and statistics, which consistently report the loss of more than 100 fire fighters in the line of duty each year in the United States and provide data about the causes and circumstances of these events.

In April 2002, the U.S. Fire Administration published *Firefighter Fatality: Retrospective Study*, which analyzed the data accumulated by the National Fire Data Center over the previous decade. The study noted that reported fire fighter line-of-duty fatalities declined from 171 deaths in 1978 to 77 deaths in 1992; however, the trend has reversed since then. Since 1993, there have been more than 100 fire fighter fatalities almost every year. Heart attacks were the leading cause of death, followed by trauma, burns, and asphyxia ▶ Figure 6-1 .

The U.S. Fire Administration has adopted a 50% reduction in fatalities over a 10-year period as a national goal. The U.S. Fire Administration study noted that during the 1990s, the number of fires declined significantly, whereas the number of fire fighter fatalities increased. The rate of fire fighter fatalities per 100,000 fire events rose by 25% in the 1990s. This trend should be of great concern to every fire officer.

When the NFPA released its annual report on fire fighter fatalities, it noted that of the 105 fire fighters who died in

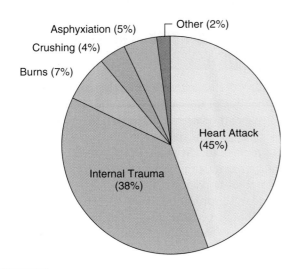

Figure 6-1   Fire fighter deaths by nature of injury—2003. Source: NFPA.

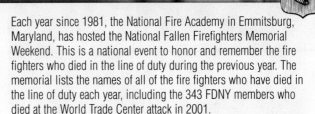

Fire Marks

Each year since 1981, the National Fire Academy in Emmitsburg, Maryland, has hosted the National Fallen Firefighters Memorial Weekend. This is a national event to honor and remember the fire fighters who died in the line of duty during the previous year. The memorial lists the names of all of the fire fighters who have died in the line of duty each year, including the 343 FDNY members who died at the World Trade Center attack in 2001.

The memorial service, like the image of a fire fighter's funeral procession, is not easily forgotten. A fallen fire fighter is honored as a hero, buried with great pomp and circumstance, and long remembered on plaques and memorials. Although we will always honor our fallen comrades, we must also remember that most fire fighter fatalities can and should be prevented. The fire service has a tremendous responsibility to its own members, to their families and friends, and to our communities to protect the lives of all fire fighters.

Whenever a fire fighter is lost, a painful and lasting impact is felt by many individuals, beginning with the members of the fire fighter's immediate family. A spouse must pick up a shattered life and try to make sense of the loss. Children will grow up without their father or mother. Parents have to bury a son or daughter. Personal friends and coworkers of the fallen fire fighter feel a similar type of loss. The loss is also felt throughout the fire department or agency and throughout the fire service whenever a fire fighter is killed or seriously injured.

A fire officer who experiences the loss of a member working under his or her supervision often feels a sense of responsibility for having allowed the loss to occur. This is one of the most difficult situations that a fire officer can experience.

2003, only 29 died on the fireground—the land or building where a fire occurs. That was the lowest number of fire fighter deaths on the fireground since the NFPA began collecting the data in 1977, and it was also the first time that fireground deaths accounted for less than 30% of the total. The data indicate that fewer fire fighters are dying on the fireground, both in actual numbers and in relative terms; however, more are being killed while responding to emergency incidents or performing duties other than fighting fires. Each time a company rolls out of the station responding to a call, the fire fighters are putting their lives at risk.

It is estimated that there are approximately 1,000 fire fighter injuries for every fire fighter fatality. The data show a wide difference when the nature of injuries is compared with the causes of fire fighter deaths. Sprains and strains account for the largest number of injuries, closely followed by cuts, bruises, bleeding, and wounds. The most common activities being performed when injuries occur are extinguishing fires and suppression support.

Heart Attack

Heart attacks are the leading cause of fire fighter line-of-duty deaths, accounting for 44% of the deaths from 1984 to 2000. As a group, fire fighters are more likely to die of a heart attack while on duty than other American workers. Although the term *occupation* is used to refer to fire fighters, the category includes a full range of affiliations, including career, volunteer, and wildland fire fighters.

This heart attack rate for fire fighters is closely related to the nature of the work, which can suddenly change from a period of relatively low activity to an episode of high stress and intense exertion. This type of sudden stress can trigger a heart attack, particularly in an individual who is predisposed to cardiovascular disease. Security guards also have a high rate of heart attacks, which represent 25% of their on-duty deaths. Construction workers, who routinely perform strenuous physical labor, and police officers, who experience similar variations between low-intensity and high-stress periods, both have lower heart attack rates than fire fighters. Among workers in all occupational categories, only 15% of on-duty deaths are caused by heart attacks.

Regular medical examinations and physical fitness programs are the most significant factors in preventing heart attacks. Studies have shown that a pre-existing heart condition was present in more than half of fire fighter heart attack deaths. Medical examinations and physical conditioning have saved the lives of many fire fighters.

Motor Vehicle Accidents

Traumatic injuries are the second leading cause of fire fighter fatalities (▶ **Figure 6-2**). Most of these deaths are caused by vehicle accidents, falls, or structural collapse of a burning building. Trauma deaths resulting from motor vehicle collisions

Figure 6-2 The largest cause of trauma deaths is motor vehicle accidents.

represented 22% of annual line-of-duty deaths in the U.S. Fire Administration retrospective study. The most common fatal motor vehicle collision scenario involves a fire fighter responding to an emergency incident in a personal vehicle. The second most common fatal collision is a tanker rollover. Tankers account for more fatal fire department vehicle collisions than all engines, ladder trucks, and aerial apparatus combined.

Burns and Asphyxiation

The third most frequent category of fire fighter fatalities includes asphyxiation and burns. In these situations, the fire itself is the direct cause of the fire fighter's death. The fatal injury can involve either burns or asphyxiation or a combination of both.

The risks of respiratory injuries and burns are inherent in a fire fighter's work environment. Firefighting is routinely performed in situations that involve high temperatures and contaminated atmospheres. Protective clothing and self-contained breathing apparatus (SCBA) are designed to reduce the risk of injury from these causes; however, their effectiveness is limited. When the capabilities of the protective ensemble are exceeded, a serious or fatal injury can occur very quickly.

Investigations into fire fighter fatalities have identified several common scenarios. Many fire fighters have died as a result of becoming lost, disoriented, or trapped inside burning buildings and exhausting their SCBA air supply. A few unprotected breaths of the atmosphere inside a burning building can be fatal. Burns most often result from sudden changes in fire conditions, such as a flashover, rollover, or backdraft, or from becoming trapped or caught in proximity to the fire. A sudden structural collapse can easily trap a fire fighter or allow a fire to spread very quickly.

Everyone Goes Home: The Fire Officer's Responsibilities

The slogan "everyone goes home" should be a constant reminder that one of the primary responsibilities of a fire officer is to ensure the safety of his or her subordinates. A fire officer must evaluate the risk factors in every situation and take action to provide for the safety of fire fighters. Safety must be fully integrated into the standard processes for conducting fire department operations.

A standard approach to safety incorporates many practices that should be fully integrated into every operational situation. When operating at emergency incidents, fire fighters must work in teams, and fire officers must maintain accountability at all times for the location and function of all members working under their supervision. Every company must operate within the parameters of a coordinated strategic plan, under the direction of the incident commander. Reliable two-way communications must be maintained through the chain of command. Adequate back-up lines must be in place to ensure that a safe exit path is maintained for crews working inside a fire building and any sudden flare-ups can be controlled. Rapid intervention crews must be established to provide immediate assistance if any fire fighter is in danger. Air supplies must be monitored to ensure that fire fighters leave the hazardous area before they run out of air. All fire fighters must watch for indications of impending building collapse. All of these issues affect the ability of fire fighters to work safely in an environment that is inherently dangerous.

Heart Attack

Heart attacks are the leading cause of death for fire fighters. Although most line-of-duty fatalities for fire fighters under the age of 35 years are from traumatic injuries, fire fighters above the age of 35 years are more likely to die from a medical cause. The National Fallen Firefighters Foundation (NFFF) Life Safety Summit noted that a disproportionate number of fire fighters over the age of 49 years die of cardiac arrest while on duty.

Safety Zone

A fire officer can demonstrate three simple habits to improve the safety of his or her crew:
- Be physically fit
- Wear seat belts
- Maintain fire company integrity at emergency incidents

Before being authorized to begin training as a fire fighter, every candidate should be examined by a qualified physician. Regular physical examinations should be scheduled for as long as the fire fighter is engaged in performing emergency duties, with an emphasis on identifying risk factors that could lead to a heart attack under stressful conditions. Fire officers should always look for indications that a member is in poor health or is unfit for duty and, if necessary, arrange for a special evaluation by a fire department physician.

The risk of heart attack is closely related to age and physical conditioning, and several actions can be taken to reduce the risk factors. Physical fitness activities should be considered an essential component of every fire fighter's training regimen. Changes in lifestyles can often reduce the risk of a fatal heart attack and can include engaging in regular physical activity, stopping smoking, lowering high blood pressure, reducing high blood cholesterol level, maintaining a healthy weight, and managing diabetes. Fire officers must understand that these changes are as important to the body as wearing full protective equipment. The fire officer should demonstrate leadership in these areas in order to increase the health and safety of the entire crew.

The International Association of Fire Chiefs, the IAFF, the NFFF, the National Volunteer Fire Council, and the NFPA have all developed resources to help the fire officer encourage healthy living. Fire officers should be well versed in methods of making positive lifestyle changes to increase fire fighter safety. NFPA 1583, *Standard on Health-Related Fitness Programs for Fire Fighters*, provides a structure and resources for a fire officer to develop a health-related fitness program. The International Association of Fire Chiefs partnered with the IAFF to develop the *Fire Service Joint Labor Management Wellness Fitness Initiative*. This initiative produced the Candidate Physical Aptitude Test, as well as a peer fitness training certificate program with the American Council on Exercise.

Regardless of what the fire department mandates, every fire officer should strive to be physically fit (▶ **Figure 6-3**).

Fitness should be a personal priority and an expression of leadership.

AED

Many fire departments have added an automatic external defibrillator (AED) to the standard inventory of every fire company as a part of their Emergency Medical Service (EMS) capability (▶ **Figure 6-4**). The AED provides an electrical shock to convert ventricular fibrillation into a proper heart rhythm. Ventricular fibrillation is the most common heart arrhythmia when a person collapses without a pulse. The sooner that an AED is used, the more likely it is that a cardiac arrest patient will survive.

Many fire departments have found that some of their first cardiac arrest survivors, after AEDs were introduced, were on-duty fire fighters. One fire officer had the disquieting experience of having the apparatus operator collapse while responding to an emergency incident. The officer had to get the engine stopped and pull the apparatus operator out of the rig, and then used the AED to save the apparatus operator's life. The apparatus operator survived and retired on disability.

Figure 6-3) The fire officer should strive to be physically fit.

Safety Zone

The NFFF summarizes the importance of physical fitness with the following items that support fire fighter maintenance:
- **Regular Medical Check-Ups.** Yes—they can be a pain, but if you do not do it for you, do it for those who need you.
- **Regular Exercise.** Even walking makes a BIG difference! Walk a mile a day and watch the changes.
- **Eat Healthy.** Think about what you are eating, and then picture operating interior at a working fire 30 minutes later. Now, what do you want to eat?

Safety Zone

Like other Americans, fire fighters often fail to recognize the signs and symptoms of a heart attack or attempt to ignore them. The belief that "it cannot be happening to me" is pervasive. Every fire fighter should be well versed in the signs and symptoms of a heart attack, and a fire officer should be constantly vigilant for these signs within the work unit. The fire officer is in a position to recognize when any of the signs are present and to encourage the individual to seek medical attention. Common signs and symptoms include:

- Chest discomfort
- Discomfort in other areas of the upper body
- Shortness of breath
- Other signs, such as cold sweats, nausea, or lightheadedness

Stress

Stress is one of the leading causes of fire fighter deaths. In many cases, the stress is the triggering factor for a heart attack. Immediate stress factors can include a wide range of forces, including intensive physical exertion, heat exhaustion or exposure to extreme cold, as well as emotional reactions to events and circumstances.

Prolonged stress can also be a contributing factor to many physical and medical conditions. Employee Assistance Programs usually include professional intervention capabilities to support fire fighters who are experiencing unusual emotional stress. These programs are covered in Chapter 8, Evaluation and Discipline.

Another form of stress is produced from exposure to an incident that causes a great deal of emotional reaction within the fire fighter. Special programs have been developed to help reduce the impact of these situations. These procedures generally begin with a critical incident stress debriefing (CISD), which should be performed only by individuals who have been properly trained. Most programs include a peer group process that provides a mechanism for the emotions to be expressed. This is often a group discussion led by a peer on the CISD team to allow those wishing to express their feelings to do so. If difficulties continue, an employee can request additional assistance, which is usually available through the CISD team. A fire officer can also make a referral for a troubled individual to obtain professional assistance.

Most fire departments have established procedures for initiating a CISD intervention when it is needed. The earlier the stress is treated, the easier the recovery. An officer should assess the need for a CISD intervention after any unusually stressful incident or situation. The situations that tend to produce more stress in workers include:

- The death of a coworker
- Mass casualty events
- Death of a child
- Death after a prolonged rescue effort
- Victim reminding one of another
- Highly dangerous event
- High media interest

Members who have been involved in a highly stressful situation may exhibit the physical signs and symptoms of stress that are identified in Chapter 8, Evaluation and Discipline, as well as others. The following are additional physical signs and symptoms that a member may be in need of assistance.

- Nausea
- Twitching
- Thirst
- Vomiting
- Chills
- Fainting
- Sweating
- Rapid heart rate

A member may also exhibit behavioral signs and symptoms. These include:

- Blaming someone
- Nightmares that do not go away
- Loss of time and place
- Flashbacks
- Guilt
- Fear
- Loss of emotional control
- Increased alcohol consumption

Motor Vehicle Accidents

The NFPA reported that in 2003, more fire fighters died while traveling to or from a fire than while actually fighting a fire; 37 fire fighters died while responding to or returning from alarms, whereas only 29 died on the fireground. In that

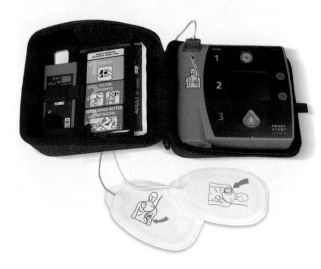

Figure 6-4 An automatic external defibrillator (AED).

year at least, motor vehicles posed a greater hazard to fire fighters than flames (▼ Figure 6-5). The NFPA notes that this was the largest number of fire fighter deaths from motor vehicle crashes since it began collecting fire fighter injury and fatality data in 1977.

The NFPA, the Federal Emergency Management Agency, and the National Occupational Institute of Safety and Health (NIOSH) have all identified motor vehicle crashes as a significant cause of fire fighter fatalities. Many of these deaths could be prevented through proper training and a change in attitudes. The injury and death statistics provide the justification for a comprehensive driving safety program.

Placing an inexperienced driver behind the wheel of an emergency vehicle places the occupants and the general public in immediate danger. The driver of an emergency vehicle is legally authorized to ignore certain restrictions that apply to other vehicles, but only when operating the emergency vehicle in a manner that provides for the safety of everyone using the roadways. Specific procedures for safe emergency response must be learned and practiced. The fire officer is responsible for ensuring that the driver consistently follows the rules of the road for emergency response.

Driver training standards are established in NFPA 1002, *Fire Apparatus Driver/Operator Professional Qualifications*. The fire officer should set high expectations for driver training and performance. The apparatus operator should be required to have nonemergency driving experience with a specific piece of apparatus before being assigned to emergency response driving duties. There are tremendous differences between the handling characteristics of fire apparatus and those of automobiles and light trucks. Stopping distances are directly related to the weight of the vehicle, and heavy fire apparatus can turn over easily in a collision. Driving a ladder truck is vastly different from driving an engine, and a tanker has a different set of handling characteristics from either of these. All of the necessary skills must be learned and prac-

ticed under controlled conditions before an operator is qualified to drive under emergency response conditions.

Requiring fire fighters to be seated and to wear seat belts is a simple rule that could prevent 10 to 15 fatalities every year. Many of the fire fighters killed in motor vehicle crashes are thrown from the vehicle. Fire officers need to ensure that all of their crew members are belted in whenever their vehicle is in motion. A fire officer who allows company members to respond unrestrained has no excuse if a fatal accident results.

The mandatory use of seat belts by fire fighters, like many other safety procedures, may require a change in the culture of some fire departments. This type of change is essential to ensure the survival of every fire fighter. A fire officer must be prepared to accept the responsibility and provide the leadership to bring about positive changes.

Fire Suppression Operations

The hazardous nature of fire suppression operations is easily recognized. As we might anticipate, considering the hazards that are present at most fires, asphyxiation, burns, or some combination of these two factors are the direct cause of many fire fighter deaths. In spite of the protection provided by protective clothing and SCBA, it is essential to remember that fire fighters routinely operate in situations where a problem, a procedural error, or an equipment failure could very quickly result in a fatality. To prevent these deaths, the fire officer must consistently implement a system of standard safety practices.

Fire officers must fully understand all local policies and procedures to guide their actions. The established standard operating procedures and safety practices should always be followed in situations that meet their criteria. If a situation does not fall under an established standard operating procedure, the fire officer must determine an appropriate course of action and give specific directives to subordinates to clearly indicate how the situation is to be handled. In these situations, the fire officer must provide an even greater level of supervision to ensure that the crew members understand and follow the plan. The fire officer must always be prepared for changing conditions and unanticipated hazards.

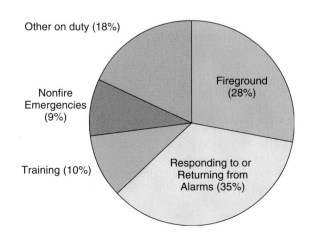

Fireground (28%)
Other on duty (18%)
Nonfire Emergencies (9%)
Training (10%)
Responding to or Returning from Alarms (35%)

Figure 6-5 Fire fighter deaths by type of duty—2003. Source: NFPA.

Safety Zone

The NFFF summarizes the issue of vehicle safety with the following items that support the driver response plan:
• It's not a race
• Safe is more important than fast
• Stop at red lights and stop signs! NO EXCUSES
• If they do not get out of your way—do not run them over! THINK and REACT CAREFULLY

Maintaining Crew Integrity

A consistent challenge for every fire officer is to maintain the integrity of the fire company under his or her command while operating at an emergency incident. The officer must know the location and function of every crew member at all times. The U.S. Fire Administration retrospective study shows that 82% of the fatal fire suppression incidents involved the death of a single fire fighter. Many of these fire fighters became lost or disoriented and died before the fire officer or incident commander was aware that a fire fighter needed help.

NIOSH has investigated fire fighter line-of-duty deaths as part of their Fatality Assessment and Control Evaluation program. The lack of an effective incident management system that includes a fire fighter accountability component is a common scenario in many of the single fire fighter deaths investigated by NIOSH.

Operating in an IDLH Environment

The Occupation Safety and Health Administration (OSHA) establishes federal workplace safety regulations in the United States, including 29 CFR 1910.134 (Respiratory Protection), which applies to the use of SCBA by the fire service. This rule applies when members are operating in an environment that is **immediately dangerous to life and health (IDLH)**. NFPA defines IDLH as any condition that would: (a) pose an immediate or delayed threat to life, (b) cause irreversible adverse health effects, or (c) interfere with an individual's ability to escape unaided from a hazardous environment. The interior of a fire building, where fire fighters are using SCBA, is considered to be an IDLH atmosphere.

The OSHA regulation establishes specific requirements for fire fighters operating in an IDLH environment. This is known as the "two-in, two-out rule" and includes:

- Having a designated officer-in-charge
- Having at least two fire fighters enter the IDLH area together and remain in visual or voice contact with one another at all times, while wearing SCBA
  ▶ Figure 6-6
- In addition, at least two properly equipped and trained fire fighters must:
  - Be positioned outside the IDLH environment
  - Account for the interior team
  - Remain capable of rescue of the interior team

The two-in, two-out rule is directly related to the concept of a **rapid intervention crew (RIC)**. NFPA 1561, *Standard on Emergency Services Incident Management System*, defines an RIC as a minimum of two fully equipped personnel, standing by on site, in a ready state for immediate rescue of injured or trapped firefighting personnel. Many departments dispatch an additional fire company to working fire incidents with the specific assignment of establishing a rapid intervention crew.

Figure 6-6 Two fire fighters should enter an area together and remain in visual or voice contact with one another at all times, while wearing self-contained breathing apparatus.

In order to ensure that the RIC can recognize in a timely manner that a fire fighter is missing, a **personnel accountability system** is needed to track the identity, assignment, and location of all fire fighters operating at the incident scene ▶ Figure 6-7 . An accountability system typically provides the following:

- A method to identify all personnel that are on the scene
- A method to identify personnel that are in the hazard area
- The process to quickly account for personnel when an evacuation order or unusual event occurs (e.g., a flashover, backdraft, or collapse)
- A process to ensure that personnel do not go unaccounted for an extended period of time
- A standardized method of notification and reaction to a Mayday event

An accountability system must function at multiple levels at an incident scene. Although every fire officer must be accountable for all assigned company members, command officers are accountable for the location and function of full companies. An officer who is assigned to manage a sector for a division at

Safety Zone

The OSHA rule includes an exception that would allow a first-arriving fire company to override the two-in, two-out rule and initiate operations in an IDLH environment if there is a chance that a victim will be rescued alive. The mere fact that there *might* be someone in danger is not sufficient justification to ignore the two-in, two-out rule; the exception exists only to avoid the situation where a procedural rule could stand in the way of saving a life.

Figure 6-7 Personnel accountability system.

an incident must know where every assigned company is operating and what they are doing at all times. The incident commander must know which companies are assigned to each sector or division.

For an accountability system to be effective, crews must routinely receive training on and follow the procedures. During an emergency situation, fire fighters will revert to habit. The proper use of an accountability system must be fully integrated as one of those habits.

Air Management

An SCBA provides a safe and reliable air supply that allows a fire fighter to operate in an IDLH atmosphere for a finite length of time. The length of time depends on the rate at which the individual consumes the available air supply, which varies considerably, depending on the user and the nature of the work that is being performed. The rated time is based on a standard consumption rate for an individual at rest and does not reflect typical experience in firefighting operations. Running out of air in an IDLH atmosphere can very quickly result in asphyxiation.

Within the past decade, many fire departments have re-evaluated how they track SCBA use during incidents. Some departments replaced their 30-minute rated SCBA air tanks with 45- or 60-minute air tanks to increase the time a fire fighter can operate in IDLH atmospheres. Other departments have established or strengthened incident management procedures to more closely follow individual fire fighter air use.

Low-pressure warning devices, which provide an indication when the remaining air supply reaches a set point, are effective only if the fire fighter is able to exit to a safe atmosphere in the time that is provided. The low-pressure alarm might not provide sufficient exit time for a fire fighter who is deep inside a burning building or unable to exit immediately.

Many departments monitor the air levels of fire fighters on entry and exit. If a fire fighter fails to exit within the expected time period, an accountability roll call is taken. Technology is evolving that will allow for the air pressure readings to be relayed through radio frequencies to a status board located with the safety officer. One individual will be able to monitor the breathing air level of all personnel on scene. This technology will also allow the fire fighters to be notified when they have to exit, based on their air consumption rate and exit time.

Teams and Tools

The minimum size of an interior work team is two fire fighters. Every team member must have full personal protective equipment, including an SCBA and a personal alert safety system (PASS) device unit. At least one member of every team should have a radio, and many fire departments provide a radio for every fire fighter.

Every interior operating team should also have a thermal imaging device. Although this technology is still new and expensive, thermal imaging devices are a critical component of a comprehensive safety program. The thermal imaging device allows crews to navigate, locate victims, evaluate fire conditions, locate hazards, and find escape routes in smoke-filled buildings or total darkness (▼ **Figure 6-8**).

A review of recent NIOSH line-of-duty death investigations shows that some departments that had thermal imaging devices did not use them in critical situations. The fire fighters did not know how to use the imager, the device was inoperable, or the device was not used. Twenty-five years ago, the same observation could have been made for SCBA, when SCBAs were left sitting in the apparatus compartments while fire fighters made interior attacks. Today, fire fighters look back and wonder how this dangerous practice could have

Figure 6-8 Thermal imaging devices are a critical component of a scene safety program.

occurred. Future fire fighters may well look back at the practice of underutilizing thermal imaging devices in the same manner. The fire officer must provide frequent training to integrate the thermal imager into interior firefighting operations.

Situation Awareness

One of the primary responsibilities of a fire officer is to maintain a continual connection between the functions being performed by the company and the overall situation. This is one of the most important reasons for establishing and maintaining an effective incident command structure at every incident. The fire officer maintains situation awareness by staying oriented, making observations, providing and receiving regular updates within the incident command system, listening to fireground radio communications, and continually assessing the risk/benefit model.

It is very easy to become distracted or preoccupied with a particular task or function and to lose track of a larger situation that is occurring and changing simultaneously. This can be a very serious problem at a dynamic emergency incident, particularly for an officer who is supervising a crew that is performing a complex task deep inside a smoke-filled building. Conditions can change very quickly, and crews might have no way of observing what is going on around them. A fire officer must always maintain the link between the immediate situation and the overall incident situation.

Risk/Benefit Analysis

Risk/benefit analysis is based on a hazard and situation assessment that weighs the risks involved in a particular course of action against the benefits to be gained for taking those risks. Fire fighters sometimes use the simple axiom, "Risk a little to save a little. Risk a lot to save a lot." However, this is a simplified description of the risk/benefit process. Risk/benefit analysis must always be approached in a structured and measured manner.

Life safety is a paramount goal, including the lives of fire fighters. If the situation requires placing fire fighters in extreme danger with little chance of success, the operation should not be undertaken. There is no justification for risking the lives of fire fighters to save property that is already lost or has no real value. Conducting an interior attack on a fire in an unoccupied abandoned building needlessly places fire fighters at risk. Similarly, entering a burning building to search for missing occupants who could not possibly still be alive cannot be justified.

Every interior fire attack operation and many other types of situations and circumstances expose fire fighters to a set of unavoidable inherent risks. These risks are managed within a measured and controlled system that is based on training, coordination, and the use of protective clothing and equipment. The nature of the mission requires fire fighters to be able to work safely in situations that are inherently dangerous. Fire officers are responsible for keeping this system in balance.

The only situation that truly justifies exposing fire fighters to a high level of risk is one where there is a realistic chance that a life can be saved. Even in this circumstance, fire fighters must use all of the resources at their disposal to limit the risks. All of the fire department's training, equipment, and systems are designed to enable fire fighters to be effective in the face of such challenging and dangerous situations.

The fire officer starts the risk/benefit analysis by preparing a **preincident plan**, which is a written document that provides information that can be used by responding personnel to determine the appropriate actions in the event of an emergency at a specific facility. Building construction, occupancy, building use, building contents, and condition of the structure are all factors that should be used to develop the risk/benefit analysis.

When operating at an emergency incident, the fire officer reviews the preincident plan and makes observations about current conditions. These two inputs are combined to produce an **incident action plan**, which is developed by the incident commander and incorporates the overall incident strategy, tactics, risk management evaluation, and organization structure for that particular situation. Incident action plans are updated throughout the incident, based on updates from operating crews and observations by the incident management team.

Incident Safety Officer

An **incident safety officer** is a designated individual at the emergency scene who performs a set of duties and responsibilities that are specified in NFPA 1521, *Standard for Fire Department Safety Officer.* The incident safety officer functions as a member of the incident command staff, reporting directly to the incident commander. The incident commander is personally responsible for performing the functions of the incident safety officer if this assignment has not been assigned or delegated to another individual.

Many fire departments assign a designated officer to respond to emergency scenes to fill this position. In the absence of a predesignated safety officer, this position may be assigned to a qualified individual by the incident commander. NFPA 1521 also specifies that the fire department must have a stan-

---

Safety Zone

The NFFF summarizes situational awareness with the following items that support the interior firefighting plan:
- Work as a team
- Stay together
- Stay oriented
- Manage your air supply
- Get off the apparatus with tools and a thermal imager for EVERY interior operating team
- Provide a radio for EVERY member
- Provide regular updates
- Constantly assess the risk/benefit model

dard operating procedure to define the criteria for the response or appointment of an incident safety officer. The same principles should be applied to any situation, whether a designated safety officer has been dispatched or an officer has been assigned to this function by the incident commander.

The fact that a safety officer has been assigned does not relieve any officer or fire fighter of the responsibility to operate safely and responsibly. The incident safety officer is an additional resource to ensure that the safety priorities of the situation are being addressed. Every fire officer shares the responsibility to act as a safety officer within his or her scope of operations.

Incident Safety Officer and Incident Management

The incident safety officer is a key component of the incident management system. The **incident management system (IMS)** (or incident command system) is the standard organizational structure that is used to manage assigned resources in order to accomplish stated objectives for an incident. Every incident requires someone, the incident commander, to be in charge at all times to coordinate resources, strategies, and tactics. This begins with the initial arriving officer, who functions as the incident commander until he or she is relieved by a higher-ranking officer. The incident management structure can become larger and more complex, depending on the nature and the magnitude of the incident.

The incident safety officer reports to the incident commander. The incident safety officer is required to monitor the scene, identify and report **hazards** (the potential for harm to people, property, or the environment) to the incident commander, and if necessary, take immediate steps to stop unsafe actions and ensure that the department's safety policies are followed. In most situations, the exchange of information between the incident commander and the incident safety officer is conducted verbally and quickly at the command post.

Safety officers have specific authority and a special set of responsibilities under the incident command system. One of the primary responsibilities of an incident safety officer is to identify hazardous situations and dangerous conditions at an emergency incident and recommend appropriate safety measures. In most cases, the safety officer acts as an observer, monitoring conditions and actions and evaluating specific situations. When an unsafe condition is observed that does not present an imminent danger, the safety officer consults with the incident commander and with other officers to determine a safe course of action.

If a situation creates an imminent hazard to personnel, the incident safety officer has the authority to immediately suspend or alter activities. When this special authority is exercised, the incident safety officer must immediately inform the incident commander of the hazardous situation and his or her actions. It is ultimately up to the incident commander to either approve or alter the action taken by the safety officer.

Qualifications to Operate as an Incident Safety Officer

NFPA 1521 outlines the criteria for an incident safety officer. Every fire officer should be trained to perform the basic duties of an incident safety officer and be prepared to act temporarily in this capacity if he or she is assigned by the incident commander.

According to NFPA 1521, the incident safety officer must be a fire department officer and as a minimum must meet the requirements for Fire Officer I as specified in NFPA 1021, *Standard for Fire Officer Professional Qualifications*. The incident safety officer also must be qualified to function in a sector officer position under the local incident management system.

The general knowledge requirements for an effective incident safety officer are as follows:
- Safety and health hazards involved in emergency operations
- Building construction
- Local fire department personnel accountability system
- Incident scene rehabilitation

Incident safety officers at a special operations incident require additional specialized knowledge and experience. Special operations are emergency incidents to which the fire department responds that require specific and advanced training and specialized tools and equipment, such as water rescue, extrication, confined space entry, hazardous materials situations, high-angle rescue, and aircraft rescue and firefighting. For example, the incident safety officer at a hazardous materials incident needs to have an advanced understanding of this situation, perhaps by training to the Hazardous Materials Technician level of NFPA 472, *Standard for Professional Competence of Responders to Hazardous Materials Incidents*. In many fire departments, specialized teams have their own designated safety specialists who are trained to work directly with the team.

Safety Zone

In order for the incident safety officer to adequately perform these duties and responsibilities at a very large or complex incident, the incident commander must define an incident action plan. The incident safety officer can then provide the incident commander with a risk assessment of incident scene operations. The incident safety officer can also develop an **incident safety plan**, which outlines the actions that are required to provide for safety at the scene, based upon the incident commander's incident action plan and the type of incident encountered. At a very large incident, the incident action plan and incident safety plan are often compiled as written documents and updated at the beginning of every work period.

VOICES OF EXPERIENCE

❝ As a fire officer, you should always remember that no call is routine. Always be on your guard, thinking about how to keep yourself and your crew safe. ❞

It was a rather cold winter's night in Philadelphia. The heaters in every home in the city were working overtime. The paramedic unit received a call from dispatch to a tiny row house in the middle of the city. A boy informed the dispatcher that his mother was having a stroke. Upon arrival, the paramedics found a severely obese woman who was disoriented. Due to the woman's physical size, the paramedics called for a "lift-assist" from an engine company.

As the paramedics prepared the woman for transport, they began to move the furniture in the small front bedroom out of the way so that the engine crew would have room to maneuver during the lift-assist. The woman's son came upstairs and suddenly began to rearrange the furniture so that the egress was blocked. This irrational behavior confused the paramedics. As they attempted to reason with the son, one of the paramedics developed a headache while the other became dizzy. As the crew from the engine company entered the very warm home, the paramedics' conditions worsened. The fire officer quickly realized that this was not a routine medical emergency. The occupants of the house and the paramedics were suffering from carbon monoxide poisoning. The fire officer ordered the immediate evacuation of the house. The engine crew had to don SCBA before re-entering the house to perform the lift-assist on the woman upstairs. Both the woman and her son were transported to the hospital for treatment.

As a fire officer, you should always remember that no call is routine. Always be on your guard, thinking about how to keep yourself and your crew safe. Being aware of the hazards and understanding the risk will improve the odds of a safe response.

William Shouldis
Philadelphia, Pennsylvania Fire Department
32 years of service

Typical Incident Safety Officer Tasks

The specific duties that an incident safety officer must perform at an incident depend on the nature of the situation. Below is a partial listing of functions that may need to be addressed at incidents:

- Ensure that safety zones, collapse zones, and other designated hazard areas are established, identified, and communicated to all members present on scene.
- Ensure that hot, warm, decontamination, and other zone designations are clearly marked and communicated to all members.
- Ensure that a rapid intervention crew is available and ready for deployment.
- Ensure that the personnel accountability system is being used.
- Evaluate traffic hazards and apparatus placement at roadway incidents.
- Monitor radio transmissions and stay alert to situations that could result in missed, unclear, or incomplete communication.
- Communicate to the incident commander the need for assistant incident safety officers because of the need, size, complexity, or duration of the incident.
- Immediately communicate any injury, illness, or exposure of personnel to the incident commander and ensure that emergency medical care is provided.
- Initiate accident investigation procedures and request assistance from the health and safety officer in the event of a serious injury, fatality, or other potentially harmful occurrence.
- Survey and evaluate the hazards associated with the designation of a landing zone and interface with helicopters.
- Ensure compliance with the department's infection control plan.
- Ensure that incident scene rehabilitation and critical incident stress management are provided as needed.
- Ensure that food, hygiene facilities, and any other special needs are provided for members at long-term operations.
- Attend strategic and tactical planning sessions and provide input on risk assessment and member safety.
- Ensure that a safety briefing, including an incident action plan and an incident safety plan, is developed and made available to all members on the scene.

Additional duties the incident safety officer must perform when fire has involved a building or buildings are as follows:

- Advise the incident commander of hazards, collapse potential, and any fire extension in such buildings.
- Evaluate visible smoke and fire conditions and advise the incident commander, tactical level management units officers, and officers on the potential for flashover, backdraft, or any other fire event that could pose a threat to operating teams.
- Monitor the accessibility of entry and egress of structures and the effect it has on the safety of members conducting interior operations.

Assistant Incident Safety Officers at Large or Complex Incidents

Some incidents, based on their size, complexity, or duration, require more than one safety officer. Assistant incident safety officers can be assigned to subdivide responsibilities for different areas and functions at incidents such as high-rise fires, hazardous materials incidents, and special rescue operations. In these cases, the incident safety officer should inform the incident commander of the need to establish a **safety unit** as a component of the incident management organization. Under the overall direction of the incident safety officer, assistant incident safety officers can be assigned to various functions, such as scene monitoring, action planning and risk management, interior operations, or special operations teams. During extended incident operations, a relief rotation can be established to ensure that safety supervision is maintained at all times.

Incident Scene Rehabilitation

Rehabilitation is the process of providing rest, rehydration, nourishment, and medical evaluation to members who are involved in strenuous or extended-duration incident scene operations (▼ **Figure 6-9**). Part of the incident safety officer's role is to ensure that an appropriate rehabilitation process is established. **Incident scene rehabilitation** is the tactical level management unit that provides for medical evaluation, treatment, monitoring, fluid and food replenishment, mental rest, and relief from climatic conditions of the incident.

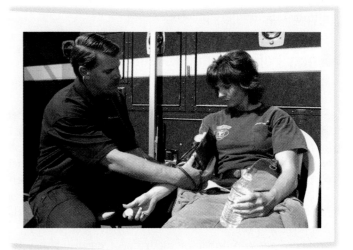

Figure 6-9 Rehabilitation is the process of providing rest, rehydration, nourishment, and medical evaluation to members who are involved in extended or extreme incident scene operations.

Safety Zone

The NFFF summarizes incident scene rehabilitation with the following items that support the fire fighter rehabilitation guidelines:

- Stop before you drop
  - Cool down when hot
  - Warm up when cold
  - Dry off when wet
- Stay hydrated
  - With noncaffeinated drinks
- Monitor vital signs

Fire fighters are very aggressive by nature. They want to do a good job and want to be where the action is occurring. This action orientation makes them very susceptible to exceeding the physical limitations of their own bodies. As a fire officer, you must constantly monitor the health and welfare of your crew. The incident safety officer is a third line of defense, after the individual and the fire officer, for ensuring that fire fighters obtain appropriate rehabilitation.

Creating and Maintaining a Safe Work Environment

Although the highest priority of a fire officer is to ensure that every subordinate member goes home alive from a call, the reality is that for every fire fighter death in the line of duty, nearly 1,000 fire fighter injuries occur. A safety program that is based on preventing fatalities is far from adequate. Every injury, or near miss, should be viewed as a potentially fatal or disabling situation, and injury prevention should be an equally important concern of the fire officer. The fire officer must model good behavior to help develop the subordinate's attitudes about injury prevention.

Safety Policies and Procedures

Most fire departments have policies that regulate safety practices at the company level. These policies are designed to address routine circumstances, and many have been developed in reaction to a previous accident or injury. The fire officer is on the front line in ensuring compliance with all safety policies. It is essential for the fire officer to fully understand each policy, to follow all safety policies and procedures, and to ensure that all subordinates fully understand and follow them. Failure to do so could have long-lasting consequences for the employee and the officer.

There are a number of methods of ensuring that fire fighters understand safety policies and procedures. Many departments require each member to sign a document to acknowledge that they have read and understand each policy and any new or amended policy. Some departments leave it

to the officer to read and explain each policy to the crew members. This can be done at the morning briefing; however, there may be little retention at this time. A more effective method would be to have the members individually read the policy and then lead a group discussion to ensure that it is understood.

One method to reinforce safety policies is to watch videos of incidents and critique them based solely on safety policies. This often reveals significant differences between the way things are supposed to be done and what actually occurs at an emergency incident. This can be a significant learning experience for everyone involved.

The fire officer must conscientiously ensure that all safety policies are followed in training activities. Training provides the luxury of time to learn and correct errors. On emergency scenes, when there is only one opportunity to do things properly, fire fighters will perform the way they have been trained.

It is impossible to write a policy to cover every conceivable hazard. For this reason, the fire officer should use good judgment to identify hazardous situations and implement mitigating measures. This might include requiring fire fighters to use safety glasses while mowing the lawn or ensuring that food preparation surfaces are properly cleaned to prevent the spread of germs. In the absence of good leadership, activities that are not regulated by an official departmental policy are likely to be ignored. The fire officer should ensure that company members understand the need for safe work practices and develop an attitude that internalizes safety, rather than relying on the fire officer to be a "safety police officer."

Emergency Incident Injury Prevention

Many of the same techniques that are used to prevent fire fighter deaths also help prevent fire fighter injuries; however, they are not necessarily the same. The following principles should be implemented to prevent injuries as well as deaths.

Physical Fitness

Fire fighters who are in good physical condition are less prone to injury, as well as being at reduced risk for heart attack. Through strength and flexibility training, the body is more resistant to sprains, strains, and other injuries and is more capable of responding to critical situations. Sprains and strains are the leading known type of injury to fire fighters. A comprehensive physical fitness program that includes cardiovascular, strength, and flexibility training is needed to reduce fire fighter deaths and injuries.

Personal Protective Equipment

The capabilities of fire fighter personal protective equipment (PPE) have improved tremendously over the past 25 years. Today, most fire departments provide their members with protective gear that meets or exceeds the requirements of NFPA standards governing personal protective equipment. The basic protective ensemble for structural firefighting includes a turnout coat and pants, boots, gloves, hood, helmet,

SCBA, and a PASS. NFPA standards define the minimum performance standards for each item.

The best protective equipment is of no use if it is not worn. At every incident, the fire officer should monitor the proper use of protective clothing and take immediate action to correct any deficiencies. Fire fighters must not be allowed to become complacent about wearing full PPE to routine calls. In some departments, fire fighters routinely enter buildings to investigate fire alarm activations and other seemingly minor situations without fully donning their PPE and sometimes without tools, hose packs, or extinguishers. This practice has proven deadly in the past when the situation turned out to be more serious than anticipated. The officer must set a good example and require every crew member to treat every call like an actual fire.

The need for protective clothing and equipment does not end when the fire is extinguished. SCBA must be used until carbon monoxide and particulate matter levels have been reduced. Coats, pants, and helmets are needed during overhaul to prevent wounds. The incident commander and the safety officer are responsible for determining whether it is safe to reduce the level of PPE that will be used at any time.

In addition to the protective clothing and equipment, fire fighters operating in hazardous areas should carry several safety-related items. A radio is needed to maintain contact, to receive instructions, or to request assistance in an emergency; at least one member of every team should have a radio, and, preferably, every fire fighter should have one. A flashlight is needed to help prevent becoming lost or disoriented and as a signal for help. A forcible entry tool should be carried to provide a method of escape in an emergency. Personal wire cutters can also prove valuable if a fire fighter becomes entangled in wires. A personal escape rope should also be carried by every fire fighter. Some fire departments issue all of these items and expect every fire fighter to be properly equipped at all times. Each item provides an additional measure of safety, and the fire officer should always set the example.

Although station uniforms are not included in most definitions of personal protective clothing, the clothing that is worn under turnout gear can provide an additional level of protection. NFPA 1975, *Standard on Station/Work Uniforms for Fire and Emergency Services* provides a set of performance requirements for station clothing. The use of clothing that meets this standard under turnout gear will further reduce the risk of burn injuries in a severe situation.

With EMS responses accounting for more than 70% of responses in many fire departments, the fire officer must also understand the requirements for medical PPE. In many cases, this requirement is as simple as wearing appropriate gloves. Today, most health care providers routinely wear gloves for all patient contact situations. When blood or other body fluids are present, eye and face protection as well as full body protection may be needed. An effective education program is essential for fire fighters to understand the importance of wearing the appropriate protection for every situation.

Safety Zone

Training facilities and training activities present a special set of safety concerns and responsibilities for the fire department. Although training facilities and exercises are often designed to replicate or simulate the conditions that can be anticipated at actual emergency incidents, the fire department has the ability to plan, manage, and conduct training activities with absolute control over the risks and safety hazards that could be encountered. The fire department is responsible for ensuring that all training facilities are as safe as it is feasible to make them. The fire department is also responsible for ensuring that all appropriate safety measures are applied to all training activities. There should be no unanticipated situations or circumstances that result in fire fighter injuries or fatalities during training.

Fire Station Safety

In addition to emergency situations, safety considerations also apply to the fire station environment. The fire station and other fire department facilities are a workplace, and the fire department is fully responsible for maintaining a safe work environment. Every fire department should have a comprehensive workplace safety program that applies to all fire department facilities. The station setting allows the fire officer to be even more proactive in enforcing safety policies than is possible at an incident scene.

Safety hazards at the fire station can have the same consequences as hazards encountered at emergency incidents. The most important difference is that the fire department has control over the fire station environment at all times and has the ability, as well as the responsibility, to identify and correct any safety problems that are present. The fire department does not have the same type of control over the conditions that are typically encountered at the scene of an emergency incident.

Clothing

Protective clothing can become contaminated at incident scenes and should never be worn in the living quarters of the fire station. Turnout gear that is known to be contaminated with fire residue, chemicals, or bodily fluids should be cleaned as soon as possible. When washing turnout gear, be sure to follow the manufacturer's recommendations. The cleaning requires the use of a commercial extractor-type washer rather than a residential washer. Many fire departments provide special washing machines for protective clothing at the fire station or send gear out for professional cleaning and decontamination. Duty uniforms should also be washed at the fire station or sent out to a commercial laundry to prevent cross-contamination of the family laundry.

Protective clothing should be inspected regularly by the fire officer, and any items that are worn or damaged should be repaired or replaced. Properly fitting gear is also important, both for comfort and for protection. Gear inspection should include an exterior evaluation while the clothing is being worn and a close examination of the condition of all interior layers.

Housekeeping

General housekeeping around the fire station is also important in injury and accident prevention. Apparatus bay floors that become slippery when wet have caused many fire fighters to slip and fall. A squeegee should be used to remove standing water, and "wet floor" warnings should be used. The same is true for interior station floors that require mopping.

Leaving equipment lying around can present a tripping hazard to fire fighters when they are in a hurry to respond to an alarm. The walking traffic flow areas should always be kept clear.

The same fire hazards that exist in other types of occupancies can be found in many fire stations. Leaving food unattended on the stove and improperly storing flammable liquids can lead to a publicly embarrassing situation. Every fire station should have the appropriate number and types of fire extinguishers, which must be properly maintained. The fire officer should ensure that working smoke alarms are present and checked regularly. Automatic sprinklers are highly recommended for all fire stations.

Food preparation activities are also responsible for many relatively minor injuries to fire fighters. Inappropriate use of kitchen knives has caused many injuries that have resulted in lost time and worker's compensation claims. In addition, proper decontamination of all food preparation surfaces is often overlooked. Regular hand washing while cooking is important to prevent the transmission of illnesses.

Many fire departments have installed diesel exhaust systems in apparatus bays to reduce the exposure of fire fighters to this contaminant. If the fire station does not have an exhaust system, the fire officer should ensure that vehicles are not left running inside the building. Daily checks should be performed outside the apparatus bay. In inclement weather conditions, some departments allow apparatus checks to be conducted inside the bay, but only with the apparatus doors fully open.

Lifting Techniques

Many back injuries are caused by improper lifting techniques. Back injuries can occur while lifting and moving patients, dragging fire hoses, setting ladders, or working at the station. No matter how they are caused, back injuries are serious and potentially career ending. Every effort should be made to prevent them.

Proper lifting techniques should always be used to reduce the risk of back injuries. A fire fighter should never bend at the waist to pick up items or patients. Instead, the fire fighter should bend at the knees and lift by standing straight up. The back should be kept in a natural position rather than locked in a hyperextended manner. When an object is too bulky, in an awkward position, or too heavy for one fire fighter, additional

help should be found. Fire officers should always reinforce the use of proper techniques and procedures to avoid injuries.

Many injuries have resulted from foolish horseplay around the station. This is an area where the fire officer must exercise supervisory control to reduce the risk of unjustifiable injuries.

Infection Control

Every fire department should have an infection control program that meets the NFPA 1581, *Standard on Fire Department Infection Control Program*. This standard identifies six components of a program:

1. A written policy with the goal of identifying and limiting exposures
2. A written risk management plan to identify risks and control measures
3. Annual training and education in infection control
4. A designated infection control officer
5. Access to appropriate immunizations for employees
6. How exposure incidents are to be handled

Proper decontamination procedures are essential after emergency medical incidents or any situation where equipment could have become contaminated. Disposable items, gloves, and equipment that has been contaminated should be disposed of in a specially marked bag that is designed for that use.

Patient compartments in ambulances should be properly decontaminated after every transport. Patient cots should also be decontaminated after every call. Equipment that is designed to be decontaminated and reused should be cleaned only in an approved decontamination sink, typically in a designated area of the fire station. Contaminated equipment should NEVER be taken into the living area of the station or cleaned in a sink where food is prepared.

Always follow written procedures and manufacturer's guidelines for decontamination of medical equipment. A 1% bleach and water solution is typically used for this purpose; however, this solution should NOT be used on turnout gear. Many of the materials commonly used in protective clothing are seriously damaged or destroyed by this type of cleaning agent.

NFPA 1581 provides very specific information on how an infection control program should be established, including equipment cleaning and storage, facility requirements, and methods of protection. The fire officer should also consult departmental guides and policies for further details.

Infectious Disease Exposure

NFPA 1581 also provides a model program for situations where a fire fighter has been exposed to an infectious or contagious disease. Most fire departments have established policies to guide postexposure procedures. The fire officer must be prepared to fulfill the duties of the initial supervisor when a member has had an infectious disease exposure.

The most important first step for any exposure is to immediately and thoroughly wash the exposed area with soap and running water. If soap and running water are not available, waterless soap, antiseptic wipes, alcohol, or other skin cleaning agents can be used until soap and running water are obtained.

The fire department infection control officer should be notified, typically within 2 hours of the exposure incident. The designated infection control officer arranges for the member who has experienced an exposure to receive medical guidance and treatment as soon as is practical, but at least within 24 hours. An exposure can be a very frightening event for a fire fighter. It is important to provide confidential, post-exposure counseling and testing. The fire fighter may not seek out this information on his or her own, so the department should proactively inform the individual of this service.

It is important to document all exposures as soon as possible using a standardized reporting form. At a minimum, the record should include the following:

- Description of how the exposure occurred
- Mode of transmission
- Entry point
- Use of personal protective equipment
- Medical follow-up and treatment

The record of an exposure incident will become part of the member's confidential health database. A complete record of the member's exposure incidents must be available to the member on request. Data on incident exposures should also be maintained by the department but without personal identifiers to allow for analysis of exposure trends and to develop strategies to prevent them.

Due to the hazardous nature of some communicable diseases, individuals who have been exposed are often required to report to the infection control officer. This information must be maintained with the strictest of confidence. It is up to the fire department physician to determine fitness-for-duty status after reviewing documentation of a member's exposure.

Accident Investigation

The fire department **health and safety officer** is charged with ensuring that all injuries, illnesses, exposures, and fatalities, or other potentially hazardous conditions and all accidents involving fire department vehicles, fire apparatus, equipment, or fire department facilities are thoroughly investigated. The initial investigation of many situations is often delegated to a local fire officer.

An **accident** is any unexpected event that interrupts or interferes with the orderly progress of fire department operations. This definition includes personal injuries as well as property damage. An accident investigation should determine the cause and circumstances of the event in order to identify any corrective actions that are needed to prevent another injury, accident, or incident from occurring.

Ensure that all required federal, state, and local documentation is complete and accurate. The result of an accident investigation should always include recommended corrective actions that are presented to the fire chief or the chief's designated representative.

Accident Investigation and Documentation

Most fire departments have established procedures for investigating accidents and injuries. The fire officer is usually responsible for conducting an initial investigation and for fully investigating many minor accidents. Accidents that involve serious injuries, fatalities, or major property damage are usually investigated personally by the health and safety officer or by other qualified individuals.

An officer could be expected to conduct the full investigation for a simple ankle sprain, a broken pike pole, or a dented rear step on the apparatus. In each case, it is the fire officer's responsibility to protect the physical and human resources of the department. To accomplish this, the officer must understand the basic principles of investigation.

An investigation normally consists of three phases:

1. The identification and collection of physical evidence
2. The interviewing of witnesses
3. The written documentation phase (at the end of an investigation)

Each step is essential for the development of a comprehensive report that can be used to initiate safer work practices and conditions. The investigation report can also provide evidence that could be used to support or refute future claims. If a department has a standardized procedure for investigating accidents, the fire officer should always use the recognized procedure.

Examination of the physical evidence includes documenting the time, day, date, and conditions that existed at the time of the incident. This is factual information that is verifiable and should not include any interpretation. The weather conditions should be noted. The scene of the accident should be fully documented, with drawings noting the locations of relevant artifacts that could provide information about how or why the accident occurred. These details are needed to help determine prevention methods.

Witnesses to an accident should also be interviewed, including all of the individuals who were directly or indirectly involved, as well as anyone who simply observed what happened. When interviewing a witness, explain the rationale for the interview and that this is a fact-finding process. Advise the person that it is okay to not know the answer to every question. Begin with having the individual give a verbal chronology of whatever occurred, being sure to not interrupt. Watch for nonverbal communication by the witness. Once the opening report has been given, ask questions to clarify the detail, but do not ask leading or misleading questions. If the level of PPE being worn at the time of the incident has not been mentioned, it is important to determine what was being worn, how it was being worn, and whether it performed as designed. Lastly, restate previous questions in a different order and from a different perspective to verify previous answers.

The last step in an accident investigation is the documentation of the findings as well as the conclusions and recommendations. Most departments have a standardized form to ensure that all information is included. The facts should be presented in a logical sequence with the results of the interview or interviews attached. A determination as to the most likely cause or causes of the accident should be included, along with a recommendation on how this type of accident could be prevented. Some departments require a determination about the likely frequency as well as the likely severity of this type of accident. This optimizes prevention efforts.

As an officer, you have a duty to perform an initial accident investigation that is fair and unbiased. Both the department and the employee have a vested interest in the report. Failure to provide a detailed and well-documented report can cause much anguish to personnel in the future. An employee who applies for a disability pension based on an accident that occurred several years earlier could be denied the pension if adequate documentation is not maintained. For this reason, most departments require a report to be filed whenever there is a potential exposure to an infectious disease, when injury has occurred to an employee or citizen, and when property damage has occurred to either private property or departmental property.

Data Analysis

The fire department is required to maintain records of all accidents, occupational deaths, injuries, illnesses, and exposures in accordance with federal, state, and local regulations. In addition to meeting regulations, these reports can assist the fire officer in identifying trends and safety-related issues.

Risk management is the identification and analysis of exposure to hazards, selection of appropriate risk management techniques to handle exposures, implementation of chosen techniques, and monitoring of results, with respect to the health and safety of members. Accident and injury reduction should be a major concern of every fire officer. The fire officer should review all injury, accident, and health exposure reports in order to identify unsafe acts and work conditions.

When reviewing an accident report, consider what was the root cause of the accident. Were unsafe acts being committed? If so, what must be done to change the behavior or attitude? If the cause was an unsafe condition, how can this condition be corrected? An evaluation of all unsafe acts and conditions and how to address each is required to prevent future occurrences. Some corrective actions can be made at the company level, whereas others require action at a higher level.

Once the cause has been identified, a report should be filed with your supervisor that outlines the problem, actions taken to correct the problem, and any additional actions that should be taken. Some departments have these reports sent directly to the department's health and safety officer.

The fire officer may be required to complete a longitudinal study of accidents, injuries, and exposures for the company. The officer must review the data to determine what areas are causing the greatest number of incidents as well as the ones that have the greatest potential for loss. The review should look for trends, such as an increase in the number of slips, trips, and falls in the engine room. This might point to the floor surface being a contributing factor, which could be corrected. A series of exposures to blood by a few individuals may indicate a need for additional training for the entire company.

Company Tips

Gordon Graham is a risk management expert. He has shared his success as a supervisor in a large police agency, where his systematic approach to risk management has reduced organizational liability and police officer injuries.

He advocates that emergency service organizations concentrate on the high-risk/low-frequency events, when the fire officer or fire fighter does not have any discretionary time to evaluate the situation or consult with others. Those events need to have clear policy and frequent training, so that the response is instinctive when a high-risk/low-frequency event is encountered with no time to think about the course of action.

Graham applied a systems approach in considering safety in public safety agencies. In 2002, he provided Graham's Rules for Enhancing Firefighter Safety (GREFS). Here is a condensed version of the rules:

- GREFS 1: People are your most critical element of the safety process. Select the best. Screen out those with a history of serious medical conditions and inappropriate vehicle operations.
- GREFS 2: Management has leadership responsibility for preventing injuries. Identify those high-risk/low-frequency events that have a probability of causing the greatest problems. Have a workable fire fighter accountability system in place.
- GREFS 3: Employees must be continuously trained to work safely.
- GREFS 4: Safety is a condition of employment. Every fire fighter has to recognize that they have an ongoing obligation to work safely. Supervisors and managers must both monitor safety performance in the workplace and follow the same rules themselves.
- GREFS 5: When safety rules are not followed, address the issue with prompt, fair, and impartial discipline.
- GREFS 6: Answer this question: "Does the benefit of fighting this fire justify the risks involved in fighting this fire?"
- GREFS 7: Women and men who work safely are more likely to be productive.
- GREFS 8: Safety extends beyond the job to be part of every person's life.
- GREFS 9: Safety is a business responsibility. Ethically, all management teams have an affirmative obligation to make sure that fire fighter safety is being taken seriously.
- GREFS 10: Most things that go wrong are highly predictable. Predictable is Preventable.

Although fire companies should ideally train on every aspect of the job, the reality is that there is usually not enough time to do it all. Risk management principles suggest that fire officers should focus attention on the high-risk activities that we perform relatively infrequently. The training helps bridge the gap for the lack of experience due to the infrequency of the activity. An example would be training on locating and removing lost or trapped fire fighters.

Postincident Analysis

One method of identifying unsafe situations is to conduct a postincident analysis. The incident safety officer provides a written report for the department that includes pertinent information relating to safety and health issues involved with the incident. This would include information about the use of protective clothing and equipment, personnel accountability system, rapid intervention crews, rehabilitation operations, and other issues that directly affect the safety and welfare of members at the incident scene.

Mitigating Hazards

The real value in a postincident analysis is the learning process that can result from the information obtained during the process. To take full advantage of the information, here are a few tips to help a fire officer who is made aware of a hazard or a dangerous situation:

1. Determine what the preferred and safer response or activity should be.
2. Develop a procedure, practice, or equipment list needed to deliver the preferred and safer response or activity.
3. Consult with supervisors for approval and formal adoption.
4. Train your peers and subordinates in the preferred and safer response or activity. Training shall address recommendations arising from the investigation of accidents, injuries, occupational deaths, illnesses, and exposures and the observation of incident scene activities.
5. Make it part of an updated standard operating procedure or directive.
6. Integrate the new response or activity into the existing continuing education and basic program.

The fire officer has three roles to provide a safe environment for the fire company and ensure that everyone goes home:

1. Identify unsafe and hazardous conditions.
2. Mitigate or reduce as many problems as possible.
3. Train and prepare for the remaining hazards.

You Are the Fire Officer: Conclusion

The most common cause of death for a fire fighter on the fireground is a heart attack, followed by trauma, asphyxiation, and burns. Heart attacks often result from stress and overexertion. Asphyxiation and burn fatalities are often caused by fire fighters becoming lost, disoriented, or trapped inside burning buildings.

The primary function of a rapid intervention crew (RIC) is to rescue any fire fighter who is lost, disoriented, trapped, or injured. The RIC should be strategically positioned, standing by ready for immediate action and monitoring fireground radio traffic.

The incident safety officer has several responsibilities, all of which are related to ensuring that operations at the incident scene are conducted safely. The most critical responsibilities include identifying imminent and potential hazards, providing a risk assessment to the incident commander, and ensuring that all safety policies and procedures are followed. The specific tasks performed by an incident safety officer depends on the situation. Typical tasks include observing the scene and operations to identify hazards, monitoring communications, establishing safety zones, and ensuring that rehabilitation is provided for fire fighters.

Wrap-Up

Ready for Review

- About 100 fire fighters die in the line of duty a year. More than 40% die of heart attacks, and more than 20% die in motor vehicle crashes.
- Ten to 15 deaths could be avoided every year if every fire fighter wore a seat belt while responding and returning to incidents.
- Almost one-third of the fire fighters who die in a motor vehicle crash are ejected from the vehicle.
- Three fire officer responsibilities for safety:
  - Be physically fit
  - Wear seat belts
  - Maintain fire company integrity at emergency incidents
- Every fire fighter operating in the interior should have a radio.
- A fire officer may be required to function as an incident safety officer.
- General knowledge requirements for the incident safety officer:
  - Safety and health hazards involved in emergency operations
  - Building construction
  - Local fire department personnel accountability system
  - Incident scene rehabilitation
- Incident safety officer monitors the scene and reports the status of conditions, hazards, and risks to the incident commander.
- A member with an infectious disease exposure must immediately and thoroughly wash the affected area.
- The infection control officer must be notified within 2 hours of the exposure.
- Infectious disease reporting information:
  - Description of how the exposure occurred
  - Mode of transmission
  - Entry point
  - Use of personal protective equipment
  - Medical follow-up and treatment
- There is an accident investigation on all occupational injuries, illnesses, exposures, and fatalities, or other potentially hazardous conditions involving fire department members and all accidents involving fire department vehicles, fire apparatus, equipment, or fire department facilities.
- The conclusion of an accident investigation includes identifying corrective actions needed to prevent another accident, injury, or incident from occurring.

Hot Terms

Accident An unplanned event that interrupts an activity and sometimes causes injury or damage. A chance occurrence arising from unknown causes; an unexpected happening due to carelessness, ignorance, and the like.

Hazards Any arrangement of materials and heat sources that presents the potential for harm, such as personal injury or ignition of combustibles.

Health and safety officer The member of the fire department assigned and authorized by the fire chief as the manager of the safety and health program.

Immediately dangerous to life and health (IDLH) Any condition that would do one or more of the following: (a) pose an immediate or delayed threat to life, (b) cause irreversible adverse health effects, or (c) interfere with an individual's ability to escape unaided from a hazardous environment.

Incident action plan The objectives reflecting the overall incident strategy, tactics, risk management, and member safety that are developed by the incident commander. Incident action plans are updated throughout the incident.

Incident management system (IMS) A system that defines the roles and responsibilities to be assumed by personnel and the operating procedures to be used in the management and direction of emergency operations; the system is also referred to as an incident command system (ICS).

Incident safety officer An individual appointed to respond to or assigned at an incident scene by the incident commander to perform the duties and responsibilities specified in NFPA 1521, *Standard for Fire Department Safety Officer.* This individual can be the health and safety officer, or it can be a separate function.

Incident safety plan The strategies and tactics developed by the incident safety officer based upon the incident commander's incident action plan and the type of incident encountered.

Incident scene rehabilitation The tactical level management unit that provides for medical evaluation, treatment, monitoring, fluid and food replenishment, mental rest, and relief from climatic conditions of the incident.

Personnel accountability system A method of tracking the identity, assignment, and location of fire fighters operating at an incident scene.

Preincident plan A written document resulting from the gathering of general and detailed data to be used by responding personnel for determining the resources and actions necessary to mitigate anticipated emergencies at a specific facility.

Fire Officer in Action

As a battalion chief, you are providing mutual aid with a company from a neighboring jurisdiction to a larger commercial structure fire as part of a second alarm. On your arrival at the scene, you stage your command vehicle and report to temporary staging at the command post. The incident commander then assigns you to the position of safety officer. Although you have completed training for safety officer, you have never filled that role at such a large incident. What are you going to do?

1. What is the role of the safety unit?
   A. Set up rehabilitation
   B. Provide support and assistance to the safety officer
   C. Work on fireground operations
   D. To be the rapid intervention team

2. What is incident scene rehabilitation?
   A. Makes sure the incident runs smoothly
   B. Provides a break for the incident commander
   C. Assesses the incident
   D. Provides treatment and rehabilitation to personnel at the scene

3. At a minimum, if personnel are exposed to an infectious disease, what should be included in the record?
   A. Type of call
   B. Demographics of the call
   C. Entry point
   D. Blood type

4. What is one of the primary functions of the incident safety officer at an emergency scene?
   A. Conduct fireground operations
   B. Operate rehabilitation
   C. Report hazards to the incident commander
   D. Supervise the rapid intervention team

Hot Terms (continued)

Rapid intervention crew (RIC) A minimum of two fully equipped personnel on site, in a ready state, for immediate rescue of disoriented, injured, lost, or trapped rescue personnel.

Rehabilitation The process of providing rest, rehydration, nourishment, and medical evaluation to members who are involved in extended or extreme incident scene operations.

Risk/benefit analysis A decision made by a responder based on a hazard and situation assessment that weighs the risks likely to be taken against the benefits to be gained for taking those risks.

Risk management Identification and analysis of exposure to hazards, selection of appropriate risk management techniques to handle exposures, implementation of chosen techniques, and monitoring of results, with respect to the health and safety of members.

Safety unit A member or members assigned to assist the incident safety officer. The tactical level management unit that can be composed of the incident safety officer alone or with additional assistant safety officers assigned to assist in providing the level of safety supervision appropriate for the magnitude of the incident and the associated hazards.

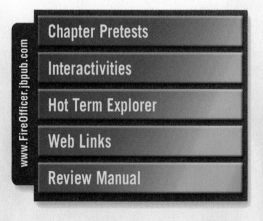

www.FireOfficer.jbpub.com

Chapter Pretests

Interactivities

Hot Term Explorer

Web Links

Review Manual

Training and Coaching

Technology Resources

www.FireOfficer.jbpub.com

- Chapter Pretests
- Interactivities
- Hot Term Explorer
- Web Links
- Review Manual

Chapter Features

- Voices of Experience
- Getting It Done
- Assessment Center Tips
- Fire Marks
- Company Tips
- Safety Zone
- Hot Terms
- Wrap-Up

NFPA 1021 Standard

Fire Officer I

4.2.3 Direct unit members during a training evolution, given a company training evolution and training policies and procedures, so that the evolution is performed in accordance with safety plans, efficiently, and as directed.

   (A) Requisite Knowledge. Verbal communication techniques to facilitate learning.

   (B) Requisite Skills. The ability to distribute issue-guided directions to unit members during training evolutions.

Fire Officer II

NFPA 1021 contains no Fire Officer II Job Performance Requirements for this chapter.

Additional NFPA Standards

NFPA 1001, *Standard for Fire Fighter Professional Qualifications*

NFPA 1403, *Standard on Live Fire Training Evolutions*

NFPA 1142, *Standard on Water Supplies for Suburban and Rural Fire Fighting*

Knowledge Objectives

After studying this chapter, you will be able to:

- Explain what accreditation means for fire fighter certification programs.
- Discuss the four-step method of job instruction training.
- Explain the difference between competence and confidence in individual skill sets.
- Identify the federal regulations that affect the training of every fire fighter.
- Describe the importance of NFPA 1403, *Standard on Live Fire Training Evolutions*.
- List the steps in developing a training program

Skills Objectives

After studying this chapter, you will be able to:

- Direct fire company members in proper completion of a prepared training evolution.

Probationary Fire Fighter Shelly is driving Engine 11 today. The last requirement of Shelly's probationary period is to become certified as an apparatus driver/operator. He has met all of the other requirements up to this point. The first month of on-the-job driver training has been a disaster for Shelly, who appears to become lost in an adrenaline haze whenever the tones go off for Engine 11. He drives erratically and has difficulty following oral directions from the officer.

This afternoon, you responded to an automatic alarm at a hotel, and Shelly drove right past the entrance. You had to advise Engine 2 to take the first-arriving assignment while you drove around the block. There are 14 hotels in your first-alarm district, and Shelly cannot seem to find any of them, in spite of the huge signs in front of each one.

Shelly will be terminated if he does not pass his apparatus driver/operator certification within the next 30 days. You need a plan to help him improve his performance.

1. What should you do when a fire fighter has a performance problem?
2. How much time should you spend with individual fire fighter performance issues?
3. How can you develop a work improvement plan?

Introduction to Training and Coaching

Training and coaching have been core fire officer tasks since the establishment of the first organized fire departments. **Training** is defined as the process of achieving proficiency through instruction and hands-on practice in the operation of equipment and systems that are expected to be used in the performance of assigned duties. Fire service training has evolved in complexity and sophistication at a rapid pace as new areas of expertise have been added to the list of services performed by fire departments.

New York City began the first organized instruction of fire officers in 1869, when Commissioner Alexander Shaler established an Officer's School. That school published the *Manual of Instructions for Commanding Officers of Engine and Hook and Ladder Companies.* Formalized instruction of FDNY fire fighters began when the department purchased the first scaling ladders in 1882. The inventor of the Hoell Lifesaving Appliance came to New York to teach scaling ladder methods and techniques. The School of Instruction was established in 1883.

The School of Instruction continued to evolve, becoming the FDNY Fire College in 1911. Four levels of instruction were conducted at the college: Officer's, Engineer's, Company, and Probationary Fireman's Schools.

Fire Fighter Certification Programs

One of the fathers of fire fighter certification training was Los Angeles Fire Chief Engineer Ralph J. Scott. Scott created a fire college in Los Angeles in 1925. The training staff researched and documented every task that a fire fighter might be required to perform. The list of almost 2000 entries evolved into a document that became known as "The Trade Analysis of Fire Fighting." While functioning as president of the International Association of Fire Chiefs in 1928, Scott convinced the U.S. Department of Vocational Education to accept this list as an official definition of fire fighter tasks.

The National Fire Protection Association (NFPA) helped standardize fire fighter training when the inaugural edition of NFPA 1001, *Standard for Fire Fighter Professional Qualifications*, was adopted by the membership in 1971. For the first time, there was a national consensus on the knowledge and skills a fire fighter should possess. This started a trend of developing national consensus standards on a wide range of fire service occupations (▶ **Table 7-1**).

Accreditation of Certification Programs

Every state or commonwealth in the United States has some type of fire service professional certification system in place. The particular programs range from local to national, depending on local history, government structure, legislation or regulation, and funding. Most of these systems are based on the NFPA professional qualifications standards.

Development of the NFPA professional qualifications standards during the 1970s created a need to validate the many different certification systems that were already in use and to establish criteria for new systems. The NFPA standards define the qualifications that an individual must demonstrate to be certified at a given level, such as Fire

Getting It Done

Training Versus Education

The relationship between training and education is often confusing. Education is the process of imparting knowledge or skill through systematic instruction. Education programs are conducted through academic institutions and are primarily directed toward an individual's comprehension of the subject matter. Training is directed toward the practical application of education to produce an action, which can be an individual or a group activity. There is an important distinction between these two types of learning.

Within the fire service, training has been considered as essential for many years. The emphasis on fire fighter education is a much more recent development. The First Wingspread Conference on Fire Service Administration, Education and Research was sponsored by the Johnson Foundation and held in Racine, Wisconsin in 1966. This conference brought together a group of leaders from the fire service to identify needs and priorities. They agreed that a broad knowledge base was needed and that an educational program was necessary to deliver that knowledge base. This became the blueprint for the development of community college fire science and fire administration programs as well as the Degrees at a Distance program. This education process built on Chief Scott's work by progressing from training into education.

In 1998, the U.S. Fire Administration hosted the first Fire and Emergency Services Higher Education (FESHE) conference. That conference produced a document "The Fire Service and Education: A Blueprint for the 21st Century," that initiated a national effort to address and update the academic needs of the fire service. Participants at the 2000 FESHE conference began work to develop a model fire science curriculum that would go from community college through graduate school. At the 2002 conference, U.S. Fire Administration Education Specialist Edward Kaplan compared the results of the FESHE effort with the Wingspread higher education curriculum. The FESHE work affirmed the soundness of the original Wingspread model, with information technology the only new knowledge item added to the curriculum.

Table 7-1 Fire Service Occupations

- NFPA 1001, *Standard for Fire Fighter Professional Qualifications*
- NFPA 1002, *Standard for Fire Apparatus Driver/Operator Professional Qualifications*
- NFPA 1003, *Standard for Airport Fire Fighter Professional Qualifications*
- NFPA 1006, *Standard for Rescue Technician Professional Qualifications*
- NFPA 1021, *Standard for Fire Officer Professional Qualifications*
- NFPA 1031, *Standard for Professional Qualifications for Fire Inspector and Plan Examiner*
- NFPA 1033, *Standard for Professional Qualifications for Fire Investigator*
- NFPA 1035, *Standard for Professional Qualifications for Public Fire and Life Safety Educator*
- NFPA 1041, *Standard for Fire Service Instructor Professional Qualifications*
- NFPA 1051, *Standard for Wildland Fire Fighter Professional Qualifications*
- NFPA 1061, *Standard for Professional Qualifications for Public Safety Telecommunicator*
- NFPA 1071, *Standard for Emergency Vehicle Technician Professional Qualifications*
- NFPA 1521, *Standard for Fire Department Safety Officer*
- NFPA 472, *Standard for Professional Competence of Responders to Hazardous Materials Incidents*

National Board on Fire Service Professional Qualifications

The Joint Council of National Fire Service Organizations created the National Professional Qualifications System (also known as the "Pro Board") in 1972. When the Joint Council dissolved in 1990, the Pro Board evolved into the independent National Board on Fire Service Professional Qualifications (NBFSPQ). The board of directors consists of representatives from national fire service organizations that have an interest in training and certification. The NBFSPQ has accredited the certification programs that are operated by 35 states, provinces, and other agencies. Accreditation is issued for certification based on 16 different standards and 67 recognized levels of fire service–related competencies.

International Fire Service Accreditation Congress

Established by the National Association of State Directors of Fire Training, the International Fire Service Accreditation Congress (IFSAC) provides accreditation to certificate-issuing entities. IFSAC also accredits fire-related degree programs at the college and university levels. The IFSAC process follows

Officer I and Fire Officer II. Accreditation establishes the qualifications of the system to award certificates that are based on the standards. A certification that is awarded by an accredited agency or institution is recognized by other accredited agencies and organizations.

Accreditation is a system whereby a certification organization determines that a school or program meets with the requirements of the fire service. A group of impartial experts is assigned to thoroughly review a given program and determine whether it is worthy of accreditation. Two organizations provide accreditation to fire service professional certification systems: the National Board on Fire Service Professional Qualifications and the International Fire Service Accreditation Congress.

the format used by higher education entities and operates through the College of Engineering, Architecture and Technology at Oklahoma State University. In 2003, there were 42 IFSAC-accredited fire service training programs and nine degree programs.

Overview of Training

Fire fighters are always training. Training is a core activity for fire departments to ensure that every fire fighter can perform competently as an individual and every fire company is prepared to operate as a high-performance team. Fire service training must anticipate high-risk situations, urgent time frames, and difficult circumstances. A wide variety of methods and practices are included in the overall category of training.

Much of the initial training for fire fighters leads to basic skill certifications, such as NFPA Fire Fighter I and EMS first responder. These certifications are often required before a fire fighter is authorized to participate in actual emergency operations. Certification usually involves a formal training program, conducted by a training academy or equivalent organization, that includes both classroom and skills practice (▼ **Figure 7-1**). The trainee must pass both skill and knowledge evaluations to be certified.

After initial certification, most fire fighters work toward achieving progressively higher levels of certification and additional specialty qualifications. Often, a fire fighter is required to achieve additional qualifications, such as Fire Fighter II, driver/operator, or emergency medical technician (EMT), within a set time period. In many fire departments, higher-level certifications, such as Fire Officer I or II, are required for promotion or advancement. Many certifications require periodic refresher or update training, and some expire after a set period unless the fire fighter completes a refresher training requirement.

Several important components of training occur at the fire station or company level under the supervision of fire officers. This type of training is often directed toward practicing basic skills and improving both individual and team performance (▶ **Figure 7-2**). Many departments have a standard set of evolutions that each company is expected

Figure 7-1 A formal training program includes both classroom and skills practice.

Figure 7-2 Training at the company level involves practicing basic skills and improving team performance.

to be able to perform without flaw, such as advancing a fire attack line over a ground ladder and into a third-floor window. Additional training at the company level often includes learning how to use new tools and equipment as well as refreshing, reinforcing, or updating knowledge and skills that are related to different aspects of the firefighting craft.

Additional company-level training often includes preincident planning and familiarization visits to different locations in a company's response area. These activities sometimes include a group of companies that would normally respond together to that location. Conducting these activities as a group is an excellent method of improving coordination between companies. Periodic multicompany drills should also be conducted for the same reason.

Fire Officer Training Responsibilities

One of the basic responsibilities of every fire officer is to provide training for subordinate fire department members. The specific training responsibilities assigned to fire officers vary, depending on the organization and the resources that are available. At a minimum, a fire officer must be prepared to conduct company-level training exercises and evolutions to ensure that the company is prepared to perform its basic responsibilities effectively and efficiently.

One of the most enduring products of the 20th century is the four-step method of skill training. This prepare-present-apply-evaluate method originated during World War I, when the armed services were teaching farmers how to fly biplanes, drive tanks, and operate ships. The process was updated and renamed as **job instruction training** during World War II,

when more than a million men and women received this type of instruction.

The four-step method is a core part of most Fire Instructor I certification programs, although most 21st century fire officers use standardized curricula and training packages, instead of starting from scratch to develop a training program. It is important to understand the process that goes into designing a new training program.

Review of the Four-Step Method

Step 1: Preparation

The fire officer or the department determines the need for instruction. Three indicators that training is needed would be a near miss, a fireground problem, or an observed performance deficiency. For example, during a structure fire, the fire officer might observe that the fire fighters are having difficulty placing a 35-foot extension ladder at an upper-floor window. The deployment of an attack line from a standpipe connection might have created a tangle of hose in the stairwell. These problems need to be addressed through additional training and practice.

The fire officer begins by obtaining the necessary material and teaching aids. If needed, the fire officer writes a lesson plan (▶ **Figure 7-3**). Components of a lesson plan include the following:

1. Break the topic down into simple units
2. Show what to teach, in what order to teach it, and exactly what procedures to follow
3. Use a guide to help accomplish the teaching objective

If this is a new lecture or topic, the fire officer should practice delivering the lesson to make sure that the important items are covered in a timely fashion. In addition, the fire officer should preview any audiovisual items and check all of the equipment that will be used during the presentation.

The final preparation activity is to check the physical environment. Make sure that you have an environment that is conducive to adult learning. That includes taking every reasonable effort to reduce distractions and student discomfort (▶ **Figure 7-4**).

Figure 7-3 Fire officers often write lesson plans in preparation for training activities.

Figure 7-4 Check the physical environment to make sure it is conducive to adult learning.

Step 2: Presentation

This is the lecture or instructional portion of the training. The objective here is to introduce the students to the subject matter, explain the importance of the topic, and create an interest in the presentation. The fire officer could be demonstrating or showing a skill or explaining a concept.

When the fire officer is teaching a skill, he or she should present it one step at a time, delivering a perfect demonstration of how it should be performed (▶ **Figure 7-5**).

When presenting a concept or idea, recommended lecture practices should be followed. The objective of a lecture is to provide knowledge and develop understanding, so the fire fighter will be able to perform the skills properly. The overall goal is to increase fire company efficiency.

The instructor should use simple but appropriate language. Begin with simple concepts and move on progressively to more complex information, relating the new material to old ideas. Lecture only on what is important at this time to achieve the teaching objective, and do not teach alternative methods. Teach in positive terms and avoid telling the fire fighter what *not* to do.

A lesson plan allows the fire officer to stay on topic and emphasizes the important points (▶ **Figure 7-6**). Increasing the number of senses engaged in the training session helps the fire fighter to retain more of the material. Audiovisual aids and training props should be used to enrich the presentation. Fire fighters retain information more effectively if they actually perform the skill in the process of learning it.

Step 3: Application

The fire fighter should then demonstrate the task or skill under the fire officer's supervision. The objective is a correct demonstration of the task, safely performed. A good reinforcing technique is to have the fire fighter explain the task while demonstrating the skill. The fire officer should provide immediate feedback, identifying omissions and correcting errors (▶ **Figure 7-7**). Success is achieved when the student can perform the task safely without input from the supervisor.

Step 4: Evaluation

There should be an evaluation of the student's progress at the end of the lesson or program. Training that is related to a certification program always includes an end-of-class evaluation. Depending on the skill and knowledge sets that are involved, the evaluation may be a written or practical examination (▶ **Figure 7-8**).

The fire officer can be certain that training has occurred only when there is an observable change in the fire fighter's performance when responding to a real situation where that task or skill is applied. For example, Fire Fighter Jones has not been coming to a complete stop at red traffic lights or stop signs when driving the engine. The fire officer provides

Figure 7-5 Fire officers should deliver a perfect demonstration of how skills are performed.

a training session to address this behavior. If, after the training session, Fire Fighter Jones always comes to a complete stop at red traffic lights and stop signs, there has been an observable change of behavior. If the behavior does not change, the training objective has not been accomplished.

Ensure Proficiency of Existing Skill Sets

The fire officer is responsible for ensuring that every fire fighter is proficient in performing a series of skill sets. This is both an individual and a company-level requirement. Some departments have a standard set of evolutions that have to be performed proficiently, either by a company or by an individual fire fighter, in which both skill and time requirements must be met. The fire officer must invest some of the available training time practicing and reviewing these standard evolutions.

It is important that some of these practice sessions be performed while the fire fighters are wearing full personal protective clothing and simulating realistic fireground situations. This may require construction of some training props, such as an assembly that allows the fire fighters to practice opening a roof using a power saw. The fire officer should always be on the alert for opportunities to acquire abandoned structures where realistic fireground skills can be practiced.

Provide New or Revised Skill Sets

On occasion, the fire officer is required to provide the initial training for a new or revised skill set. This is often related to a new device that has been acquired, such as a new type of

breathing apparatus or a thermal image camera. The fire fighters need to become familiar with the new equipment and proficient in its use, especially during emergency procedures. In the case of a thermal imaging camera, the fire fighters have to learn how it works and how to maintain it, as well as how to incorporate its use into fireground operations. This type of training could also be required in order to introduce a change in standard operating procedures. Sometimes procedures are changed or additional training is mandated in response to a near miss that caused a problem to be recognized.

Teaching new skills takes more time than maintaining proficiency of existing skills. The fire officer should obtain as much information as possible about the device or procedure, especially identifying any fire fighter safety issues. The emphasis of fire station–based training should be on the safe and effective use of the device or procedure. The fire officer should plan to spend a couple of training periods developing competency and showing how it relates to existing procedures. Avoid spending more than 15–20 minutes on any lecture or video presentation; the adult learner drifts away if the presentation takes more time.

Ensure Competence and Confidence

The fire officer works as a **coach** when providing training to an individual or a team. After team members have learned the basic skills and can appropriately demonstrate them, the coach has to work with them to build competence and confidence. The coach has to provide the guidance that advances them from being capable of performing the basic required skills to being able to perform them effectively, efficiently, and consistently.

Tools for the Instructor's Toolbelt
By David Hall

Rationale: At some point in their career in the fire service, most everyone is going to be required to teach others about a topic. To effectively do this, the instructor must understand the basic principles of learning and have an understanding of the material that they are going to present. This course will give an overview of those principles and then focus on two methods that may be particularly effective in teaching many fire fighting topics.

Purpose: The purpose of this course is to provide participants with the knowledge required to analyze the content to be presented, the target audience, and the desired outcomes, and be able to select the appropriate teaching method. The course will also teach the instructor how to effectively utilize games and case studies to present material appropriate for these methods.

Objectives:
At the conclusion of this course, the participant will be able to:
1. Describe how to choose the right instructional tool
2. Identify rote information, hard skills, and soft skills
3. Describe how to evaluate the audience
4. Identify the outcomes of a course
5. List the advantages of using games
6. Demonstrate the use of the "Bingo" game
7. Demonstrate the use of the "Generic Game Board"
8. Demonstrate the use of a case study
9. Develop a case study without external resources
10. Develop a case study using external resources

Qualifications for Attendance: This course is designed for all members of the fire service that instruct other members. The course is primarily aimed at individuals that are the Fire Service Instructor I or greater. This course will also be particularly beneficial to fire officers that instruct at the company level.

Summary of Subject areas:	Hours:
Introduction and Welcome	.50
Choosing the Right Tool	.50
Games	1.50
Case Studies	1.50
Summary and Evaluation	.50
TOTAL	4.5

Figure 7-6 A sample lesson plan.

Figure 7-7 The fire officer should provide immediate feedback to the fire fighter.

Figure 7-8 A student's progress is evaluated through a written or practical examination.

Many fire fighter tasks involve psychomotor skill sets. Psychomotor skill levels fall into four categories. The levels can be described using the following example of a new driver/operator who is required to know, from memory, every address in the fire district.

- **Initial.** The driver/operator knows the main streets and has a basic understanding of how the street grid and numbering system works. The fire officer has to help by reading the map when responding in subdivisions and office parks.
- **Plateau.** The driver/operator can drive to more than 85% of the streets in the company's response area without assistance from the officer. At this level, the driver/operator is *competent*.
- **Latency.** The driver/operator can remember the route to an area of the district where the engine company has not had a response for several months.
- **Mastery.** The driver/operator can easily drive to any address in the district and knows at least two or three alternative routes to each area. The driver knows all of the subdivisions, the layout of every office park, and the locations of all hydrants and fire protection connections. At this level, the driver/operator is *confident* in knowledge of the district.

To bring fire fighters up to the confident or mastery level, the fire officer must work every day to reinforce his or her skills. Many fire fighter skill sets are infrequently used, yet when that skill set is needed, the fire fighter has to deliver a near-perfect performance under urgent or critical conditions.

It is dangerous to ask a fire fighter to deliver a rusty skill set in a critical emergency situation.

The fire officer needs to provide enough repetition and simulations to maintain fire fighter confidence in seldom-used skill sets (▶ **Figure 7-9**). The continuing expansion of fire fighters as all-hazard mitigation specialists increases the fire officer challenge. Fire fighters must become proficient as first responders to terrorist incidents involving chemical, biological, or nuclear elements.

When New Member Training Is On-the-Job

Many departments require that a new fire fighter attend a recruit school and obtain NFPA Fire Fighter I certification before responding to emergencies. Fire officers have special responsibilities when operating with inexperienced fire fighters and fire fighters in training. At the very first meeting with the new fire fighter trainee, the fire officer should explain the procedures in the fire station when the company receives an alarm, assign a senior fire fighter to function as a mentor, and describe any restrictions that are placed on fire fighters in training.

Skills That Must Be Immediately Learned

Three federal regulations affect every fire fighter. These are usually some of the very first topics that are covered in any emergency service training program:

- **Bloodborne pathogens.** Even "suppression only" fire fighters are at risk of being exposed to blood and other bodily fluids. The Occupational Safety and Health Administration (OSHA) has issued regulation 29 CFR 1910.1030: "Occupational Exposure to

Figure 7-9 Skills practice may require training props.

Bloodborne Pathogens." This requires all fire fighters to be trained about the department's exposure control plan, the personal protective equipment used by the fire fighter, and the reporting requirements if there is an exposure. Usually, this training takes about 4 hours.

• **Hazardous materials awareness.** Every public safety member as well as a large group of local government employees is required to have training at the hazardous materials awareness level. OSHA regulation 29 CFR 1910.20: "Hazardous Waste Operations and Emergency Response (HAZWOPER) Training" describes these requirements. The awareness level is generally an 8-hour program that provides a brief overview of hazardous materials and describes the NFPA 704 marking system, the Department of Transportation (DOT) placard system, how to use the DOT Emergency Response Guide, and general precautions that the fire fighter must take in responding to incidents that could involve hazardous materials.

• **SCBA fit testing.** OSHA regulation 29 CFR 1910.134: "Respiratory Protection Training" requires that anyone who uses respiratory protection during job tasks must be provided with appropriate training, be fit-tested for a mask, and be subject to a health monitoring program. This training is usually part of the Fire Fighter I training program.

In addition to the federally required training, the fire department needs to provide emergency scene awareness training to reduce the risk that the fire fighter in training will be injured at the emergency scene. This training is often a responsibility of the fire officer and would include such items

as how to avoid being struck by a car when operating on an interstate highway. The fire officer should spell out the expected behavior of the trainee when operating at emergencies. This may include assigning the trainee to shadow, or work with, a senior fire fighter on the team.

The expected behavior falls into three areas: responding to alarms, on-scene activity, and emergency procedures. Responding to alarms includes the trainee behavior that should occur when an alarm is received at the fire station. Be specific about the appropriate actions necessary when a fire fighter is preparing to respond to an alarm. Emphasize the importance of not running in the station, donning protective clothing, and always wearing a seat belt when riding in a fire vehicle.

On-scene activity includes clearly defining the expected location, activities, and behavior of the trainee at an emergency incident. Make sure that the trainee knows the safe locations when working at an incident that is on a roadway. Identify off-limits activity. The fire officer should know all of the restrictions that apply to each individual and ensure that the trainee is equally aware of those limitations. A trainee who has not completed SCBA training and fit testing must not wear breathing apparatus or enter an immediately dangerous to life and health environment. If the trainee is a teenager, additional restrictions might apply; allowing a 15-year-old to operate inside a burning building is prohibited in many jurisdictions.

Finally, the officer should clearly identify the trainee's expected behavior when operating under an emergency situation, such as a Mayday (fire fighter down or lost) or a potentially violent situation. The fire officer should ensure that the trainee will be in a safe location and stay out of the way until the emergency situation is controlled.

Skills Necessary for Staying Alive

Once the "skills that must be immediately learned" are covered, the fire officer should concentrate on the skills that the fire fighter in training needs to know in order to stay alive. Most states and commonwealths require the trainee to pass both a knowledge and a skill test for Fire Fighter I before engaging in interior structural firefighting. Some communities start with the trainee limited to performing "outside only" activities at fire scenes. Suggested topics include:

1. Fireground tasks that emphasize teamwork and require mastering the location of all of the equipment on the apparatus:
   a. Supply line evolutions
   b. Ropes and knots
   c. Laddering the fire building
   d. Lights, fan, and power deployment
2. Crashes and medical emergencies
   a. CPR and semiautomatic defibrillator training
   b. Outside circle activities on a crash extrication
   c. Helicopter landing zone procedure
   d. Assisting the paramedics on a medical emergency

The trainee obtains experience responding to emergencies and functioning in an "outside only" role while obtaining appropriate training in order to qualify for Fire Fighter I certification. Some fire departments do not allow trainees to respond to any emergency incidents until they have achieved the initial certification level.

Live Fire Training

It is vital that the fire fighter in training receive appropriate training before any live fire evolutions are considered. It is unacceptable to injure trainees during a live fire training session. NFPA 1403, *Standard on Live Fire Training Evolutions,* provides essential information for any type of live fire training session.

Student Prerequisites

Before participating in any live fire training evolutions, the student must have received training to meet the Fire Fighter I performance objectives from the following sections of NFPA 1001, *Standard for Fire Fighter Professional Qualifications.*
- Safety
- Fire behavior
- Portable extinguishers
- Personal protective equipment
- Ladders
- Fire hose, appliances, and streams
- Overhaul
- Water supply
- Ventilation
- Forcible entry

In addition to the prerequisite training, the trainee must be equipped with full protective clothing, a personal alarm device, and self-contained breathing apparatus that is compliant with the relevant NFPA standard.

Fire Officer Preparation Responsibilities

NFPA 1403 provides detailed instructions on how to conduct a safe live fire training evolution under five different scenarios: acquired structures, gas-fired training center buildings, non–gas-fired training buildings, exterior props, and exterior class B fires. The fire officer should read the standard and, if available, consult with a training officer who has had training in delivering live fire training evolutions.

There are general points that apply to all live fire evolutions. The fire officer who is the instructor-in-charge develops a training action plan that includes:
- Preburn plan, including communications, personnel accountability, rehabilitation, and building evacuation procedures. The instructor-in-charge must remain mindful of severe weather conditions that increase the risk of injury or illness.
- Calculate the needed water supply. NFPA 1142, *Standard on Water Supplies for Suburban and Rural Fire Fighting*, provides a reference for this subject. Each attack and back-up hoseline shall be capable of flowing a minimum of 95 gallons per minute. The water supply resources should be capable of delivering 150% of the minimum needed water supply at the training site from two separate sources.

Safety Zone

Training Injury Patterns

About 10% of fire fighter line-of-duty deaths occur in training exercises. This is an unacceptable situation. Most training deaths and serious injuries occur during live fire training and involve similar patterns:
1. Missing, incomplete, or inadequate preparation of the drill site
2. Inadequate training of instructors and assistant instructors
3. Inadequate orientation or training of recruits. Live fire evolutions are not the place to conduct or complete initial SCBA training. The students should be competent in SCBA use and emergency procedures *before* participating in a live fire training exercise.
4. Inadequate planning of the training evolutions by the lead instructor
5. Nonexistent incident management system, or no designated safety officer
6. No provision of rehabilitation or EMS services at the drill site

VOICES OF EXPERIENCE

" Effective training cannot occur without effective coaching. "

Training and coaching fire fighters is similar to training and coaching athletes. As a trainer, you have to plan your training based on what skills your team needs to improve. You also need to train individual skills in order to improve your team's overall abilities.

For example, a baseball coach notices a continued pattern of errors on fielding ground balls by his team during games. A good coach will see that and realize that his team needs to spend time in practice working on fielding ground balls. Likewise, a fire officer who notices that his fire fighters have been spending a great deal of time getting hoses reloaded on the trucks after calls would realize that his team needs to spend time practicing loading hoses and would integrate it into their training plan. Both coaches should end up helping their teams improve by targeting their training.

Effective training cannot occur without effective coaching. Handing someone a manual or setting them down in front of a training video is ineffective training without the coaching to support it. Imagine Michael Jordan's coach handing him a book about basketball and saying, "Here. Read this." Take the time and put forth the effort to be a coach as well as a trainer.

Bart Ridings, MSgt
Air National Guard Fire Department
182nd Airlift Wing, Peoria, Illinois
19 years of service

Getting It Done

Performance in Context

Whole-skill training or training in context are effective tools for helping experienced fire fighters improve their group performance at standard evolutions. Athletes use visualization and sequential task mastery to develop consistent and high-level performances. Members who prepare for the Fire Fighter Combat Challenge, Transportation Emergency Rescue Committee (TERC) Extrication Challenge, or other timed skill events also use training in context to improve performance.

- Arrange to have a dedicated ambulance or emergency medical services unit assigned on site during the training evolution.
- Establish a designated and appropriately equipped rest and rehabilitation area.
- Inspect the structure to ensure it is safe for use in live fire evolutions. Determine the required class A materials to be used to provide the fire load. Watch out for excessive fire load and for materials that could lead to early and unwanted flashover.
- Assign dedicated positions of safety officer and ignition officer.
- Assign instructors to each functional crew. Maximum size of a functional crew is five trainees. An additional instructor is assigned to each back-up hoseline and functional assignment. All instructors have full protective clothing and SCBA. Weather, duration of the evolution, and size of the trainee group determine whether additional instructors are required.
- Conduct a preburn briefing session and walk-through for instructors, safety crew, and trainees. Identify evacuation routes during the walk-through. All facets of each evolution are discussed and assignments made before the fire is lit.

Prohibited Live Fire Training Activities

NFPA 1403 identifies activities that are absolutely prohibited. These prohibitions are the direct results of investigations of prior live fire evolutions that resulted in fire fighter deaths or serious injuries.

- The first rule is that no live "victims" may be used in live fire training evolutions. *Never.*
- The second prohibited activity is that flammable or combustible liquids cannot be used as fuel.
- Acquired structures have been the source of many line-of-duty deaths. There must be no more than one fire evolution in an acquired structure at a time.

Developing a Specific Training Program

On occasion, the fire officer may need to develop a specific training program that is not covered by an existing certification training program or prepared lesson plan. This may be a work improvement plan or training related to a new device or procedure. NFPA Fire Service Instructor II covers the development of a training program in more detail. Here is an overview of the five steps.

Needs Assessment

The fire officer must first confirm that there is a need for a training program. Some performance problems may be better solved through an engineering solution.

Example: The effective use of personal alert safety systems (PASS) devices was a problem. Many fire fighters were operating in an immediately dangerous to life and health environment while their PASS device was not armed. Despite training programs and, in some departments, progressive discipline, the devices remained unarmed. An engineering solution has resolved this performance issue. The PASS device is now integrated into the SCBA and is armed every time the high-pressure hose is charged.

Form Objectives

Training has occurred when there is an observable change in behavior. Identify the specific behavior you want the fire fighters to exhibit after the training. The desired behavior could be that fire fighters will demonstrate the proper procedures for deploying a ground ladder.

The description must include the conditions under which the behavior will be demonstrated. For example, if an expectation is that a ground ladder will be deployed while fire fighters are wearing full protective clothing and SCBA, then that condition must be part of the expected fire fighter behavior.

The final part of the objective is the measure of performance. Fire fighter evolutions are timed events, so the measure of performance is usually described in terms of the time required to properly complete a task.

Develop Training Program

Various methods exist for developing the training program. If training is needed for a new device, the manufacturer or vendor may have a training package available. If training is needed for a new procedure or company evolution, the department may have a template the fire officer can use.

The fire officer needs to consider how the training will be delivered. Will it require a skill drill using multiple companies or can the skill be practiced by an individual fire fighter?

Program Delivery

New programs should have a pilot class or trial run before finalization. This provides the opportunity to tweak the program, identify any problems, and correct any unforeseen issues.

Although the use of a lesson plan is part of the four-step method, it is important that the fire officer develop lesson plans for every training program. A good lesson plan:

- Organizes the lesson
- Identifies key points
- Can be reused
- Allows others to teach the program

Evaluation

Have you accomplished the change of behavior? Was the training program worth the fire fighters' or instructor's time? Was the instructional method appropriate for the learning objective? What can the instructor-developer do to improve the training program?

Many successful national fire service training programs, like the "Saving Our Own"™ fire fighter rescue class or the "Car Busters" extrication seminars, have been developed by fire officers and fire fighters. They have the perspective, the understanding, and the need to develop and share vital emergency activity training.

You Are the Fire Officer: Conclusion

You have a closed-door meeting with Shelly. "This is your last probationary period examination. I know you've been a little nervous, but I think your driving skills will dramatically improve when you know where you are going."

You go on to describe how Shelly's workday will progress until he completes his apparatus driver/operator training. He will be driving the rig every day and will be working with a senior driver. Shelly will use flash cards, flip chart maps, driving drills, and map drawings in order to learn station 11's district.

- 0700: Line up, flash-card pop quiz
- 0730–0830: Apparatus and equipment check
- 0830–1030: Physical training, station maintenance, and pumping or maneuvering drill in rear of station 11
- 1030–1100: Street drill with flip chart maps (Shelly and senior driver)
- 1100–1200: Driving drill (find 10 locations in the district) and lunch run
- 1300–1700: Scheduled company training, productivity, and flash-card pop quiz
- 1700–1900: Dinner and station clean-up
- 1900–2000: Shelly works on drawing personal map of district

You tell Shelly that you will meet with him every third workday in the evening to review his progress. "I think you've got what it takes to be a valuable member of this department." You further say, "I think you know how important it is for you to be able to drive safely and get where you're going. I want you to succeed, and that's why I'm taking these steps to help you." You suggest that Shelly invest as much off-duty time as possible to learn the district. You both know Shelly's job depends on it.

Wrap-Up

Ready for Review

- Education is the process of imparting knowledge or skill through systematic instruction. Training is directed toward the practical application of education to produce an action, which can be an individual or group activity.
- The NBFSPQ has accredited the certification programs that are operated by 35 states, provinces, and other agencies. Accreditation is issued for certification based on 16 different standards and 67 recognized levels of fire service–related competencies.
- The four-step method is a core part of most fire instructor I certification programs.
- The components of the four-step method are:
  - Preparation
  - Presentation
  - Application
  - Evaluation
- When presenting tasks, break the topic down into simple units.
- Ensuring proficiency of existing skill sets includes having fire fighters practice their craft in full personal protective clothing and perform fireground tasks, such as cutting a ventilation hole.
- In-station lectures should be no more than 15 or 20 minutes long.
- The fire officer has to provide the guidance that advances fire fighters from being capable of performing the basic required skills to being able to perform them effectively, efficiently, and consistently.
- In psychomotor skill development,
  - At the plateau level, fire fighter is competent.
  - At the latency level, the fire fighter can recall skills steps without significant effort.
  - At the mastery level, the fire fighter is confident in his or her ability to perform the skill.

- The three federal regulations that require fire fighter training are:
  - OSHA regulation 29 CFR 1910.1030: "Occupational Exposure to Bloodborne Pathogens."
  - OSHA regulation 29 CFR 1910.20: "Hazardous Waste Operations and Emergency Response (HAZWOPER) Training."
  - OSHA regulation 29 CFR 1910.134: "Respiratory Protection Training."
- Live fire training must always comply with NFPA 1403.
- A good lesson plan:
  - Organizes the lesson
  - Identifies key points
  - Can be reused
  - Allows others to teach the program

Hot Terms

Accreditation A system whereby a certification organization determines that a school or program meets with the requirements of the fire service.

Coach To provide training to an individual or a team.

Education The process of imparting knowledge or skill through systematic instruction.

Job instruction training A systematic four-step approach to training fire fighters in a basic job skill: (1) prepare the fire fighters to learn, (2) demonstrate how the job is done, (3) try them out by letting them do the job, and (4) gradually put them on their own.

Training The process of achieving proficiency through instruction and hands-on practice in the operation of equipment and systems that are expected to be used in the performance of assigned duties.

As the fire officer, one of your quarterly tasks is to review the training requirements distributed by the training officer and schedule those into your quarterly schedule. When scheduling training, you must also plan on delivering the training provided and making any adjustments as necessary to meet the needs of your shift.

1. Fire fighter tasks involve the use of different psychomotor skills. These skills fall into four categories. What is one of those categories?

   A. Preparation

   B. Competence

   C. Plateau

   D. Intrinsic

2. There are three federal regulations that affect every fire fighter. What is one of them?

   A. Fire Fighter I

   B. Rescue Systems I

   C. Confined space awareness

   D. SCBA fit testing

3. What is one step in the Whole Teaching Process?

   A. Application

   B. Readiness

   C. Eager to learn

   D. Operation is explained

4. Part of the quarterly training includes a review of the new SCBA recently purchased by the department. You are expected to develop on-going refresher training on the new equipment. What should be one of the five steps to develop a new refresher course?

   A. Performance

   B. Form objectives

   C. Whole skill

   D. Create proper mind-set

Evaluation and Discipline

Technology Resources

www.FireOfficer.jbpub.com

- Chapter Pretests
- Interactivities
- Hot Term Explorer
- Web Links
- Review Manual

Chapter Features

- Voices of Experience
- Getting It Done
- Assessment Center Tips
- Fire Marks
- Company Tips
- Safety Zone
- Hot Terms
- Wrap-Up

NFPA 1021 Standard

Fire Officer I

4.2 **Human Resource Management.** This duty involves utilizing human resources to accomplish assignments in accordance with safety plans and in an efficient manner. This duty also involves evaluating member performance and supervising personnel during emergency and nonemergency work periods, according to the following job performance requirements.

4.2.4 Recommend action for member-related problems given a member with a situation requiring assistance and the member assistance policies and procedures, so that the situation is identified and the actions taken are within the established policies and procedures.

(A)*Requisite Knowledge. The signs and symptoms of member-related problems, causes of stress in emergency services personnel, and adverse effects of stress on the performance of emergency service personnel.

(B) Requisite Skills. The ability to recommend a course of action for a member in need of assistance.

4.2.5 Apply human resource policies and procedures, given an administrative situation requiring action, so that policies and procedures are followed.

(A) Requisite Knowledge. Human resource policies and procedures.

(B) Requisite Skills. The ability to communicate orally and in writing and to relate interpersonally.

Fire Officer II

5.2 **Human Resource Management.** This duty involves evaluating member performance, according to the following job performance requirements.

5.2.1 Initiate actions to maximize member performance and/or to correct unacceptable performance, given human resource policies and procedures, so that member and/or unit performance improves or the issue is referred to the next level of supervision.

(A) Requisite Knowledge. Human resource policies and procedures, problem identification, organizational behavior, group dynamics, leadership styles, types of power, and interpersonal dynamics.

(B) Requisite Skills. The ability to communicate orally and in writing, to solve problems, to increase teamwork, and to counsel members.

5.2.2 Evaluate the job performance of assigned members, given personnel records and evaluation forms, so each member's performance is evaluated accurately and reported according to human resource policies and procedures.

(A) Requisite Knowledge. Human resource policies and procedures, job descriptions, objectives of a member evaluation program, and common errors in evaluating.

(B) Requisite Skills. The ability to communicate orally and in writing and to plan and conduct evaluations.

Knowledge Objectives

After studying this chapter, you will be able to:

- Describe the special requirements for supervising a probationary fire fighter who has a structured in-station training program.
- Describe how to use a performance log or T-account to document fire fighter work performance.
- List the activities associated with a mid-year review.
- Describe the requirements of an advanced notice of a substandard employee evaluation.
- Describe the concept of progressive discipline.
- List and describe the components of a written reprimand.
- Describe the services available through an employee assistance program (EAP).

Skills Objectives

After studying this chapter, you will be able to:

- Use a performance log or T-account.
- Issue an oral reprimand, warning, or admonishment when a fire fighter demonstrates a substandard or unacceptable behavior.

You cringe when the fire station telephone rings at 6:55 A.M. It is Fire Fighter Franks, reporting that he will be about 90 minutes late to work this morning. This is supposed to be the first day that Franks will be working at Engine 5, after requesting the transfer to work at the fire station nearest his home. You had just finished reviewing his personnel file when he called. This will be the third time Franks has been late in the past 7 months, as well as the third time he has violated the Work Hours regulation by not calling in before 6:00 A.M. This offense should result in 48 hours of leave without pay, according to the regulation. His next offense will result in termination.

You had planned to spend this first day with Franks outlining your expectations and reviewing his progress in his probationary fire fighter in-station training program. It appears that Franks is a couple of months behind schedule in the program and that his previous supervisor disregarded some of the requirements of the program.

1. How should you handle a fire fighter who has an existing disciplinary issue?
2. What should you do about this morning's late reporting to work?
3. Should you be involved in resolving off-duty problems that affect on-duty fire fighter performance?

Introduction to Evaluation and Discipline

Evaluation and discipline are two essential components of a fire fighter's work performance development. The fire officer plays a key role in the development and success of the fire fighters who are under the officer's command. For probationary fire fighters, their first fire officer sets the stage for a 20- or 30-year career.

Supervision of fire fighters requires that the fire officer provide regular evaluations to provide feedback on progress and problems. Regular evaluations of employees are a standard part of the officer's job and should be approached in a standard, professional manner.

The officer is responsible for supervising the employee's performance and behavior on the job. The officer is also responsible, to a certain extent, for supervising off-the-job behavior, where it reflects on the fire department. Performance and behavior are two different issues, but they are linked in the sense that the officer has to monitor, evaluate, and deal with both types of problems.

Discipline can be either positive or negative. Positive discipline is intended to help the employee recognize problems and make corrections to improve performance or behavior. Negative discipline is stronger and is intended as punishment for unsatisfactory performance, unacceptable behavior, or failure to respond appropriately to positive discipline. Different types and levels of discipline are appropriate for different problems.

Discipline should also be progressive, moving from positive to negative and from minor to major, depending on the nature and the seriousness of the problem and the employee's efforts to correct continuing problems. Progressive discipline could begin with verbal counseling, then move on to a verbal reprimand with a written note to be put in the employee's file, then to a formal written reprimand, and finally to suspensions, reductions in pay, or demotion. Dismissal is considered as the ultimate level of negative discipline for an employee.

Some problems are so serious that the earlier steps in progressive discipline might be bypassed. Most fire departments have a list of offenses that are considered to be so serious that they can lead to immediate termination. All of these steps and actions should be spelled out in written policies; however, a fire officer still has to use a significant amount of judgment and discretion when performing evaluation and disciplinary functions.

Evaluation

One of the fire officer's primary supervisory responsibilities is to conduct regular evaluations of fire fighter performance. Evaluations are performed to ensure that each fire fighter knows the performance that is expected while on the job and where he or she stands in relation to those expectations. This process helps the fire fighter set personal goals for professional development and performance improvement and provides positive motivation to perform at the highest possible level (▶ Figure 8-1).

Most career fire departments require a supervisor to conduct an annual performance evaluation for each assigned employee. The annual performance evaluation is a formal written documentation of the fire fighter's performance during the rating period. This permanent personnel record is

Figure 8-1 The evaluation process helps the fire fighter set personal goals for professional development and performance.

important to both management and the fire fighter, but it cannot replace the ongoing supervisory actions that maintain and improve employee performance throughout the year.

In a volunteer fire department, there should be an equivalent form of periodic evaluation for each member by a higher-ranking individual. The documentation procedures for a volunteer organization are often less structured; however, every member deserves some type of regular report card. The responsibilities of a supervisor are just as serious in a volunteer fire department as a full-time career organization.

Starting the Evaluation Process With a New Fire Fighter

Fire officers have a special responsibility when starting an evaluation process with a probationary fire fighter. The fire officer is helping to shape that individual's fire department career or volunteer experience. The officer will be shepherding the probationary fire fighter through the first real-life emergency experiences, providing feedback and guidance. Most fire fighters have vivid memories of their first fire officer. They remember and respond to the expectations established by that officer. That first fire officer creates a foundation for a fire fighter's entire career.

New fire fighters start out with wide variations in the range and depth of their skills. In some volunteer departments, probationary members can start to ride the apparatus with only 16 hours of life-safety orientation. Many fire departments require fire fighters to complete a recruit school, where they achieve NFPA 1001 Fire Fighter I certification before they are authorized to respond to any alarms. It is the fire officer's responsibility to determine each individual's skills, knowledge, aptitudes, strengths, and weaknesses and then set specific expectations for each new fire fighter.

Recruit Probationary Period

Most fire departments have a structured in-station training program that is part of the recruit fire fighter's probationary period, after the completion of basic training. Regardless of the new fire fighter's level of certification or pre-employment experience, the fire officer is responsible for evaluating each individual during the probationary period. In career fire departments, there is usually a classified job description that specifies all of the required knowledge, skills, and abilities that a fire fighter is expected to master within a specified time period in order to complete the probationary requirements. Volunteer fire departments should have an equivalent set of performance expectations for a fire fighter to successfully complete the probationary period. The fire fighter should be provided with a copy of the specific requirements to use as a checklist ▶ **Figure 8-2** .

During the probationary period, which is generally 6 to 12 months, the recruit fire fighter is expected to obtain experiences and demonstrate competencies related to the fire fighter's classified job description. This period often includes rotation assignments on different type of companies and specialty units operated by the department, such as an engine company, a ladder company, an ambulance, and even the 911 center. Competencies may include demonstrating all of the skill sets required for NFPA 1001 Fire Fighter II certification.

Structured probationary programs usually require the fire officer to complete a monthly evaluation of each probationary fire fighter. This evaluation typically assesses the probationary fire fighter's progress in four areas:

- Fire fighter skill competency, including proficiency as an apparatus operator
- Progress in learning job-specific information not covered in basic training, such as the local fire prevention code and the department's rules, regulations, and standard operating procedures
- Progress in learning the fire company district, including both streets and target hazards
- Performance of other job tasks, such as recording deliveries, performing housework, conducting in-station tours, and completing reports

City of Charlottesville Fire Department

Fire Fighter Job Description

General Definition of Work
The fire fighter performs responsible service work in fire profession and prevention; does related work as required. Work is performed under the regular supervision of a company and/or shift commander.

Typical Tasks
Responds to alarms, drives and operates equipment and related apparatus and assists in the suppression of fires, including rescue, advancing lines, entry, ventilation and salvage work, extrication, and emergency medical care of victims.
Performs cleanup and overhaul work, establishes temporary utility services.
Assists in maintaining and repairing fire apparatus and equipment, and cleaning fire stations and grounds.
Checks fire hydrant flows.
Makes fire code inspections of business establishments and prepares pre-fire plans.
Responds to emergency and non-emergency calls, pumps out basements, inspects for gas leaks, secures vehicle accidents, inspects chimneys, etc.
Participates in continuing training and instruction programs by individual study of technical material and attendance at scheduled drills and classes.
Conducts station tours for the public, school, and community demonstrations and programs.
Backs up for dispatching personnel, monitors alarm boards, receives and transmits radio and telephone messages.
Performs related tasks as necessary.

Knowledge, Skills, and Abilities
General knowledge of elementary physics, chemistry, and mechanics; general knowledge of technical firefighting principles and techniques, and principles of hydraulics applied to fire suppression; general knowledge of the street system and physical layout of the city; general knowledge of emergency care methods, techniques, and equipment; ability to understand and follow written and oral instructions; ability to establish and maintain cooperative relationships with fellow employees and the public; ability to keep simple records and prepare reports; possess a strong mechanical aptitude; ability to perform heavy manual labor; skill in operation of heavy emergency equipment.

Education and Experience
Any combination of experience and training equivalent to graduation from high school.

Special Requirements
Possession of a valid driver's permit issued by the Commonwealth of Virginia.

Future Requirements at Three Years of Service
NFPA 1001   Fire Fighter Level II
NFPA 1002   Driver/Operator Certification
Commonwealth of Virginia Emergency Medical Technician or greater

Figure 8-2 Sample fire fighter job description.

In volunteer fire departments, it is not uncommon for different officers or individuals to certify that a probationary fire fighter has met the requirements for different components of the program. One officer should be specifically assigned to oversee the progress of each individual probationary member to ensure that the overall program requirements are being accomplished successfully.

The fire fighter probationary period is an important foundation for a long career. The fire officer has a special opportunity to prepare future departmental leaders by providing a comprehensive and effective probationary period.

Annual Evaluations

The personnel regulations in most career fire departments require that every employee who has completed the initial probationary period will receive an annual written evaluation from his or her immediate supervisor. These annual evaluations become a formal part of the employee's work history.

In many organizations, the fire officer will receive an annual evaluation form from the human resources or personnel division a month or two before the employee's annual evaluation is due to be submitted. Some organizations expect the officer to keep track of the dates and submit a completed form when the evaluation for each employee is due. In general, the annual evaluation process requires four steps. First, the supervisor fills out a standardized evaluation form. The form asks the supervisor to evaluate the subordinate on a number of knowledge, skills, and abilities appropriate for the subordinate's rank and classified job description. There is usually an opportunity for the supervisor to add comments in a narrative section. In some systems, the supervisor's manager is required to review the evaluation before it is issued to the subordinate.

The subordinate is then allowed to review and comment on the officer's evaluation. The subordinate has an opportunity to provide feedback or additional information that would affect the performance rating.

The third step is a face-to-face feedback interview between the supervisor and the subordinate to discuss the evaluation. This is the opportunity to clarify points and review the performance over the last rating period. A goal of the feedback interview is that both the supervisor and the subordinate understand the results of the evaluation.

The final step, usually completed at the end of the feedback interview, is to establish goals for the subordinate to accomplish during the next evaluation period. These are usually specific, measurable goals that allow the subordinate to improve on the knowledge, skills, and abilities of the current job or prepare for the next higher job level.

Conducting the Annual Evaluation

The actual completion of the annual evaluation forms should be viewed as a formality. The fire officer should be providing continual evaluation and feedback to the fire fighter throughout the year, so the information that goes onto the form should not come as a surprise. Unless there is a problem with the employee's performance, the scheduled evaluation should be used primarily as an opportunity to discuss future goals and objectives.

The methods outlined in the following section are designed to track an employee's performance and progress throughout the year. Using the information gathered from a year of entries in a performance log or T-account, the fire officer can provide an objective and well-documented annual performance evaluation. This documentation is essential, regardless of whether the fire fighter receives an "outstanding" rating or an "unsatisfactory" rating.

Keeping Track of Every Fire Fighter's Activity

In a **performance log**, the fire officer maintains a list of the fire fighter's activities by date, along with a brief description of performance observations (▼ **Figure 8-3**).

The **T-account** is a slightly more sophisticated documentation system, similar to an accounting balance sheet listing credits and debits. A single-sheet form is used to list the assets on the left side and the liabilities on the right side, appearing like a letter "T" (▶ **Figure 8-4**).

Using either method, the fire officer compiles an extemporaneous "when, what, and how" record of each fire fighter's work history throughout the evaluation period. This record can be a powerful evaluation and motivational tool.

Establishing Annual Fire Fighter Goals

The fire officer should require all fire fighters who have completed probation to identify three work-related goals that they want to achieve during the next evaluation period. This provides focus beyond the minimum day-to-day activities

---

March 11, Incident 3467. Trouble positioning apparatus to hook up soft suction and hydrant. Charged wrong attack line. First structural fire as engine driver.

Figure 8-3) Sample notation in a performance log.

+	−
March 11, Incident 3467. Rescue of elderly female from hallway outside burning bedroom in a one-story, single-family dwelling.	March 11, Incident 3467. Had to return to Engine 4 to retrieve forcible entry tools and firefighting gloves during initial actions.

Figure 8-4 Sample notation in a T-account.

and helps fire fighters prepare for future promotional examinations or assignments. Examples of individual fire fighter goals might include:

- Enrolling in and completing a "Building Construction" class at the local community college
- Becoming qualified as an aerial operator
- Learning how to use the computer-aided design software in order to develop tactical preincident plans

The fire officer should track the progress toward these goals during the evaluation period, recording progress in a T-account or a performance log.

Informal Work Performance Reviews

Most civil service and personnel regulations require only that an annual evaluation be performed for each employee; however, many departments encourage fire officers to conduct an informal performance review with each fire fighter every 2 or 3 months. During these informal sessions, the fire officer can review the T-account or performance log with the fire fighter and see what the officer can do to assist the fire fighter in meeting the established work-related goals. The fire officer and fire fighter can identify situations or work conditions that impede progress toward the goals. For example, a fire fighter who is spending 70% of the time assigned to the detail out of the station might have difficulty becoming qualified as a back-up aerial operator. They could discuss changing the goal or adjusting duty assignments so that the fire fighter would have more time at the station to work on meeting the goal.

Mid-Year Review

The mid-year review is a more formal review session. The officer should have each fire fighter write a self-evaluation in preparation for the session. This allows the fire fighter to focus on the job description and the personal goals that were set at the beginning of the year. The officer and the fire fighter should review this document together and discuss how well expectations are being met.

If necessary, the officer can identify resources or assistance that would help the fire fighter meet the mutually established goals. The officer might suggest meeting with another fire fighter who is experienced at working with the preincident planning software or point out that only one semester remains to enroll in the building construction course at the college during the evaluation period.

This is also a time when the personal goals can be adjusted because of changes in the work environment. It would be unrealistic to maintain the goal to attend a building construction class at the community college if the department has selected the fire fighter to attend paramedic school for the next 6 months.

Advance Notice of a Substandard Employee Evaluation

A fire fighter who is not meeting expectations should know there is a problem long before the annual evaluation. The subordinate should be given adequate time to change the behavior or improve the skill, particularly if it is jeopardizing a scheduled pay raise or continuing employment.

Many municipalities require a supervisor to provide a formal notification to an employee who is likely to be evaluated below the level necessary to obtain a pay increase or a "satisfactory" rating. The notice must identify the aspect of the job performance that is substandard and what the employee must do to avoid receiving a substandard annual evaluation. A common practice is to require such a notification 10 weeks before the annual evaluation is due.

If the employee receives a substandard annual evaluation, then the municipality might require a work improvement plan. The plan should cover a specific period of time, such as 120 calendar days, that would be designated as a special evaluation period. The employee is expected to fully participate in a work improvement plan during this special evaluation period. He or she is given an opportunity to demonstrate the desired workplace behavior or performance no later than the end of the special evaluation period.

The following is an example of the human resource division requirements that could be applied to a work improvement plan:

- The work improvement plan shall be in writing, stating the performance deficiencies and listing the improvements in performance or changes in behavior required to obtain a "satisfactory" evaluation.
- During the special evaluation period, the employee shall receive monthly progress reports.

- If, at the end of the special evaluation period, the employee's performance rating is "satisfactory" or better, the time-in-rank pay increase will start at the first pay period after the work improvement period.
- If, at the end of the special evaluation period, the employee rating remains unsatisfactory, then no time-in-grade pay increase will be issued. In addition, the supervisor will determine whether additional corrective action is appropriate.

Regardless of the outcome, a special evaluation period performance evaluation is filled out and submitted as a permanent record in the employee's official file.

In most organizations, the employee's performance is officially evaluated during the annual review. Conducting informal bimonthly or quarterly reviews and midyear reviews enables a fire officer to identify an underperforming fire fighter early in the annual evaluation period. The advanced-notice procedure should provide the fire fighter with enough time to change the behavior or improve the necessary skills before the formal evaluation is scheduled to occur. If the employee has still not improved, a substandard evaluation would not be a surprise to either the fire fighter or the fire officer's supervisor.

Four Weeks Before the End of Annual Evaluation Period

The fire fighter should be prompted to conduct another self-evaluation approximately 4 weeks before the official annual evaluation is due. The goal of this self-evaluation is to identify how well the fire fighter has met the organizational expectations and personal goals for the year. The fire officer should review this document and provide feedback. Together, the fire officer and fire fighter develop the final, formal evaluation report for this rating period, which includes developing three new work-related personal goals for the next evaluation period.

Completing annual evaluation reports is an important fire officer responsibility. It is an opportunity to identify and evaluate the work performed by subordinates. The fire officer should recognize outstanding accomplishments, encourage improvement, and, in some cases, identify those who should be considering another career option.

The fire officer who maintains a performance log or T-account, conducts bimonthly informal reviews, conducts a midyear review, and has the fire fighter complete a self-assessment 1 month before the annual evaluation will have a wealth of information to present an accurate, well-documented, and comprehensive annual evaluation.

Evaluation Errors

Using a T-account or performance log to assemble a detailed list of work behavior observations over the evaluation period provides excellent background when the fire officer is preparing a

fire fighter's evaluation. Evaluation is a largely subjective process that is vulnerable to unintentional biases and errors.

Leniency or Severity

In a variation of the central tendency, called leniency or severity, some fire officers tend to rate all of their fire fighters either higher or lower than their work performance deserves. Leniency reduces conflict. A positive evaluation is likely to make the evaluation a more pleasant experience and avoids confrontation. Leniency is sometimes found when the fire officer is required to conduct a face-to-face meeting with the fire fighter to review the evaluation.

Some fire officers lean in the opposite direction and rate all fire fighters with "needs improvement" or "unsatisfactory." Some officers think that low ratings will cause fire fighters to be motivated to work harder. Newer fire officers sometimes make this mistake when they have not received adequate training in preparing performance evaluations.

Personal Bias

Personal bias is an evaluation error that occurs when the evaluator's personal bias skews the evaluation such that the classified job knowledge, skills, and abilities are not appropriately evaluated. Fire officers must not allow an evaluation to be slanted by race, religion, gender, disability, or age.

Recency

Recency is an evaluation error in which the fire fighter is evaluated only on recent incidents rather than on all of the events that occurred throughout the evaluation period. Fire fighters who are aware of this tendency are on their best behavior in the weeks leading up to the fire officer's evaluation.

Central Tendency

Most evaluation systems involve some type of rated scale, ranging from unsatisfactory to outstanding. A fire officer demonstrates a **central tendency** when a fire fighter is rated in the middle of the range for all dimensions of work performance.

A central tendency evaluation provides little value to the fire fighter or the evaluation process. Being rated as "OK in all areas" is not very informative or helpful.

Frame of Reference

In a **frame of reference** evaluation error, the fire fighter is evaluated on the basis of the fire officer's personal ideals instead of the classified job standards. For example, a fire officer who spent hours every day fixing, improving, and polishing the engine when he was an apparatus driver/operator might issue "unsatisfactory" or "needs improvement" ratings to apparatus operators who are meeting all of the departmental standards but are not meeting his personal ideals.

Halo and Horn Effect

Like life, fire fighter's performances are not just black and white. When the fire officer concentrates on only one aspect of the fire fighter's performance, which is either exceptionally good or bad, and applies that perception across the board to all aspects of the individual's work performance, this is called the **halo and horn effect**.

Contrast Effect

Contrast effect is an evaluation error that can occur when the fire officer compares the performance of one subordinate with the performance of another subordinate instead of against the classified job standards.

Providing Feedback After an Incident or Activity

Performance evaluation should be a continual supervisory process, not a special event that is performed only when a scheduled rating has to be submitted. Regular feedback from the fire officer should keep fire fighters aware of how they are doing, particularly after incidents or activities that present a special challenge.

Performance feedback is most effective when delivered as soon as possible after an action or incident. That means providing essential feedback before the ashes get cold after a struc-

ture fire. The fire fighter is intensely aware of the event and wants to know how well he or she performed. The fire officer needs to be ready to provide specific information in order to recognize or improve fire fighter performance (◄ Figure 8-5).

Discipline

Discipline is a moral, mental, and physical state in which all ranks respond to the will of the leader. The fire officer builds discipline by training to meet performance standards, using rewards and punishments judiciously, instilling confidence in and building trust among team leaders, and creating a knowledgeable collective will.

Within the fire department, discipline divides into positive and negative sides. Positive discipline is based on encouraging and reinforcing appropriate behavior and desirable performance. Negative discipline is based on punishing inappropriate behavior or unacceptable performance. Both positive and negative discipline can be applied to a full range of activities, including emergency incidents and administrative functions.

As a general rule, positive discipline should always be used before negative discipline. Progressive discipline refers to starting out to correct a problem with positive discipline and then increasing the intensity of the discipline if the individual fails to respond to the positive form, perhaps by using mild negative discipline. Increasingly, negative discipline might have to be used in a situation in which an individual fails to respond in an appropriate manner to correct the problem. There are exceptions to this rule; some actions or behaviors are so unacceptable that they must result in immediate negative discipline.

Positive Discipline: Reinforcing Positive Performance

Positive discipline is directed toward motivating individuals and groups to meet or exceed expectations. The key to positive discipline is to convince them that they want to do better

Figure 8-5 Performance feedback should be delivered as soon as possible after an action or incident.

Safety Zone

Immediately Correct Unsafe Conditions

Fire fighter safety is one of the primary responsibilities of a fire officer. Although negative feedback should be issued to an individual in private, the fire officer must always correct unsafe conditions as soon as they are noticed. There is no excuse for allowing an unsafe action or situation to occur without taking corrective action, even if that means shouting an order at a crowded fireground or issuing a direct instruction over the radio. Once the incident is under control, the fire officer needs to follow up with a private face-to-face meeting with the fire fighter or fire fighters who created or ignored an unsafe condition.

and are capable and willing to make the effort. A fire officer provides positive discipline by identifying weaknesses, setting goals and objectives to improve performance, and providing the capability to meet those targets.

The starting point for positive discipline is to establish a set of expectations for behavior and performance. Once the expectations are known, there must be a consistent and conscientious effort to meet them. The expectations have to apply to the entire team as well as to each individual team member. Positive discipline is reinforced by recognizing improved performance and rewarding excellent performance (▼ **Figure 8-6**).

If an individual fire fighter's performance needs improvement in a particular area, the fire officer should coach that person, providing guidance and extra opportunities to correct the problem. Sometimes, simply pointing the individual in the right direction and offering encouragement can achieve the objective. In many cases, the officer can arrange for another fire fighter to work with the individual to correct the problem.

Teamwork is a key factor for fire companies. In addition to each individual having the required knowledge and skills, the company must be able to work efficiently and effectively as a team. All of the company members have to work, learn, and practice together to become capable and confident. The fire officer has to provide the leadership to make it happen.

An officer sets the stage for positive discipline by setting clear expectations and by "walking the talk." Fire officers are working supervisors; they supervise while directly participating in firefighting activities and performing nonemergency duties. The officer should demonstrate a personal commitment to the department's goals, objectives, programs, rules, and regulations by participating in all of the activities that are expected of fire fighters, such as physical fitness training and regular company drills. Fire fighters can gauge an officer's level of commitment by observing the characteristics of self-discipline (▼ **Figure 8-7**).

Competitiveness can also be used as a stimulant in positive discipline, particularly at the company level. Most fire fighters are naturally competitive, and most fire companies work hard to prove that they can be better, faster, more skillful, or more impressive than a rival company. The officer has to point that competitive energy in a positive direction, making sure that the ultimate objective is high performance.

Empowerment

Empowerment is one of the most effective strategies of positive discipline. Fire fighters often complain that they have little control over their work environment; they are told where to go, what to do, and when to do it. Fire officers can help make fire fighters feel stronger by learning how to control their own destiny. An officer who identifies an area where improvement is needed can often empower fire fighters to correct the problem on their own. It is important for the officer to identify the target and provide the resources, but doing the work on their own and demonstrating their capabilities can be a very positive motivator for the fire fighters.

Providing information to help fire fighters learn more about "the system" can support the empowerment process.

Figure 8-6 Positive discipline is reinforced by recognizing improved performance and rewarding excellent performance.

Figure 8-7 Fire officers set the stage for positive discipline by "walking the talk."

VOICES OF EXPERIENCE

❝ When conducted fairly and consistently, most subordinates and peers will respect the manner in which you handle these situations. ❞

Conducting employee performance appraisals and administering discipline is one of the most unpleasant aspects of the fire officer's job. Often this is the result of many factors such as poor appraisal instruments, the lack of adequate training, and the close personal bonds that develop in a firehouse setting. The fire officer's responsibility is to fairly evaluate the current situation, compare the fire fighter's performance to the job standards, and then take corrective actions if required. A successful fire officer takes this responsibility seriously and understands that it is in the employee's best interest, as well as the fire department's, to do so. Giving performance appraisals that are too glowing or do not differentiate between good performers and poor performers only serves to pacify the underperformers and never challenges them to excel. It also serves to de-motivate those who work hard to excel when their efforts are not formally recognized.

Some pointers that may help a new officer deal effectively with performance appraisals and employee discipline:

- Make notes throughout the year of both good and bad behaviors by your employees.
- Discuss deficient performances immediately when they occur.
- Get additional training in conducting performance appraisals and the disciplinary process.
- Call a seasoned officer whom you respect for advice if you are unsure how to handle a situation.
- If you have a Human Resources department, seek their advice when you believe a situation could result in disciplinary action.
- Ask yourself if your actions are fair to the employee and fair to your employer.
- Complete a draft of the performance appraisal and then mull it over for a shift or two prior to completing the final version.
- When meeting to discuss the appraisal or discipline, be prepared, clear, even tempered, and unapologetic.

When conducted fairly and consistently, most subordinates and peers will respect the manner in which you handle these situations. Most employees want to please their supervisor and the performance appraisal gives them the understanding of what you expect. For those instances where an employee's actions require discipline, remember that it is the employee that has determined the actions you must take.

David Hall
Springfield, Missouri Fire Department
22 years of service

The first component can be described as "Local Government 101." This information helps fire fighters learn more about how the fire department and local government work. It could include reviewing the approved budget for the next fiscal year or discussing which functions are performed by each part of the organization. The more fire fighters understand about how the organization and local government work, the more they can feel a sense of participation.

Figure 8-8 Fire fighters should practice working with hoselines and tools in simulated fireground scenarios while wearing full personal protective clothing.

The second component could be described as "Success 102." This information would identify the tools that others have used to achieve success within the fire department. If some companies are consistently recognized for positive performance, what are they doing right? When a few individuals achieve high scores on every promotional examination, what is their secret method? Do fire fighter self-study groups improve promotional examination results? Answering these questions may take some research by the fire officer, such as identifying the individuals or groups that have performed well in the promotion process and asking them how they prepared. An officer can often help fire fighters succeed by identifying successful practices.

Negative Discipline: Correcting Unacceptable Behavior

Whereas positive discipline is directed toward encouraging desirable behavior and high performance, negative discipline is aimed at discouraging unacceptable behavior and poor performance. Negative discipline is a stronger force that sometimes needs to be applied to varying degrees.

Sometimes, an effort that begins with positive discipline (motivation, training, and coaching) evolves into negative discipline because of the continuing inability or unwillingness of a fire fighter to meet the required performance or behavioral expectations. If an individual does not respond to positive efforts to correct a problem, the next logical step is to punish continuing unsatisfactory performance. **Progressive negative discipline** moves from mild to more severe punishments if the problem is not corrected. The ultimate goal is still to improve fire fighter performance. In an extreme case, in which progressive discipline fails to solve the problem, it may become necessary for the organization to terminate an employee who is ineffective and unwilling to improve.

The disciplinary process is designed to be fair and well documented. Typical steps in a progressive negative discipline system could be to:

- Counsel the fire fighter about poor performance and ensure that he or she understands the requirements. Ascertain whether there are any issues contributing to the poor performance that are not immediately obvious to the supervisor. Resolve these issues, if possible.
- Verbally reprimand the fire fighter for poor performance.
- Issue a written reprimand and place a copy in the fire fighter's file.
- Suspend the fire fighter from work for an escalating number of days.
- Terminate the employment of a fire fighter who refuses to improve.

Some employee behaviors require the fire officer to immediately implement negative discipline. This is a common policy when there is willful misconduct, as opposed to inadequate performance. Personnel regulations usually provide a list of behaviors that will lead to immediate negative discipline, such as:

- Knowingly providing false information affecting an employee's pay or benefits or in the course of an administrative investigation
- Willfully violating an established policy or procedure
- Being convicted of a criminal offense that affects the ability of the employee to perform his job
- Displaying insubordination
- Behaving in a careless or negligent way that leads to personnel injury, property damage, or liability to the municipality
- Reporting to work when under the influence of alcohol or a controlled substance
- Misappropriating fire department property or funds

As the penalty increases, negative discipline requires the participation of higher levels of supervision. A fire officer is usually encouraged to consult with his or her supervisor before issuing any formal negative discipline. Here is an example of how the negative discipline process might proceed:

- **Informal oral or written reprimand.** Issued by a fire officer. Remains at the fire station level. Expires after a time period no longer than 1 year.
- **Formal written reprimand.** Initiated by a fire officer (usually after consulting with his/her supervisor). Some organizations require a battalion chief or other command officer to issue a written reprimand. A copy of the letter goes into the individual's official personnel file; however, it expires after a set time period, usually 1 year.
- **Suspension.** May be initiated and/or recommended by a fire officer. The suspension notice is usually issued by a battalion chief or a higher-level command officer after consultation with the fire chief or designate. The record of a suspension remains permanently in the employee's official file. Occasionally, the record is removed after a grievance, arbitration, civil service hearing, or lawsuit.
- **Termination.** Although a termination is usually recommended by a lower-level command officer, the fire chief issues the formal termination notice after consulting with the personnel office.

Oral Reprimand, Warning, or Admonishment

An **oral reprimand, warning, or admonishment** is the first level of negative discipline, considered "informal" by many organizations. By informal, it means that this discipline stays with the fire officer and does not become part of the employee's official record. For example, a new fire officer observes that the apparatus driver does not comply with the department regulation requiring that units responding to an emergency come to a complete stop when encountering a red traffic light. When this occurs, the fire officer should have a private face-to-face meeting with the apparatus driver. In this meeting, the fire officer would determine why the driver did not stop at a red

traffic light and would clearly state the policy and expectation for all subsequent responses. If the fire officer determines the reason was willful noncompliance with the department regulation, he or she would use the following steps:

- Tell the driver that the officer expects compliance with the regulation requiring that rigs stop at red lights before proceeding through an intersection. Provide the driver with a printed copy of the regulation.
- Inform the driver that this is a verbal reprimand and that the fire officer will maintain a written record of the reprimand for whatever time period is required by the authority having jurisdiction. (A common practice is to maintain the written record for 365 days. At the end of that period, if there have been no further problems related to the same individual or issue, the paper record is removed and destroyed.)
- Inform the driver that continued failure to comply with the regulation will result in more severe discipline.

In most situations, these actions correct the fire fighter's behavior. If the fire fighter continues to have difficulty complying with the regulation, this is the first step in progressive negative discipline.

Informal Written Reprimand

Some fire departments require the fire officer to use a standard form when issuing an oral reprimand, warning, or admonishment. The form covers the three items described earlier and provides a space for the employee to respond to the reprimand. The form ensures that the fire officer covers all of the requirements of an informal reprimand. In addition, the form allows the fire fighter to clearly understand that this is a disciplinary issue and to have an opportunity to respond. This type of contemporaneous record is often valued by the personnel office if the issue becomes a grievance or results in a civil service hearing.

Formal Written Reprimand

A **formal written reprimand** represents an official negative supervisory action at the lowest level of the progressive discipline process. Even when the department requires a command-level officer to issue a formal written reprimand, the document is often prepared by a fire officer. The fire officer is the closest person to the issue and will be involved in the remediation effort. The fire officer should consult the personnel regulations and departmental guidelines when preparing a written reprimand. In departments in which a fire officer can issue a written reprimand, the fire officer's supervisor should review and approve the document before it is issued to the fire fighter.

The written reprimand should contain the following:

- Statement of charges in sufficient detail to enable the fire fighter to understand the violation, infraction, conduct, or offense that generated the reprimand.
- Statement that this is an official letter of reprimand and will be placed in the employee's official personnel folder.

- List of previous offenses in cases in which the letter is considered a continuation of progressive discipline.
- Statement that similar occurrences could result in more severe disciplinary action, up to and including termination.

A written reprimand starts the formal paper trail of a progressive disciplinary process (▼ **Figure 8-9**). If the behavior is not repeated, most written reprimands are removed from the files after a designated period of time, usually 1 year after the reprimand is issued.

This written reprimand has a wide potential audience. If the reprimand is appealed or becomes part of a larger dis-ciplinary action, the fire chief, labor representatives, civil service commissioners, and attorneys read the reprimand. Fire officers must focus on the work-related behaviors and clearly explain the behavior or action that generated the reprimand.

Suspension

Suspensions are the next step of a progressive negative discipline path but have many different forms. A **suspension** is a negative disciplinary action that removes a fire fighter from the work location and prohibits the fire fighter from performing any fire department duties. A disciplinary suspension usually

---

Metro City Fire Department

Memorandum

To: Fire Fighter Joseph Knowles   March 4, 2005
        Engine 6, Green Platoon

From: Lieutenant Peter Schwartz
        Engine 6, Green Platoon

Subject: Written Reprimand: Driving through red light intersections

On March 2, 2005, at 20:12 hours you were driving Engine 6 and responding to a reported house fire at 9604 Daylilly Drive, incident #3411. You were responding north on Waterway Boulevard and approaching Edgerly Street when the traffic signal turned red for Waterway. While you slowed down, you did not come to a complete stop for the red traffic signal, as required in Standard Operating Procedure 3.4.1.b: *Emergency Vehicle Response*.

This is the fourth time you have failed to come to a complete stop when encountering a red light intersection when responding to an emergency.

Feb 11:   Verbal reprimand by Lt. Schwartz for incident #1123
Dec 24: Counseling by Captain Ortega, Engine 4, for incident #0034
Oct 09:   Counseling by Lt. Schwartz for incident #2173

Continued failure to come to a complete stop at red light intersections may result in more severe proposed discipline, including suspension or termination. This reprimand will remain in your permanent personnel folder for 365 days.

Chapter 8 of the Metro City Personnel Regulations explains your rights and options if you wish to appeal this discipline.

I have read and received a copy of this written reprimand:

_____
Fire Fighter Joseph Knowles   Date

cc: Personnel File
       Battalion Chief George Anders
       3rd Battalion, Green Platoon

Figure 8-9 Sample written reprimand.

results from a willful violation of a policy or procedure or another specific act of misconduct. For career fire fighters, suspension is a personnel action that places an employee on a leave-without-pay (LWOP) status for a specified period. For volunteer fire fighters, a suspension means that they are not allowed to respond to emergencies and, in some cases, are prohibited from entering the fire station or participating in other fire department activities. Suspensions usually run from 1 to 30 days.

A fire officer must usually recommend a suspension action to a higher-level officer and provide the required documentation. In some organizations, a fire officer has the authority to immediately suspend a fire fighter for the balance of a work period, pending a formal disciplinary action. Depending on the nature and the severity of the offense, career fire fighters can also be suspended with pay or placed on restrictive duty while an administrative investigation is being conducted. <u>Restrictive duty</u> is usually a work assignment that isolates the fire fighter from the public, often an administrative assignment away from the fire station environment.

In a few situations, the municipality can suspend a fire fighter without pay before an investigation is completed. This can occur when the employee is being investigated by the fire department or a law enforcement agency for an offense that is reasonably related to fire department employment, or when an employee is waiting to be tried for an offense that is job related or a felony.

Termination

<u>Termination</u> means that the organization has determined that the employee is unsuitable for continued employment. In general, only the top municipal official, such as the mayor, county executive, city manager, or the civil service commission, can terminate an employee. Many senior municipal officials delegate this task to the agency or department heads, such as the fire chief. Terminations are usually high-stress events, with labor representatives, the personnel office, and the fire chief's office all involved in the process.

Predetermined Disciplinary Policies

Some common disciplinary issues may have a predetermined policy for specific offenses. For example, here is how one department deals with fire fighters who are late reporting to work, without mitigating circumstances:

- **First offense.** Oral admonition and LWOP covering the period of time employee was late. Admonition remains active for 366 days with immediate supervisor.
- **Second offense** (within 366 days of first offense). Written reprimand and LWOP for the period of time employee was late.
- **Third offense** (within 366 days of second offense). Suspension for 1 workday (8 hours for day-work staff or 12 hours for shift workers).

- **Fourth offense** (within 366 days of third offense). Suspension for 4 workdays (32 hours for day-work staff, 48 hours for shift workers).
- **Fifth offense** (within 366 days of fourth offense.) Suspension for 10 workdays (80 hours for day-work staff, 120 hours for shift workers).
- **Sixth offense** (within 366 days of fifth offense.) Proposed termination.

Alternative Disciplinary Actions

Depending on the nature and the severity of the offense, a variety of penalties can be imposed, such as:

- **Extension of a probationary period.** If the fire fighter is in a probationary period, as occurs with a recruit or after a promotion, the probationary period can be extended until the work performance issue is resolved.
- **Establish a special evaluation period.** An incumbent fire fighter might be given a <u>special evaluation period</u> to resolve a work performance/behavioral issue. For example, a fire fighter who fails a required recertification examination could be placed in a special evaluation period until he or she is recertified.
- **Involuntary transfer or detail.** An <u>involuntary transfer or detail</u> occurs when a fire fighter is transferred or detailed to a different or a less desirable work location or assignment.
- **Make financial restitution.** For example, the fire fighter might be required to pay the insurance deductible after a property damage incident. One department has $2,500 deductible insurance policy on fire and EMS vehicles and equipment. If the fire fighter is judged to be responsible for a department vehicle crash, he or she is required to repay the deductible amount. The restitution payment period can take up to a year through payroll deduction.
- **Loss of leave.** A fire fighter might lose annual or compensatory leave. This is equivalent to the practice of paying a cash fine.
- **Demotion.** A <u>demotion</u> occurs when an individual is reduced in rank, with a corresponding reduction in pay. Demotions are more common in the supervisory ranks.

Predisciplinary Conference

In most cases, a predisciplinary conference or hearing must be conducted before a suspension, demotion, or involuntary termination can be invoked. The degree of investigative effort and the opportunities for fire fighter response before these punishments are issued are set higher than for less severe levels of discipline. Most fire departments require a formal disciplinary hearing before a suspension is issued, in order to provide an opportunity for the fire fighter to formally respond to the charges.

Known as a **Loudermill hearing**, this process resulted from a 1985 U.S. Supreme Court decision in *Cleveland Board of Education vs Loudermill*. The Board of Education hired James Loudermill as a security guard in 1979. Loudermill received a letter from the Board on November 3, 1980 dismissing him because of dishonesty in filling out his employment application. He had not listed a 1968 felony conviction of grand larceny on his application. The U.S. Supreme Court combined the Loudermill case with a similar appeal filed by Richard Donnelly, a Parma Board of Education bus mechanic fired in August 1997 after failing an eye examination.

The court ruling indicated that a **pretermination hearing**, including a written or oral notice, in which the employee has an opportunity to present his or her side of the case, and an explanation of adverse evidence were essential to protecting the worker's due process rights. The hearing is a check against a possible mistaken decision, to determine whether there are reasonable grounds to believe that the charges against the employee are true and whether they support the proposed action.

Not all disciplinary actions require a Loudermill hearing before suspension. The rules vary from state to state. In *Gilbert vs Homar*, the U.S. Supreme Court ruled in 1997 that East Stroudsburg University did not violate the constitutional due process rights of a university police officer when it immediately suspended him without pay after learning he had been charged with a drug felony.

A predisciplinary hearing can be conducted by a disciplinary board, by the fire chief or another ranking officer, or by a hearing officer. The designated individual or board reviews the case and makes a recommendation to the fire chief. Before the hearing, the fire fighter receives a letter that outlines the offense and the results of the investigation. The proposed duration of the suspension without pay could also be stated in the letter.

Here is a typical example of the process:

- Fire officer investigates alleged employee offenses promptly and obtains all pertinent facts in the case (time, place, events, and circumstances) including, but not limited to, making contact with persons involved or having knowledge of the incident.
- Fire officer prepares a detailed report outlining the offense, the circumstances, the individual's related prior disciplinary history, and recommended disciplinary action. The report is submitted to a higher-ranking officer in the chain of command.
- Fire department representative consults with the Human Resources Director or his/her designee if necessary when suspensions are contemplated.
- Disciplinary board hearing is scheduled, and an advance notice letter is prepared.
- Disciplinary board considers the charges and hears the employee's response.
- Disciplinary board makes a recommendation to the fire chief, who issues the final decision.

At the hearing, the accused individual has an opportunity to refute the charge or present additional information about the mitigating circumstances. In most career fire departments, a union representative is present at the hearing to advise the individual. An alternative or lower level of discipline might be proposed, based on past department practices. The disciplinary board makes a final recommendation to the fire chief.

The fire fighter might or might not have the ability to appeal the disciplinary action through the grievance procedure, depending on the personnel rule of the organization. The final resolution of a disciplinary action usually resides with the civil service commission or the city personnel director. Some municipalities use arbitration to resolve these issues.

Documentation and Record Keeping

Municipal personnel rules usually require all of the official records of an employee's work history to be in a secured central repository. This repository includes all of the official documents accumulated during employment, including:

- Hiring packet (application, Candidate Physical Aptitude Test score, and medical examination results)
- Tax withholding, I-9 status, and insurance forms
- Personnel actions (changes in rank and pay)
- Evaluation reports (probationary, annual, and special)
- Grievances
- Formal discipline

Some fire departments maintain a second personnel file at fire headquarters. That file may include letters of commendation, transfer requests, protective clothing record, work history, copies of certifications, and other fire department specific information that does not need to be kept in the secured personnel file.

Employee Assistance Program

An **employee assistance program (EAP)** is designed to deal with issues such as substance abuse, emotional or mental health issues, marital and family difficulties, or other difficulties that affect job performance. EAPs helps the employees cope with underlying issues that might be affecting workplace performance. Fire department EAPs are comprehensive programs that deal with a wide range of issues that can affect fire fighters. When an EAP is available, it is a resource that fire fighters can turn to when in crisis.

One important characteristic of an EAP is the ability to maintain the value to the organization of highly trained emergency service professionals. Earlier in this chapter, a six-step progressive discipline process, followed by one department to handle chronic tardiness, was discussed. If the fire fighter is unable to correct this behavioral problem, and the process is followed, then termination is inevitable. The underlying problem could be an off-the-job issue that the individual cannot solve without assistance. Termination of an otherwise good employee would be a tremendous waste in resources because it could cost the department up to $70,000 to find and train a replacement. If EAP involvement can help solve the problem, it is worth the effort.

Consider some of the reasons a fire fighter would continue to report late for work:

- Child or elder care issues; employee is late due to unanticipated coverage problems
- Family crisis, such as a divorce proceeding or a dying family member
- Alcoholism or substance abuse
- Coming from a second job that is needed to handle a financial crisis, such as a child with a significant health problem not covered by insurance, a crushing debt, or gambling losses
- Employee is suffering from a psychological condition or chemical imbalance

Often, these situations cause stress for the employee. A referral to the EAP may assist in addressing the issues. For an EAP to be successful, the fire officer must be able to recognize stress in an employee. Signs that may be noticed at work could include absenteeism, unexplained fatigue, memory problems, irritability, insomnia, increased use of products with caffeine/nicotine, withdrawal from the crew, resentment towards management/coworkers, stress related illnesses, moodiness, or weight gain or loss. Although any of these individually may not indicate the employee is stressed, multiple signs may indicate that the fire officer should ensure that the employee is aware of the EAP.

If the stressful situation goes unresolved, it can seriously affect performance on the job. Obviously, an employee that is absent reduces the efficiency of a fire company. However, more subtle are the symptoms of stress such as physical exhaustion from stress related fatigue or lack of sleep. The fatigue or the indecision caused by it can have serious safety consequences to every person on scene. It may also affect the group dynamics of the crew. A crew member that is stressed may not intend on taking it out on the crew at work, but the tension created is often very evident to the rest of the company.

The goal of the EAP program is to provide counseling and rehabilitation services to get the employee back to full productive duty as soon as possible. Fire department EAP programs have been successful in lowering employee turnover and reducing absenteeism, tardiness, accidents, and injuries. In addition, there are fewer employee grievances and severe disciplinary actions when an EAP program is in place.

Successful EAP programs place a high value on confidentiality and require that fire fighters enter the program voluntarily. Although a fire officer can recommend or suggest that a fire fighter consider seeking assistance from an EAP program, the fire officer cannot know the details of any fire fighter/EAP interaction. The fire officer's focus is on the fire fighter's job performance.

You Are the Fire Officer: Conclusion

You arrange to have a private face-to-face meeting with Fire Fighter Franks once he reports to work. You review the Work Hours regulation and show where he is in the progressive negative discipline procedure, pointing out that he will receive 48 hours of leave without pay and could lose his job if he is late one more time. You ask Franks what is causing his continuing difficulty in reporting for work.

Franks explains that he is recently separated from his wife and is a single parent. Sometimes, the babysitter does not show up on time. This morning she called at the last minute to say that she was sick, and Franks had to find a substitute. He lost track of time and apologizes for calling 5 minutes before he was supposed to be on duty at the fire station.

You appreciate Franks' situation, but you explain the vital importance of Franks' showing up on time or, failing that, calling in sick no later than 6:00 A.M. You provide him with information about the EAP and suggest that it may be able to provide some helpful ideas. You conclude the meeting by reinforcing the repercussions if Franks shows up or calls in late a fourth time.

Wrap-Up

Ready for Review

- Fire fighter evaluations are an ongoing process throughout the year. The annual performance evaluation is a formal written documentation of the fire fighter's performance during the rating period.
- New fire fighters should receive a copy of the classified fire fighter job description.
- The fire officer is responsible for overseeing any structured probationary fire fighter in-field training and experience program.
- Performance feedback is most effective when delivered as soon as possible after an action or incident. It should be done in private and specific enough for the fire fighter to understand what needs to improve.
- Fire officers should maintain an extemporaneous performance log or T-account on every fire fighter under the officer's command.
- Informal work performance reviews should be scheduled every 2 months with each fire fighter under the fire officer's command.
- A formal midyear review includes a fire fighter prepared self-evaluation that is reviewed by the fire officer.
- Some municipalities require that an employee receive a formal notification if the annual evaluation is below "satisfactory." Usually issued 10 weeks before the annual evaluation is due.
- Four weeks before the end of the evaluation period, the fire fighter completes another self-evaluation that is reviewed by the fire officer. Together, they develop the final, formal evaluation report for the rating period.
- Progressive discipline is a process for dealing with job-related behavior that does not meet expected and communicated performance standards.
  - Positive discipline makes fire fighters strong and confident.
  - Negative discipline is aimed at correcting individual behavior.
- An oral reprimand, warning, or admonishment is an "informal" negative discipline, which stays at the fire officer level and expires after a year.
- Written reprimand requires review by the fire officer's supervisor, and is placed in the fire fighter's personnel folder for a specified period of time. This starts the paper trail of progressive discipline.
- Suspension is a personnel action that places a career fire fighter on a leave-without-pay (LWOP) status for a specified period of time.
- Negative disciplinary actions may include:
  - Financial restitution
  - Extension of a probationary period
  - Involuntary transfer or detail
  - Loss of annual or compensatory leave
  - Establishment of a special evaluation period
  - Demotion
- An EAP provides counseling and rehabilitation services to get the fire fighter back to full duty as soon as possible.

Hot Terms

Central tendency An evaluation error that occurs when a fire fighter is rated in the middle of the range for all dimensions of work performance.

Contrast effect An evaluation error in which a fire fighter is rated on the basis of the performance of another fire fighter and not on the classified job standards.

Demotion A reduction in rank, with a corresponding reduction in pay.

Discipline A moral, mental, and physical state in which all ranks respond to the will of the leader.

Employee assistance program (EAP) An employee benefit that covers all or part of the cost for employees to receive counseling, referrals, and advice in dealing with stressful issues in their lives. These may include substance abuse, bereavement, marital problems, weight issues, or general wellness issues.

Formal written reprimand An official negative supervisory action at the lowest level of the progressive disciplinary process.

Frame of reference An evaluation error in which the fire fighter is evaluated on the basis of the fire officer's personal standards instead of the classified job description standards.

Halo and horn effect Evaluation error in which the fire officer takes one aspect of a fire fighter's job task and applies it to all aspects of work performance.

Involuntary transfer or detail A disciplinary action in which a fire fighter is transferred or assigned to a less desirable or different work location or assignment.

Loudermill hearing A predisciplinary conference that occurs before a suspension, demotion, or involuntary termination is issued. Term refers to a Supreme Court decision.

Oral reprimand, warning, or admonishment The first level of negative discipline. Considered informal, it remains with the fire officer and is not part of the fire fighter's official record.

Performance log Informal record maintained by the fire officer listing fire fighter activities by date and with a brief description. Used to provide documentation for annual evaluations and special recognitions.

Personal bias An evaluation error that occurs when the evaluator's personal bias skews the evaluation such that the classified job knowledge, skills, and abilities are not appropriately evaluated.

Pretermination hearing An initial check to determine if there are reasonable grounds to believe that the charges against the employee are true and support the proposed termination.

Progressive negative discipline A process for dealing with job-related behavior that does not meet expected and communicated performance standards. Discipline increases from mild to more severe punishments if the problem is not corrected.

Recency An evaluation error in which the fire fighter is evaluated only on recent incidents rather than on the entire evaluation period.

Fire Officer in Action

As a new fire officer, you are beginning to work on your first round of annual performance evaluations. You have never before given a performance evaluation and the entire process seems a little overwhelming. The biggest concern is that you want to ensure that you complete them correctly and conduct the process in a fair manner while providing valuable feedback to your personnel.

1. What is one common error that can occur with employee evaluations?
   A. Central tendency
   B. Discipline
   C. Empowerment
   D. Providing feedback

2. An oral reprimand is considered part of what type of discipline?
   A. Positive
   B. Constructive
   C. Demonstrative
   D. Negative

3. If a fire officer was evaluating a fire fighter that was also a close friend, what could be construed by others in regard to the evaluation process?
   A. Central tendency
   B. Halo and horn effect
   C. Contrast effect
   D. Personal bias

4. When should correction of an unsafe act be made?
   A. After training
   B. After the incident
   C. Immediately
   D. Next shift

Hot Terms (cont'd)

Restrictive duty A temporary work assignment during an administrative investigation that isolates the fire fighter from the public and usually is an administrative assignment away from the fire station.

Special evaluation period A designated period of time when an employee is provided additional training to resolve a work performance/behavioral issue. The supervisor issues an evaluation at the end of the special evaluation period.

Suspension A negative disciplinary action that removes a fire fighter from the work location; he or she is generally not allowed to perform any fire department duties.

T-account A documentation system, similar to an accounting balance sheet, listing credits and debits, in which a single sheet form is used to list the assets on the left side and liabilities on the right side, appearing like a letter "T."

Termination The organization has determined that the member is unsuitable for continued employment.

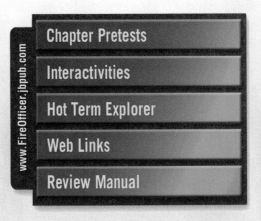

Leading the Fire Company

NFPA 1021 Standard

Fire Officer I

4.2.1 Assign tasks or responsibilities to unit members, given an assignment at an emergency operation, so that the instructions are complete, clear, and concise; safety considerations are addressed; and the desired outcomes are conveyed.

(A) Requisite Knowledge. Verbal communications during emergency situations, techniques used to make assignments under stressful situations, and methods of confirming understanding.

4.2.2 Assign tasks or responsibilities to unit members, given an assignment under nonemergency conditions at a station or other work location, so that the instructions are complete, clear, and concise; safety considerations are addressed; and the desired outcomes are conveyed.

(A) Requisite Knowledge. Verbal communications under nonemergency situations, techniques used to make assignments under routine situations, and methods of confirming understanding.

(B) Requisite Skills. The ability to issue instructions for frequently assigned unit tasks based on department policy.

Fire Officer II

5.2.1 Initiate actions to maximize member performance and/or to correct unacceptable performance, given human resource policies and procedures, so that member and/or unit performance improves or the issue is referred to the next level of supervision.

(A) Requisite Knowledge. Human resource policies and procedures, problem identification, organizational behavior, group dynamics, leadership styles, types of power, and interpersonal dynamics.

Knowledge Objectives

After studying this chapter, you will be able to:

- Describe leadership styles.
- Describe how to motivate.
- Describe leadership in routine situations.
- Describe emergency scene leadership.
- Describe the fire officer challenges in the 21st century.
- Compare and contrast the fire station as a municipal work location versus a fire fighter home.

Skills Objectives

There are no skills objectives for this chapter.

You Are the Fire Officer

You are a newly promoted captain and are assigned to the legendary "Three House." Most of the fire fighters who previously worked at Engine and Tower 3 have retired or were recently promoted. Your eight-person work group includes a new lieutenant, a new apparatus driver/operator, and three fire fighters with less than 2 years experience. Truck 3's apparatus operator has 15 years working at Station 3. The senior fire fighter on Engine 3 has been there for 7 years and is ranked #2 on the lieutenant's eligible list.

The Three House had a reputation of aggressive firefighting, accumulating many unit citations and individual commendations. The workload remains, but the crew is new. It is your job to lead them.

1. How can you develop trust with a new work force?
2. How can you develop the new work group into an effective firefighting work force?

Leadership

<u>Leadership</u> is a complex process by which a person influences others to accomplish a mission, task, or objective and directs the organization in a way that makes it more cohesive and coherent. An alternative way to describe leadership is the art of getting someone else to do what you want them to do but they do it because they want to do it. A person carries out this process by applying his or her leadership attributes (belief, values, ethics, character, knowledge, and skills). Although your position as a fire officer gives you the authority to accomplish certain tasks and objectives in the organization, this power does not make you a leader—it simply makes you the supervisor. Bosses tell people to accomplish a task or objective, whereas leadership makes them *want* to achieve high goals and objectives. Your goal should be to lead, not just to be a boss.

The Fire Officer as a Follower

Leaders can be effective only to the extent that others are willing to accept their leadership. This is sometimes described as <u>followership</u>. It is self-evident that the leader cannot lead if others will not follow. An effective leader uses persuasiveness and motivation to overcome resistance. In some cases, for a variety of reasons, others simply refuse to follow.

The fire officer has to be both a leader and a follower. This means that the officer leads the fire company to achieve the goals and objectives that have been established by the department or the jurisdiction. The officer has to follow leadership that comes from a higher level, even if it is not always pointing in a direction where the officer would prefer to go. In many cases, the officer is the messenger who has to deliver unpopular news to the company members and then provide the leadership to ensure compliance with instructions that came from a higher level. Chapter 3, *Fire Fighters and the Fire Officer,* discussed how the fire officer handles an unpopular order.

Followership is particularly important for a fire officer because subordinates are always aware of what the fire officer does. If the fire officer demonstrates selective following of orders from the fire chief, deciding which ones to follow and which ones to ignore, the officer sends a clear message to the company members: It is acceptable for the fire fighters to be selective in following orders from their fire officer.

Leadership Styles

There are many ways to look at leadership, including how to describe a leader, what leaders do, and how leaders act. One approach to better understanding a leader is to look at the different styles of leadership. Originally, most believed that the style a leader used did not change; however, today most agree that an effective leader changes the style of

Company Tips

Transitioning From Participatory to Autocratic

The increase in specialized and highly technical emergency services creates a unique challenge for the fire officer. The best way to size up a complicated situation and develop an incident action plan is to use the democratic style, using the knowledge and experience of all responders to develop the best approach. Once the operation begins, the fire officer needs to assume the autocratic role to ensure the safe execution of the plan.

leadership based on the specific situation. Three leadership styles are traditionally identified: autocratic, democratic, and laissez-faire.

Autocratic

This iron-hand approach is used when the fire officer needs to maintain high personal control of the group. The fire officer is telling subordinates what to do and is expecting immediate and complete adherence to the issued instructions. This style is required in two situations. The first situation occurs when the fire company is involved in a high-risk, emergency scene activity, such as conducting a primary search during a structure fire (▼ **Figure 9-1**). There is no time for discussion, and this is not the situation to experiment with alternative approaches.

The second situation occurs when the fire officer needs to take immediate corrective supervisory activity, such as during a "control, neutralize, command" response to a confrontation. In this case, the officer has to be firmly in control of the situation.

Democratic

This consultative approach uses all of the ingenuity and resourcefulness of the group in determining how to meet an objective or complete a task (▶ **Figure 9-2**). The officer should use the democratic style when planning a project or developing the daily work plan of the company. This approach can also be used in some low-risk emergency scene operations.

Specialized and highly technical fire companies often use the democratic approach when faced with a complex or unusual emergency situation. They depend on the skills and experience of the individual team members to analyze the situation, consider the alternatives, and develop the incident

> ## Company Tips
>
> A goal of an effective fire officer is to push decision making to the lowest possible level. Because a fire officer gains experience as a leader and builds confidence and trust with a team, many routine activities can be delegated. For example, a fire fighter could be assigned to manage in-station training activities, another could oversee routine station maintenance, and a third could keep preincident plan files updated. The fire officer continues with the morning line-up and evening check-up, but the individual fire fighters perform their assignments semiautonomously. This allows the officer to focus on activities that require his or her personal attention.

action plan. Execution of the plan often involves an autocratic command style.

Laissez-faire

This free-reign style moves the decision making from the fire officer to the individual fire fighters. The fire officer depends on the fire fighters' good judgment and sense of responsibility to get things done within basic guidelines. This is an effective leadership style when working with experienced fire fighters and when handling routine duties that pose little personal hazard.

Motivation

One of the key components of leading is the ability to motivate. Chapter 4, *Understanding People,* introduced some of the background for motivation. The fire officer must be able to apply methods to inspire subordinates to achieve their

Figure 9-1 An autocratic approach to leadership is required in a high-risk, emergency scene activity.

Figure 9-2 The democratic approach uses the resourcefulness of the group in determining how to meet an objective.

maximum potential. Although each method has a slightly different focus, the overriding principles are:

- Recognize individual differences
- Use goals
- Ensure goals are perceived as attainable
- Individualize rewards
- Check for system equality

Reinforcement Theory

Probably the best-known motivational theory is the reinforcement theory. This theory suggests that behavior is a function of its consequences. To motivate employees to perform, the officer must provide reinforcers to encourage the employee to act in the desired manner. Reinforcement must immediately follow an action in order to increase the probability that the desired behavior will recur. There are four types of reinforcers:

- **Positive reinforcement**—giving a reward for good behavior
- **Negative reinforcement**—removing an undesirable consequence of good behavior
- **Extinction**—ignoring bad behavior
- **Punishment**—punishing bad behavior

Whereas positive and negative reinforcement increase the likelihood of good behavior, extinction and punishment decrease the likelihood of bad behavior. Using punishment and extinction does not guarantee good behavior but rather only reduces the specific bad behavior. It may be replaced with a different undesirable behavior.

Most fire officers quickly learn to use positive reinforcement. It is easy to pat someone on the back for a good job. Most people are also familiar with and may use punishment in order to correct an employee who does not perform as desired. Most parents can identify with using extinction. Frequently, this strategy is used when a child is acting up. The parent simply ignores the behavior in order to decrease the chance of the child doing it again.

Negative reinforcement is the reinforcer that is often underutilized. This involves removing the negative consequences of good performance. For example, if a company completes their yearly inspection list early, many departments increase the number of inspections they are required to complete the following year. In effect, the crews are punished for good behavior. Negative reinforcement would remove the undesirable consequence.

Motivation-Hygiene Theory

One method of motivation was described by Frederick Herzberg. This theory breaks the motivational process into two parts: hygiene factors and motivation factors. Hygiene factors are conditions that are external to the individual, such as pay and work conditions. Motivation factors are the individual's internally determined motivators, such as the desire for recognition, achievement, responsibility, and advancement.

Hygiene factors do not motivate individuals, but if the person is not satisfied with any of the external conditions, he or she will not be motivated. On the other hand, the motivation factors *do* motivate. In order to achieve maximal performance, the officer must ensure that the fire fighters are satisfied with their work environment, pay, supervision and company policy and then motivate them by giving recognition for performance, provide promotional opportunities, and allow for added responsibilities.

Probably the most significant lesson to be learned from this theory is that employees who are dissatisfied with external conditions will not be motivated to achieve maximal performance for the company. Alternatively, once fire fighters are satisfied with the external conditions, improving these will not increase performance. Because fire fighters are people, no two will consider the same conditions to be at the same satisfaction level. The officer must consider the individual's perception. To determine this, the fire officer must have open and honest communication with the individual fire fighters.

Goal-Setting Theory

Another method of motivating fire fighters is by the use of goal setting, which relies on the natural competitiveness of people. In this theory, the key to motivation is for the officer to set specific goals that will increase performance. The goals must be difficult, but attainable. If they are easy to achieve, there is little motivation to work hard. If they are perceived to be impossible to achieve, the fire fighter will not try hard because failure is inevitable. The fire fighter must honestly believe that performing well can result in success.

Clear, specific goals are essential for motivation. Vague goals, such as "get to the apparatus as quick as you can," do not give the fire fighter clear enough expectations to determine whether they succeeded. Instead, a goal might be to "reduce your turnout time to less than 1 minute." The fire fighter in the latter example would be much more motivated than the fire fighter in the former example.

The most significant lesson in the goal-setting theory that a fire officer can use to motivate fire fighters is to carefully consider what actions are needed to improve the organization and then to develop clear, specific, and challenging but attainable goals for the individual fire fighter.

Equity Theory

This motivational process suggests that employees evaluate the outcomes that they receive for their inputs and compare them with the outcomes others receive for their inputs.

Outcomes range from pay and benefits to recognition, achievement, and promotion. Inputs include educational level, performance level, risks taken, and special skills.

This theory supports why most people understand why the fire chief is paid more than a fire fighter or why a fire fighter is paid more than a dishwasher at a restaurant. Conversely, according to this theory, most fire fighters would not understand why they are paid in the lowest one third in a survey of comparable cities whereas the fire chief's pay is in the top one third in the survey.

If a fire officer wants to motivate the fire fighter, he or she must determine where the fire fighter believes an inequity exists. Although the fire officer would likely not have the ability to change the organization's pay structure, he or she may be able to provide other outcomes that are desired by the fire fighters. This might include a more flexible time trade policy or recognition.

Expectancy Theory

This motivational theory is based on the premise that people act in a manner that they believe will lead to an outcome they value. According to expectancy theory, there are three considerations that the fire officer must address in order to motivate the individual:

- The employee's belief that his or her effort will achieve the goal
- The employee's belief that meeting the goal will lead to the reward
- The employee's desire for the reward

Using this method, the fire officer might indicate to the fire fighter that if he or she has a turnout time of less than 1 minute for the next month, the officer will make sure that this performance is reflected on the fire fighter's evaluation, which is used for the upcoming promotional process in which the fire fighter was interested.

Leadership in Routine Situations

Most fire officer leadership activity is directed toward accomplishing routine organizational goals and objectives in nonemergency conditions. This includes being well prepared to perform in emergency situations. The fire officer accomplishes most of these goals and objectives through the efforts of fire fighters. The role of the officer involves influencing, operating, and improving to ensure that those efforts achieve the desired results.

Historically, fire officers used an autocratic style of leadership in both emergency and nonemergency duties. The officer decided what was to be done and who was to do it, and the fire fighters responded without question. Today, the work force is very different. All employees demand to be included in the decision-making process. Effective fire officers today provide a more participative form of leadership in routine activities.

Although the fire officer is commonly given specific assignments that the company must complete, there can be much discretion about when, how, and by whom activities are carried out. Many officers have found it particularly effective to sit down with their crews at the start of the shift and cover the assignments, if any, that the battalion chief has indicated need to be completed for the day. He or she may also outline the status of other duties that are long term that will need to be completed but are not due as of yet. An open question is then addressed to the crew as to what they think should be done, including any areas that they believe should be addressed. Depending on the outcome of the discussion, the officer then decides or has the group decide the plan for the day on how to accomplish the activities.

Some tasks are to be completed every day. In most departments, station cleaning is performed daily, making a trip to the grocery store is required, and equipment must be checked. The fire officer may decide that these duties are easiest if they are preassigned. Typically, in conjunction with the crew, a determination is made about what duties will be completed every day based on the position of the individual. The driver is usually assigned to check all equipment and clean the apparatus bay. The rookie fire fighter may be assigned to the bathroom and kitchen detail, whereas the senior fire fighter is assigned to sweeping and vacuuming floors.

Emergency Scene Leadership

A core responsibility of a fire officer is to handle emergencies effectively. Chapter 15, *Managing Incidents,* covers structural components of incident management in more detail. This chapter examines some of the leadership issues that apply specifically to a fire officer.

A fire officer always has direct leadership responsibility for the company that he or she is commanding. The first-arriving fire officer has additional important responsibilities to establish command of the incident and to provide direction to the rest of the responding units. The first-arriving fire officer also has leadership responsibilities that relate to the communications center.

VOICES OF EXPERIENCE

" Leadership is not awarded— it is earned. "

Being promoted to fire officer does not automatically make you a leader. Leadership is not awarded—it is earned. Leadership is earned on a daily basis, while directing routine tasks and while managing a three-alarm fire.

Leaders do not always have titles. Fire fighters can become leaders in the minds of their company because they have proven themselves to be "go-to" people. If there is a problem that needs a solution, the "go-to" fire fighter always has an answer. If 110% needs to be given, the "go-to" fire fighter gives 125%. The "go-to" fire fighter does not have to have rank, because the "go-to" fire fighter has earned the trust of the company and has obtained the company members as followers.

Many people think that management and leadership are one in the same. They could not be more wrong. Management is given. Leadership is influence that is continuously earned.

Matthew B. Thorpe
City of King, North Carolina Fire Department
12 years of service

Methods of Assigning Tasks

The fire officer's primary responsibility is for the team of fire fighters under his or her direct supervision. Those company members must function as a single team. The fire officer is responsible for their safety, their actions, and their performance at the incident scene. The fire officer should develop a consistent approach to emergency activities, in which all of the department's operating procedures are followed.

Most departments have some form of standard operating procedures (SOPs) that specify what is expected of each company and what to do in a given situation. Some are very specific and detailed, whereas others are others are more general. For departments with very detailed procedures, the task that each fire fighter is to perform may already be assigned. The SOPs may indicate what tools the fire fighter is to carry based on whether the company has three or four persons. In these cases, the fire officer is responsible for ensuring that the fire fighters know and follow the policy.

For departments that have broad standard operating procedures or none at all, the fire officer has two choices in assigning tasks: preassigning them or assigning them as needed on the scene. The advantage of preassignment is that it relieves the officer of having to make as many decisions during the emergency. The advantage of assigning tasks on the scene is that it reduces unnecessary efforts.

When giving assignments on the scene, the fire officer must be clear and concise and ensure that the fire fighter understands. In this situation, the fire officer usually uses an autocratic style of leadership because the emergency scene does not usually allow for participative decision making. For this method of operation to be effective, the fire fighter must clearly understand the reason for the different leadership styles before the incident. One method that may help a fire officer make assignments under pressure is to develop a standard method of handling situations. For example, if you need to assign someone to rescue, always consider first whether you have a rescue company available; if not, then a truck company and lastly an engine company can be used. This allows a sequential thought process to develop, making assignments easier.

A method of assigning tasks that allows for more participation in the process yet reduces the number of decisions the officer must make on scene is the broad standard operating procedures, in which the fire officer discusses with the crew the various "routine" emergencies that they respond to. The officer and crew can discuss the needs of the incident and what must be done to mitigate it. Finally, the officer and crew decide who will be responsible for doing what.

For example, the crew may decide before the incident that on any medical call that is potentially a code, the fire fighter is responsible for checking for respirations and securing an airway and then providing ventilation. The driver is responsible for checking for a pulse and uncontrolled bleeding and then providing compressions. The officer is responsible for setting up and using the semiautomatic external defibrillator. This plan ensures that the crew covers all the basics without fumbling and waiting to be directed.

Critical Situations

Dangerous situations can develop very suddenly during an emergency scene operation. A fire officer must always be prepared to implement the autocratic leadership style when immediate action is required. Orders to evacuate a building and a fire fighter Mayday are two critical events in which strong autocratic leadership would be essential.

Occasionally, the fire company encounters a dangerous emergency scene situation that requires immediate action. When command issues an evacuation order, the fire company should immediately leave the fire building. Evacuation may mean that the company leaves a fire hose in the building and the fire fighters quickly leave the building.

Safety Zone

Practice Like You Play

One of the keys to a consistent performance at emergency incidents is to "practice like you play." Every training activity and every response should be approached with the same degree of professionalism and strict adherence to standard operating procedures. This approach is essential to ensuring that the company performs as expected when the situation demands maximum performance. The fire officer should ensure that fire fighters completely and properly use personal protective gear and follow standard operating procedures at every incident.

Most structure fire responses involve relatively minor situations, such as activated fire alarms and fires that involve less than a room and contents. To prepare the fire company for larger fires and critical situations, each response should require the same consistent approach. Below are some possible examples.

- All crew members don their protective clothing, sit down, and fasten their seat belts before the apparatus moves, and they remain seated and belted as long as the apparatus is moving.
- The first-arriving engine company always lays a supply line.
- The first-arriving ladder company positions and prepares to deploy the aerial.
- Crews entering the building wear full protective gear, including SCBA; carry their tools; and perform their assignments for a working fire.

If there is any doubt about whether there is a fire in the structure, all operations should be performed with the assumption that there is. Without creating a customer service problem, the fire company should search the entire building, including attics and basements. This allows each fire fighter to develop awareness of the built environment. More importantly, when the company encounters a working fire, the standard fireground tasks have been practiced and are familiar.

The fire officer is responsible for accounting for the fire fighters under his or her supervision. The fire officer provides a head count to the sector command officer or the incident commander.

A Mayday requires a complex response from the company operating within a hot zone or burning structure. The first obligation is to maintain radio discipline so that command can determine the Mayday location and situation. The second obligation is to maintain company or sector integrity. It is an understandable desire for every fire fighter to rush to assist the fire fighter in trouble. The fire officer needs to maintain company integrity to facilitate the Mayday rescue, continue to fight the fire, or control the hazard. Changes in assignment come from the sector incident commander.

The fire officer may directly encounter a dangerous condition, such as a collapsing building, a person with a gun, or a careening vehicle. The officer must make an immediate, clear, and autocratic command to protect the fire fighters under his or her supervision by removing them from the danger area. Then, the fire officer needs to inform command.

After every incident, the fire officer should briefly review the event while still at the scene or as soon as possible after returning to quarters. This is an opportunity to clarify issues and answer questions. The fire officer can reinforce good practices and immediately identify any unacceptable performance (▼ Figure 9-3).

The Dispatch Center

One of the responsibilities of the first-arriving fire company-level officer at an incident scene is to concisely describe conditions to the dispatch center (► Figure 9-4). The content of the initial radio report needs to meet the departmental operating procedures. In addition, this verbal picture establishes the tone of the incident. A fire officer who provides a calm and complete description of the situation on arrival demonstrates leadership and ability to control the action that will occur. An officer who gives the impression of confusion, indecision, or uncertainty fails the first test of leadership.

Without a specific requirement, a radio report would include:

- Identification of the company arriving at the scene
- A brief description of the incident situation. That may include providing information about building size, occupancy, hazardous chemical release, and multiple-vehicle accident
- Obvious conditions: working fire, multiple patients, hazardous materials spill, or dangerous situation
- Brief description of action to be taken. "Engine 2 is advancing an attack line into the first floor."
- Declaration of strategy to be used: offensive or defensive
- Any obvious safety concerns
- Assumption, identification, and location of command
- Request or release resources as required

A calm, concise, and complete radio report helps the dispatch center, the chief officer, and other responding companies to understand the situation at hand. It is important to use radio terminology that meets the department's operating procedures and is clear to everyone who is listening. Close compliance with the communications standard operating

Figure 9-3 After every incident, the officer should briefly review the event while still at the scene.

Figure 9-4 The first-arriving fire company-level officer needs to concisely describe conditions to the dispatch center.

procedures is an important component of effective fire-ground command.

A visit to the dispatch center is a valuable experience for every fire officer. It is important to understand the dispatch center's operating conditions and how dispatchers depend on fire officers at the incident scene to provide good information.

Other Responding Units

The first-arriving fire officer provides leadership and direction to responding units by implementing the local jurisdiction's incident management plan. It is important to recognize the leadership component of incident command. To effectively manage an incident, the fire officer must demonstrate the ability to take control of the situation and provide specific direction to all of the units that are operating and arriving. This requires mastery of the autocratic style of leadership.

Fire Officer Challenges in the 21st Century

The fire officer works in a dynamic environment with changing conditions and evolving organizational needs. The basic leadership concepts remain, but the environment is continually changing. The fire officer faces two challenges: the fire station as a work location and the special challenges of leading a volunteer fire company.

Fire Station as Municipal Work Location Versus Fire Fighter Home

Most fire fighters do not think of the fire station simply as the place they go to work. The fire fighters in a fire station work rotating shifts together and prepare and share meals, often spending days of dull routine, punctuated by episodes of intense excitement. These factors create a powerful and special community among fire fighters that is difficult to compare with any other workplace environment. The fire station becomes a "home away from home," and the fire fighters become a type of extended family.

Although this workplace environment produces a special type of bonding among fire fighters, it can easily produce a variety of productivity problems, as well as behavioral and personality traits that are often associated with a dysfunctional family. The resulting situations can be quite different from the problems that occur in a "normal" work environment, and the behavioral issues are often more severe. Case law and administrative actions reinforce the expectation that the fire station is essentially a place of business, subject to the same rules and expectations as any other workplace. They expect the fire fighters who are in the station at 3 A.M. to behave the same way the accounts receivable staff behaves at 3 P.M.

Getting It Done

Look Official When Performing Formal Fire Officer Duties

Many departments provide two different uniforms that can be worn by the fire officer. The station work uniform could be a golf shirt, or "job" shirt. The more formal class A uniform generally includes a white shirt with insignia, badge, and tie.

One way that the fire officer can ensure that the fire fighters understand when he or she is performing an official supervisory task is to wear the formal uniform shirt. If the officer always puts on a dress shirt and tie to conduct employee evaluations or counseling sessions and to issue commendations or disciplinary actions, the serious and official nature of these activities is clearly visible. This increases the fire officer's effectiveness by alerting the fire fighters that a formal supervisory task is at hand (▼ **Figure 9-5**).

A fire officer must balance the expectations of the employer with the realities of a fire station work environment and the desire to create an effective team. A certain amount of spirited behavior can be healthy, as long as some basic rules are observed. The fire officer has to maintain order and ensure that whatever occurs can be explained and would be acceptable to a rational observer.

The fire officer should establish some house rules. Two general rules for nonemergency activities are as follows:

1. Do not compromise the ability of the fire company to respond to emergencies in its district.
2. Do not jeopardize the public's trust in the fire department.

Responding to emergencies and preserving the public trust are the two most important values that every fire fighter

Figure 9-5 Wearing a uniform alerts the fire fighters that a formal supervisory task is at hand.

needs to respect. These still provide wide latitude for team-building extracurricular activities.

Leadership in the Volunteer Fire Service

Fire officers leading volunteer fire companies have to rely on their leadership skills even more than their municipal counterparts. Pride and personal commitment are the only forces that compel a volunteer to remain active and loyal to the organization—there is no biweekly paycheck that compels a volunteer to tolerate an unpleasant situation. The volunteer fire officer must pay attention to the satisfaction level of every member and be alert for issues that create conflict or frustration.

The relationship between the fire officer and fire fighters is more like that of a mom-and-pop family business than that of a municipal agency. Effective leadership is often the strongest force that influences their performance and commitment to the organization. If the negative aspects outweigh the positive aspects, an individual can simply step aside or drop out.

Some of the unique issues that a volunteer fire officer must consider are as follows:

- Personal and family issues often affect volunteer availability. Changes in employment, family situations, or child or elder care profoundly affect the time a volunteer is able to devote to the fire department.
- Extensive training requirements can have an impact on volunteer availability. It is easy to assign a municipal employee to attend a 40-hour training class that occurs during the regular workweek. It is difficult to accomplish the same training when the opportunities are restricted to evenings and weekends.
- Interpersonal conflicts can exist between members. Conflicts can drive away members and erode fire company preparedness. The volunteer officer must act quickly when such problems are identified. The rights and reasonable expectations of individual members must be considered, without compromising on the mission or the good of the organization.

One of the issues that is a special concern in many volunteer organizations is the political balance that results from electing officers. In many volunteer organizations, the officers are elected by the members, either annually or biannually. A volunteer fire officer may have to decide between the right decision or policy and a more popular option. In a perfect world, the selection of the most appropriate policy would earn the respect of the voting membership; however, the world is not perfect. In many cases, a conscientious volunteer officer has to use strong leadership and the courage of conviction to implement an unpopular policy.

Summary

Leadership requires the fire officer to provide purpose, direction, and motivation to fire fighters. These ingredients are produced through a complex process that has to maintain balance between a wide range of factors. The fire officer applies personal attributes to the leadership process. This is a lifelong endeavor as the fire officer develops and refines the belief, values, ethics, character, knowledge, and skills that make up the sum of a person.

You Are the Fire Officer: Conclusion

You have lunch with the recently retired captain. He suggested having frequent hose-and-ladder drills to hone the crew's ability to advance hoselines and throw ladders. He said that the Three House secret has always been superior training. They trained every day in the basic fire suppression skills. He provided a name of a contractor who was redeveloping a housing complex in Station 3's district. That contractor would allow the Three House to practice their craft on the buildings scheduled for demolition. He also noted that they took pride in getting out of the station quickly on structure fire calls. He finished lunch with the recommendation that every little win should be recognized.

Wrap-Up

Ready for Review

- Effective leaders are also good followers, supporting the fire department leadership.
- Three situational leadership styles are:
  - Autocratic
  - Democratic
  - Laissez-faire
- Dangerous emergency scene situations, such as an evacuation order or a fire fighter Mayday, require an immediate and autocratic response from the fire officer.
- Leadership in routine situations is accomplished through influencing, operating, and improving.

- Fire officer house rules are as follows:
  - Do not affect the ability of the fire company to respond to emergencies in their district.
  - Do not jeopardize the public trust of the fire department.

Hot Terms

Followership Leaders can be effective only to the extent that followers are willing to accept their leadership.

Leadership A complex process by which a person influences others to accomplish a mission, task, or objective and directs the organization in a way that makes it more cohesive and coherent.

Your company responds to a serious car accident with multiple injuries. Two of the three occupants perish from their injuries. The third occupant, an infant, is airlifted to the nearest trauma facility and is not expected to live. Your company takes the incident very hard as several members of your company have children of their own. On the way back to the station everyone is very quiet and there is not much discussion for the rest of the afternoon.

1. Because fire stations are like second homes, what are the best ways to deal with issues that may arise on or after stressful incidents so they do not become issues that affect the mood and bond of the crew?

   A. Not bringing them up is the best way, talking about things that have obviously bothered the crew is just another way to add tension

   B. The shift should make the decision. Everyone has different ways of dealing with stress and the crew being a tight-knit group should respect that some individuals will want to talk about the incident while others will not

   C. By being forced to talk about the issues in a roundtable-like discussion

   D. Ignoring the issues if they are brought up

2. What is one of two general house rules that should be followed around the station?

   A. Don't jeopardize public trust

   B. It's okay to horseplay after the chief leaves

   C. Give people space

   D. Do everything as a group

3. The fire officer works in a dynamic environment with changing conditions. What is one of the two challenges that fire officers deal with today?

   A. Dealing with the chief

   B. Fire station as a work location

   C. Personal strength

   D. Integrity

4. What is one consideration that a volunteer fire officer must think about?

   A. Volunteers are unmotivated

   B. Personal and family issues often affect volunteer availability

   C. Personnel expect more

   D. Personnel are not as well trained

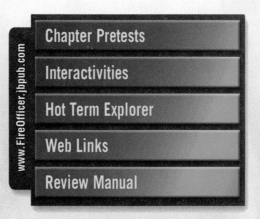

Chapter Pretests

Interactivities

Hot Term Explorer

Web Links

Review Manual

NFPA 1021 Standard

Fire Officer I

4.3 **Community and Government Relations.** This duty involves dealing with inquiries of the community and projecting the role of the department to the public and delivering safety, injury, and fire prevention education programs, according to the following job performance requirements.

4.3.1 Initiate action on a community need, given policies and procedures, so that the need is addressed.

(A) Requisite Knowledge. Community demographics and service organizations, as well as verbal and nonverbal communication.

(B) Requisite Skills. Familiarity with public relations and the ability to communicate verbally.

4.3.3 Respond to a public inquiry, given policies and procedures, so that the inquiry is answered accurately, courteously, and in accordance with applicable policies and procedures.

(A) Requisite Knowledge. Written and oral communication techniques.

(B) Requisite Skills. The ability to relate interpersonally and to respond to public inquiries.

4.3.4 Deliver a public education program, given the target audience and topic, so that the intended message is conveyed clearly.

(A) Requisite Knowledge. Contents of the fire department's public education program as it relates to the target audience.

(B) Requisite Skills. The ability to communicate to the target audience.

Fire Officer II

5.3 **Community and Government Relations.** No additional requirements at this level.

5.4 **Administration.** This duty involves preparing a project or divisional budget, new releases, and policy changes, according to the following job performance requirements.

5.4.4 Prepare a news release, given an event or topic, so that the information is accurate and formatted correctly.

(A) Requisite Knowledge. Policies and procedures and the format used for news releases.

(B) Requisite Skills. The ability to communicate orally and in writing.

Knowledge Objectives

After studying this chapter, you will be able to:

- Discuss community demographics.
- Discuss risk reduction.
- Discuss public inquiries.
- List four objectives of a public safety education program.
- Describe Fire Prevention Week.
- Describe the National Fire Protection Association (NFPA) Risk Watch program.
- Describe the goal of a Community Emergency Response Team.
- Describe developing a local public education program.
- Describe the three steps NFPA recommends in developing a relationship with the media.
- Define proactive media communications.
- List the four NFPA guidelines to use when conducting an interview with the media.

Skills Objectives

After studying this chapter, you will be able to:

- Develop a public education program using a five-step program.
- Prepare a media release that conforms to local format.
- Conduct a media interview as the fire department representative.

You Are the Fire Officer

You are tired and sooty after 3 hours spent fighting an outside fire that ignited the exterior siding of a house in the suburbs of Metro City. The homeowner was astounded that the fire burned so quickly, unaware that a 4-year drought had created a tinderbox situation in the community. You took the time to explain why it is important to cut a clear area to provide proper clearance between the house and the brush, while asking yourself how many more homes will suffer a similar fate before their owners get the message.

As you arrive at the fire station, you notice a private vehicle parked at the front door. A citizen approaches as soon as you climb down from the cab and asks if you can take care of her problem right away. She has been waiting for 2 hours for someone to help her install a child car seat. She had an appointment and is unhappy because no one was at the station when she arrived.

Before you can ask your driver/operator, who is also the designated car seat technician, to help the citizen, you see the local day care center minivan arriving for a scheduled station tour. You look at the two adults and six preschoolers in the van and see how excited they are to visit a fire station.

1. What are your options for dealing with the station tour at this point?
2. How do you respond to the citizen who is unhappy about having to wait 2 hours for your return?
3. What will you do with your visitors if you get another alarm within the next 30 minutes?
4. The departmental regulations require you to be clean and dressed in uniform when conducting public education activities. At this moment you are wet and dirty; what should you do?

Introduction to Working in the Community

Fire departments generally have a close relationship with the communities that they protect. Historically, volunteer fire departments were established within local communities and were closely associated with the local area. During the 1800s, fire stations were frequently used for public meetings and assemblies, and membership in the local volunteer fire company was often a steppingstone for politicians and power brokers.

In many small towns, this type of close relationship still exists between the fire department and the community. Even in many cities where the 20th century saw the transition to career firefighting, the bonds between local communities and their fire fighters have remained strong. The apparatus doors of many fire stations are usually left open, and the proverbial outside bench is still a popular location for fire fighters to maintain close contact with the neighborhood.

The situation is quite different in some cities, where the benches have been moved inside and the apparatus doors are always closed. This is a reflection of fire companies' isolation from their communities. Urban decay and civil unrest have changed the nature of the relationship between fire departments and their communities in hundreds of cities.

Even in large cities, the fire station continues to be widely viewed as an important community member. Citizens tend to think of the fire department in relation to the local fire station. The financial crises of the 1980s showed the value of strong community ties, as residents in New York City, Baltimore, and other cities rallied to keep their local fire stations open in the face of budget cuts, measuring their personal perceptions of safety by the proximity of the closest fire station.

The trend of local government in the early 21st century is to move toward more community-based local government. Next to schools, the fire department is generally the most decentralized and community-based function of local government. Many cities are establishing their neighborhood fire stations as a primary point of contact for various local government services.

The fire officer in each fire station is the official fire department representative in the local community and also ensures that the community's needs are being addressed by the department. The fire company is also a neighbor to the community. This chapter provides the fire officer with information about how to learn more about the community, how to provide fire and life-safety education to meet local needs, and how to work with the news media to disseminate information clearly and appropriately.

Understanding the Community

It is important for a fire officer to understand the characteristics of the local community. Each community has special needs and has different characteristics that should be considered in relation to every service and program provided by the fire department. In most cases, the most significant information comes from understanding who lives or works in the community. The fire officer should work on developing a good understanding of the population and demographics of the local community. Fire officers should take the time to learn about the characteristics of the population in the particular areas where the company responds.

The federal government undertakes a nationwide census once every decade. This information is readily available and provides an excellent starting point to begin an analysis of the local community. The census collects and identifies a massive amount of information about the **demographics** of the nation. Demographic data describe the characteristics of human populations and population segments. The 1990 and 2000 census data can be analyzed with the use of sophisticated database software tools and digitized mapping to profile many characteristics of local populations down to the neighborhood level.

One event of the information age, coupled with the business boom of the 1980s, was the explosion of demographic analysis techniques. Demographic data are often used to identify and analyze consumer markets, to help retailers predict what consumers will buy, and to give advertisers insight into the messages that will be most effective in particular markets. Politicians make extensive use of demographic data to predict the voter response to policies and messages in each area. Messages and programs are fine tuned to reach certain segments of the population based on detailed analysis of what different types of people like, what they value, what they believe, and where they live. The same approach can be used to ensure that the fire department is delivering the appropriate services and information to the local community.

Fire departments must be able to communicate effectively with the citizens they serve and understand the cultural factors that influence particular behaviors. Information can be used to identify groups with special needs to improve the delivery of emergency services. Demographic information is especially valuable in producing fire prevention and public education programs to meet the needs of diverse communities. An effective program should be designed to meet the needs of the particular community, and the message has to be formulated to reach the target audience.

America is a land of immigrants that continues to become more multiethnic with each census. There is a continual flow of new immigrants into the country from many different parts of the world, bringing with them a tremendous variety of languages, cultures, religions, traditions, and beliefs. Since 1970, the number of immigrants living in the United States has tripled. There are school districts within major cities, such as New York, Los Angeles, Chicago, and Washington D.C., where the students speak more than 100 different languages.

One dimension of demographics is the identification of communities where people share the same cultural background

Company Tips

Demographics can assist the fire department in planning for new safety programs, such as child safety seat installations. In the late 1990s, public safety officials in many communities identified a need to provide public education and assistance regarding proper child car seat installations. The sudden popularity of this service overwhelmed some fire departments that were not adequately prepared.

The proper installation of a child car seat is complex. A 40-hour class is required for a public safety official to achieve certification as a child car seat technician. It makes sense to provide this training to fire fighters in the stations, where a high demand for the service can be anticipated.

The demand for this type of service is predictably concentrated in areas where there are families with automobiles and young children. The fire department can use demographic information to identify the fire stations that serve communities where there are concentrations of young families who would take advantage of the service.

Fire officers can then survey those communities to identify the days and times when the demand for the service would be highest. To meet the public's need, specific days and times might be designated to install child safety seats. Dedicated staffing could be assigned on those days to ensure that a certified child car seat technician would be available to perform the installation, regardless of the emergency workload. This information can be disseminated in press releases and posted on the department's website and on signs at every fire station (▶ Figure 10-1).

Company Tips

Se Habla Espanol?

In some areas, fire fighters have to be able to communicate in a second language to effectively provide services to their communities. In the southwestern states, there are many communities where Spanish is the primary language, and it is not unusual to encounter individuals or whole families that cannot speak English. Many fire departments encourage their members to learn additional languages, and some even provide special classes or offer pay incentives for bilingual fire fighters. At least one major department has experimented with a language immersion program, making Spanish the primary language at a designated fire station. The fire fighters assigned to that station answer the telephone, speak to each other, and handle emergencies while speaking Spanish.

City of Hazelwood

415 Elm Grove Lane
Hazelwood, MO 63042-1917
Phone: 314-839-3700
Fax: 314-839-0249

Hazelwood Fire Activities   **02/02/2003**

The Hazelwood Fire Department would like to remind the residents of our community that we continue to serve you with our new child car safety seat installation program. This is a free service residents can use to get information and assistance with learning how to properly install your child's car seat. Please understand that it is important to call ahead to make an appointment with the car seat installer for the day you would like to come in. Unfortunately, our work schedule does not always allow for "walk-ins" to be immediately accommodated. We almost always have at least one certified car seat installer on duty at any given time, so it is likely that you may be able to come in the same day as you call. It just may not be at the same time as your call or random visit. Mornings are our busiest time, but we are willing to be very flexible in scheduling installations including evenings, weekends, etc. We want you to be sure that your child is as safe as he or she can possibly be while riding in your car. And we are willing to do whatever we can to achieve that goal. Thank you.

If you would like to schedule a car seat installation appointment, call the Hazelwood Fire Department at 731-3424.

© City of Hazelwood.

Figure 10-1 Description of a car seat installation program.

and language. Within a small neighborhood, there may be thousands of people who speak one particular language and share a culture that is totally different from that of the adjoining neighborhoods. Highly diverse populations can be found in many suburban communities as well as inner cities. This type of diversity presents tremendous challenges to the fire service in trying to meet the needs of different groups.

Demographic information allows the fire officer to identify the characteristics of a community before delivering a safety message. Safety information should be applicable to the particular community and delivered in a format that the

community can understand and act on. The format and the method of delivery should be different when fire fighters are meeting with a community of new immigrant families who understand very little English versus a community of college students living in fraternity houses.

Both emergency service and public education efforts have to be fine-tuned to identify the needs and meet those needs in particular communities. Fire fighters may encounter languages, cultures, and traditions that present barriers to delivering emergency services or providing fire prevention and public education. The fast and loud aspects of a typical Western world response to a fire emergency could be overwhelming, terrifying, or embarrassing to the person who needs assistance. The customs and traditions of another culture may clash with the invasive, brash, and blunt team approach to a cardiac arrest. A shop owner may be angered by the invasion of a fire company into her shop to perform a fire safety inspection. Cultural sensitivity is required to help the customers appreciate the services the fire department is providing.

Risk Reduction

The fire service has a long and rich history of striving to reduce the risk of fire in the community. This has been achieved through emergency response, fire safety education, adoption of fire codes, and enforcement of those codes. However, most fire departments have recognized that their primary goal is to save lives and property in general, not just from fires. This broader vision has been evident as the fire service has evolved from responding only to fires and fire alarms to responding to extrications, drownings, heart attacks, strokes, hazardous materials releases, trench collapses, building collapses, high-angle rescues, swift water rescues, underwater and ice rescues, as well as weapons of mass destruction incidents.

The fire service is called to respond to each of these types of incidents. With this broadened scope, the fire service has taken on the role of prevention and mitigation of these types of incidents, much like early fire prevention efforts. The best method of preventing fire injuries and deaths is by preventing the fire or reducing the severity of the fire. The same is true for the other types of incidents to which the fire department is called. Progressive fire departments realize that it is the fire department's responsibility to save the lives and the property of its citizens from all types of incidents through prevention.

This change in thinking often requires a shift in the culture of the fire department. Many fire fighters' attitude is that they became a fire fighter to fight fires—not to check car seats. The attitude must be developed that the job of the fire fighter is to save lives and property from every type of incident to which we might be called. All too frequently, this could be a motor vehicle accident in which a child is severely injured or killed because of the lack of proper restraints. The link must be made between the incidents to which the fire department responds and preventing those types of incidents.

The fire officer is at the front line of developing this attitude. Over time, the fire company will reflect the attitude of the fire officer. When the fire company observes the fire officer making an effort to prevent all types of injuries and deaths, the fire fighters will notice and attitudes will change. When incidents are encountered that could be used to reinforce the need for risk-reduction programs, the fire officer should use them as examples. For example, the fire company responds to a child

Getting It Done

Risk-Reduction Programs

There are a wide range of risk-reduction programs currently being used in departments across the country. Each is designed to meet a community need.

- Public access to automated external defibrillators
- Bicycle helmet programs
- Residential sprinkler education
- Residential smoke alarm programs
- Antismoking education
- Carbon monoxide poisoning prevention
- Creating disaster resistant communities
- Immunization programs
- Teen driving
- Water awareness programs
- Heart attack survival kits
- Injury reduction program for seniors

If you would like more information on these or other risk-reduction programs, please request "Leading Community Risk-Reduction Innovative Approaches" from the National Fire Academy.

who dies from head injuries in a bicycle accident that could have been prevented by wearing a helmet. Real life can provide powerful lessons as to why the fire department has a responsibility to create prevention programs.

There are two levels of needs that should be addressed: those that are systemic and those that are individual. Systemic needs are needs that the fire department can address through the development of programs to eliminate or reduce risk, such as a bicycle helmet safety program directed toward children. Often, these programs are a community-wide response and involve many organizations and agencies. Many of these needs are identified at the departmental or community level rather than the fire company level; however, most fire officers can readily identify these problems. Fire officers may identify systemic needs from the calls to which the fire company responds. If a fire officer identifies preventable incidents, the fire officer should report this to fire administration for consideration.

The fire officer is also in the unique position to identify the needs of the individual. The fire department is one of the few community resources that regularly respond into the private homes of the citizens. The fire officer has the capability to recognize community needs first. These needs can vary from reporting signs of child abuse to heating assistance programs. Typically, individual needs are those that have already been identified by the community, and programs are already in place to address the need.

Although most fire officers may intuitively recognize needs, often they do not know what community assistance is available. This requires the fire officer to proactively learn what community programs are available. Many programs are run by the health department and the department of social services. Often, these departments know what assistance programs are available for low-income families and the elderly. The area safety council and hospitals are valuable resources for information on the prevention of accidents and injuries. The police department and the highway department have a vast amount of information on vehicle accident and injury prevention. Churches and civic groups also have many programs to assist in community needs.

Each community has a varying level of support for addressing community needs. A fire officer needs to seek out this information before he or she needs it. Often, brochures are a good method of having the information available when it is needed; this allows the fire officer to address the need on the scene. For example, during a call, the fire officer notices a pool with an unlocked gate. Discussing the issue with the resident and leaving a pool safety information packet may prevent the fire department from responding to a child drowning at the location. Other issues may not be able to be addressed on the scene. An elderly couple who lives alone and is unable to care for themselves will likely need to be turned over to the department of aging for a follow-up visit.

It is important to follow departmental policies for these situations. Sometimes, the fire officer is allowed to call directly to the agency that can provide the assistance. Fire officers are likelier to take action in fire departments that empower the fire officer to initiate action directly.

Responding to Public Inquiries

Sometimes, the public makes an inquiry to the fire officer, for example, for general information, such as a request for a description of the services the fire department or city provides or a request to remove a cat from a tree. Both of these situations have the potential to leave the citizen satisfied or for the citizen to go away with a negative view of the fire department.

When a citizen makes an inquiry to a fire officer, the fire officer must treat all requests professionally and with respect. Even though you may find the request less than critical, the citizen believes that it is valid and important. Failure to approach the request with sincerity can have a lasting negative impact.

Every effort should be made by the fire officer to answer each question fully and accurately. The fire officer may not know all the information that the individual wants to know. For example, the fire officer may not know where the closest flu shot clinic is located. When this occurs, the fire officer should seek out the information immediately rather than just expressing that he or she does not know. A phone call or two can usually lead to the answer and leaves the citizen with a positive image of the fire department.

Some requests that citizens make may not be within the fire officer's authority. In these situations, the fire officer

Getting It Done

"My Cat Is Stuck in the Tree. Can You Get it Down?"

Many citizens still have the image of the fire fighter rescuing a cat from a tree. Over the years, many fire departments have decided to not provide this service. Recently, there has been a trend to reconsider this position in light of the risk-reduction philosophy as well as from a public relations standpoint. First, if the cat does not come down, the owner experiences a loss. Second, if the resident attempts to retrieve the cat, he or she can be placed in grave danger.

One method that has been used is to evaluate the situation. Most cats are perfectly able and will come down in a matter of time. An empathetic explanation to the owner may allay concerns. If the cat is actually stuck in the tree, such as between two branches and has remained there for several days and is in jeopardy of dying, the fire officer may consider the situation and determine whether the fire crew can retrieve the cat in a reasonably safe manner. The fire officer must balance the safety of the fire crew against the seriousness of the situation, the risk to the citizen from the fire department's inaction, and the potential benefits of a positive result.

should respectfully provide a method of moving the request to the level where it can be resolved. If time allows, the best method is to get the citizen's contact information, write up a summary of the discussion, and forward it to the appropriate fire chief in fire administration. The fire officer should also follow up to ensure that the citizen is contacted in a reasonable amount of time. Some citizens may prefer to contact fire administration themselves. In these situations, the fire officer should give the citizen specific information on whom to contact. The fire officer should also contact the fire chief before the citizen does to provide him or her with information on the situation.

Many fire departments have specific policies that address how citizen inquiries are to be handled. When the fire department has a policy, it is important that the fire officer understand and follow these policies in all cases. Failure to do so may lead the citizen to believe that he or she is being treated unfairly. It can also leave the fire officer open to disciplinary action.

Public Education

Almost every fire department has some involvement in public fire safety education. The goals, objectives, content, and delivery mechanisms for public education programs vary greatly among fire departments, depending on their resources and circumstances. Some large fire departments have staff bureaus that specialize in the development and delivery of public education programs. Other fire departments adopt

Figure 10-2 Stop, Drop, and Roll teaches young children what to do if their clothing catches fire. © National Fire Protection Association.

national programs or adapt a program that was developed by some other department. In many cases, a pubic education program is developed at the local level to meet local needs. Public fire safety education programs include:

- Learn Not to Burn®
- Risk Watch®
- Stop, Drop, and Roll (▲ **Figure 10-2**)
- Change Your Clock—Change Your Battery
- Fire safety for babysitters
- Fire safety for seniors
- Wildland fire prevention programs

The delivery of public education programs often involves fire suppression companies and depends on fire officers to transmit the message to the intended audience. Fire officers can also be assigned to positions that involve specific responsibilities for public education program planning and development. Every fire officer should understand some of the basic principles of public education programs.

The fundamental goal of a public safety education program should be to prevent injury, death, or loss due to fire or other types of incidents. A public safety education program can have four objectives:

- Educate target audiences in specific subjects in order to change their behavior
- Instruct target audiences on how to perform specific tasks, such as Stop, Drop, and Roll or operate fire extinguishers
- Inform large groups of people about fire safety issues
- Distribute information on timely subjects to target audiences

An educational presentation is successful when it causes an observable change of behavior. The fire officer should have a specific goal in mind when educating the public, just as there should be a goal when providing training for fire fighters (▼ **Figure 10-3**).

National and Regional Public Education Programs

The NFPA, the United States Fire Administration, and other specialized associations and industry groups have developed national and regional public education programs that can be used by fire officers to meet local community needs. Some of these programs are designed for general outreach, such as Change Your Clock—Change Your Battery and fire prevention week themes. Programs targeted toward particular problems and population groups are also available.

Fire Prevention Week

The history of Fire Prevention Week has its roots in the Great Chicago Fire, which began on October 8, 1871 but continued into and did the most damage on October 9, 1871. In just 27 hours, this tragic conflagration killed 300 people, left 90,000 homeless, and destroyed more than 17,400 structures.

Figure 10-3 The goal of public safety education programs is to prevent injury, death, or loss due to fire.

VOICES OF EXPERIENCE

❝ Perhaps the greatest difficulty is dealing with the increase in responsibility and greater concern for the image of the fire department. ❞

Rising through the ranks can be a difficult proposition for a fire fighter, especially after attaining the rank of fire officer. After many years being a subordinate, the individual now finds himself or herself in the position of supervisor and leader. Perhaps the greatest difficulty is dealing with the increase in responsibility and greater concern for the image of the fire department.

In 1997, I found myself in a unique position—we were having our 75th anniversary celebration at the firehouse, located in a quiet suburban Philadelphia community. The fire fighters became a little loud towards the end of the evening. This offended some of the neighbors. While their complaints were legitimate, I began to think, "Don't these people know we are celebrating 75 years of volunteer service to the community?" Additionally, one of the complainants was an individual who was notorious for complaining about everything about the firehouse—from the siren to the flashing lights, it seemed as if everything bothered him.

My first thought was to just dismiss the complaints; however, I realized that the people expressing their concerns were our neighbors and had probably overlooked other times when there was some excessive noise. Perhaps they were unaware of the celebration that was going on. Also, as the assistant chief, I had an obligation to the company to protect its image not only with the neighbors but also with the community as a whole. My response was a prompt letter thanking the neighbors for expressing their concern and reassured them that we would make every effort in the future to alleviate excessive noise.

Edward J. Wurster
Bucks County Public Safety Training Center, Doylestown, Pennsylvania
19 years of service

On the 40th anniversary of the Great Chicago Fire, the Fire Marshals Association of North America (now known as the International Fire Marshals Association) decided that the date of this occurrence should be observed in a way that would keep the public informed about the importance of fire prevention.

President Woodrow Wilson issued the first National Fire Prevention Day proclamation in 1920. National Fire Prevention Week has been observed on the Sunday-through-Saturday period in which October 8 falls, every year since 1922. In addition, the President of the United States has signed a proclamation announcing the national observance of Fire Prevention Week every year since 1925. NFPA has

officially sponsored Fire Prevention Week since the observance was first established.

Risk Watch®

Risk Watch is the first comprehensive injury prevention program available for use in schools. Developed by the NFPA, with cofunding from the Lowe's Home Safety Council and in collaboration with a panel of respected safety and injury prevention experts, Risk Watch gives children and their families the skills and knowledge they need to create safer homes and communities.

Risk Watch is a school-based curriculum that links teachers with community safety experts and parents. The curriculum is divided into five age-appropriate teaching modules (pre-K/kindergarten, grades 1–2, grades 3–4, grades 5–6, and grades 7–8), each of which addresses the following topics:

- Motor Vehicle Safety
- Fire and Burn Prevention
- Choking, Suffocation, and Strangulation Prevention
- Poisoning Prevention
- Firearms Injury Prevention
- Bike and Pedestrian Safety
- Water Safety
- Natural Disasters

Community Emergency Response Team

The Community Emergency Response Team (CERT) concept was developed and first implemented by the Los Angeles City Fire Department (LAFD) in 1985. The CERT program

Figure 10-4 Audiovisual aids, handouts, or souvenirs should be available for planned visits. © National Fire Protection Association.

helps citizens understand their responsibilities in preparing for disaster and increases their ability to safely help themselves, their families, and their neighbors in many types of situations.

The concept of the program was to provide basic training to residents and employees in local communities, as well as to government workers, that would allow them to function effectively during the first 72 hours after a catastrophic event. Experience had shown that in the event of an earthquake or similar event, the emergency services would be overwhelmed with serious incidents and unable to respond promptly to every problem. The CERT program was developed to train citizens to help themselves.

CERT groups can provide immediate assistance to victims in their area and collect disaster intelligence that assists professional responders with prioritization and allocation of resources after a disaster (▶ **Figure 10-5**). The training also teaches the CERT members how to organize spontaneous volunteers who have not undergone the training.

The Whittier Narrows earthquake in 1987 underscored the area-wide threat of a major disaster in California as well

as the effectiveness of the CERT program. As predicted, during that event, there were so many individual incidents over such a wide area that fire department responders could not quickly get to every location where assistance was needed. Where the CERT program had been implemented, the value of having trained groups of citizens was clearly demonstrated. As a result, the LAFD Disaster Preparedness Division was established, and the CERT program was expanded.

The Emergency Management Institute and the National Fire Academy adopted and expanded the CERT materials, believing them to be applicable to a wide variety of situations. This training was made available nationally by the Federal Emergency Management Agency (FEMA) in 1993. Since the September 11, 2001 terrorist attacks, many more fire departments have started similar CERT training programs in their communities. In 2003, FEMA reported that CERT programs had been established in more than 340 communities in at least 45 states.

CERT Course Schedule

The 20-hour CERT course is delivered in the community by a team of first responders. The instructors should complete a CERT Train-the-Trainer Program, conducted by their State Training Office for Emergency Management, in order to learn the training techniques that are still used successfully by the LAFD.

The CERT training for community groups is usually delivered in seven 2.5-hour sessions and delivered one evening per week. The FEMA version of the training consists of the following:

- **Session 1, *Disaster Preparedness*:** Addresses the different hazards to which people are vulnerable in their community. Materials cover actions that participants and their families can take before, during, and after a disaster. As the session progresses, the instructor begins to explore an expanded response role for civilians to become disaster workers. The CERT concept and organization are discussed, as are applicable laws governing volunteers in that jurisdiction.
- **Session 2, *Disaster Fire Suppression*:** Briefly covers fire chemistry, hazardous materials, fire hazards, and fire suppression strategies. The thrust of this session is the safe use of fire extinguishers, sizing up of the situation, control of utilities, and extinguishing of a small fire.
- **Session 3, *Disaster Medical Operations, Part I*:** Participants practice diagnosing and treating airway obstruction, bleeding, and shock by using simple triage and rapid treatment techniques.
- **Session 4, *Disaster Medical Operations, Part II*:** Covers evaluating patients by performing a head-to-toe assessment, establishing a medical treatment area, performing basic first aid, and demonstrating medical operations in a safe and sanitary manner.

Figure 10-5 CERT groups can provide immediate assistance to victims and assist professional responders with prioritization and allocation of resources.

Getting It Done

Catastrophic Planning

Some fire departments have developed media communication/public education plans to assists their communities in responding to natural and man-made catastrophes. For example, the San Francisco Fire Department accepted the services of a retired local television anchorperson to develop preassembled public safety messages and citizen instructions in the event of a wide variety of major emergencies.

Fire departments have to be prepared to respond to dynamically changing conditions. For example, national concerns, such as the 2001 anthrax scare or the 2003 severe acute respiratory syndrome (SARS) epidemic, required fire departments to quickly develop specific public information messages, using breaking information updates from the Centers for Disease Control and Prevention. The sniper attacks in the Washington, D.C. area in 2003 required more than a dozen federal, state, and local public safety agencies to coordinate their efforts to provide a consistent message to the general public.

- **Session 5, *Light Search-and-Rescue Operations*:** Participants learn about search-and-rescue planning, size-up, search techniques, rescue techniques, and most important, rescuer safety.
- **Session 6, *Disaster Psychology and Team Organization*:** Covers signs and symptoms that the disaster victim and worker might experience. It addresses CERT organization and management principles and the need for documentation.
- **Session 7, *Course Review and Disaster Simulation*:** Participants review their answers from a take-home examination. Finally, they practice the skills they have learned during the previous six sessions in disaster activity.
- **Optional Session 8, *Heart Attack Rescue Training*:** LAFD added an eighth class that provides additional training on airway management, choking emergencies, and a Heartsaver-level cardiopulmonary resuscitation training (only adult unwitnessed scenario).

Maintaining CERT Involvement

CERT members should receive recognition for completing their training. Some communities issue identification cards, vests, and helmets to graduates.

It is important to keep the graduates involved and practiced in their skills after they have completed the basic training course. Trainers should offer periodic refresher sessions to reinforce the basic training. CERT teams can sponsor events such as drills, picnics, neighborhood clean-ups, and disaster education fairs, which keep them involved and trained.

Locally Developed Programs

Like the CERT program, many regional and national public education programs started within local fire departments. Brush abatement programs for wildland fire areas and fireplace ash disposal programs in the northern states are two examples of regional programs that meet specific needs.

In large fire departments, the development of new public education programs is usually assigned to an individual or a work group specializing in this area. In a smaller department, a fire officer could be given this assignment as a special project.

The five step planning process can help local fire departments create and develop programs. This process was created by the U.S. Fire Administration in the 1970s and updated in 2003 by FEMA.

Identify the Problem

The first task is to identify a fire or life-safety problem. In many cases, a public education project is initiated to address a problem that demands attention, such as an increasing number of fire deaths in a community. The process sometimes works in the opposite direction, with the decision to undertake a public education effort being made first, before the objective is identified. In either case, the fire officer should attempt to clearly identify the problem before developing the program.

The fire officer should look at local emergency incident data to identify the types of events that generate deaths or serious injuries as well as the most frequent causes of fires. The analysis should consider both the frequency and the severity of different risks. For example, the fire data for a particular city might indicate an average of 100 food-on-the-stove fires and 25 bedroom fires a year. The food-on-the-stove fires result in three serious burn injuries each year and one fatality every 3 years. The bedroom fires cause an average of three fatalities and six serious injuries annually. This analysis indicates that bedroom fires are the more critical life-safety problem in this community. There are more kitchen fires than bedroom fires, but the bedroom fires cause more deaths and serious injuries. (Chapter 6, *Safety and Risk Management,* discusses the concept of hazard and risk in more detail.)

Data from another city might point out that the total number of fire deaths is lower than the number of young children who drown in backyard swimming pools in an average year. A public education project that is directed toward the drowning problem would be a higher priority than a program designed to prevent fire deaths.

Getting It Done

Engineering, Education, or Enforcement?

- **Fireplace ashes:** Fire departments in cold climates often experience large-loss fires due to the improper disposal of fireplace ashes. In many cases, the ashes are placed in combustible containers and left in a garage or on a deck. The smoldering ashes eventually ignite the container and cause a fire. An engineering solution for this problem might be for the fire department to provide free metal ash cans to the community as part of a seasonal fire safety education program.
- **Dead smoke detectors:** Many fire departments encounter single-station smoke alarms with dead or missing batteries in residences. Some departments make it a practice to check the smoke alarm during every fire company response to a residential occupancy in the community. After stabilizing the initial incident, the fire officer may request permission to check the smoke alarms in the home. A simple engineering solution can quickly solve this problem. Every fire company carries a supply of nine-volt batteries that are used to replace any dead or missing batteries. Some fire departments even issue complete smoke alarms to their fire companies with instructions to install one in any home that does not have a functioning smoke alarm. The fact that the fire department takes this problem so seriously conveys an added public education lesson.
- **Brush abatement:** Failure to maintain a clear area around a building is a common cause of structure fire losses in wildland fires. Los Angeles County uses all three factors—education, engineering, and enforcement—in efforts to reduce the impact of brush fires on the built environment. Education starts with seasonal public safety messages through the media in multiple languages. Engineering includes a description of the size of the required clear area and a program to certify brush clearance contractors. Enforcement includes local government inspection of properties, as well as issuing of fines and work orders if the brush has not been cleared.

Public education programs are, in many cases, developed in response to a single, high-profile incident. Sometimes, the high-profile incident directs public attention toward a problem that has been ignored in the past and creates an opportunity to provide valuable public education. Unfortunately, in other cases, this type of program is more reactive than effective and causes resources to be wasted on low-priority problems.

Select the Method

After identifying the problem, the fire officer should try to select the most effective method to address it. Public life-safety education is not necessarily the best or the only response for every type of problem. The fire officer should try to determine whether an engineering solution or an enforcement solution would be more efficient.

Design the Program

When designing a public education program, the fire officer should review the ABCD of course preparation, just as in fire fighter training:

- Identify the *audience*.
- Explain the desired *behavior* the student should demonstrate after training.
- Under what *conditions* will the student perform the task?
- What *degree* of proficiency is expected?

The fire officer has to identify the specific objectives of the educational program. The program format should be matched to the message, the audience, and the available resources. Demographic information assists in describing the audience.

The development process should include delivery of a pilot class to a target audience to see how well it works. Most new programs benefit from revision after the pilot is evaluated.

Implement the Program

Implementation occurs when the program is actually delivered to the target audience. Timing is often an important consideration when a public education program is implemented. For example, a fire safety program for university students would be most effective during the first couple of weeks in the fall semester, in order to reach all of the new residents. The objective could be to reach every on- and off-campus student housing location at the start of the academic year.

Evaluate the Program

There are two types of evaluation. Immediate feedback is obtained from the students at the conclusion of the training, usually from a short survey form. Immediate feedback can help the fire officer determine whether the specific objectives were met by the presentation. This type of immediate feedback is most effective in evaluating the mechanics of the class presentation, such as interaction between the speaker and the audiovisual media.

A longer-term evaluation should focus on the effectiveness of the program in relation to the desired effect. Success is accomplished when the students actually demonstrate the desired behavior in the anticipated situation. For example, the goal of a program at a university could be to reduce the number of malicious false alarms from pull stations. To evaluate the effectiveness of the program, the number of malicious false alarms in a time period before the training would be compared with the number during an equivalent period after the training.

Media Relations

Most fire departments have frequent interactions with the local news media, including during emergency incidents, in which reporters want to obtain information about a situation that has just occurred or is still occurring, as well as in newsworthy situations that are related to other fire department programs or activities. In addition, in many situations, the fire department has information or an important message to communicate to the public.

Significant emergency incidents are high-profile news events that capture public attention. The local print, radio, and television media can be expected to appear at major emergency incidents or to call the fire department seeking information. Many other fire department activities, ranging from inspection and public education programs to promotions, retirements, awards ceremonies, or delivery of a new fire truck, also generate media interest.

The news media have an important and legitimate mission to obtain and report information to the general public. The fire department is an organization that performs a public function, operates in the public view, and is usually supported by public funds. It is always in the best interest of the fire department to maintain a good relationship with the local media. Providing information to the news media should be a normal occurrence.

Although the news media have a legitimate purpose in seeking information, there are limitations on the information that can be released in many circumstances. The fire officer should know the applicable departmental procedures and guidelines and how to respond to a request for information that the fire officer is not authorized to release. The fire officer should understand that the interaction that occurs

between the fire department and the news media is likely to have a direct impact on the department's public image, reputation, and credibility. The fire department depends on public trust and confidence to be able to perform its mission.

The Fire Department Public Information Officer

Some fire departments have a full-time or part-time public information officer (PIO) who functions as the media contact person and the source of official fire department information. Where this position exists, most interactions with the news media should go through this individual. Other officers have limited responsibilities and opportunities to interact with the news media. There may be occasions when the PIO refers a reporter to another fire officer to obtain information about a particular subject or situation.

Many smaller departments rely on the local fire officer or a staff officer to function as the PIO and interact with the local newspaper, radio, and television media as the need arises. In these situations, the individual should be guided by the same basic principles as a regularly assigned PIO.

This section presents basic recommended practices for a PIO who has to establish and maintain an ongoing relationship with the local media. This aspect is particularly important when the fire department has information that it needs to release to the general public. Reaching out to a community is much easier for an organization that has already established solid contacts and relationships with the local media (▼ **Figure 10-6**).

The NFPA provides a "Get Your Message Out" media primer to help the fire service publicize community outreach campaigns and other events associated with public safety

(**Figure 10-6**) Reaching out to a community is easier with solid contacts and relationships with members of the media.

education. The basic concepts are valuable for any fire officer who has to be prepared to interact with media representatives. The NFPA publication recommends three steps when working with the media:

- Build a strong foundation
- Use a proactive outreach
- Use measured responsiveness

Step One: Build a Strong Foundation

Look at your relationship with the media as a business arrangement: the media have something you need (an audience and the means to communicate with them), and you have something they need (news and information). If you work collaboratively, you will both achieve your objectives. If you expect too much or if you do not anticipate the media outlet's needs, someone is going to end up disappointed, and that could be counterproductive.

The most important media asset you will ever have is a good relationship with the media. Members of the local media need to be familiar with you and need to be comfortable working with you; they must have confidence in the information you provide and, most importantly, they need to trust you. Whatever else you do, always tell the truth to the media, even if the truth is "I do not know" or "I cannot release that information at this time" or "It is premature to answer that question." Do not guess, speculate, or lie.

Do not wait until you need a reporter to establish media contacts; lay the groundwork in advance for a collaborative working relationship. Build a list of local media contacts and keep it up-to-date. Obtain the names, titles, and contact information for key individuals. Learn who has responsibility at the newspapers, radio stations, and television stations.

Find out about the deadlines for different news outlets and how to reach reporters with late-breaking information. Television reporters need to have the information before news time, and print reporters need it before the paper is printed.

Be a consumer of your local media. Read the local newspapers, watch television, and listen to the radio (▶ **Figure 10-7**). Pay attention to the kinds of stories they tend to cover, the angles they use, and the personal style of individual reporters. If a reporter does a good job reporting a fire department story, make a thank-you call. If a story contains inaccurate information, make a call to provide the correct information, keeping the tone positive.

Make sure that the news representatives know how to contact you whenever they need information. If you are not going to be available, make sure that someone else is equally accessible. Your objective is to make sure that the news media representatives know that your department is interested in working with them and in responding to their inquiries, that you are the person to contact, and that they can have confidence in any information you give them.

Figure 10-7 Be a consumer of your local media.

Step Two: Proactive Outreach

If you have the resources, a proactive media communications plan can be much more beneficial than a reactive approach. Simply responding to media inquiries as they come to you is a reactive strategy. You should be as helpful as possible, providing information within the resource capabilities of your department.

Being proactive means that in addition to being responsive, you actively look for opportunities to use the media to communicate your department's objectives and mission. The media need factual, timely stories to report, and the fire department generates broad public interest. You can help them by providing announcements of newsworthy issues, topics, and events.

Once the working relationship has been established, reporters begin to look to you for story ideas, resources, and quotes. You can call a reporter or editor and suggest a story that could be interesting to him or her. Learn the types of stories that are most interesting to different reporters and call that person when you have something that fits. They are not going to automatically accept every story idea, but they will listen to what you have to suggest.

When you have information that you want to distribute to several news outlets, use a press release or hold a press conference. A press conference is a staged event in which you are inviting the news media to come and hear something important that you have to announce. Press conferences should be reserved for topics and situations that definitely have broad interest; on these occasions, you know that the media will want to attend and ask questions.

Step Three: Measured Responsiveness

When dealing with the news media, remember that one of your responsibilities is to present a positive image of the fire department. In certain situations, this is difficult, particularly if something negative has occurred. It is no use denying that something has happened when the facts are clearly evident. Sometimes, the best that you can do is to be honest and factual to maintain the credibility of the organization.

Be wary of situations in which someone could be trying to misrepresent a situation in a manner that reflects badly on the department. The reporter who comes to you may have already been given a highly inaccurate or slanted viewpoint on a story by some other source. Do not allow yourself to be placed in a situation in which you make a statement that is inaccurate, misleading, or damaging to the fire department. Make sure that your response is accurate and that the information is appropriate for release before you react.

Press Releases

A press release is used to make an official announcement to the news media from the fire department PIO. It could be an announcement of a special event, a promotion or appointment, an award ceremony, a fire station opening, a retirement, or any similar occurrence that the department wants to make known to the public. A press release could also provide information about a program, such as the launching of a new juvenile fire setter counseling program or the fact that the annual report has been released and statistics show a 34% decrease in residential fires since a major public education program was implemented.

A press release should be dated and typed on department stationery, with the PIO contact information at the top. The release should be as brief as possible, ideally one to two pages, using an accepted journalistic writing style. Make the lead paragraph powerful to entice the reporter to read on and answer the basic "who, what, when, where, and why" questions (▶ **Figure 10-8**).

Depending on the preference of your local media contacts, you can mail, fax, or e-mail press releases. Reporters and editors are likelier to read your releases if they receive only those that pertain to their particular beat or to topics in which they are interested. Personalize the release to an individual at each destination by name and title. For example, a release announcing the official opening of a new fire station could go to the assignment editor at a television station and to the newspaper reporter who usually covers the fire department. A release on cooking fire safety could be directed to the food editor.

Company Tips

Grammar Counts

Press releases are public documents that are read by reporters, the public, and local elected officials. The presentation and the content both reflect on the image of the fire department. Invest in a dictionary, thesaurus, and grammar book or style guide to improve your written presentation skills. Check the spelling, particularly names, and be sure that ranks, titles, and telephone numbers are correct.

The Fire Officer as Spokesperson

Every fire officer should be prepared to act as a spokesperson for the fire department. Even if the department has a PIO who is the designated official spokesperson, there are likely to be occasions when other fire officers have to use these skills. Every fire officer should know the basic guidelines that apply to an official spokesperson.

The most common situation in which a fire officer would appear as an official spokesperson is an interview. NFPA provides these guidelines for the fire officer conducting an interview with the media:

- Be prepared
- Stay in control
- Look and act the part
- It is not over until it is over

Be Prepared

Before agreeing to be interviewed, make sure that you are authorized to speak on the subject as a departmental spokesperson and be sure that you have the appropriate information. Determine the reporter's story angle and what he or she already knows about the topic. Ask with whom he or she has already spoken or plan to contact and what he or she hopes to learn from you.

If you are uncomfortable with the situation angle or do not have the information, you do not have to agree to be interviewed. You should be polite but firm if you decline a request for an interview. If possible, refer the reporter to someone else who can provide the information.

Be on time for an agreed-on interview, but do not begin the interview until you are ready. The reporter probably has a deadline that you should try to respect, but not at the expense of your readiness.

When preparing for an interview for which you want to deliver a specific message, determine no more than three key message points in advance and practice saying them in varied ways. Learn to use wording that emphasizes that what you are saying is important, such as, "The number-one thing to remember is ___" or "More than anything else, people should realize that ___." Practice your talking points

MEDIA RELEASE

LAS VEGAS FIRE & RESCUE
DATE/TIME: 061003/13:00:00 PDT
RELEASE: 03-085   Contact: Tim Szymanski
Telephone: 229-0145

Smoke Alarm Alerts Sleeping Family About Fire In Home

Newly installed batteries in a smoke alarm alerted a sleeping family to a fire in their northwest Las Vegas home Tuesday morning. The two adults and three small children escaped the burning home without injury. The cause of the fire has been ruled accidental.

Fire fighters were dispatched to 6001 Carmel Way at 7:53 A.M. While en route to the scene, fire fighters could see a column of smoke indicating they had a working fire. Fire fighters arrived to find flames coming out a side window of the one-story, wood frame/stucco house. They were able to douse the flames in less than five minutes.

The fire gutted one bedroom, caused heavy damage to two other bedrooms, and there was heavy smoke damage throughout the rest of the house. Damage was estimated at $60,000.

There were two adults and three small children asleep in the house when the smoke alarm activated and woke them. They found the hallway full of smoke and the adults evacuated the children and alerted neighbors of the fire. The mother returned to a rear bedroom with a fire extinguisher and found the mattress on fire. She said the flames were too intense and had to retreat.

Fire investigators determined that one of the children was playing with a lighter and started the fire.

The American Red Cross is assisting two adults and four children that live in the house. The woman stated that she just replaced the battery in her smoke alarm this past weekend.

Figure 10-8   Example of a press release.

with a colleague or a tape recorder. Do not memorize your message points but be very familiar with them. Check any statistics, trends, or other information you would not know on a casual basis.

Stay in Control

Be cooperative, but stay focused on your key message points. Listen carefully to the reporter's questions and ask for clarification if you do not fully understand a question. If you need a few seconds to think through your answer, take the time to formulate your answer. If you do not know the answer, say so and offer to find the information and get back to the reporter. Avoid jargon and highly technical language that confuses or distracts your audience. When you have answered the question, stop talking.

Remember that the reporter came to you looking for information and expertise. Be confident and authoritative, steering the interview in the direction that you want it to go. Showcase your message points and your department's mission and work.

Look and Act the Part

When doing a television interview, make sure your appearance is clean, tidy, and professional. Wear your uniform properly, and show pride in the department that you are representing (▼ Figure 10-9). Adopt the appropriate demeanor for the subject. If you are on the scene of a fire fatality, a somber yet authoritative tone is appropriate. If you are conducting a television interview from a fire station open house, a more enthusiastic tone is acceptable.

It Is Not Over Until It Is Over

When speaking to a reporter, do not assume that anything you say is "off the record." Assume that everything you say could be quoted, including conversations before and after the actual interview. When you are wearing a microphone, anything you say can be overheard and recorded.

After the interview, double-check to ensure that the reporter has the correct name and spelling for you and your department and accurate information about any program or event that you are promoting. Leave your business card to ensure that the proper name, title, and organization attribution is recorded.

Summary

The fire officer is an ambassador of the fire department. Formal and informal educational sessions can have a tremendous impact on community safety. A fire officer's challenge is to be able to meet the community needs and respond to emergency incidents in a manner that creates community goodwill and reduces fire-related deaths and injuries.

Figure 10-9 Make sure your appearance is clean, tidy, and professional, and wear your uniform properly.

Assessment Center Tips

Television Interview Behavior Tips

The following recommendations from the NFPA will help if part of the assessment center process includes conducting a television interview.

Look into the reporter's eyes, not into the camera. Keep your gaze steady and avoid rolling your eyes, blinking excessively, or closing your eyes when you are thinking about your answer. If you are standing, do not sway. Plant your feet about 18 inches apart, and keep your hands down to your sides or clasped together loosely in front of you. Do not put your hands in your pockets. If you are sitting, ask for a stationary chair. If you must sit in a swivel chair, plant your feet to help you keep the chair still. Sit with your knees together and your feet flat on the floor or your ankles loosely crossed. Keep your hands folded comfortably in your lap. Do not clench your fists, crack your knuckles, pick your nails, or play with your earrings. These are not only distracting behaviors; they signal to viewers that you are nervous, which can be interpreted as a lack of confidence about the subject. (Imagine a close-up camera angle of your clenched fists or fidgety fingers.)

Wrap-Up

Ready for Review

- Demographics describe a community through characteristics of human population (age, race, income, education).
- Community characteristics might call for an innovative public safety response, such as a Spanish-language immersion course at the fire station.
- Effective public fire and life-safety education programs must result in the desired change in behavior.
- Public safety education programs should educate, instruct, inform, and distribute information.
- NFPA's Risk Watch is an age-appropriate, school-based curriculum linking teachers with community safety experts and parents.
- The Los Angeles Fire Department developed Community Emergency Response Teams in 1985 to train citizens to assist each other in the first 72 hours after any calamity that would delay the response of public safety agencies, such as an earthquake.
- Fire officers should use the identify-select-design-implement-evaluation method to develop local public education programs.
- Media relations require the fire officer to build a strong foundation, outreach proactively, and provide measured responsiveness.
- The fire officer should be prepared, remain in control, and have appropriate demeanor/appearance when participating in a media interview.

Hot Terms

Demographics The characteristics of human populations and population segments, especially when used to identify consumer markets. Generally includes age, race, sex, income, education, and family status.

Show and tell Activities in which the fire and rescue department is called to present a program at a school or public gathering on short notice, resulting in the closest fire company taking their apparatus to a location and presenting a program.

You Are the Fire Officer: Conclusion

Apologize to the mother who waited 2 hours for you to return to the fire station. Explain to her that you were out fighting a fire and that the driver/operator will install the child safety seat as soon as the apparatus is ready for the next emergency. Have the senior fire fighter assist in getting the engine ready to reduce the delay. Offer to reschedule the installation if this arrangement is inconvenient for her.

Have the fire fighters clean up quickly and prepare for the tour while you greet the children and teachers from the day care center. You can start the tour while the fire fighters put on clean uniforms. Apologize for your appearance and explain that you look the way you do because you have been fighting a fire. When the fire fighters return, you can excuse yourself briefly to get cleaned up. After the fire fighters finish the tour, rejoin the group and help the fire fighters demonstrate the Stop, Drop, and Roll technique to the children.

Once the engine is ready, the car seat has been installed, and the day care center group has left, you should plan a follow-up to the neighborhood where the fire occurred. The goal is to make sure that the rest of the community is complying with the brush abatement ordinance and is prepared for another drought season.

Fire Officer in Action

Your engine company is preparing to clear from a double fatality motor vehicle accident when you notice a news van arrive on scene. The news crew seems to rush over to you before the van even stops moving and you are immediately flooded with questions.

1. The reporter immediately inquires, "Was anybody hurt?" You respond with:

   A. "I'm not sure yet, let me get back to you."

   B. "There were a few minor injuries, but everyone seems to be okay."

   C. "Unfortunately there were two fatalities, but I cannot release any further information at this time."

   D. "So far we have two fatalities, some alcohol seems to be involved, and the victims were identified as. ..."

2. When approached by a reporter or an onlooker on the scene of an emergency, an appropriate response would be:

   A. "We are very busy here. We do not have time for your questions!"

   B. "I am not sure what happened, but when I find out I will let you know."

   C. "Our crew is busy mitigating the scene at this time. If you have any questions please feel free to stand back at a safe distance and we can arrange for someone to answer your questions at a later time."

   D. "This is a dangerous area for you to be in! You need to leave!"

3. What are some steps you can take to make yourself a better candidate for a public address situation on an emergency scene?

   A. You can make notes and memorize them before speaking to any member of the public, to ensure you will not say anything that you do not mean to.

   B. You can do some homework and research on questions you think will be asked.

   C. Being stern and professional is always the best approach for public speaking; showing emotion or sympathy is not tolerated.

   D. The public should always be addressed by only your department's PIO officer; you are not qualified to speak in these situations.

4. The reporter addresses you on the scene of the accident and asks if he can ask you a few questions "off the record." You should:

   A. Agree and proceed to speak candidly about the accident with the reporter, being as helpful as possible.

   B. Agree, with the limitation that he does not use your name in the article.

   C. Tell the reporter that you cannot comment at this time, but you invite him to go out for a cup of coffee so that you can give him the information he wants at a later time.

   D. Inform the reporter that anything you happen to say is "on the record" and you are not at liberty to make personal comments or judgments on anything involving the accident, reminding the reporter that you are representing your department at all times.

Chapter Pretests

Interactivities

Hot Term Explorer

Web Links

Review Manual

Handling Problems, Conflicts, and Mistakes

Technology Resources

www.FireOfficer.jbpub.com

- Chapter Pretests
- Interactivities
- Hot Term Explorer
- Web Links
- Review Manual

Chapter Features

- Voices of Experience
- Getting It Done
- Assessment Center Tips
- Fire Marks
- Company Tips
- Safety Zone
- Hot Terms
- Wrap-Up

NFPA 1021 Standard

Fire Officer I

4.3.2. Initiate action to a citizen's concern, given policies and procedures, so that the concern is answered or referred to the correct individual for action and all policies and procedures are complied with.

(A) Requisite Knowledge. Interpersonal relationships and verbal and nonverbal communication.

(B) Requisite Skills. Familiarity with public relations and the ability to communicate verbally.

4.4.1. Recommend changes to existing departmental policies and/or implement a new departmental policy, so that the policy is communicated to and understood by all members.

(A) Requisite Knowledge. Written and oral communication.

(B) Requisite Skills. The ability to relate interpersonally.

Fire Officer II

5.4 **Administration.** This duty involves preparing a project or divisional budget, news releases, and policy changes, according to the following job performance requirements.

5.4.1. Develop a policy or procedure, given an assignment, so that the recommended policy or procedure identifies the problem and proposes a solution.

(A) Requisite Knowledge. Policies and procedures and problem identification.

(B) Requisite Skills. The ability to communicate in writing and to solve problems.

Knowledge Objectives

After studying this chapter, you will be able to:

- Describe how to manage conflict.
- Describe how to deal with citizen complaints.
- Describe how to recommend policies and policy changes.
- Describe how to implement policies.
- Describe the difference between customer service and customer satisfaction.

Skills Objectives

After studying this chapter, you will be able to:

- Develop a policy or procedure.

You Are the Fire Officer

You are in the fire station preparing to give an in-station drill when you hear people shouting. Moving toward the front driveway of the station, you hear a crash, followed by more shouting. The engine has apparently rolled down the driveway and crashed into a large passenger vehicle in front of the station. Two of the fire fighters are helping the elderly driver out of the vehicle. You hear the driver/operator calling dispatch and requesting police and an EMS unit.

You are relieved to see that the vehicle driver appears to be uninjured. However, your relief quickly turns to dismay as some of the fire fighters begin to laugh as they look at the crushed passenger doors of the vehicle. Just when you think the situation cannot get any worse, you realize that the driver is the city council chairperson.

1. How should you handle this situation?
2. What steps should be immediately taken?
3. What steps should be considered after initial issues are addressed?

Introduction to Problems

In general, a problem is the difference between the actual state and the desired state. If the way we park at a scene places us at greater risk than is necessary, it is a problem. If a council person is hit by a runaway apparatus, it is a problem. In each case, there is a discrepancy between the actual state and the desired state.

A fire officer must be prepared to deal with several different types of problems. Fires and emergency incidents present a very special category of problems that call for highly specialized problem-solving skills. Aside from emergency incidents, many other types of situations require a fire officer to apply more conventional problem-solving skills and techniques. These situations can include a wide range of supervisory, management, and administrative activities in which the fire officer is directly responsible for solving the problem, as well as problems that can be solved only at a higher level. They can also include nonemergency situations that involve individuals or organizations outside the fire department.

Decision-making skills are required whenever the fire officer is faced with a problem or any type of situation that requires an action or a response. Promotional examinations routinely evaluate the ability of fire officer candidates to make good decisions and exercise good judgment. A fire officer who is called upon to make a job-related decision must always determine the appropriate decision for the organization. The fire officer is the front-line representative of the organization and must be guided by organizational values, guidelines, policies, and procedures, even if that decision does not coincide with the fire officer's own opinion or personal preferences.

The fact that a fire officer has to act in the best interests of the fire department in solving problems and making decisions does not mean that all other values and considerations are ignored or overruled. A problem can present several possible solutions, some better than others and some more desirable to one set of interests than another. In many cases, there is a reasonable solution that serves multiple interests and satisfies multiple concerns, whereas in other cases, one concern has to prevail over all others. Problem-solving techniques are designed to identify and evaluate the realistic potential solutions to a problem and determine the best decision.

Complaints, Conflicts, and Mistakes

Complaints, conflicts, and mistakes are special categories of problems. One of the most critical areas in decision making is how to deal with situations that involve conflicts or complaints.

Definitions

- A **complaint** is an expression of grief, regret, pain, censure, or resentment; lamentation; accusation; or fault finding.
- A **conflict** is a state of opposition between two parties. A complaint is often a manifestation of a conflict.
- A **mistake** is an error or fault resulting from defective judgment, deficient knowledge, or carelessness. A mistake can also be a misconception or misunderstanding. Mistakes happen; the issue is how to deal with a mistake, or the perception of a mistake, when someone complains to the fire officer about it.

It is inevitable that a fire officer will have to deal with conflict and complaints from time to time. Sometimes, a fire officer has to make a decision or enforce a policy that is not popular with the crew members. A citizen could be less than satisfied with something the fire department has done or can be simply unhappy with a situation. People misbehave and make mistakes. Disagreements and differences of opinion occur, and it is not possible to make everyone happy all of the time. A fire officer must be prepared to deal with all of these situations in a professional manner, acting as the official representative of the fire department. Dealing appropriately with problems and conflicts requires maturity, patience, determination, and courage (▼ Figure 11-1).

The types of problems that a fire officer could be expected to encounter can be divided into four broad categories:

1. In-house issues include situations, decisions, or activities that occur at the fire officer's work location and within the direct scope of supervisory responsibilities. An example might be a complaint about the assignment of duties to different individuals within a fire station. Most of these conflicts begin and end at the company officer level.

2. Internal departmental issues involve fire department operational policies, decisions, or activities that go beyond the fire station level. An example in a career organization could be a conflict about where a reserve ladder truck is housed and which company is responsible for maintaining it. In a volunteer system, it could be a dispute between companies over the boundaries of their first-due response areas. The resolution usually requires action by command officers at a higher level on the organizational chart.

3. External issues include any fire department activities that involve private citizens or another organization. Examples might include an engine backing into a parked car or a citizen complaint about a fire fighter making an inappropriate remark during a medical assist response. NOTE: External issues require the fire officer to perform one additional task early in the

conflict resolution process: making sure that the fire officer's supervisor is not surprised. It is poor form for the battalion chief to learn about a fire department incident from the local media.

4. High-profile incidents can involve any type of issue that is likely to become a major one for the organization because of its nature. An example of such an incident might include a fire fighter arrested while on duty. The department must take immediate actions to respond to these events. Senior fire administrators often become directly involved in the situations or keep a close watch on how they are handled.

A problem should generally be solved at the lowest possible level within an organization. A fire officer is expected to be able to manage problems up to a certain level. At the same time, the fire officer should quickly recognize problems that can be handled only at a higher level and make the appropriate notifications without delay. If there is any doubt, it is a wise policy to discuss the situation with the next higher level officer in the chain of command. Asking is always better than making a major error.

General Decision-Making Procedures

A fire officer is called upon to make many different types of decisions about a wide variety of subjects. At the fire officer level, most of the problems are fairly uncomplicated, although they are not necessarily easy to solve. As the fire officer moves up through the ranks, the number of decision-making situations increases exponentially, and the problems are often much more complex.

The technique described in the following section is a systematic approach to high-quality decision making. This five-step process can be applied to many different types of problems:

1. Define the problem
2. Generate alternative solutions
3. Select a solution
4. Implement the solution
5. Evaluate the result

Although the five-step technique appears to be designed for situations where there is plenty of time to go through the steps in a logical sequence before making a decision, the same basic approach is used for emergency incidents. When the situation calls for a rapid response, a fire officer should be able to move through the steps very quickly because training and experience have prepared that individual to very quickly identify the pertinent problem, move on to the realistic solutions, and then select the best option.

Define the Problem

The first step in solving any problem should be to closely examine and carefully define the problem. A well-defined problem is one that is half solved. Tremendous time and

Figure 11-1 Dealing appropriately with problems and conflicts requires maturity, patience, determination, and courage.

effort can be wasted in efforts to solve a problem before the problem itself is clearly identified. By practicing the activities listed in this section, the fire officer improves his or her ability to detect and define any departmental problem.

Pay Attention

Tom Peters used the phrase "management by walking around" to describe a practice he observed in the successful businesses he profiled in the 1982 bestseller *In Search of Excellence*. An effective manager should know what is going on within the organization and address most issues before they become major problems. The best way to prevent major problems is to deal with issues before they reach the crisis stage (▼ **Figure 11-2**).

Ask Basic Questions

Peter Drucker is a famed author, educator, and management consultant. He has worked with private, public, and non-profit organizations to improve their management skills. He served as a Professor of Management at the Graduate Business School of New York University for more than 20 years. However, he is best known for his more than 30 books on management and as an editorial columnist for the *Wall Street Journal* for 20 years.

Peter Drucker encourages managers to question the value of each organizational activity, usually once a year. What may have been a vital activity last year may be of minimal importance this year. Most fire departments have a strong inclination to keep on doing the same things the same ways and exhibit strong cultural resistance to this type of critical self-analysis. The fire officer should make a conscious effort to identify activities that can be changed, improved, or updated. If there is a better way to do something,

give it appropriate consideration. If we are doing something that is no longer worth doing, why not use the time to do something more productive?

How Quickly Do You Get Bad News?

Few things damage a new fire officer's career more effectively than not finding out about something that is not going well until it is too late to fix it. Effective fire officers create a work environment that encourages subordinates to report bad news immediately. The sooner the fire officer receives the bad news, the quicker he or she can implement corrective action. Fire fighters should be encouraged to tell their fire officer whenever an event or performance is out of balance with expectations (▼ **Figure 11-3**). This should also include reporting personal injuries or broken equipment immediately, regardless of the time of day.

This principle also applies to bad news during an emergency. Fire officers who create a barrier to receiving administrative bad news can suffer catastrophic results on the emergency scene. From the fire fighter's perspective, the fire officer who does not appreciate hearing bad news in the fire station will probably not want to hear it at the emergency scene either. There are many cases on record where a fire fighter was afraid to point out a problem that could have saved a life or prevented a serious injury.

Fear Versus Trust

The goal of every fire officer should be to foster a trusting relationship with the employees. Employees who do not trust their boss or each other are unlikely to make good decisions when faced with a problem. Employees who do not feel that their input is valuable will stop passing vital information to the fire officer. Effective problem solving requires good information.

Figure 11-2 Fire officers must pay attention to what is going on within the organization and deal with problems before they reach the crisis stage.

Figure 11-3 Fire fighters should be encouraged to report problems immediately.

Many fire officers believe that because they have been promoted, they can identify the problem. This is not necessarily true. The only way the fire officer can best define the problem is with the best information. Eliminating sources of valuable information because of fear and mistrust will cause the fire officer to make an inaccurate assessment.

Generate Alternative Solutions

Involve Anyone with Direct Knowledge of the Problem

The best people to solve a problem are usually those who are directly involved in the problem. The fire officer who is struggling with a new incident management clipboard in the rain has a valuable perspective on what might improve the clipboard's performance in inclement weather. The fire fighters who make ventilation openings in roofs probably have some good ideas about how it could be accomplished more efficiently. Company-level problems are most likely to be solved by involving the members of the company.

Brainstorming

Brainstorming is a method of shared problem solving in which all members of a group spontaneously contribute ideas. A typical fire company is a good size and a natural group for brainstorming (▼ **Figure 11-4**). Fire fighters are usually very adept at solving problems.

The following eight steps will assist the fire officer when brainstorming alternative solutions:

1. Using a flip chart, white board, or chalkboard, write out the problem statement. Everyone should agree with the words used to describe the problem.

2. Give the group a time limit to generate ideas. Although the scope of the problem is a factor, 15 to 25 minutes seems to work for groups of four to 16.
3. The fire officer should function as the scribe. The scribe writes down the ideas and keeps the group on task.
4. Tell everyone to bring up alternative solutions. At this point, there is no commenting on ideas. All suggestions are welcome.
5. Once the brainstorming time is up, have the group select the five ideas they like the best.
6. Write out five criteria for judging which solution best solves the problem. Criteria statements start with "should." For example, "It should be legal," or "It should be possible to complete in 6 months."
7. Have every participant rate the five alternative solutions, using a 0 to 5 scale. The value 5 means that the solution meets all of the criteria, and 0 means that the solution does not meet any of the criteria.
8. Add up the scores for each idea. The idea with the highest score is the one that provides the best solution for your problem.

There are some constraints to brainstorming. The process assumes that the problem statement is accurate and the criteria are valid. On occasion, the best solution that comes out of a fire station may crash at headquarters because of a criterion or restriction that is unknown to the fire officer, such as a new federal or state regulation.

Should the Fire Chief Participate?

Fire company group dynamics are different when the fire chief is in the fire station. Whether intentional or accidental, the presence of a senior command officer significantly influences a group that is brainstorming alternative solutions. The verbal and nonverbal signals sent by the chief officer are likely to influence the group, and the participants self-censure their suggestions and limit the range of possible and plausible solutions because of how they think the chief will react to them. The best ideas occur when they come from the fire fighters directly affected by the problem or issue. Each layer of hierarchy added to the group reduces the range of options.

Do Fire Fighters Feel Comfortable Sharing Ideas?

Fire fighters are naturally competitive. Even when the decision group is restricted to the fire fighters who are directly affected, such as in determining how fill-in engine drivers will be trained and used, they may be reluctant to share ideas. The fire officer's role should be to encourage group participation by creating a positive and nonhostile work environment.

Is the Process Legitimate?

A legitimate problem-solving process has to be reasonable and based on logic and organizational values. Fire fighters should be able to anticipate that their decision will be implemented if it meets the criteria. Going through a process that results in no change or provides no feedback to the fire company members is the quickest way to destroy fire fighter participation in the decision-making process. A legitimate process reinforces the trust between the fire company and the fire administration.

Select a Solution

You have defined the problem, generated solutions, and ranked them based on criteria. One factor in deciding on the best solution is the core value system of your department. For example, if participation in local neighborhoods is one of the core values, a solution that increases the fire department's involvement in local neighborhoods would be preferred.

Implement the Solution

Once the decision has been made, the solution still has to be implemented. The implementation phase is often the most challenging aspect of problem solving, particularly if it requires the coordinated involvement of many different people. One of the reasons for involving as many players as possible in making the decision is to capture their commitment to the plan when it is time for implementation.

Consider a plan for reducing the number of fire fatalities in residential occupancies. After careful analysis, the decision has been made that the best strategy would be a community outreach program to check every smoke detector in every multiple-family dwelling in a community over a 4-month period. The plan supports the fire department's mission statement and reflects the organization's values. To accomplish this objective, each fire company will have to invest 2 hours every evening for the next 4 months.

Before this plan can be implemented, a buy-in is required. The plan has to be "sold" to the people who will actually take action, particularly to those who were not involved in the decision-making process. Rearranging schedules to do something different for 2 hours every evening involves a significant behavioral change, even if it is for a limited time and for a good cause. It would be possible to simply order everyone involved to "just do it"; however, this is probably not the best implementation strategy for a program that involves extensive public contact. Willing participation almost always works out better than involuntary compliance.

Who Does What When?

To avoid confusion, the fire officer must clearly assign tasks to individuals or teams. The fire officer benefits by using a project plan, which lists tasks, responsibilities, and due dates (▶ **Figure 11-5**). Most of the projects supervised by a company-level officer do not require a sophisticated planning system. A simple project control document can be used to divide a project into segments, with milestones to identify progress. Complex and long-range tasks may require a formal project management plan and a designated coordinator,

Fire Marks

New York City hired the RAND Corporation in the late 1960s to perform operations research on various city agencies and make recommendations to Mayor Lindsay. The goal was to deliver city services more efficiently and effectively. As a result of this project, RAND published groundbreaking research relating to fire department deployment analysis. RAND published the research and New York City publicized the successes.

One of the high-profile projects was a computer simulation model that was used to predict the FDNY workloads in the Bronx. Based on the model, a policy change was implemented to reduce the response to a pulled fire alarm street box to a single-fire company. In the high-activity areas, the data showed there was a less than 1% probability that a pulled fire alarm box, with no other information, would turn out to be a serious fire. Although the initial response to pull boxes was reduced, the full first alarm assignment was dispatched immediately if there was a second call reporting a fire at the same location.

Other fire departments read the reports and mimicked the dispatch and alarm assignments that the FDNY had developed. One small city fire department, which was similar in density and demographics, adopted the reduced response to pulled fire alarm street boxes with disastrous results. Several situations occurred in which a single company responded and encountered a serious fire, sometimes requiring civilian rescue.

The fire officers in this city revisited the problem-solving process. They identified one very significant difference in the response data for their city. Going through records in the fire alarm dispatch office, they discovered that about 30% of the serious fires were reported only through a pulled fire alarm street box, with no telephone call or other notification. They also obtained information from the local telephone company showing that many of the small apartments in the city lacked residential phone service. Senior staff quickly eliminated the reduced response, restoring the response of six companies to a street box alarm.

particularly if they require the coordinated activity of multiple agencies.

No Deadline Means No Implementation

An implementation plan must include a schedule to ensure that the goals are met. Deadlines focus effort and help prioritize activities. If the solution requires activity by other organizations, such as changing a local ordinance or submitting a budget request, then the fire officer must determine the time it will take to accomplish that task and incorporate that time into the schedule. The schedule is valuable only if it is followed and someone ensures that the deadlines are met.

Plan B

Many problems remain unsolved, long after the problem has been clearly defined and a good solution has been selected. The problem is not truly solved unless the solution is imple-

Orchard Valley Fire Station 712
Fiscal Year 2005 Capital Improvement   **3/1/2006 Status**
Training Room

Goal		Due Date	
Paint, carpet, and repair fire station training room		5/1/2006	

Task	Assigned to	Due Date	Status
Volunteer and career staff select colors and décor	Capt Smyth	1/4/2006	Done
Order paint and wallpaper	Treasurer Allen	1/20/2006	Done—at fire station
Order lumber and equipment for media center	Technician Gold	1/20/2006	Done—at fire station
Order new furniture	Capt Smyth	2/1/2006	Done—3/21/06 delivery
Order carpet	Treasurer Allen	2/15/2006	Done—to install 3/20/2006
Build media center (built in-house)	Technician Gold	3/13/2006	Done
Remove old furniture, tables, and carpet	B shift	3/16/2006	
Strip wall coverings and prepare to paint	A shift	3/17/2006	
Install media center	Technician Gold	3/18/2006	
Paint and wallpaper party	C shift and volunteers	3/18/2006	
Vendor installs carpet		3/20/2006	
Move new furniture into training room	A shift	3/21/2006	

Figure 11-5 Example of a project control document for a fire station work project.

mented. There are many reasons why good solutions are not implemented, including cases where the required approvals cannot be obtained or the necessary resources are not available. Quite simply, the organization might not have the capability to solve the problem or to implement the solution that was selected.

Fire officers should consider a "plan B" if the original solution cannot be implemented. Plan B could be an extended implementation schedule, a modified plan, or a completely different solution to the problem.

Evaluate the Results

After implementing the solution, the fire officer must assess whether or not it produced the desired results. Evaluation should be a standard part of the process of any problem-solving activity. The nature of the evaluation depends on the complexity of the problem and the solution; in most cases, an initial evaluation should be performed immediately after implementation, and then follow-up evaluations should be performed at regular intervals.

Determining whether the solution actually solved the problem requires some type of measurement that compares the original condition with the condition after implementation.

For example, do the new hose loads really result in quicker deployment of attack lines? The answer to this question requires data on how long the old way took versus how long the new way takes. The evaluation should also look out for situations where the original problem is solved, but another unintended and equally bad situation is created. If the hose is deployed more quickly but it comes out twisted and kinked, the negative could outweigh the positive.

Change the Plan If Necessary

If necessary, the fire officer needs to be prepared to adjust the plan or re-evaluate the original decision. Many problems are solved in stages, with gradual progress being made toward a solution. In spite of the analysis, plan B may turn out to be a better choice. Changing a plan should not be viewed as failure.

Feedback

Part of the evaluation process is to go back and listen to the people who identified the original problem and ask for feedback (▶ **Figure 11-6**). In the example illustrated in the box, the fire chief listened to the fire officers and dispatchers and learned that sending a single fire company to pulled fire alarm street boxes was providing a lower level of service and producing undesirable results. The policy was quickly changed.

Figure 11-6 Feedback is an important part of the evaluation process.

Managing Conflict

One of the factors that distinguishes a fire officer from a fire fighter is the responsibility to act as an agent of the formal organization. A fire officer is the official first-level representative of the fire department administration when dealing with subordinates and enforcing policies and procedures. This responsibility places the fire officer in the first position to deal with a wide variety of problems, including situations that potentially involve conflict, emotions, or serious differences of opinion.

Situations that involve conflicts and grievances require an additional set of skills that go beyond the general problem-solving model. The general model is designed to focus on solving the problem itself. In conflict situations, the issues are often much more complicated and sensitive. There could be a relatively simple problem that becomes complicated by the ways that different individuals react to it or to each other. In some cases, the problem is centered on the relationship between individuals or groups and is played out in relation to other issues.

Personnel Conflicts and Grievances

The close living relationships within a fire company can produce a variety of tensions, anxieties, and interpersonal conflicts, in addition to the common types of conflicts in most workplaces. One of the most difficult situations for a fire officer is an interpersonal conflict or grievance within the company or directly involving a company member.

A fire officer might be faced with any one of four different types of internal situations that can originate within the fire station. A fire fighter might come to an officer with a complaint about:

- A coworker (or coworkers)
- The work environment, including the fire station, apparatus, or equipment
- A fire department policy or procedure
- The fire officer's own behavior, decisions, or actions

Company Tips

New York City Mayor Ed Koch was famous for asking voters "How am I doing?" This is a very good question for fire officers to the people who work around them.

In each of these situations, the fire officer is the individual's first point of contact with the formal organization. The official response to the problem begins when the officer becomes aware that a problem exists. The relationship of the fire officer to the conflict and the complainant makes a significant difference in the role that the officer can play in resolving the conflict.

Fire officers at higher levels or performing other assignments within the organization must also be prepared to deal with problems that involve conflict. Their relationship to the individuals involved in the problem is likely to be different; however, their responsibility to officially represent the formal organization is the same.

Conflict Resolution Model

The conflict resolution model is a basic approach that can be used in situations where interpersonal conflict is the primary problem or a complicating factor.

Listen and Take Detailed Notes

The first phase of the conflict management template is to obtain as much information as possible about the problem. The fire officer should encourage the complainant to explain the situation completely. If the details are even slightly complicated, the fire officer should take notes **▼ Figure 11-7**.

Figure 11-7 The fire officer should take notes if the details of a complaint are even slightly complicated.

Company Tips

A fire officer has a different relationship with the company members than most supervisors have with their subordinates. Spending full shifts together in a fire station and engaging in emergency operations tend to create close relationships, which can be either an advantage or a disadvantage in different situations. The fire officer is likely to become aware of most internal problems at an early stage, so the opportunity for preventive action and early intervention can be a positive factor. On the other hand, the fire officer can become too close to some problems to deal with them effectively.

The kitchen is one of the most important rooms in the fire station. Given enough coffee and time, it is in this room that members who are present will discuss and attempt to solve all of the department's problems. This is a great place for a fire officer to learn about the issues and concerns that are on the minds of fire department members. It is also a good opportunity for the fire officer to informally release information and measure reactions to different issues. If the fire officer has important official information to share with the company members, it should not be slipped into casual conversation.

Although large quantities of information are exchanged in the kitchen, it should not substitute for the official communications mechanisms or the formal chain of command. Information that comes from a fire officer, even if it is intended as informal conversation, can be interpreted as a statement of official fire department policy. In the same way, the fire officer cannot deny official knowledge of information that was discussed at the kitchen table. Although a fire officer is entitled to have a personal opinion, it is unprofessional and confusing to express mixed messages.

Some fire officers make a practice of meeting informally, one-on-one, with every fire fighter every 3 to 6 months. These sessions help the fire officer understand what is going on with each person in the fire company. Because these discussions are private, the fire fighter may be comfortable expressing opinions or concerns that may not be shared at the kitchen table with everyone. These informal sessions can help the fire officer become aware of potential conflicts and problems before they become an issue. The fire officer can be proactive and take actions to resolve issues at the earliest opportunity.

The person who is making a complaint has a certain perspective on the situation. Whether you agree or disagree with that person's perspective, an important starting point is to find out what the complainant thinks about the situation.

Active Listening

When dealing with an individual who is expressing a concern or a problem, the fire officer should focus on active listening. Engaged or active listening is the conscious process of securing all kinds of information through a combination of listening and observing. The listener gives the speaker full attention, being alert to any clues of unspoken meaning while also listening intently to every word that is spoken. The fire officer should be aware of nonverbal clues that may indicate agreement, dissatisfaction, anger, or other emotions. Often, these nonverbal clues provide great insight as to the disposition of the speaker. The fire officer actively seeks to keep the conversation open and satisfying to the speaker, showing an interest in feelings and emotions as well as raw information (▶ **Figure 11-8**).

Paraphrase and Feedback

The first objective should be to understand the issue and why the individual is complaining. After listening, the fire officer should be able to paraphrase the complaint and recite it back to the complainant. Paraphrasing the issue and receiving feedback from the complainant accomplishes two goals: The fire officer finishes this phase with a good understanding of the issue from the complainant's perspective, and the complainant feels that the fire officer really listened.

Do Not Explain or Excuse

In situations where the complaint is directly related to actions taken or policies enforced by the fire officer, it is understandable that the fire officer would want to immediately respond to the complaint. In this type of situation, it is particularly important to listen and process the information before deciding on an appropriate response. In many situations, a reflex explanation or excuse gives the individual an additional reason to complain. If the complainant feels strongly enough to complain about something the officer has done, that officer's explanation is probably not going to solve the problem.

Figure 11-8 Give the speaker your full attention.

VOICES OF EXPERIENCE

❝ It is the fire officer's responsibility to address these issues within the standard operating procedures of the department to make certain that all parties receive adequate information and resolution. ❞

Upon arriving at the medical office complex, the fire and EMS crew proceeded into a physician's office to find a short-of-breath middle-aged woman. The physician stated that the woman could have a spontaneous pneumothorax and needed to be transported to the local Emergency Department. The physician left the room and the woman was moved to a stretcher, 100% oxygen applied, and taken to the ambulance. Upon entering the ambulance, the paramedic reassessed lung sounds and placed the patient on the oxygen saturation monitor. The woman stated that her breathing was getting worse and she was now speaking one or two word sentences. It is the opinion of the paramedic that the woman met the SOP for chest needle decompression. This was performed and the woman showed marked improvement. Transport was complete with no further changes.

Upon arrival at the Emergency Department, the receiving physician received the paramedic's report and promptly became very upset with the paramedic. According to the sending physician, there was no need for the procedure. In front of the patient and the family members, the receiving physician stated that the paramedic had acted inappropriately.

The paramedic immediately contacted his EMS Shift Commander who notified the EMS Division Chief. Following a SOP regarding complaints by the medical community, the EMS Division Chief contacted the Medical Director. The appearance of the EMS Shift Commander and the EMS Division Chief at the Emergency Department insured that all concerned understood the seriousness of the concerns raised by the physician. It was determined that the paramedic made a mistake, re-education of the paramedic would immediately be implemented. If the paramedic's action was deemed appropriate, the results of the investigation would be conveyed directly to the complaining physician to avoid ambiguity and bad feelings.

Frequently, managing complaints in the Fire/EMS service is dependant upon open, honest communication that recognizes that we all can make mistakes. It is the fire officer's responsibility to address these issues within the standard operating procedures of the department to make certain that all parties receive adequate information and resolution.

Gregg Lord
Cherokee Georgia County Fire-Emergency Services
26 years of service

Investigate

An **investigation** is a detailed inquiry or systematic examination. All complaints should be investigated, even if the foundation for the complaint appears to be weak or nonexistent. Fire department procedures should dictate who will conduct the investigation, depending on the nature of the complaint and the relationship of the individuals who are involved. Sometimes, the fire officer who received the complaint is assigned to conduct the investigation; however, a fire officer who is directly or personally involved in the problem should never be involved in conducting the investigation. Sensitive matters require an appropriate level of investigator.

The purpose of the investigation is to obtain additional information, beyond the initial information and perspective that came from the complainant. The investigator must be impartial in gathering and documenting information. The information could come from other individuals, departmental regulations, or incident-specific data. When investigating a human resource conflict, such as a payroll or work assignment issue, the fire officer might have to refer to specific departmental directives and regulations.

The product of an investigation is a report, which is provided in an appropriate format for the fire officer's immediate supervisor (▶ **Figure 11-9**). A complete investigative report has three objectives:

1. The report must first identify and clearly explain the issues.
2. The report should then provide a complete, impartial, and factual presentation of the background information and relevant facts.
3. The conclusion should be a recommended action plan, which is based on and supported by the information.

Take Action

Once the investigation is completed, the fire officer presents the findings and recommended action to a supervisor at a higher level. There are four possible responses:

1. **Take no further action.** The investigation concludes that the complaint was unfounded or requires no further action. If the complaint was related to an earlier decision or action, that original decision is affirmed. The response should include the reasons why no further action is recommended.
2. **Recommend the action requested by the complainant.** In this case, the investigation concludes that the complaint was justified and the requested action is the best solution to the problem.
3. **Suggest an alternative solution.** The investigation concludes that some alternative action or policy is the best solution to address the complainant's concerns. For example, a citizen might complain that the fire

truck blocks several spaces in the parking lot at the local gym and does not want the fire fighters to go there. The fire officer meets with the citizen, points out that benefits of fire fighters using the gym, and proposes a more appropriate parking space for the fire truck. The compromise is acceptable to the citizen and to the department.

4. **Refer the issue to the office or person that can provide a remedy.** Other members of the fire department or some other municipal agency can provide the relief the complainant seeks. Grievance procedures require that the employee start with the immediate supervisor for all complaints. If the employee is not satisfied with the response at that level, the grievance can then be taken to a higher level. If the problem involves a paycheck deduction issue, it will probably have to be resolved by the payroll clerk or human resources office. The fire officer's duty is to refer the complaint to the appropriate person.

Follow-up

For many of the conflicts, the fire officer needs to follow-up with the complainant to see whether the problem is resolved.

Emotions and Sensitivity

Fire fighters are passionate about their profession and are deeply concerned about many issues that affect the job. They live, breath, eat, and sometimes dream about fire department operations. Because of this, they often are very emotional when making a complaint. Michael Taigman, a writer, teacher, and consultant on emergency medical services quality issues, provides a conflict resolution model that is especially effective when emotions are high. It was effective for Mike Taigman when working with employees during a stressful creation of a paramedic ambulance service under a tight schedule. He continued to refine the model while working as a consultant for other emergency service organizations. The model follows four steps:

1. Drain the emotional bubble.
2. Understand the complainant's viewpoint.
3. Help the complainant feel understood.
4. Identify the complainant's expectation for resolution.

Step 1: Drain the Emotional Bubble

The body reacts to emotional conflict or stress the same way it does when you are the first-arriving fire officer at a greater-alarm structural fire. Both situations result in a dumping of adrenaline to prepare the body to fight or run away. The same adrenaline-induced red haze that reduces fire fighter effectiveness at emergencies also happens to ordinary citizens. It tends to bring complaints to the surface and impedes

Metro City Fire Department
Internal memorandum

Date: November 1, 2006

To: Deputy Chief Bruce Appleton, Operations Division

Thru: Battalion Chief Jane Stapleton, Battalion 2

From: Captain Stan Holtz, Engine 7

Subj: Citizen Complaint: slow response of Engine 7 to fire emergency.
Inquiry 2006-71   Incident #0429
October 28, 2005   1437 Richland Street

Mrs. Cindi Creswell, 1437 Richland St., (555) 555-1267, is the owner of the fire building and made the 9-1-1 calls. Mrs. Creswell told Captain Holtz that it took 20 minutes for the fire department to arrive after she called 9-1-1. Captain Holtz said that he would investigate. Captain Holtz notified Battalion Chief Jane Stapleton on October 28.

Investigation:

According to the computer aided dispatch records:

21:35:12: 9-1-1 receives telephone call reporting odor.
21:37:20: Engine 7 dispatched to 1437 Richland Street.
21:38:40: Engine 7 marks EN ROUTE to call.
21:39:07: Second 9-1-1 call "Where is fire department?" Caller reports basement is on fire.
21:39:29: Engines 11, 3, and 9; Truck 11 and Battalion 2 dispatched for a reported structure fire.
21:39:40: Dispatch advises Engine 7 of upgraded alarm and report of basement fire.
21:42:24: Engine 7 marks ON SCENE and reports "Two-story, single-family frame with smoke showing from basement windows."

Engine 7 was on the scene five minutes and four seconds from dispatch. There was no delay in turnout (80 seconds) or response.

Action Taken:

Located Mrs. Creswell at the Shady Acres Hotel, (555) 555-9767. Chief Stapleton and Captain Holtz met with Mrs. Creswell on October 30 and reviewed dispatch and response information.

Follow-Up:

No further action is recommended.

Figure 11-9 Example of a fire officer investigation report.

resolution of conflicts. Mike Taigman, a former paramedic, says that the adrenaline fills up the prefrontal lobes of the neocortex of the brain. That creates an emotional bubble that interferes with the ability of the complainant to hear the fire officer or consider any response to the issue.

Mr. Taigman recommends listening deeply, actively, and empathetically in order to drain this emotional bubble. This is not an easy task for the fire officer, but writing detailed notes and not explaining or excusing helps the process. The fire officer asks questions and encourages responses, draining the emotional bubble by allowing the complainant to completely express grief, regret, pain, censure, or resentment.

This discussion should be in private and should have a couple of ground rules. There should be no physical contact. If this discussion is between a fire fighter and a fire officer, avoid personal attacks and concentrate on the work issues.

Step 2: Understand the Complainant's Viewpoint

The initial complaint or behavior may be a sign or symptom of a larger problem. By draining the emotional bubble through active listening, the fire officer may identify the root cause or issue of the complaint.

Internal conflicts, grievances, or issues occasionally suffer from long memories. It may require some investigating to understand the current issue, which may be related to something that happened years ago or may be wrapped up in history and tradition.

Step 3: Help the Complainant Feel Understood

The "listen and take detailed notes" part of the basic conflict management template includes the recommendation to paraphrase and feed back what you heard from the complainant. Some issues are resolved if the complainant feels that the fire officer understands the issue, conflict, or problem.

Safety Zone

Aggression, Anger, and Acting Out

Some people have trouble handling their anger. Some have very abrasive personalities, others launch vicious verbal attacks, and a few may physically act out their feelings. The fire officer needs to monitor the situation and the complainant. If, in attempting to drain the emotional bubble, there is a rise in rage, then take a time-out.

If this situation involves a fire fighter, follow your organization's procedure on workplace violence or hostility. If the conflict is with a civilian at an emergency scene, ask for police assistance. The police have much more training in handling angry and potentially violent citizens.

Step 4: Identify the Complainant's Expectations for Resolution

By this final step, the complainant has drained the emotional bubble, has been able to describe what is going on, and feels that the fire officer understands the issue, problem, conflict, or grievance. The fire officer should now ask what the complainant expects the department to do to resolve this issue. If the problem is an internal grievance, this is where the employee should be asked to describe the desired action.

Citizen Complaints

The fire officer represents the department in dealing with citizens, public or private organizations, and other governmental agencies. A fire officer could be faced with three different types of citizen complaints. A citizen might complain about:

1. The conduct or behavior of a fire fighter (or a group of fire fighters)
2. The fire company performance or service delivery
3. Fire department policy

Sometimes, a citizen may want to express an alternative viewpoint on an issue and try to see whether there is a resolution that is mutually agreeable to the department and the citizen. At other times, the citizen is making a formal complaint. When this occurs, the fire officer is functioning as the official recipient of the complaint. In some cases, the fire officer is the subject of the complaint. The first role of the fire officer is to respond to a complaint in a professional manner that effectively obtains the needed information and does not make the situation worse. The methods outlined in resolving conflict within the company also apply to a citizen complaint.

The fire officer must take notes and function as an active listener. By listening attentively and taking detailed notes, the fire officer is demonstrating that the complaint is officially considered to be important and is receiving the fire officer's full attention. If the immediate response to a conflict is an explanation or excuse, the complainant will feel that the fire officer is not paying attention, does not care about the issue, or has something to hide. The person who had a complaint will likely stop providing information and feel even more strongly that the complaint was justified.

On some citizen complaints, the fire officer may be able to resolve the problem. On other issues, such as a complaint about using a siren at night, the fire officer should explain the rationale for the use of the siren at all times as well as the laws that regulate its use. Be empathetic and listen to the problem without interrupting. Frequently, these kinds of complaints are a method of venting frustration by the citizen rather than a real expectation that it will change. Allowing the citizen to express the frustration is important to resolving the issue.

If the fire officer does not have the authority to make a decision on the issue or the citizen is dissatisfied with the officer's decision, he or she should ask whether the citizen would like the officer to refer the concern to a higher level. If the citizen would like further action, the fire officer should determine the appropriate organizational level where the decision can be made. Generally, the best method to determine this is to consider the scope of the complaint. If the complaint is strictly an operational issue, it should be sent to the chief of operations. If it is an issue about codes, it would be sent to the chief fire marshal. If the issue affects all areas of the department, it would be forwarded to the chief of the department. Complaints about personnel should be forwarded to the supervisor of the individual that is involved for action.

All relevant facts should be identified and forwarded, along with the details of the complaint. In some departments, the proper procedure is to follow the chain of command. In other departments, fire officers are encouraged to forward the information directly to the decision maker in order to ensure prompt attention to the matter. Even in these circumstances, the fire officer should inform his or her supervisor about the situation. If there is any doubt about how to proceed, discuss the issue with your supervisor before taking any action.

Policy Recommendations and Implementation

Because the fire officer is the level of management that is regularly in direct contact with the fire fighters as well as the citizens, the fire officer is often in the best position to recommend new departmental policies. This change could be the result of a citizen or employee complaint or an identified problem. The problem could be anything that creates a disparity between the actual state and the desired state. For example, not having a standardized location for all equipment on the apparatus reduces efficiency on incident scenes; therefore, it is a problem.

Recommending Policies and Policy Changes

The fire officer must understand the procedure for adopting new policies within the department. Although many fire fighters may believe that the procedure simply consists of getting the chief to put his name on it, usually it entails much more. Many departments have a policy on how to implement a new policy. If the department has such a policy, the fire officer should follow it.

The most common method used by fire officers to get new policies adopted or to cause changes to existing policies is by outlining the problem that they have identified to their supervisor with the recommendation that "someone ought to do something about this." This places all responsibility on the supervisor to come up with a solution and get the department to adopt it. The benefit that this method is easy for the officer is far offset by its ineffectiveness. This is why most of the time, nothing changes, and if it does, the change may create more problems than it solved.

To successfully recommend a new policy or a change to an old policy, the fire officer should carefully identify the problem and develop documentation to support the need for a change. This could be with statistical measures, anecdotal evidence, or both. The fire officer's opinion seldom carries enough weight to cause a change because other officers may have different opinions. One effective technique is to discuss the situation with other fire officers initially to determine how widespread the perceived problem is.

Once the supporting evidence has been developed that there is a widespread problem, the problem-solving techniques outlined earlier in the chapter should be used to develop and choose the best alternative. At this point, the fire officer is ready to write out a proposal to administration. The problem should be carefully outlined, along with the proposed policy change that will resolve it. Resources that will be needed for the solution must also be identified, including financial and time commitments. The benefits of the solution should be identified, as should any potential negative affects and how they will be addressed.

Once a written proposal is developed, the fire officer should approach his or her superior if that person has not already been part of the development. Along with the recommendation, the policy should be presented to the appropriate officer whose scope is to oversee the area affected by the policy. It is critical to get the agreement of this fire officer, because without this recommendation, the policy will likely not be implemented.

With this officer's recommendation, the proposal is usually forwarded to the chief of the department for review. Usually, a review committee composed of senior staff officers evaluates the proposal and makes changes and accepts or rejects the proposal. Once it is accepted, a draft policy is usually developed in the proper format. The draft policy is distributed to all personnel for review and comment. After a comment period, all comments are addressed, and a final policy is signed by the Chief and distributed to the department.

Implementing Policies

Like the recommendation process, the process for implementing new policies varies widely. The fire officer should follow departmental procedures. In their absence, the officer may use the following methods to improve their communication and understanding.

Because policies are the backbone of order for the fire department, every individual should understand the policies that affect him or her. For this to occur, the fire officer must take responsibility to ensure that the fire fighters are informed about the policy and take the time to understand it. The fire officer must also ensure that he or she follows all policies. Failure of the officer to follow all policies undermines their importance, and fire fighters will develop the attitude that they can choose which policies they will follow.

When a new or amended policy is distributed, the fire officer should ensure that it is communicated to the subordinates. Frequently, this occurs at the morning roll call. At the beginning of each shift, the new policies should be discussed. The fire officer should point out the points that are most relevant, particularly those areas that change the current practice rather than formalize the current practice. Reading through a full policy may not draw the attention to the subtle changes that have occurred—the fire officer should point these changes out. For example, if a revised driving policy is distributed that now requires every apparatus to come to a complete stop at all red signals and stop signs rather than slow to a speed so as to avoid an accident, the fire officer should specifically point this change out.

The fire officer should require that all fire fighters read the policy and sign off that they understand the policy. This helps ensure accountability. Employees tend to follow policies more closely if they believe they will be held accountable for them. To make sure that employees understand the policy, the fire officer should give situations covered by the policy and ask them to apply the policy. Many departments also require that a copy of the policy be posted on the bulletin board for a period of time for employees who are not working when it is published. Some also require a notation in the station log book that the policy was distributed.

The fire officer should evaluate the employee's actions against the policy. If a policy is violated, the fire officer should review the policy with the employee involved. For repeated violations or a violation of a safety policy, disciplinary action should be considered. Employees should never get the feeling that it is acceptable to violate policies.

Lastly, a regular review of policies should occur. Selecting policies that are not routinely encountered and testing fire fighters' knowledge about them will help keep everyone up-to-date.

Customer Service Versus Customer Satisfaction

Customer service is a term that public safety has adopted from the retail business. A focus on customer service fixes problems, straightens out procedural glitches, corrects errors of omission (or commission), and provides information.

Customer satisfaction focuses on meeting the customer's expectations. Fire departments meet customers during one of the worst days of their lives. They did not start the day planning to crash the car, ignite a fire in the trash can, or have trouble breathing. Good customer service requires sensitivity on the part of every person involved.

Although there is generally little competition for others to provide public safety services, the importance of satisfied customers is vital. Creating satisfied customers is one of the most important activities that a fire officer can do to improve fire department status within the community (▶ Figure 11-10).

Complainant Expectations

When dealing with a civilian, the fire officer should ask what could resolve the situation. The fire officer needs to take the response from the complainant in order to resolve the problem.

In some cases, the resolution may be as simple as acknowledging that the complainant has had a bad day or a bad experience with the department. In other cases, it may be as extreme as recommending that the department terminate the employee or employees involved. Unless it is within the scope of the fire officer's authority, the fire officer should make no promises or imply that certain actions will be taken in the discussion with the complainant. In many cases, the complainant's expectations must be relayed to the fire officer's supervisor as a part of resolving the issue. If the proposed resolution involves discipline, the fire officer is also obligated to protect employee privacy and civil service due process (▶ Figure 11-11).

Keep Complainant Informed

If the fire officer needs to do research, consult with others, or obtain direction from supervisors, he or she should keep the complainant informed during the process. This communication

(**Figure 11-10**) Creating satisfied customers is one of the fire officer's most important activities.

Metro City Fire Department

Headquarters: 100 Public Safety Lane

January 11, 2005

Mister Marvin Bryant
3314 Enzo Terrace
Metro City

Mr. Bryant:

We have completed our internal investigation in response to your property damage claim that you filed with the city on December 29th.

Around 6:30 AM on December 26, 2005, Engine 6 backed into your automobile while it was parked in front of your townhouse.

When you went to Fire Station 6 later that morning to file a complaint, Captain Dale Smith made untruthful statements when he stated that the pumper had been in the fire station all morning.

I apologize for the behavior of Captain Smith. The department will be taking appropriate disciplinary measures.

Mr. Sam Hall from the Mayor's staff is the city representative who will be coordinating the repairs to your vehicle. I understand he has already contacted you and arranged for a rental replacement until your car is repaired.

Sincerely,

Carl Cosgrove

Fire Chief
Metro City

Figure 11-11 Example of a letter to a civilian in response to a complaint.

demonstrates that the fire officer is not ignoring the issue, and it educates the complainant on the process.

Follow-up

A common citizen complaint is that local government is unresponsive. By following up with the complainant, the fire officer reinforces the impression that the complainant's issue is important. This is especially important if the fire officer has referred the issue to another individual, agency, or organization.

Follow-ups may be inappropriate if the fire officer was the subject of the complaint or the conflict was handled at a higher supervisory level. The fire officer may consider consulting with a supervisor before conducting a follow-up.

You Are the Fire Officer: Conclusion

You first make sure that the driver is evaluated by EMS and confirm that police are en route. You notify the chief that the engine has been involved in a crash and is unavailable for emergency response. Do not forget to mention that the city chairperson was the driver of the passenger vehicle. You assign the laughing fire fighters to a task that moves them away from the incident.

You meet with the driver of the vehicle to apologize for the accident and explain what you are doing to remedy the situation. While speaking with the driver, you perform active listening and allow the driver to express her feelings. You do all you can to assist her in making notifications or arranging for assistance.

Once the driver is properly situated, either transported to the hospital or talking with the police officer investigating the accident, you begin the administrative paperwork associated with an apparatus accident. A key point of the investigation is determining why the engine rolled down the driveway. Was it due to mechanical failure, operator error, or procedural failure? A long-term goal is to identify the cause of the accident and take appropriate actions to avoid a repeat of the conditions that lead to this mishap.

Wrap-Up

Ready for Review

- The fire officer has to deal with conflict, emotions, or serious differences of opinion.
- Situations that involve conflicts and grievances require an additional set of skills.
- The fire officer takes or recommends four actions after completing an investigation:
    - Take no further action
    - Recommend the action requested by the complainant
    - Suggest an alternative solution
    - Refer the issue to the office or person that can provide remedy
- A citizen might complain about:
    - The conduct or behavior of a fire fighter
    - The fire company performance or service delivery
    - Fire department policy
- Because the fire officer is in regular direct contact with both fire fighters and citizens, he or she is often in the best position to recommend new departmental policies.

- A fire officer must understand the procedure for adopting new polices within the department.
- The process for implementing new policies varies.
- Customer service is a important part of customer satisfaction.

Hot Terms

Brainstorming A method of shared problem solving in which all members of a group spontaneously contribute ideas.

Complaint Expression of grief, regret, pain, censure, or resentment; lamentation; accusation; or fault finding.

Conflict A state of opposition between two parties. A complaint is a manifestation of a conflict

Investigation A systematic inquiry or examination.

Mistake An error or fault resulting from defective judgment, deficient knowledge, or carelessness. A misconception or misunderstanding.

The fire chief issued a memo a week before regarding the upcoming budget process. He has asked the company officers responsible for different areas of the budget to get together and work on an operations budget that needs to be submitted to him by the end of the month. As the senior company officer, you organize a meeting and plan to leave the meeting with a draft budget plan to submit to the chief.

1. What is a method of shared problem solving in which members of a group spontaneously contribute ideas?

- **A.** Active listening
- **B.** Brainstorming
- **C.** Open input
- **D.** Analysis

2. What is one step that will assist in brainstorming?

- **A.** Function as the scribe
- **B.** Filter the ideas
- **C.** Limit input
- **D.** Ensure that the process is autocratic

3. What is the problem with having the chief participate in brainstorming?

- **A.** People get angry
- **B.** Group dynamic changes and people do not feel comfortable
- **C.** The subject changes
- **D.** Fire fighters feel too comfortable

4. How do you select the best solution?

- **A.** Pick out of a hat
- **B.** Forward to the chief for discussion
- **C.** Open to voting by the whole department
- **D.** Rank and select based on which one meets core values of the department

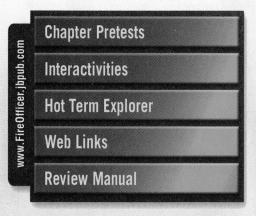

www.FireOfficer.jbpub.com

Chapter Pretests

Interactivities

Hot Term Explorer

Web Links

Review Manual

www.FireOfficer.jbpub.com

Preincident Planning and Code Enforcement

EXIT

Technology Resources

www.FireOfficer.jbpub.com

- Chapter Pretests
- Interactivities
- Hot Term Explorer
- Web Links
- Review Manual

Chapter Features

- Voices of Experience
- Getting It Done
- Assessment Center Tips
- Fire Marks
- Company Tips
- Safety Zone
- Hot Terms
- Wrap-Up

NFPA 1021 Standard

Fire Officer I

4.6* **Emergency Service Delivery.** This duty involves supervising emergency operations, conducting preincident planning, and deploying assigned resources in accordance with the local emergency plan and according to the following job performance requirements.

4.6.1 Develop a preincident plan, given an assigned facility and pre-planning policies, procedures, and forms, so that all required elements are identified and the approved forms are completed and processed in accordance with policies and procedures.

(A) Requisite Knowledge. Elements of the local emergency plan, a preincident plan, basic building construction, basic fire protection systems and features, basic water supply, basic fuel loading, and fire growth and development.

(B) Requisite Skills. The ability to write reports, to communicate orally, and to evaluate skills.

Fire Officer II

5.5 **Inspection and Investigation.** This duty involves conducting inspections to identify hazards and address violations and conducting fire investigations to determine origin and preliminary cause, according to the following job performance requirements.

5.5.1 Describe the procedures for conducting fire inspections, given any of the following occupancies, so that all hazards, including hazardous materials, are identified, approved forms are completed, and approved action is initiated:

1. Assembly
2. Educational
3. Health care
4. Detention and correctional
5. Residential
6. Mercantile
7. Business
8. Industrial
9. Storage
10. Unusual structures
11. Mixed occupancies

(A) Requisite Knowledge. Inspection procedures; fire detection, alarm, and protection systems; identification of fire and life safety hazards; and marking and identification systems for hazardous materials.

(B) Requisite Skills. The ability to communicate in writing and to apply the appropriate codes.

5.6 **Emergency Service Delivery.** This duty involves supervising multiunit emergency operations, conducting preincident planning, and deploying assigned resources, according to the following job requirements.

Additional NFPA Standards

NFPA Fire and Life Safety Inspection Manual
NFPA 1, *Uniform Fire Code*
NFPA 10, *Standard for Portable Fire Extinguishers*
NFPA 12, *Standard on Carbon Dioxide Extinguishing Systems*
NFPA 101, *Life Safety Code*
NFPA 220, *Standard on Types of Building Construction*
NFPA 291, *Recommended Practice for Fire Flow Testing and Marking of Hydrants*
NFPA 704, *System of the Identification of Hazards of Materials for Emergency Response*
NFPA 1620, *Recommended Practice for Preincident Planning*
NFPA 2001, *Standard on Clean Agent Extinguishing Systems*

Knowledge Objectives

After studying this chapter, you will be able to:

- Discuss how to develop a preincident plan.
- Understand built-in fire protection systems.
- Understand fire code compliance inspections.
- Identify the five types of building construction as used in the fire prevention code.
- Prepare for an inspection.
- Describe general inspection requirements.

Skills Objectives

After studying this chapter, you will be able to:

- Create a preincident plan.
- Demonstrate how to conduct an inspection.

The fire company crew members drop their jaws. This is their first visit to the Towne Centre complex site since construction started. The architect shows you and your crew the plot plans. Four city blocks of mixed retail, office, and parking, plus a half-dozen residential buildings, from four to 30 stories high. The senior fire fighter looks at you and says, "We never had anything like this before in our district. Think of the preincident planning we need to do!"

You realize that most of this project will also require regular fire safety inspections and fire prevention code enforcement.

1. What is the best way to perform a preincident plan for a complex facility?
2. How can you prepare fire fighters for fire inspection duties?

The Fire Officer's Role in Community Fire Safety

A fire officer should always function as a partner with the community in a continuing effort to create a fire-safe community. Fire departments perform a range of functions that are related to fire prevention, risk reduction, preincident planning, and public education. Fire officers play multiple roles in relation to properties within their communities, which generally include:

1. Identifying and correcting fire safety hazards—code enforcement
2. Developing preincident plans
3. Promoting fire safety—public education

The assignment of responsibilities to perform these functions varies considerably in different fire departments. In most areas, fire inspectors and fire officers working in staff assignments perform fire inspections and code enforcement duties (◄ Figure 12-1). Fire suppression companies are usually involved in preincident planning and also perform code enforcement inspections in many jurisdictions. Public education activities are often preformed by a combination of staff personnel and fire companies.

Even where the role of fire companies does not include code enforcement, fire officers and fire fighters should conduct regular visits to properties to develop preincident plans and look for potential hazards. A fire officer should always try to reduce the impact of any potential fire emergency that could occur, including identifying and correcting conditions that could lead to the ignition of fires or increase the risk to citizens and fire fighters if a fire does occur.

Developing a Preincident Plan

A **preincident plan** is described by NFPA 1620, *Recommended Practice for Preincident Planning* as a document developed by gathering general and detailed data used by responding personnel to determine the resources and actions necessary to mitigate anticipated emergencies at a specific facility (► Figure 12-2). Many fire departments conduct code enforcement inspections and develop preincident plans at the same time, whereas others divide the work into two separate activities.

Figure 12-1 Fire officers often conduct preincident planning and fire inspection activities during the course of their workday.

Tactical Priorities

Address: 1500, 1510, 1520		
Occupancy Name:		
Preplan #: 02-N-01	Number Drawings: 1	Revised Date: 12/2002
District: E275A	Subzone: 60208	By: ACEVEDO

Rescue Considerations: Yes ( ) No (X)

Occupancy Load Day:	Occupancy Load Night:
Building Size:	Best Access:

Knox Box: NONE	Knox Switch: NONE	Opticom: NONE
Roof Type: X	Attic Space: Yes ( ) No ( )	Attic Height: X

Ventilation Horizontal:	Ventilation Vertical:

Sprinklers: Yes (X) No ( )	Full (X) Partial ( )
Standpipes: Yes (X) No ( )	Wet ( ) Dry ( )
Gas: Yes ( ) No ( )	Lpg ( )

Hazardous Materials: Yes (X) No ( )

DIESEL GENERATORS

1,000 GALLON TANKS

BATTERY ROOM

Firefighter Safety Considerations:
ELEVATOR PIT

Property Conservation And Special Considerations:
VENTILATION: AUTOMATIC SMOKE REMOVAL SYSTEM

3 OFFICE BLDGS; 2 PARKING STRUCTURES

6 FLRS - 1230 W. WASHINGTON ST.
4 FLRS - 1500 N. PRIEST DR.
4 FLRS - PARKING GARAGE

Figure 12-2 An example of a preincident plan.

The original purpose of a preincident plan was to provide information that would be useful to the fire department in the event of a fire at a high-value or high-risk location. Today's preincident plans often include information that could apply to a variety of potential situations that could occur at the location. Facilities that store or handle hazardous materials are required to submit information to the fire department and the local emergency planning council (LEPC). Since 2001, many jurisdictions and businesses have included potential terrorist activities in preincident planning.

The most essential reason for developing a preincident plan is to obtain information before an incident occurs that would be difficult to obtain during the incident. The investment that is made in the preincident phase results in a more valuable and effective response during an incident. When an incident occurs, that information is immediately available to the incident commander and other responding fire officers. Ideally, the individual preparing the plan should try to include any information that could be valuable during an incident.

The second purpose of a preincident plan is to identify in advance the strategies, tactics, and actions that should be taken if a predictable situation occurs. The preincident plan allows for the incident to be thought out and practiced in advance. Preincident plans can be particularly useful at the company level for practicing initial operations for buildings in their district.

When the fire officer is assessing the potential situations that could affect a facility during emergency conditions, the following factors need to be evaluated:

- Construction
- Occupant characteristics
- Fire protection systems
- Capabilities of public or industrial responding personnel
- Availability of mutual aid
- Water supply
- Exposure factors
- Access
- Utility cutoff locations

During this assessment, the fire officer also evaluates the department's ability to respond and control an incident at the facility. For example, a small fire department would probably need to call for mutual aid for a serious fire in a large building. The anticipated mutual aid requests can be planned in advance of the incident.

A Systematic Approach

To ensure a systematic approach that collects all of the required data, the fire officer should use a standardized method for each preincident plan. NFPA 1620 provides a six-step method of developing a preincident plan:

1. Evaluate physical elements and site considerations.
2. Evaluate occupant considerations.
3. Evaluate fire protection systems and water supply.
4. Evaluate special hazards.
5. Evaluate emergency operation considerations.
6. Evaluate special or unusual characteristics of common occupancy.

Step 1: Evaluate physical elements and site considerations

The first step is to evaluate the physical elements and site considerations. The building's size and dimensions, including overall height, number of stories, length, width, and square footage, should be determined and included in the plan. Connections between buildings and distances to exposures should be noted. Access routes and points of entry should be clearly indicated. The preincident plan should identify concealed spaces and windows that can be used for ventilation, rescue, or both.

The preincident plan should include detailed information about the construction of the roof, floor, and walls, and their structural integrity. The focus is on factors that could lead to collapse of building components or spread of fire or toxic gases. Any factor that would affect the ability of responding fire fighters to enter and safely perform interior operations should be highlighted. All indications of deterioration that could lead to premature structural collapse or other conditions that could affect the spread of fire or toxic gases should be noted. Conditions that affect the safety of fire operations on the roof should also be noted.

This part of the preincident plan also includes locations of electrical power and domestic water shutoffs, heating-ventilating-air conditioning controls, as well as any fuels, compressed or liquefied gases, or steam. Elevator information should include the number, location, floors served, type of elevator, and type of recall or override service.

This section of the preincident plan should also include information on conditions that would hamper the access of responding personnel, including weight-restricted bridges, low overhead clearances, and roads subject to natural or man-made blockages. Security information, such as presence of fences and guard dogs, should be included in the preincident plan. The fire officer must also determine the environmental impact of any potential contaminants that could be released.

During a test of the preincident plan, the fire officer should document any interference or poor coverage of the communications system that will be used during the emergency. For example, when conducting a preincident plan of a high rise, be sure to test your portable radios to ensure that you can contact dispatch when in the inner core of the building and when below grade. Often, these areas are prone to cause "dead" spots in the radio system. A method to overcome these deficiencies should be developed. It might include moving from a frequency that is repeated to one that is simplex, in which the portable radios work in a line of sight. It could also be resolved by having a mobile repeater respond to the scene or by having a fixed antenna system installed in the building.

Step 2: Evaluate occupant considerations

Life safety considerations are always a priority. The preincident plan should allow responding emergency service personnel to assist in rapid and safe evacuation of a facility or, alternatively, support protection in place. Protecting in place would require the officer to determine what areas within the structure are resistant to the magnitude of fire that would be expected to occur. For example, in a high-rise building of fire-resistive construction, the officer might determine that the most likely fire would be contained to the floor of origin. With this information, the officer could determine that only the fire floor occupants would need to be evacuated. The occupants on all other floors would be instructed on the proper method of sealing their doors and windows and closing air handling system vents and then to remain in their rooms.

If the preincident plans determine that the occupants should be removed, the plan should identify how that will be accomplished. It could specify that the second- and third-arriving companies would immediately begin removing occupants, first from the fire floor and then from the floor immediately above. It would also identify the escape routes that should be used to evacuate occupants.

Information should be gathered on the number of occupants and their ages, their physical or mental conditions, and whether they are ambulatory or nonambulatory. The information should include hours of operation, occupant load, and location of occupants. Each of these has a bearing on how many companies would be needed and what they can expect for removing occupants.

The preincident plan should note the location of exits and any special locking conditions, such as a delayed release or stairwell unlocking system. The fire officer should review the facility's emergency action and evacuation plans and ensure that the preincident plan reflects coordination between the facility staff and the fire fighters. This might include a meeting between the fire officer and the staff of a nursing facility in which they predetermine where occupants will be moved by the staff. This assists the fire department in evaluating whether further removal would be needed, and if this is the case, this allows for the prompt request for additional units. Both the emergency action plan and the preincident plan should identify the locations of any occupants who need assistance to safely evacuate the building, noting the locations of equipment such as stair chairs, stretchers, and lifts.

The preincident plan should contain provisions for sheltering evacuated occupants. When large numbers of people are relocated, they quickly need basic services, such as food, water, and shower facilities. The preincident plan should identify where occupants would be sheltered and how these needs would be met. The public school systems often work closely with the fire department for these types of situations. Identifying the nearest location and the contact person in advance makes an evacuation go much smoother.

An accountability system for determining the location and the number of occupants should be established, particularly for locations where the number of occupants varies with the facility's operations and time of day, is also needed. The Red Cross can be a valuable asset in assisting the fire officer with these determinations.

Some facilities cannot be completely evacuated in a short time, such as high-rise, health care, or detention facilities. In those occupancies, the preincident plan could be to defend in place, providing areas within the structure that are designed to provide occupant protection during a fire. These facilities are normally fire-resistive construction, which allows occupants to remain protected from the fire for significant periods of time. However, for this system to be effective, the products of combustion must be segregated from the occupants. This is normally accomplished by moving those in most peril to areas away from the fire. Fire doors are then closed, and the individual rooms can also be sealed with wet towels to prevent smoke from filtering into the room. The heating-ventilating-air conditioning is often designed to shut down and then work to eject the smoke from the structure. This is a common practice in hospitals. When a fire alarm occurs, the patients of that wing are moved into adjoining wings, which are separated by fire walls. The doors are closed, and when the incident is stabilized, the patients are moved back to their rooms.

Step 3: Evaluate fire protection systems and water supply

The required water flow should be determined by evaluating the site in terms of size of the building or buildings, contents, construction type, occupancy, exposures, fire protection systems, and any other features that could affect the amount of water needed to control or extinguish the fire. The adequacy of available water for sprinkler systems, inside and outside hose streams, and any other special requirements or needs should be considered when a site is evaluated for its fire loss potential.

The preincident plan should identify the water supply that is available. The officer needs to identify the location of the closest fire hydrants, their flow rates, and how the distribution system that feeds them is configured. Some fire hydrants may have only single or double 2½-inch outlets, whereas another might have double 2½-inch outlets and a steamer connection. A department that routinely uses a large-diameter hose might want to opt for the closest hydrant with a steamer port. The distribution system that feeds the hydrants is also important to understand. A dead-end main or a main that feeds the fire protection system should be avoided. The ideal hydrant would be fed from a large main that is part of a grid that allows water to flow from several directions. A water supply test should be conducted in accordance with NFPA 291, *Recommended Practice for Fire Flow Testing and Marking of Hydrants*. When the demand exceeds the available supply, the preincident plan should address an appropriate response to mitigate the deficiency. This might include a water shuttle operation or use of relay operations. Some sites may have a private water system, including their own water tower that can be used if it does not interfere with the fire protection system.

The preincident plan should identify the location and details of every fire department connection, fire pump, standpipe system, and automatic sprinkler system (► Figure 12-3). Smoke management and special hazard protection systems should also be detailed on the preincident plan. Finally, the preincident plan should include data on the protective signaling system.

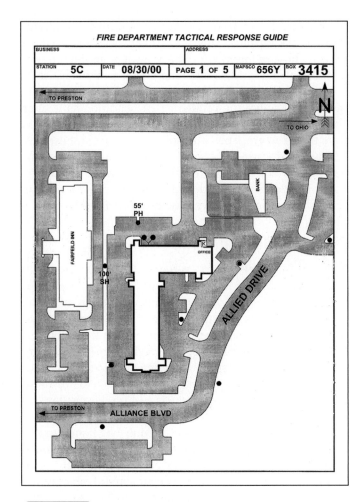

Figure 12-3 A diagram showing hydrant locations around a building.

Figure 12-4 Document any special hazard in the facility.

document emergency operating procedures and identify personnel that can provide technical assistance during emergency incident mitigation.

The preincident plan should include instructions for de-energizing high-voltage electrical systems and isolating and securing mechanical systems, or it should indicate where the instruction can be found. These systems can be shut down to protect emergency responders from the hazards of electrocution, mechanical entrapment, or other perils. Digital cameras can be used to collect this type of information for training and familiarization.

Step 5: Evaluate emergency operation considerations

The preincident plan should address a standard response to an incident at the facility or site. The initial emergency scene incident action plan should be based on the priorities of life safety, scene stabilization, and incident mitigation. The preincident plan should address the implementation of an appropriate incident management system on the arrival of the first emergency units. The planned location of the incident command post and emergency operations center, if provided, should be noted.

One consideration that must be evaluated is the assessment of the fuel loading. Fuel load refers to the total quantity of all the combustible products that are within a room or space. The fuel load determines how much heat and smoke will be produced by a fire, assuming that all of the combustible fuel in that space is consumed. The size and the shape of interior objects and the types of materials used to create them have a tremendous impact on the objects' ability to burn and on their rate of combustion. The same factors influence the rate of fire spread to other objects. Even the arrangement of the furnishings within a room can make a difference in the way a fire burns.

Step 4: Evaluate special hazards

The fire officer should document any special hazard in the facility to ensure that responding personnel are aware of these hazards, appropriate response considerations are made, and proper resources are activated during an emergency ▶ **Figure 12-4**). Special hazards include flammable or combustible liquids, explosives, toxic or biological agents, radioactive materials, and reactive chemicals or materials. During the site survey, the maximum inventory values of hazardous materials and highly combustible products should be requested. The preincident plan should include a method to obtain the current inventories' material safety data sheets at the time of the emergency. Contact information for the hazardous materials coordinator should be on the preincident plan.

Some occupancies contain very specialized operations, processes, and hazards that can pose unique challenges in an emergency. The preincident plan should identify any area of the occupancy that contains gases or vapors that could present a hazard to emergency responders. This includes confined spaces, inert atmospheres, ripening facilities, and special equipment treating atmospheres. The preincident plan should

Consideration must also be given to fire growth and development. Fires develop in phases: ignition, growth, fully developed, and decay. The ignition phase of a fire is the starting point of a fire. A fire in this phase usually involves only the object of origin. A preincident plan may identify what would be required to handle a fire. For example, a loaded stream-type extinguisher may be required for the combustibles present in the building. A fire in the growth phase involves other objects in the fire. The preincident plan might indicate that a growth-stage fire would require a minimum of a 1¾-inch handline. A fire in the fully developed phase has already flashed over. Depending on occupancy and construction factors, the preincident plan might recommend that a fire in this stage be fought defensively on arrival, or it might indicate that an attack with large handlines could be initiated. When the fire has consumed all of the oxygen but has retained the heat and has fuel, it has entered the decay phase. Plans would indicate how the building would be ventilated when this situation is encountered.

The preincident plan should also identify how the fire will progress. Open pipe chases, chimneys, elevator shafts, and balloon construction all lend to vertical fire spread. Large open areas may conceal the magnitude of a fire that has only light smoke showing from the outside yet allows for horizontal fire spread. False ceilings and cocklofts may also conceal horizontal fire spread and therefore should be identified.

If the facility has an evacuation plan or an emergency operations plan, the fire department's preincident plan should be fully coordinated with the internal plan. The facility should provide the fire department with an on-site liaison as soon as command is established.

Step 6: Evaluate special or unusual characteristics of common occupancy

Particular hazards for each occupancy group should be identified. This might include reviewing hazards that are specific to a restaurant, such as to ensure that the hood and duct system is checked for grease buildup. A more detailed discussion of specific hazards is presented later in the chapter under "Selected Use Group Specific Concerns." After gathering data on the specific facility, the fire officer should research the applicable codes, standards, and other information sources to identify the common and unusual hazards that are associated with that occupancy group. For example, multiple fatality fires are the major risk factor that should be considered in assembly occupancies.

Effective preincident plans for simple sites or facilities or plans with simple objectives can be developed with minimal amounts of data. Additional data are required for more complex sites or facilities, locations with numerous potential hazards, plans with more complex objectives, and potential incidents with greater risks. All data that might be useful should be collected, with the understanding that it can be filtered out later if it is not needed in the final plan.

Data collection forms can be developed to aid in the efficient and consistent collection of information for preincident plans. When developing a data collection system, the intended audience for the final preincident plan should be considered. A consensus should be reached regarding the types of information that are needed and the level of detail that is useful during an incident. These considerations should govern the data collection effort.

Several factors, such as available resources, time, proprietary information, and privacy concerns, often constrain the collection of data. The fire officer might have to prioritize the data collection effort to obtain the most critical and potentially useful information. An incremental process can be used, in which initial preincident plans are developed and issued (in lieu of having none) and subsequently revised and enhanced.

As an example, a local municipality could begin by preparing simple preincident plans for all of the hospitals in the community. As additional resources become available, the plans can be enhanced. This approach might be preferable to expending all of the available resources to prepare a complex and comprehensive preincident plan for only one hospital, while leaving the other hospitals without any plan.

Putting the Data to Use

The goal before the incident is to develop a written plan that would be valuable to both the owner and the fire department if an incident occurs at that location. The preincident plan should provide critical information that could be advantageous for responding personnel, presented in a format that can be easily deciphered under emergency conditions. Many fire departments maintain their preincident plans in electronic form instead of printing out hard copies. The electronic files can be easily updated and quickly transmitted to responding units when an incident occurs.

Data storage systems allow the information to be automatically retrieved when the dispatch system processes an alarm for the location. The information can be printed at the fire station, or on-board mobile data equipment can provide access to the information from a command post or while en route to the incident. Additional detailed information, such as building plans and fire alarm drawings, can be kept in a lock box or other secured area at the site.

For this to occur, the data that have been collected must be organized into usable information. This usually requires the officer to draw a plot plan, which is a drawing of the property as seen from a view from above. The plot plan includes the property lines and locates the building within those boundaries; it is similar to what a surveyor would draw. The plot plan should show the relationship of the building to other buildings, streets, hydrants, utility controls, and other features. The plot plan visually represents these objects, allowing the officer to quickly identify relevant facts.

Some departments may also have a drawing of the interior of the buildings. Like the plot plan, the floor plan allows the officer to quickly identify considerations for a fire attack. The floor plan is a drawing of the interior of the structure and is similar to an architect's blueprints. The floor plan notes stairwell locations, elevators, standpipe connections, hazardous material storage areas, alarm panel locations, and points of entry. The floor plan allows the officer to show arriving companies a general layout of the interior of the structure.

A report in a written format also accompanies most preincident plans. The written report identifies the address, type of occupancy, construction type, size of the occupancy, and required fire flow in a section. In another section, the report identifies the resources available, including responding units and water sources and availability. Another section includes any special hazards or considerations, such as fire department connection location, utility control locations, hazardous material storage, and rapid-entry key box locations. It is important that these are in a standardized format with standardized information to allow for quick access by the officer, whether the report is in an electronic form or a hard copy.

Understanding Fire Codes

Fire officers are often involved in performing inspections to enforce a fire code (or fire prevention code). A fire prevention code establishes legally enforceable regulations that relate specifically to fire safety, although some fire codes also include related subjects, such as regulation of hazardous materials.

Codes can be adopted by different jurisdictions. A state fire code applies everywhere in the state, whereas a locally adopted code can only be enforced within that particular jurisdiction. In many cases, there is a state or provincial fire code that sets a minimum standard, and local jurisdictions have the option of adopting more stringent requirements.

Fire code requirements are often adopted or amended in reaction to fire disasters. This is known as the **catastrophic theory of reform**. Many egress code requirements can be traced back to disasters, such as the 1903 Iroquois Theatre (Chicago: 602 dead), the 1911 Triangle Shirtwaist Company (New York: 146 dead), the 1942 Cocoanut Grove Nightclub (Boston: 491 dead), the 1944 Ringling Brothers-Barnum and Bailey Circus (Hartford, Connecticut: 168 dead), the 1977 Beverly Hills Supper Club (Southland, Kentucky: 164 dead), the 1980 MGM Grand Hotel/Casino (Clark County, Nevada: 84 dead), the 1990 Happy Land Social Club (Bronx, New York: 87 dead), and the 2003 The Station Nightclub (West Warwick, Rhode Island: 100 dead).

Authority having jurisdiction is a term used in NFPA documents to refer to "an organization, office, or individual responsible for enforcing the requirements of a code or standard, or for approving equipment, materials, an installation, or a procedure." The authority having jurisdiction for a state fire code is usually the state fire marshal. In the case of a provincial fire code, the authority having jurisdiction would be the provincial fire marshal or fire commissioner. The local

fire chief, fire marshal, or code enforcement official would be the authority having jurisdiction for a local fire code.

The regulations contained in a fire code are enforced through code compliance inspections. Depending on the jurisdiction, these inspections could be conducted by the state fire marshal's office, by the local fire department, or by code enforcement officials who might or might not be a part of the fire department. The authority having jurisdiction delegates the power to enforce the code to the fire officers, inspectors, and other individuals who actually conduct inspections.

Building Code Versus Fire Code

It is important to differentiate between a building code and a fire code. Building codes and fire prevention codes both establish legally enforceable minimum safety standards within a state, province, or local jurisdiction. A building code generally contains regulations that apply to the construction of a new building or to an extension or major renovation of an existing building, whereas a fire code applies to existing buildings and to situations that involve a potential fire risk or hazard. For example, the building code might require the installation of automatic sprinklers, a fire alarm system, and a minimum number of exits in a new building; the fire code would require the building owner to properly maintain the sprinkler and alarm systems and to keep the exits unlocked and unobstructed at all times when the building is occupied. Sometimes, a fire code includes certain requirements that apply to new buildings that are beyond the scope of the building code, such as a regulation requiring the installation of fire lanes and hydrants.

State Fire Codes

Most US states and Canadian provinces have adopted a set of safety regulations that apply to all properties, without regard to local codes and ordinances. Where there is a state or provincial fire code, it is generally the minimum legal standard in all jurisdictions within that state or province. The state or provincial fire marshal usually delegates enforcement authority down to local fire officials.

Most states allow local authorities the option of adopting additional regulations or a more restrictive code. A few states

Safety Zone

NFPA 101® *Life Safety Code®* is a model code document that contains requirements specifically related to protecting the lives of building occupants, particularly exit details. When NFPA 101 is adopted by a jurisdiction, it can be applied to both new and exiting buildings. In some cases, the requirements in the *Life Safety Code* make allowances for exiting buildings where the exits do not comply with the standards for newly constructed buildings.

have adopted **mini/max codes**, which means that local jurisdictions do not have the option of adopting more restrictive regulations.

The *Fire Protection Handbook* identifies seven different organizational patterns for state fire marshal organizations in the United States. The state fire marshal may work in:

- The department of insurance
- The department of public safety
- A separate government department
- A regulatory agency
- The state police
- A cabinet-level official
- The state fire commission (▶ **Table 12-1**)

Local Fire Codes

At the local level, fire and safety codes are enacted by adopting an **ordinance**, which is a law enacted by an authorized subdivision of a state, such as a city, county, or town. The local jurisdiction adopts an ordinance that establishes the fire code as a set of legally enforceable regulations and empowers the fire chief or the local fire marshal to conduct inspections and take enforcement actions. This authority can then be delegated to fire officers, fire inspectors, and other individuals within the fire department.

Model Codes

Model codes are documents developed by a professional code-making organization, such as the NFPA, and made available for adoption by authorities having jurisdiction. *The National Building Code* was first published by the National Board of Fire Underwriters in 1905. A model code is generally developed through a consensus process using a network of technical committees. Most jurisdictions use model codes developed by the NFPA and the International Code Council.

Most states and local jurisdictions adopt a nationally recognized model code, with or without amendments, additions, and exclusions. Some jurisdictions choose to maintain their own independent codes. A complete set of model codes includes a building code, electrical code, plumbing code, mechanical code, and fire prevention code. The primary advantages of a model code are that the same regulations apply in many jurisdictions, and all of the requirements are coordinated to work together without conflicts.

If a model code is adopted by a local jurisdiction, it may occur in one of two ways. **Adoption by reference** occurs when the jurisdiction passes an ordinance that adopts a specific edition of the model code. A local jurisdiction could adopt *NFPA 1™ Uniform Fire Code™ (2005)* by reference. The requirements specified in NFPA 1 then become local requirements that can be enforced by designated local officials. A fire officer would need to obtain a copy of the 2005 edition of *NFPA 1, Uniform Fire Code* to read the specific requirements.

Table 12-1   State Fire Marshal Organizational Patterns*

Under Department of Insurance

• Alabama	• New Mexico
• Florida	• North Carolina
• Georgia	• Tennessee
• Idaho	• Texas
• Mississippi	

Under Department of Public Safety

• Alaska	• Minnesota
• Colorado	• Missouri
• Connecticut	• Nevada
• District of Columbia	• New Hampshire
• Iowa	• Oklahoma
• Kentucky	• Utah
• Maine	• West Virginia
• Massachusetts	

Under a Separate Government Department

• Indiana (State Emergency Management)	• South Carolina (Department of Labor)
• Louisiana (Department of State)	• South Dakota (Department of Community Affairs)
• Michigan (Department of Consumer and Industry Services)	• Virginia (Department of Housing and Community Development)
• Montana (Justice, Attorney General)	• Wisconsin (Fire Investigation, Justice, Attorney General)
• Kansas	• Wyoming

Under a Regulatory Agency

• Arizona (Department of Building Fire Safety)	• Wisconsin (Other functions, Department of Commerce)
• Vermont	

Under State Police

• Arkansas	• Oregon
• Maryland	• Pennsylvania
• Michigan (Fire Investigation)	• Washington

Under Cabinet-Level Official

• California	• New York (Secretary of State)
• Illinois	• North Dakota (Attorney General)
• Nebraska	• Ohio (Department of Commerce)
• New Jersey (Department of Community Affairs)	• Rhode Island (Governor's Office)

Under State Fire Commission

- Delaware

*Courtesy of the National Fire Protection Association.

Adoption by transcription occurs when the jurisdiction adopts the entire text of the model code and publishes it as part of the adopting ordinance. For example, a city would copy the language of the code and include it in its entirety within the ordinance. A fire officer would only need to read the city's ordinance to read the specific requirements rather than having to find the appropriate code book.

The fire officer must know which code and which annual edition is used by the local jurisdiction. Although the model code process updates the code every 2 to 4 years, the authority having jurisdiction must specifically adopt the new edition of a model code before it becomes legally enforceable. Different codes or different editions of the same code might apply to different occupancies. Some communities adopt selected portions of one or more model codes and defer to a state code or locally written ordinances to cover other issues.

Retroactive Code Requirements

In general, the regulations that applied to a particular building at the time it was built remain in effect, as long as it is occupied for the same purpose. The fire officer may have to determine the specific code document (name and year) that was in effect when the building was built to determine whether a building is still in compliance with the code requirements that apply. Most codes include provisions that can be applied to buildings that were constructed before a code was adopted.

If a building is remodeled or extensively renovated, or if the occupancy use changes, most codes specify that all of the current requirements of the code must be applied; otherwise, new code requirements, adopted after a certificate of occupancy has been issued, do not apply unless specific language is included in the adopting ordinance. For example, Scottsdale, Arizona established a mandatory requirement for sprinkler systems in all new residential construction in 1986. The ordinance did not require retroactive installation of sprinkler systems in residential structures built before that date.

On occasion, a state or local authority having jurisdiction passes a code revision that is specifically identified as applying retroactively to all affected occupancies. After the 1980 Las Vegas MGM Grand fire that killed 84 and injured 679, the state of Nevada required fire suppression systems to be installed in all existing casinos and hotels. Many cities have required retroactive installation of sprinklers or fire alarm systems in high rises, nursing homes, or other occupancies that have a high life-safety risk. Rhode Island and neighboring Massachusetts adopted retroactive sprinkler requirements for nightclubs in response to the 2003 fire in The Station nightclub that killed 100 and injured more than 200 patrons.

Understanding Built-in Fire Protection Systems

Next to access and egress issues, the status of the built-in fire protection features is the second most important reason for a fire company to perform an ongoing code compliance inspection. The ability of occupants to safely egress a structure in the event of a fire is essential for their safety, but ensuring that egress routes comply with the codes does not have any effect on the fire itself. Built-in fire protection systems are designed as tools to assist fire fighters in fighting a fire. For example, ensuring that exit lights are properly working assists occupants in escaping during a fire, whereas the standpipe system enables the fire fighters to more quickly extinguish a fire. The officer must have an understanding of how systems work and why the code requires them.

The codes often allow more flexibility in the design of a building because built-in fire protection systems are included. For example, if an automatic sprinkler system is installed in a building, the builder has more flexibility in terms of the building's dimensions and construction materials. A building with a sprinkler system can be larger and taller, the travel distance to exits can be longer, and the access for fire apparatus could be limited. The building may have sprinklers because the design does not allow aerial apparatus to be positioned on all sides. All of these trade-offs depend on a properly functioning sprinkler system.

If there is a fire in the building, you are depending on the built-in fire protection systems to assist you. The dependability of these systems to fire fighters' safety is of prime importance. An inspection is the best method of ensuring the systems will work as designed if the need arises. Most importantly, the officer must ensure that the system is turned on. Industrial Risk Insurers is a large provider of insurance for industrial properties. In their insured properties over many years, the biggest cause of large-loss industrial fires in properties with sprinkler systems was the fact that systems had been shut down for various reasons and not properly restored. As a result, Industrial Risk Insurers promotes the RSVP (Restore Shut Valve Promptly) program.

In the 1980s, the fire departments in Los Angeles, Phoenix, and Fairfax County, Virginia each started a fire protection sys-

Safety Zone

Operating With Built-in Fire Protection Systems

A turned-off automatic sprinkler system and improperly maintained standpipe system contributed to the death of a lieutenant and two fire fighters from FDNY Ladder 170 on December 18, 1998. They were the first-arriving ladder company to a 10th floor apartment fire in a 5-year-old city-owned high rise for the elderly. There was a report of an occupant trapped. The fire fighters from Engines 257 and 290 were having difficulties connecting to the standpipe system because of the insulation encasing the hook-ups. Before the attack lines were in place, a 29-mile-per-hour wind gust blew a fireball into the public hallway, killing Ladder 170's inside team.

An inspection of the standpipe system before the fire should have identified the problems that would have been encountered during a fire, and actions could have been required that would have allowed the fire companies to quickly deploy the attack lines.

tem testing program. All three jurisdictions encountered significant failures in built-in fire protection systems and created a retesting program to reduce the failure rate.

Water-Based Fire Suppression Elements

Automatic sprinkler systems, standpipe systems, and fire pumps are the three primary components of water-based fire protection systems. An **automatic sprinkler system** consists of a series of pipes with small discharge nozzles (sprinklers) located throughout a building. When a fire occurs, heat rising from the fire causes one or more heads to open and release water onto the fire. When the water starts flowing, a water flow alarm is activated (▼ **Figure 12-5**). This alarm may be monitored by a central station alarm service, and/or on-site safety/security service. Some systems are "local alarms" and sound a gong or bell only at the outside of the building (▶ **Figure 12-6**).

Depending on the usage (and the weather), automatic sprinkler systems may be wet pipe, dry pipe, deluge, or pre-action. In a wet-pipe system, there is normally water in all of the pipes throughout the system. When a sprinkler head opens, water is discharged immediately. In general, wet-pipe systems require less maintenance than dry-pipe systems. Because of the faster reaction, fewer sprinkler heads are activated to control most fires.

Dry-pipe systems are used in locations where a wet-pipe system would be likely to freeze during cold weather, such as unheated storage facilities and parking garages. Instead of water, the pipes are filled with compressed air or nitrogen until a sprinkler head opens. When the air pressure drops, the dry-pipe valve opens and water is released into the system. Dry-pipe systems require higher maintenance because activation of the sprinkler system requires the entire sprinkler system to be drained. Sometimes, an antifreeze solution is added to the water in a wet-pipe system to protect an unheated area, (e.g., a freezer or a loading dock), instead of installing a dry system.

Deluge systems are special versions of wet- or dry-pipe systems in which large quantities of water are needed to quickly control a fast-developing fire. Deluge systems are most often found in ordnance plants, occupancies with flammable liquid hazards, and aircraft hangers. All of the sprinkler heads are open and ready to discharge water as soon as the control valve opens. The system uses smoke, flame, or fire detectors to sense a fire and trigger the system.

Pre-action sprinkler systems are similar to dry-pipe systems, with a separate detection system to trigger the dry-pipe valve and fill the sprinkler pipes with water. At this point, it becomes equivalent to a wet system. A pre-action system is designed to reduce the risk of water damage due to accidental sprinkler discharge or a broken pipe.

Some sprinkler systems are designed to discharge protein or aqueous film-forming foam as the extinguishing agent.

Standpipe systems provide the ability to connect fire hoses within a building. Standpipes are an arrangement of piping, valves, hose connections, and allied equipment that allow water to be discharged through hoses and nozzles to reach all parts of the building. Like automatic sprinkler systems, you may find dry standpipes in unheated areas. Standpipes are subdivided into three classes based on their expected use:

- Class I provides 2.5-inch hose outlets, intended for use by fire department or fire brigade members trained in the use of large hose streams.

Figure 12-5 The basic components of an automatic sprinkler system.

Sprinkler heads
Piping
Control valve
Water supply

Figure 12-6 Some systems are "local alarms" and sound a gong or bell only at the outside of the building.

Getting It Done

Fire Sprinkler Activations

In most buildings with automatic sprinklers, a fire alarm system is also present. In many cases, smoke detectors and local fire alarm pull stations are also connected to the system. Such sophisticated alarm systems are often monitored by a central fire alarm service that calls in the fire alarm event to the local 911 center. Most of the notifications from fire alarm systems are nonfire events. Some of these events are caused by situations that appropriately trigger an alarm, such as cigarette smoke or burnt food. Many others are due to other causes, such as defective or dirty smoke detectors. The fire officer must remain vigilant because the same notification process is used when a waterflow alarm occurs.

A waterflow alarm means that water has moved past a flow switch that is located inside the water supply pipe and set off the alarm. Non-fire events that set off the waterflow alarm often include public water supply surges, sprinkler heads broken accidentally, and burst pipes. Many older buildings have a single waterflow device for the entire structure, so the problem could be anywhere in the building.

A trouble alarm in a monitored fire alarm system is like a warning light in a car: All that it is telling you is something is not right. Old and inadequately maintained fire alarm systems that detect a fire or water-

flow alarm may default into a trouble alarm after the reset button is pushed. Many communities encourage or require that a log book be left in the buildings with large and sophisticated alarm systems, usually in the fire control room. Both the building maintenance department and the fire department make entries in this log book with every activated fire alarm. Fire officers who work in districts with many monitored buildings develop a detailed knowledge of the features and problems in each fire alarm system. Although the code may specify a specific type of system, these fire officers appreciate that every building has unique characteristics.

If the fire alarm system resets without going back into alarm or trouble, a surge in the municipal water supply may have tripped the waterflow alarm. If the alarm system properly resets and does not reactivate within 1 or 2 minutes, you would not need to make a detailed floor-to-floor search.

However, if you respond back to a second waterflow activation within a couple of hours, assume that water is flowing somewhere. Many communities require that a building maintenance representative respond to every fire alarm activation. If you are making a return visit, make sure that a building representative responds.

- Class II provides 1.5-inch hose coupling with a pre-connected hose and nozzle in a hose station cabinet. The hose is designed for occupant use.
- Class III provides both 1.5- and 2.5-inch connections. The 1.5-inch connection may have a preconnected hoseline to be used by the occupants until the fire department arrives.

During a code compliance inspection, you should pay particular attention to the condition of the standpipe system; it is one of the few built-in fire protection devices that is specifically designed to help you fight a fire. Fire officers should identify fire hose access and ensure proper calibration of pressure regulating devices. During the 1991 high-rise office fire at One Meridian Plaza in Philadelphia, fire fighters discovered that the pressure-regulating devices were improperly set, and only a weak flow could be obtained from each outlet. A pressure-regulating device limits the discharge pressures from standpipe hose outlets. They are usually found on the lower floors of high-rise standpipes.

Fire Pumps

Fire pumps increase the water pressure in standpipe and automatic sprinkler systems (► Figure 12-7). They are designed to start automatically when the water pressure drops in a system or a fire suppression system is activated. A field inspection generally is confined to a visual inspection to confirm that the pump appears to be in good condition and is free of physical damage. The fire officer should also ensure that the fire pump has passed the annual performance test.

Special Extinguishing Systems

Generally, a fire officer encounters four different types of special extinguishing systems: carbon dioxide, dry or wet chemical, Halon, or foam. In general, the ongoing code compliance inspection consists of visual inspection, looking for physical damage and confirming that the safety seals are intact, as well as verification of the documentation for required tests and inspections. The safety seal is placed on any handle that would activate the system. If the safety seal is broken, the system could have been discharged. Note these specialized systems in the preincident plan.

Carbon Dioxide

Carbon dioxide systems are fixed systems that discharge carbon dioxide from either low- or high-pressure tanks, through a system of piping and nozzles, either to protect a specific device or process (e.g., a printing press) or to flood an enclosed space. You may recall from your fire fighter training that carbon dioxide extinguishes fire by displacing oxygen. The gas is heavier than air, so it settles in low spaces. Fixed systems are generally required to comply with NFPA 12, *Standard on Carbon Dioxide Extinguishing Systems*. Your preincident plan should require the use of SCBA if a fixed system has activated.

Dry or Wet Chemical

Fixed chemical extinguishing systems discharge a chemical extinguishing agent through a system of piping and nozzles. Either fusible links, or local pull stations, or both, are used to discharge the chemical. These systems can be found protect-

Figure 12-7 A fire pump may be needed to maintain or increase water pressure in standpipe and automatic sprinkler systems.

ing commercial cooking devices, as well as industrial processes where flammable or combustible liquids are used. During a field inspection, observe the conditions of the nozzles. Many have a protective cap to protect the nozzle from becoming obstructed with cooking grease.

The newer wet chemical systems are preferred for protecting cooking equipment. The wet chemical agent reacts with hot grease to form a foam blanket, reducing the release of combustible vapors. The foam blanket cools the grill and reduces the possibility of a rekindle. The old-style dry chemical systems left a residue that was very difficult to clean up.

Both systems are activated in one of two ways: (1) a fusible link that melts on flame contact or (2) a manual pull station. Activation of the system also turns off the cooking device by closing the cooking fuel valve or turning off the electricity.

Halon

Halon 1301 was the extinguishing agent of choice for fire protection in computer rooms and to protect electronic equipment from the 1960s to the 1990s. Based on the weight of the agent, Halon 1301 is about 250% more efficient than carbon dioxide for extinguishing fires. Unfortunately, it also contributes to depletion of the ozone layer in the stratosphere. The 1990 Clean Air Act Amendments implemented the Montreal Protocol to phase in restrictions on the manufacture and import of Halon. As of January 1, 2000, Halon cannot be manufactured or imported into the United States, but systems may still be recharged using "banked" agent. You may encounter some existing fixed Halon systems, but new installations are extremely rare because of the high cost of the agent.

NFPA 2001, *Standard on Clean Agent Extinguishing Systems* covers the use of alternative agents that have been developed to replace Halon systems protecting electrical and electronic telecommunications systems. Most of these systems are designed to flood a room or an enclosure and, like Halon, involve some degree of toxicity. The room or enclosure must be sealed, and the occupants have to leave before the agent is discharged. The system can be automatically or manually fired; both methods include a pre-alert warning for the occupants to leave the room before the agent discharges. The preincident survey should identify the chemical used, the duration of the discharge (10 seconds to 1 minute), the enclosure protected, the automatic activation sequence, and the location of the manual activation station.

Foam Systems

There are several types of fixed foam systems for different hazards. A low-expansion foam system is most likely to be used to protect hazards involving flammable or combustible liquids, such as gasoline storage tanks. These systems discharge foam bubbles over a liquid surface to create a smothering blanket that extinguishes the fire and suppresses vapor production. High-expansion foam is used in areas where the goal is to fill a large space with foam in order to exclude air and smother the fire.

Fire Alarm and Detection Systems

Many occupancies require fire alarm systems. A fire alarm system consists of devices that monitor for a fire and notify the appropriate personnel. Manual fire alarm boxes, smoke detectors, or heat detectors may activate the fire alarm system. The alarm may also be activated by the fire protection system, such as waterflow in a sprinkler system. Once activated, the system notifies the appropriate personnel, including the building occupants, with audible and visual signals and these may be transmitted outside the building and to the fire department or a monitoring firm. Like other fire protection systems, fire alarm systems require regular maintenance and operation.

Understanding Fire Code Compliance Inspections

The objective of a fire code compliance inspection is to determine whether an existing property is in compliance with all of the applicable fire code requirements. Some codes call this a maintenance inspection. The basic goal is to observe the housekeeping to ensure that there are no fire hazards and confirm that all of the built-in fire protection features, such as fire exit doors and sprinkler systems, are in proper working order.

The fire department's authority and responsibilities for conducting code compliance inspections are usually included in the ordinance that adopts the local fire code. The responsibility for code enforcement is usually assigned to the fire chief or fire marshal. The fire chief can delegate the

VOICES OF EXPERIENCE

❝ As a fire officer, you should always take the opportunity to truly learn a structure. ❞

Preincident planning gives fire officers an opportunity to not only meet with the people in the community, but to see what is waiting to harm fire fighters if a structure experiences a hazardous event. Experienced fire officers can shed light on unique structural features that were present during the construction of the structure. Interior walls, single or multiple roofs, fire department connections and piping, fire control panels, and void spaces are all readily visible when the structure is under construction. Over time and through different owners, the structure may change in appearance and structural stability. That is why it is important to revisit a structure and make certain that your preincident plan is up-to-date.

As a fire officer, you should always take the opportunity to truly learn a structure. Studying the structure in ideal conditions can prepare you to command and control an incident in fire conditions. Preplanning gives confidence to the fire officer when dealing with incident management, rapid intervention, fire fighter progress reports, and Mayday operations. Areas of egress, plans of operation, safety or danger zones can be identified before the fire and keep everyone safer.

Gerard P. Forte
City of Palm Coast, Florida Fire Department
15 years of service

responsibility for code enforcement activities to different individuals or units within the department.

In many fire departments, fire officers and fire inspectors who are assigned to a fire prevention bureau or code enforcement division conduct inspections of specific types of properties. In departments where fire companies conduct code compliance inspections, there is usually an individual or a group assigned to coordinate their activities and provide technical assistance.

Fire Company Inspections

Because preventing a fire is a basic goal of every fire department, conducting fire inspections is an essential skill that every fire officer must posses. The purpose of conducting fire inspections is to identify hazards that exist and ensure that the violations are corrected. This process helps enable the fire department to be proactive in preventing fires. One method of maximizing the number of inspections is for them to be conducted by the fire companies.

Inspections conducted at the fire company level have been an activity advocated by the National Fire Protection Association since the first *NFPA Quarterly* was published more than a hundred years ago. Local fire companies have the ability to become familiar with their response areas and take actions to prevent the start of fires and, if a fire starts, minimize the fire spread and destruction of the neighborhood.

Before conducting an inspection, it is important to understand the scope of code enforcement authority that is delegated to a fire officer in your jurisdiction. There is wide diversity in the scope and authority of inspections conducted by fire companies. In some areas, fire suppression companies are authorized to enforce all fire code **regulations**; one metropolitan city authorizes fire officers to issue corrective orders that would require the installation of automatic sprinklers in a business under renovation. In other fire departments, fire suppression companies are assigned to inspect only certain types of occupancies, whereas the fire marshal's office or fire prevention division inspects other types of occupancies. Some fire departments can only provide safety recommendations to citizens and property owners, whereas others can issue citations and compliance orders. The fire officer needs to determine the source and the scope of his or her code enforcement authority before conducting any fire safety inspection, as well as which particular codes are used by the jurisdiction.

Outside of fire emergencies, a fire officer generally cannot enter private property without the permission of the owner or occupant. Most fire codes include a section that authorizes code enforcement officials to enter private properties at any reasonable time to conduct fire and life-safety inspections; however, the scope of this authority usually depends on the type of property involved. In most cases, the permission of the owner or occupant is required for entrance into a dwelling unit, whereas access to public areas is less restricted. The fire code often contains a section that, if necessary, allows for the issuance of a court order requiring the owner or occupant to allow the fire department agent to enter the occupancy to conduct an inspection.

Classifying by Building or Occupancy

Many of the code requirements that apply to a particular building or occupancy are based on its classification. The codes classify a building by construction type, occupancy type, and use group. You need to know how to classify a building before you can determine the code requirements that apply to it. In addition to these three classifications, local zoning ordinances often regulate where certain types of occupancies or buildings are permitted.

Construction Type

The building itself is classified by **construction type**, which refers to the design and the materials used in construction. NFPA 220, *Standard on Types of Building Construction,* addresses building construction and firefighting. The type of building construction has a significant influence on firefighting strategy. Construction type is a fundamental size-up consideration when firefighting strategy for a burning structure is being determined.

The most commonly used model codes subdivide construction into five basic types:

- **Type I: Fire resistive**. The construction elements are noncombustible and are protected from the effects of fire by encasement, using concrete, gypsum, or spray-on coatings (**▼ Figure 12-8**). Depending on the model code used, type I is divided into subtypes, based on the level of fire protection provided. The

Figure 12-8) A type I building.

level of protection is described by the number of hours a building element can resist the effect of fire.

- **Type II: Noncombustible.** The structural elements can be made from either noncombustible or limited combustible materials (▼ **Figure 12-9**). Like type I, this type also has subdivisions of, based on the level of fire resistance. Although the buildings are assembled from noncombustible components, the structural elements have limited or no fire resistance. A type IIA structural frame is expected to resist fire for 1 hour. The structural frame in a type IIB building is not expected to resist the effects of fire. A strip shopping center with cinderblock walls, unprotected steel columns, and steel bar joists supporting a steel roof deck is an example of a type IIB building.
- **Type III: Limited combustible (ordinary).** The exterior load-bearing walls of the building are noncombustible masonry (▶ **Figure 12-10**). A **masonry wall** may consist of brick, stone, concrete block, terra cotta, tile, adobe, or concrete. The interior structural elements may be combustible or a combination of combustible or noncombustible. Like types I and II, there are different levels of fire protection for type III buildings. The structural frame of a type IIIA building is protected, which means that it is encased in concrete, gypsum, or spray-on coatings and is expected to have a fire-resistive rating of 1 or 2 hours. A type IIIB structural frame is unprotected and has no fire resistance rating.
- **Type IV: Heavy timber.** The exterior walls are noncombustible (masonry), and the interior structural elements are unprotected wood beams and columns with large cross-sectional dimensions (▶ **Figure 12-11**). Mill construction, which was used in many New

Figure 12-10 A type III building.

England textile buildings that were built in the 1800s, is an example of heavy timber construction. Mill construction features massive wood columns and wood floors.

- **Type V: Wood frame.** The entire structure may be constructed of wood or any other approved material (▶ **Figure 12-12**). Sometimes, a masonry veneer is applied to the exterior, but the structural elements are wood frame.

It is often difficult to determine the type of construction from the exterior of a building. Many low-rise offices, apartment buildings, and residences appear to be ordinary construction; however, they are actually wood frame buildings with a brick, stone, or masonry veneer on the exterior walls. This type of wall covering does not contribute to the strength or fire resistance of the building.

Figure 12-9 A type II building.

Figure 12-11 A type IV building.

Figure 12-12   A type V building.

Figure 12-13   Example of a business occupancy.

Occupancy and Use Group

The <u>occupancy type</u> refers to the purpose for which a building or portion of a building is used or is intended to be used. There are hundreds of different occupancy classifications, such as restaurant, dining hall, tavern, and bar. The code requirements that apply to a particular occupancy are determined by its <u>use group</u>. Occupancies are classified into use groups based on the characteristics of the occupants, the activities that are conducted, and the risk factors associated with the contents.

Assembly

An assembly occupancy is an occupancy used for the gathering of people for deliberation, worship, entertainment, eating, drinking, amusement, or awaiting transportation. The classification may be further divided into more specific types of assemblies. Examples of assembly type occupancies include:

- Churches
- Taverns or bars
- Nightclubs
- Basketball arenas
- Restaurants
- Theaters

Business

A business occupancy is an occupancy used for account and record keeping or transaction of business other than mercantile (▶ **Figure 12-13**). Examples of business occupancies include:

- Dental offices
- Banks
- Architect offices

- Hair salons
- Colleges and universities
- Doctors' offices
- Investment offices
- Insurance offices
- Radio and television stations

Educational

An educational occupancy is an occupancy used for educational purposes through the 12th grade. Typically, this designation refers to schools. Educational occupancy may also cover some day care centers for children older than $2\frac{1}{2}$ years.

Industrial

An industrial occupancy is an occupancy in which products are manufactured or in which processing, assembling, mixing, packaging, finishing, decorating, or repair operations are conducted (▶ **Figure 12-14**). This would include occupancies such as:

- Automobile assembly plants
- Clothing manufacturers
- Food processing plants
- Cement plants
- Furniture production facilities

Health Care

A health care occupancy is an occupancy used for purposes of medical or other treatment or care of four or more persons, where such occupants are mostly incapable of self-preservation due to age, physical or mental disability, or security measures not under the occupants' control. This would include hospitals and nursing homes.

Figure 12-14 Example of an industrial occupancy.

Detention and Correctional

A detention and correctional occupancy is an occupancy used to house four or more persons under varied degrees of restraint or security, where such occupants are mostly incapable of self-preservation because of security measures not under the occupants' control. This could include prisons, jails, and detention facilities.

Mercantile

A mercantile occupancy is an occupancy used for the display and sale of merchandise (▼ **Figure 12-15**). This group includes the following:

- Retail stores
- Convenience stores
- Department stores
- Drug stores
- Shops

Residential

A residential occupancy is an occupancy that provides sleeping accommodations for purposes other than health care or detention and correctional. There are five subcategories:

- One- and two-family dwelling units: buildings that contain nor more than two dwelling units with independent cooking and bathroom facilities
- Lodging or rooming houses: buildings that do not qualify as one- or two-family dwellings that provide sleeping accommodations for a total of 16 or fewer people on a transient or permanent basis, without personal care services, with or without meals, but without separate cooking facilities for individual occupants
- Hotel: buildings under the same management in which there are sleeping accommodations for more than 16 persons and primarily used by transients for lodging with or without meals
- Dormitory: buildings in which group sleeping accommodations are provided for more than 16 persons who are not members of the same family in one room, or a series of closely associated rooms, under joint occupancy and single management, with or without meals, but without individual cooking facilities
- Apartment building: buildings containing three or more dwelling units with independent cooking and bathroom facilities

Storage

A storage occupancy is an occupancy used primarily for the storage or sheltering of goods, merchandise, products, vehicles, or animals. Examples include:

- Cold storage plants
- Granaries
- Lumber yards
- Warehouses

Figure 12-15 Examples of mercantile occupancies.

Mixed

A mixed-use property is a property that has multiple types of occupancies within a single structure. An example would include an old commercial building that has been renovated to include multifamily residential loft apartments on the second and third floor while there is a bakery on the first floor.

Unusual

In addition to the previous occupancies, some do not fit neatly into the other categories. These occupancies can be placed in a miscellaneous category that represents unusual structures, such as towers, water tanks, and barns.

You can find the use group classifications in the building code or NFPA 101, *Life Safety Code* (▼ **Table 12-2**).

Once the use group classification is known, the code provisions that relate to a building's allowable height, floor area, and construction type can be referenced in the building code. The building code (or NFPA 101, *Life Safety Code*) also includes the requirements for the number of exits, the maximum travel distance to an exit, and the minimum width of each exit. Requirements for the installation of fixed fire protection systems in new buildings are also included in the building code. Additional requirements for fire protection systems are sometimes included in the fire code, particularly for existing buildings.

One of the first steps of a code compliance inspection is to confirm that the occupancy is still being used for the approved purpose. For example, if a rental space in a shopping center has been converted from a clothing store to a restaurant or bar, the exit requirements and the built-in fire protection system requirements could be different.

NFPA 704 Marking System

Many of the occupancies may have sufficient quantities of hazardous materials to require the use of a standardized marking system that notifies responders to the hazards associated with materials they might encounter. The most recognized standard is the marking system outlined in NFPA 704, *System for the Identification of the Hazards of Materials for Emergency Response*. The marking system consists of a color-coded array of numbers or letters arranged in a diamond shape.

Each color represents a specific type of hazard. Blue represents health hazards, red represents flammability hazards, and yellow represents the material's reactivity hazard. Within each color diamond, there is a number from zero to 4 that represents the relative hazard of each. A zero means that material poses essentially no hazard, whereas a four indicates extreme danger. For example, on the exterior of the building, there might be a marking with a 2 inside the blue area, a 4 inside the red area, and a zero inside the yellow area. This means that inside the building, there are materials that pose a moderate health hazard, a severe flammability hazard, and no real reactivity hazard. The last quadrant of the diamond is white and is used to represent special hazards. This could include letters or numbers. For example, it might include a placard that indicates to the responder that there is material inside that is water reactive.

This system also requires labels to be affixed to containers inside the structure to indicate the hazards of the substance. This might include labels such as "corrosive," "flammable," or "poison." When completing an inspection, the officer must understand the requirements of NFPA 704 and when it applies. In general, the system requires a 704 marker at each entrance to the building, on doorways to chemical storage areas, and on fixed storage tanks.

Table 12-2 Classification of Use Groups

Major Use Classification	Occupancy Categories
Assembly	Theaters, auditoriums, and churches
	Arenas and stadiums
	Convention centers and meeting halls
	Bars and restaurants
Health care	Hospitals
	Nursing homes
Detention and correctional	Prisons
	Penitentiaries
Mercantile	Retail stores
Business	Offices
Industrial	Factories
Storage	Warehouses
	Parking garages
Educational	Schools
Residential	Homes
	Apartments
	Dormitories
	Hotels

Preparing for an Inspection

Preparation is required before a fire officer conducts a fire inspection for the purpose of code enforcement. The assumption in this section is that the fire officer has the responsibility, authorization, and training necessary to conduct such an activity. This section covers how to prepare to conduct such an inspection.

Fire officers have a special responsibility to demonstrate a professional concern about fire safety and knowledge of the local fire prevention code. The fire officer should approach the owner or occupant as a well-prepared and professional ally who is concerned about the safety of lives and property. This begins with preparing for the inspection by reviewing the applicable code provisions and any information on previous inspections that can be found in the occupancy file.

For the business owner or occupant, a visit by the local fire company is a major and disruptive event. Some jurisdictions send a preinspection or self-survey form to the business or property a few weeks before an official fire department inspection will occur.

Reviewing the Fire Code

Before conducting an inspection, you should review the specific sections of the code that apply to the specific property. In some cases, you discover unanticipated situations during an inspection that require you to research an additional section of the code. The *NFPA Fire and Life Safety Inspection Manual* is a valuable reference. This book provides general information about a large variety of processes, equipment, and systems.

Review Prior Inspection Reports, Fire History, and Preincident Plans

If possible, review the file of past fire safety inspection reports for the property. You may see a trend or a chronic problem in that particular occupancy. A facility that always requires two follow-up inspections to correct the violations that are noted during inspections might indicate that the fire officer needs to turn up the salesmanship to ensure quicker compliance with the fire prevention code. For example, when calling to schedule the inspection, the fire officer might take the opportunity to review the violations noted during the previous inspection and encourage the owner to check those items.

Look at the fire history for the occupancy. If a steakhouse restaurant has a history of a flue or hood fire every 4 to 6 months, you should plan to concentrate on the grill and hood system during your inspection. You should think about what you can recommend to reduce the frequency of accidental fires.

Bring a copy of the preincident plan with you. This is an excellent time for you to update the information, such as names and phone numbers. You can also use the preincident plan to identify any modifications and additions that have been made to the occupancy.

Coordinate Activity With the Fire Prevention Division

In departments where there is both a fire prevention division and a company-level inspection program, coordination is required. Some jurisdictions assign a list of occupancies to be inspected by each company on a monthly or quarterly basis. In buildings where a process, storage, or occupancy is required to have an annual fire prevention permit, the local fire com-

pany's **ongoing compliance inspection** could be scheduled for 6 months after the fire prevention division issues the permit.

The authority having jurisdiction usually determines the frequency of inspections for each type of occupancy. For example, day care facilities might be required to have an annual inspection from the fire marshal's office, whereas business occupancies are inspected by a local fire company once every 2 or 3 years. The fire officer at the local fire company is expected to follow the inspection schedule and coordinate with the fire marshal's office.

Arrange a Visit

It is professional and good business practice to contact the owner or business representative to schedule a day and a time for the fire safety inspection. Many businesses are cyclical, and some days or months are more difficult to accommodate a fire department inspection than others. For example, April 14th may not be the best time for an inspection of an accounting firm's office because it is the peak of their annual work cycle.

In some cases, a time that is inconvenient for a business is an important time in terms of fire safety. Between early November and January 1, many retail stores are packed with extra stock for the holiday buying season. Some retail businesses earn 30% or more of their annual revenue at this time of year. A fire safety inspection of a store in an enclosed mall on December 1 may reveal boxes stored from floor to ceiling, obstructing the sprinkler heads. Cardboard boxes may be pushed up tight against the electrical panel, instead of maintaining a 30-inch clearance. The electrical panel may be hotter than usual because of all of the extra power needed to run the holiday displays. Because the store is short staffed, the trash may have piled up in the storage room and have blocked the rear emergency exit.

Assemble Tools and References

Once you have reviewed the information on the occupancy to be inspected and the applicable fire prevention codes, you are ready to perform the inspection. In addition to your knowledge, you will need to bring some equipment to assist

---

Company Tips

Neatness Counts

It might be a good idea to make sure that your fire helmet is relatively clean. A "salty" helmet with a heavy buildup of carbon and other products of combustion can create a poor impression. If your hand gets dirty when you touch your helmet, it is time to clean it. While you are at it, make sure that the fire fighters under your supervision appear neat, clean, and professional.

you during the visit. You should have the following tools with you:

- Pen, pencil, and eraser
- Inspection form
- Clipboard. Consider using the type of clipboard that law enforcement and EMS professionals use; it would allow you to store the completed forms in a safe place in case you need to respond to an emergency during the code enforcement period.
- Graph paper and a ruler
- Digital camera with flash attachment
- Coveralls
- Fire department business cards
- Reference code books

Plan to wear your safety shoes and bring your fire helmet, eye protection, and protective gloves if you will be inspecting an area that is under renovation or inspecting an occupancy that requires such protective equipment.

Conducting the Inspection

Some departments require that the entire fire company perform the fire inspection together. Business owners and managers sometimes complain about the disruption that occurs when four to six fire fighters come into a business. Portable fire radios are blaring, and, sometimes, the fire fighters seem to be more interested in sightseeing or shopping than identifying fire safety hazards. Other departments deploy the fire company in two or three teams to conduct parallel inspections in adjacent occupancies. Make sure that you know what procedure and practice your organization requires.

Conducting a fire inspection should be approached in a systematic manner. The fire officer should begin every inspection in the same manner. The first step should be to circle the area as you park the apparatus to get a general overview of the property. Second, meet the property owner or manager to let them know that you have arrived and will begin the inspection. Third, begin the inspection at the exterior of the building and work systematically throughout the inside of the building, beginning in the lowest level and working up. Fourth, conduct an exit interview with the contact person. Last, write a formal report on the inspection.

General Overview

As you approach the occupancy to be inspected, make an effort to circle the property and observe all four sides of the building. You are looking for any obvious access or storage problems and any new construction since your last visit, and you are confirming the location of hydrants, sprinkler/standpipe hook-ups, and other outside features with your preincident survey sheet.

Park the fire apparatus in a location that does not disrupt the business and allows the fire company to respond if they are dispatched to an emergency. Some departments require

that one fire fighter remain with the apparatus to both listen to the radio and protect the rig from vandals or thieves. In general, the apparatus should not be parked in a fire lane. It is hard to convince the business owner to comply with the fire prevention code when the fire engine is violating the code by sitting in a fire lane when there is no emergency.

Meet With the Representative

Enter the business through the main door and make contact with the appropriate representative. If you have called ahead to schedule this visit, you already have the designated representative's name and office location. Introduce your crew and briefly explain the goal of this visit. This is a great time to review and update all of the contact names, phone numbers, and information found in your preincident survey sheet. Ask to have a representative with the appropriate access cards and keys accompany the inspection team.

Inspecting From the Outside in, Bottom to Top

A fire company-level ongoing compliance inspection is designed to confirm that all of the built-in fire protection systems are in place and fully operational and to make sure that

Company Tips

Conducting Inspections and Surveys

One fire officer has a four-step system that meets the organization's need for fire company inspections and preincident surveys and helps his fire company focus on what is important.

Step 1

Schedule the inspections based on use group or occupancy. For example, all of the service stations and auto repair facilities are scheduled for November.

Step 2

Have an in-station drill to review the fire prevention code sections that apply to the types of items that would be found during a typical code compliance inspection in that type of occupancy. In October, the fire officer schedules an in-station drill to review the fire code requirements for a service station or automotive repair shop and also delegates to a senior fire fighter the responsibility to schedule the planned inspections in the district.

Step 3

Break the fire company into two-person inspection teams. In general, one of the team members would focus on the code enforcement, and the other would update the preincident survey form.

Step 4

The fourth and final step is for the company to get together and review their findings. This would be the time to discuss any changes that are required on incident action plans, based on the inspections. For example, if a local auto repair shop had added a vehicle spray booth, the plan for that occupancy should be revised.

the area is free of avoidable accidental fire ignition sources. This begins with a walk around the exterior of the premises. Look to ensure that the address is present and properly identified on both the front and the rear of the building. Consider whether extension cords are being used improperly and electrical boxes are open. If the building has a fire department connection, is it unobstructed from view, is it open for foreign objects to be inside the piping, and are the threads in workable condition? The fire company must always verify that all means of access and egress are clear and in proper operating order; there is no more important reason for this type of inspection. Exit problems require immediate correction.

If the building requires any special markings, such as NFPA 704 markings, are they present? Fire companies routinely respond to dumpster fires that are both intentionally and unintentionally set. The fire officer should ensure that the location of dumpsters does not present a fire hazard. Outside storage buildings should be checked for compliance, particularly with the storage of hazardous materials.

Once the exterior has been evaluated, the officer should begin the interior evaluation in the basement. Often, utility rooms, fire pumps, fire protection system control valves, back-up generators, and laundry rooms are found there. The importance of a properly working fire protection system cannot be overstated. These areas are also susceptible to improper storage of combustibles near electrical panels, open junction boxes, improperly stored hazardous material, and blocked exits.

After the basement has been inspected, the fire officer should systematically work through the building, checking for fire safety issues. Like the exterior and the basement, the ability of occupants to quickly exit a building is a primary concern. This may include inventory stock that is blocking exits, locked doors, and exit and egress lights that are not working.

Looking for conditions that are prone to starting fires is also important: open electrical wiring, use of extension cords, and storage of combustibles too close to heat sources. Consideration should also be given to the storage of flammable liquids, as well as the use of candles, portable heating units, and fireplaces or wood stoves.

Lastly, consider items that would assist in extinguishing a fire once it has started. This would include the clearance between sprinkler heads and objects, the condition of smoke and heat detectors, and the operability of fire extinguishers. The fire officer should check to ensure that fire doors are not propped open and are in operable condition. If the structure is equipped with a commercial kitchen, the hood and duct system and fire suppression system should be checked for compliance.

Exit Interview

It is important to wrap up your ongoing compliance inspection/preincident survey by meeting with the owner or designated representative to review what was found and issue any required correction orders. Remember that one of your roles is to be a partner with the business to reduce the risk of acci-

dental fire ignition or fire spread. Generally, once this is complete, a written report needs to be completed, with one copy going to the occupant.

Writing the Inspection/Correction Report

There are wide varieties of inspection/correction reports. Most fire departments have developed standardized forms for the report. Many use a check-off system allowing the officer to put a check next to the corresponding deficiency. This allows the citation to have the appropriate code without having to look up each violation. If violations are not corrected on a reinspection, a notice of hazard is generally typed up rather than reported on a standardized inspection form.

In general, the report needs to identify the address and business name with the appropriate contact information. You should clearly describe any needed corrections and quote the appropriate sections of the code or ordinance. Some communities have developed report forms that list the most commonly occurring violations.

You should review this report with the owner or representative. If violations are found during the inspection, you have to explain to the responsible individual what needs to occur to correct the problem. You retain the original report for follow-up purposes. The owner or representative should sign the form and retain a copy of the report. The report should then be forwarded to the appropriate division of the department; usually, this is the fire prevention division. The records are filed for future reference if needed, and the need for follow-up visits is documented.

Life-threatening hazards, such as locked exits, must be corrected immediately. Less critical issues can be corrected within a reasonable time period, generally 30 to 90 days. You should arrange for any needed follow-up inspections to verify that the corrections have been made. Here is the procedure used by one fire department for non–life-threatening code enforcement issues:

- **Fails first fire company–level inspection**. Schedule follow-up inspection in 30 days.
- **Fails second fire company–level inspection**. Follow-up inspection in 15 to 30 days.
- **Fails third fire company–level inspection**. Issue referred to the fire prevention division for resolution. Depending on the nature of the issue and the history of this occupancy, the fire prevention division contacts the owner and makes an inspection within 2 to 30 days.
- **Fails first fire prevention division inspection**. Notice of violation or correction order issued by the fire marshal. Time to comply with notice or order is determined by the situation and by local regulations. Owner can file for appeal or variance within 10 days of first fire prevention division inspection.
- **Fails second fire prevention division inspection**. Second notice of violation or correction order issued

by fire marshal. Generally, time to comply is shorter. Owner is warned of more severe consequences if the issues remain unresolved.

- **Fails third inspection**. Follow-up inspection in days; building representative may be subject to misdemeanor charge with potential fines, or jail, or both. Occupancy may be required to cease operations until matter is resolved.
- **Fails fourth inspection**. Follow-up inspection in 24 hours. Fines and legal action are initiated, with assessment for every 24 hours that the facility is in violation.

General Inspection Requirements

The fire code includes general fire and life-safety requirements that apply to every type of occupancy. These items should be checked during every inspection. The general requirements include properly operating exit doors and unobstructed paths to fire exits. The general requirements also require built-in fire protection systems, such as automatic sprinklers and fire detection systems, as well as portable fire extinguishers, to be regularly inspected and properly maintained in operational condition. The fire code also includes general precautions that are required in the use or storage of hazardous, flammable, or combustible goods. Some fire codes include topics relating to emergency planning and conducting of fire drills, as well as features like rapid-entry key boxes for fire department use.

Access and Egress Issues

As the inspection progresses, the officer must take into account the type of occupancy that is being inspected for special considerations; however, some general conditions are considered for most occupancy types. The most important issue is access and egress. The fire department must ensure that there is sufficient means of egress for the occupants. These provisions of the fire code are often violated because improper storage causes the exits to be obstructed.

Proper storage practices mean that the goods for sale or the items used by the business or industrial process are safely stored in accordance with the fire prevention code. In the example of a retail store during the holidays, the fire hazard is greatly increased when the store is jammed to the ceiling with stock. In some cases, a fire company may discover a very hazardous condition due to excessive, illegal, or dangerous storage practices. This situation requires immediate corrective action and, if available, assistance from the fire prevention division or the fire marshal's office.

Exit Signs and Egress Lighting

Equally important in occupant evacuation are exit signs and egress lights. Exit signs are those that indicate the direction of exit to occupants during a fire. Egress lights light the path

of the exit. These are particularly important when the occupants may be unfamiliar with the location of exits other than the main entrance, such as at theaters and nightclubs. Often, officers find exit lights that are burned out or that the back-up battery no longer works. The same is true for egress lights. The occupant is required to maintain documentation of monthly checks performed on these systems.

Portable Fire Extinguishers

Almost every building you inspect has portable fire extinguishers, which are provided for the occupants to control incipient fires. The fire prevention code describes requirements for the size, type, and locations of extinguishers needed in various occupancies, usually referring to NFPA 10, *Standard for Portable Fire Extinguishers* for the details. During an ongoing compliance inspection, you should verify that the extinguishers are in place and are of the appropriate size and type for the hazard. In addition to visually inspecting for physical damage, look to see that the instructions are visible, the safety or tamper seal is present, there is current inspection and testing documentation, and the pressure gauge is in the normal range.

Inspecting Built-in Fire Protection Systems

The readiness of the built-in fire protection systems is essential to both occupant and fire fighter safety. *Fire Protection Systems: Inspection, Test and Maintenance Manual* published by the NFPA provides both guidance for initial acceptance testing of built-in fire protection systems and the required periodic inspection, testing, and maintenance. Each chapter includes inspection forms that can be used for 20 different types of built-in fire protection systems. In general, the codes require annual testing, inspection, and maintenance. The officer should check to see that all fire department connection caps are in place, and that they are unobstructed and accessible. For sprinkler systems, extra heads and a changing wrench should be readily available. The officer should ensure that all control valves are in the open position and are locked or have a supervisory alarm to protect the system from being accidentally shut down.

It is important that properly trained personnel perform any test on the fire protection system. When three fire departments started an aggressive retest program, some embarrassing situations arose. Using a fire department engine to charge a standpipe system blew one standpipe right out of the building wall. Another pressure test filled a mercantile basement with hundreds of gallons of water. For the purposes of retesting, one of the departments replaced the fire engine with an air compressor. There are issues with liability and repair costs when the local fire company is operating built-in fire protection systems when there is no fire emergency. In some retesting programs, the fire prevention bureau requires a licensed fire protection contractor to perform the retest, and the contractor submits a certified report.

Electrical

Many fires occur because of faulty electrical wiring. While inspecting an occupancy, ensure that no combustibles are stored around the electrical panels and that the panel covers have not been removed. In addition, the use of extension cords is often abused. Extension cords are designed to be used with items on a temporary basis, such as a drill or a fan. They are not designed to be used as permanent wiring, such as with a refrigerator or lighting. Indications that the wiring is being used in a permanent manner are situations in which the cord is run through a wall. The officer should also check for multiple extension cords being used in sequence and for open electrical boxes for outlets, switches, or junctions. The use of open splices in electrical wiring is also unacceptable.

Special Hazards

The nature of an industrial process by itself may be hazardous. For most processes with very high hazards, the occupancy may be required to have a **fire prevention division or hazardous use permit**. This is a local government permit that is renewed annually after the fire prevention division performs a code compliance inspection. The local fire company may be the first to discover a high-risk occupancy that has not been inspected by the fire prevention division, like the Los Angeles fire company that discovered a fireworks warehouse during a routine inspection. This is especially true if your district includes industrial parks where the buildings were built on speculation, known as "on spec" or "spec built." The property owners often construct buildings that are generic spaces and rent out space to a variety of tenants. This type of space often has a high turnover of occupants with different fire risk factors. A space that was used for storing building materials last year may be used for assembling computer hardware equipment this year.

Hazard Identification Signs

When required, visible hazard identification signs meeting NFPA 704 conditions should be placed on stationary containers and above-ground tanks and at entrances to locations where hazardous materials are stored, dispensed, used, or handled. Individual containers, cartons, or packages must be conspicuously marked or labeled. Material safety data sheets must be readily available on the premises for regulated material.

Selected Use Group Specific Concerns

Each use group classification involves special concerns and considerations in addition to the general concerns. Some of the more significant considerations are described in the following section.

Assembly

The primary concern for a code compliance inspection in a public assembly occupancy is to ensure that all of the access and egress pathways are clear and in good order (▼ Figure 12-16). The number and the size of exits should have been approved before the notice of occupancy was given to the business.

During a company inspection, the primary concern is to ensure that the exits are not blocked or locked and are in good working order and that there has not been a change in the layout. The code also specifies when panic hardware must be installed and the direction of swing of the doors. Like the number of exits, these are not usually an issue unless remodeling has occurred. Storage of combustibles under exit stairs is usually not permitted.

Exit lighting and egress lights are an important consideration for assembly-type occupancies. Typically, occupants are not familiar with exit locations, so methods to direct occupants to the exits are essential. Because of the large numbers of people in these types of occupancies, they require battery back-up or a back-up generator to provide illumination during a power failure. Evacuation plans may also be required.

The officer should check for the use of flame-resistant material in curtains, draperies, and other decorative materials. Other considerations include the storage of hazardous materials.

Figure 12-16 All access and egress pathways must be clear.

Fire extinguishers may be required and should be readily available. Commercial exhaust hood and ducts also need to be inspected. The fire officer should check for unvented, fuel-fired heating equipment. These occupancies are also prone to propping open fire doors and smoke barriers.

Business

The primary concern for a code compliance inspection in a business occupancy is to make sure that all of the access and egress pathways are clear and in good order. This includes exits that are blocked with office furnishings. Exit egress lights also tend to be problematic. Most occupants are familiar with building design, but often, fire protection equipment, such as fire extinguishers, is not maintained properly.

These occupancies are notorious for inappropriate use of electrical cords. In older buildings, open electrical boxes and spliced wiring tend to be a problem. These occupancies also tend to store small amounts of flammable liquids and other hazardous materials inappropriately.

Educational

A single school fire produced the swiftest example of the catastrophic theory of reform. The Our Lady of Angels School burned on December 1, 1958. Despite heroic efforts by the Chicago Fire Department, 95 people died, most of them children. NFPA's Percy Bugbee reported that within days of the disaster, state and local officials ordered the immediate inspection of schools. The NFPA partnered with Los Angeles Fire Marshal Raymond Hill to conduct 172 fire tests to develop new fire safety regulations for schools. *Operation School Burning* provided smoke and heat data that guided the technical committee in drafting new standards. Within 1 year, it was reported that major improvements had been made in 16,500 schools throughout the country.

Like assembly-type occupancies, exit paths are essential for occupant safety. Fire officers should ensure that doors are not blocked or locked, preventing the safe exit. When inspecting these occupancies, the fire officer must also ensure that built-in fire protection systems are in working order and are properly maintained. These facilities should have a fire evacuation plan, which should be up-to-date and practiced. Employees should also be trained in fire emergency procedures.

Factory Industrial

Factory industrial occupancies include many of the same hazards as business ones. They also may have processes that are particular to that type of factory; these should be identified for specific fire hazards. Factories sometimes have combustibles that are improperly stored. They are also prone to having fire protection systems that have been shut off or are

improperly maintained. The officer may also find fire doors that are propped open.

Hazardous

Hazardous occupancies are at a great risk for fire. Many of the same considerations apply as for factory industrial occupancies. In addition, special consideration should be given to preventing fires from starting. Codes specify what items and quantities can be stored in a facility. They also specify the markings that are required. Signs should be posted that prohibit open flames and smoking. Fire doors and smoke barriers should be in working order. Fire protection systems should be regularly inspected and maintained. Fire emergency plans should be in place, along with evacuation plans which should be practiced.

Health Care

Institutional occupancies include hospitals, nursing homes, and similar occupancies where the occupants are likely to require special assistance to evacuate. Built-in fixed fire protection systems, including automatic sprinklers, smoke detection systems, and sophisticated alarm systems, are required to provide additional evacuation time. The fire officer must ensure that these systems are in working order and are properly maintained. Fire detection systems should also be checked for proper maintenance and operation. These facilities should have a fire evacuation plan that should be up-to-date and practiced. Employees should also be trained in fire emergency procedures.

Mercantile

Mercantile occupancies include retail shops and stores selling stocks of retail goods. Stand-alone 24-hour convenience markets, department stores, and big-box home improvement and discount stores are all mercantile occupancies. There is a higher-than-average number of fire fighter line-of-duty deaths in mercantile fires. Like in the assembly classification, the occupants of mercantile occupancies may not be aware of the location of exits. Ensuring that exits are properly marked and lit is a prime concern.

Housekeeping is also a concern in these occupancies. Frequently, exits and aisles are blocked with merchandise, sometimes to the point of covering sprinkler heads, standpipe connections, and fire extinguishers. Exits may also be locked. The storage of flammable liquids might also be noted.

Residential

When the fire officer is inspecting residential units, only common areas can be inspected unless otherwise requested by the occupant. Hallways, utility areas, entryways, common laundry rooms, and parking areas are all open for inspection. Residential concerns vary by specific occupancy type; however, in general, exits are a concern. Fire doors are routinely propped open and may have items stored in the path. Exit and egress lighting are often not maintained properly.

Fire protection systems also need to be checked for adequate maintenance and operation. Vandals often remove caps and place objects inside standpipes. Valves may be closed, preventing the system from operating. Smoke alarms may have been removed. Hose cabinets may have damaged threads.

Preventing fires from igniting is a prime concern. Poor housekeeping, such as bags of trash in hallways, are invitations for fire setters. Frequently, managers of garden apartments store mowers and gasoline under stairways both inside and outside the structure.

Special Requirements

This class has a wide variety of hazards. Before inspecting a special use occupancy, the officer needs to review applicable code requirements that are specific to the occupancy type.

Detention

It is very important to ensure that the built-in fire protection systems in detention facilities are in good working order because of the inability of occupants to protect themselves from fires. The detection and alarm systems ensure that a fire is quickly detected while it is in its ignition phase and that the appropriate authorities are notified, including the fire department. If the system is inoperable, a fire may develop to a point at which the occupants are not able to be moved by on-scene workers. Check the records for proper testing and maintenance of these structures.

A working sprinkler system is essential to preventing the fire from growing. These systems not only protect the occupants but also assist the fire department in extinguishing the fire. Because of the security measures of the facility, the fire department is often delayed in accessing the seat of the fire. Like the alarm systems, be sure to check for the required testing and maintenance records. The inspector should also ensure that all valves are in their proper position. The inspector should also verify that the fire department connection is clear and easily accessible and that the threads match those of the hose used by the department.

Many of these facilities have standpipe systems because of the travel distance to the exterior. Often, they have "house lines," which are preconnected $1\frac{1}{2}$-inch hoselines attached to the standpipe connection. These lines are used by workers on scene and may be in a locked metal cabinet. When inspecting one of these facilities, be sure to take a coupling to attach to the standpipe connections to ensure that the threads are compatible with those used by the fire department.

These facilities also generally have fire extinguishers located throughout. The inspector should ensure that the extinguishers have undergone their monthly and annual tests

as required. Like the standpipe connections, these may be locked in a metal cabinet; however, they should be properly marked.

Typically, these facilities do not have a great deal of combustibles, and the housekeeping is normally excellent. The exception to this is in the areas where detainees are not permitted, such as the mechanical rooms. These areas tend to collect combustibles and may even have improperly stored chemicals.

Storage

Storage facilities often have sprinkler and standpipe systems, which should be checked for proper inspection and maintenance records. Frequently, this has not been completed as required. The inspection should also include a thorough check of the access to the fire department connection and the clearance for each sprinkler head. Storage facilities regularly stock material too close to the sprinkler heads.

These facilities generally have fire extinguishers, which need to be monitored for their monthly and annual checks. They may have been removed and set on the ground or may be damaged from fork lifts.

These facilities may also store hazardous materials. Often, these materials are stored for only brief periods of time. The fire officer should determine whether the proper NFPA 704 markings are in place. Check with management to determine the normal amount of materials that are on site as well as the maximum amount that could be present.

These facilities often block access and egress paths. The need to use space often causes the occupants to partially block exits with pallets of material. The side exit doors may also be found locked, and the exit and egress lights may not work and must be checked.

Unusual

Unusual occupancies can have a wide variety of hazards with no one hazard that is customary in all of them. The normal inspection concerns should be considered, with particular attention being paid to housekeeping, exit and egress routes, and built-in fire protection systems.

Mixed

Frequently, the officer finds buildings with multiple occupancy types inside. Each is considered individually in its requirements, but the building must meet the most stringent requirement of all of the occupancies inside. This might include a building that has a fireworks storage area in one unit and a barbershop in another. Although each occupancy must meet the codes that apply to that specific occupancy, the building in general must meet the most stringent standard. In this case, it would be the fireworks storage.

You Are the Fire Officer: Conclusion

You start a resource file on the Towne Centre project, using a copy of the basic site plan and the preliminary information provided by the developer. You call the fire prevention division to consult with the person performing the plan review for the project. You assign each fire fighter to research the special or unique characteristics of one type of occupancy that will be constructed. The driver-operator is assigned the task of identifying the fire flow requirements and evaluating the water supply and hydrant locations.

Wrap-Up

Ready for Review

- NFPA 1620 provides a six-step method of developing a preincident plan.

  Step 1: Evaluate physical elements and site considerations.

  Step 2: Evaluate occupant considerations.

  Step 3: Evaluate fire protection systems and water supply.

  Step 4: Evaluate special hazards.

  Step 5: Evaluate emergency operation considerations.

  Step 6: Evaluate special or unusual characteristics of common occupancy.

- The state or commonwealth determines the range and scope of local community fire code enforcement. The local community adopts an ordinance or regulation to establish a fire code.

- Automatic sprinkler systems come in wet pipe and dry pipe versions.

- There are three classes of standpipes, based on the size of the hose coupling outlets and the intended use of the standpipe.

- There are five building construction classifications.

- Every inspection is completed when the fire officer conducts an exit interview with the owner or representative and leaves a copy of a fire inspection report.

Hot Terms

Adoption by reference Method of code adoption in which the specific edition of a model code is referred to within the adopting ordinance or regulation.

Adoption by transcription Method of code adoption in which the entire text of the code is published within the adopting ordinance or regulation.

Authority having jurisdiction An organization, office, or individual responsible for enforcing the requirements of a code or standard, or for approving equipment, materials, an installation, or a procedure.

Automatic sprinkler system A system of pipes with water under pressure that allows water to be discharged immediately when a sprinkler head operates.

Catastrophic theory of reform When fire prevention codes or firefighting procedures are changed in reaction to a fire disaster.

Construction type The combination of materials used in the construction of a building or structure, based on the varying degrees of fire resistance and combustibility.

Fire prevention division or hazardous use permit A local government permit renewed annually after the fire prevention division performs a code compliance inspection. A permit is required if the process, storage, or occupancy activity creates a life-safety hazard. Restaurants with more than 50 seats, flammable liquid storage, and printing shops that use ammonia are three examples of occupancies that may require a permit.

Masonry wall May consist of brick, stone, concrete block, terra cotta, tile, adobe, precast, or cast-in-place concrete.

Mini/max codes Codes developed and adopted at the state level for either mandatory or optional enforcement by local governments; these codes cannot be amended by local governments.

Model codes Codes generally developed through the consensus process with the use of technical committees developed by a code-making organization.

Occupancy type The purpose for which a building or a portion thereof is used or intended to be used.

Ongoing compliance inspection Inspection of an existing occupancy to observe the housekeeping and confirm that the built-in fire protection features, such as fire exit doors and sprinkler systems, are in working order.

Ordinance A law of an authorized subdivision of a state, such as a city, county, or town.

Preincident plan A written document resulting from the gathering of general and detailed data to be used by responding personnel for determining the resources and actions necessary to mitigate anticipated emergencies at a specific facility.

Regulations Governmental orders written by a governmental agency in accordance with the statute or ordinance authorizing the agency to create the regulation. Regulations are not laws but have the force of law.

Standpipe system An arrangement of piping, valves, hose connections, and allied equipment installed in a building or structure, with the hose connections located in such a manner that water can be discharged in streams or spray patterns through attached hose and nozzles, for the purpose of extinguishing a fire, thereby protecting a building or structure and its contents in addition to protecting the occupants. This is accomplished by means of connections to water supply systems or by means of pumps, tanks, and other equipment necessary to provide an adequate supply of water to the hose connections.

Use group Found in the building code classification system whereby buildings and structures are grouped together by their use and by the characteristics of their occupants.

Your administration has just announced that they will be instituting a new company-level fire inspection program in a month. There is a lot of dissent in the room from some of the other officers, but you look forward to the opportunity to get out and do this level of inspection. The fire marshal's office has been too overwhelmed with the amount of new construction and cannot keep up with the inspection schedule, prompting the new program. In the next month, the fire marshal will be conducting training with each of the shifts to ensure that the program is carried out properly.

1. What is type II construction type?

    A. Fire resistive

    B. Limited combustible

    C. Heavy timber

    D. Noncombustible

2. What is occupancy type?

    A. Who lives there

    B. What a building is used or intended to be used for

    C. What processes occur at a building

    D. Development

3. What is one of the major use classifications?

    A. Retail stores

    B. Hospitals

    C. Residential

    D. Arena

4. What tools should you have with you to complete the inspection?

    A. Hose

    B. Digital camera with flash attachment

    C. NFPA 1901

    D. PPE

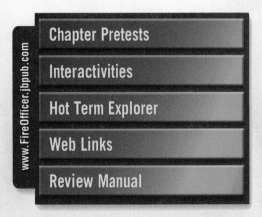

www.FireOfficer.jbpub.com

- Chapter Pretests
- Interactivities
- Hot Term Explorer
- Web Links
- Review Manual

Budgeting

NFPA 1021 Standard

Fire Officer I

4.4.3. Prepare a budget request, given a need and budget forms, so that the request is in the proper format and is supported with data.

(A) Requisite Knowledge. Policies and procedures and the revenue sources and budget process.

(B) Requisite Skill. The ability to communicate in writing.

Fire Officer II

5.1.2 **General Prerequisite Skills.** Intergovernmental and inter-agency cooperation.

5.4 **Administration.** This duty involves preparing a project or divisional budget, news releases, and policy changes, according to the following job performance requirements.

5.4.2. Develop a project or divisional budget, given schedules and guidelines concerning its preparation, so that capital, operating, and personnel costs are determined and justified.

(A) Requisite Knowledge. The supplies and equipment necessary for ongoing or new projects; repairs to existing facilities; new equipment, apparatus maintenance, and personnel costs; and appropriate budgeting system.

(B) Requisite Skill. The ability to allocate finances, to relate interpersonally, and to communicate orally and in writing.

5.4.3. Describe the process of purchasing, including soliciting and awarding bids, given established specifications, in order to ensure competitive bidding.

(A) Requisite Knowledge. Purchasing laws, policies, and procedures.

(B) Requisite Skill. The ability to use evaluative methods and to communicate orally and in writing.

Knowledge Objectives

After studying this chapter, you will be able to:

- List the supplies and equipment necessary for ongoing and new projects, repairs to existing facilities; new equipment, apparatus maintenance, and personnel costs; and appropriate budgeting system.
- Understand purchasing laws, policies, and procedures.

Skills Objectives

After studying this chapter, you will be able to:

- Prepare a budget.

As commander of Engine 4 in the high-rise district, you are running more than 16 alarm system activations in every 24-hour period. Tower 4 is running 10 additional alarm activations a day. The two fire trucks have rapidly aged because of the wear and tear of this extraordinary workload. You have made many suggestions and proposals to address this problem and keep the companies available for true emergency calls. Last year, the department was unsuccessful in getting approval to establish Quint 4, a third fire suppression unit that would share the workload. One of your radical ideas got the attention of the command staff.

You suggested the creation of a two-person Alarm Unit 4 that would be the primary responder to most of the activated smoke detector and manual fire alarm systems in the district. Alarm Unit 4 would be dispatched on alarms that come from a fire protection system monitored by a central fire alarm service without a report of actual smoke or fire. One of the team members would be a fire inspector empowered to issue repair orders, citations, and warrants. The other would be a fire suppression officer.

The command officers meet with you to discuss this proposal. The chief fire marshal is enthusiastic about the proposal, stating that the current prevention staff did not have the time to focus on the issue of defective alarm systems. Three years ago, the city passed a faulty alarm ordinance, but the budget director never authorized the staffing that was needed to implement the regulation. The fire operations chief was willing to support the alarm unit as long as the team was provided with clear operating guidelines and the unit was equipped with personal protective equipment and the appropriate tools to investigate an activated fire alarm.

You are excited to hear that the chiefs were willing to try out your idea. However, excitement quickly turns to dread when the fire operations chief tells you to submit a proposed budget for the alarm unit before the end of the month. You have never developed a budget and have a hard time balancing a checkbook.

1. What do you have to consider when preparing a budget?
2. How do you prepare a budget proposal?

Introduction to Budgeting

A **budget** is an itemized summary of estimated or intended revenues and expenditures. **Revenues** are the income of a government from all sources appropriated for the payment of public expenses. **Expenditures** are the money spent for goods or services. Every fire department has some type of budget that defines the funds that are available to operate the organization for a particular period of time, generally 1 year. The budget process is a cycle:

1. Identification of needs and required resources
2. Preparation of a budget request
3. Local government and public review of requested budget
4. Adoption of an approved budget
5. Administration of approved budget, with quarterly review and revision
6. Close out of budget year

Budget preparation is both a technical and a political process. The funds that are allocated to the fire department define the services that the department would be able to provide for that year. The technical part relates to calculating the funds that would be required to achieve different objectives, whereas the political part is related to elected officials making the decisions on which programs should be funded among numerous alternatives.

For most cities, the fire department is one of the larger components of the overall budget, along with schools, police, and public works. The elected officials have to set tax rates to generate the revenue and then allocate the available funds among all of the different organizations and programs supported by local government. Federal, state, and local regulations all influence the budget process.

The Budget Cycle

Every organization has a budget cycle, typically 12 months for most government agencies. The budget document describes where the revenue comes from (input) and where it goes (output) in terms of personnel, operating, and capital expenditures. Annual budgets usually apply to a **fiscal year**, such as starting on July 1 and ending on June 30 of the following year. The budget for fiscal year 2008, referred as FY08, would start on July 1, 2007 and end on June 30, 2008. In many cases, the process for developing the FY08 budget would start in 2006, a full year before the beginning of the fiscal year (▶ **Table 13-1**).

The timeline in Table 13-1 shows that there is a long period from the time a fire officer makes a budget request to the time the money is available to spend. Consider the timeline for a replacement pumper. If the initial request is submitted in August 2006 and it is approved at every stage of the process, the funding is not authorized until July 2007. Then, the department can issue a request for proposals from apparatus manufacturers to build the vehicle. It then takes a couple of months to evaluate the submissions and award a contract, and the time until delivery could be 6 months. The replacement pumper that was requested in August 2006 may not be in service before mid-2008.

Base and Supplemental Budgets

Most municipal governments incorporate the concept of a **base budget** in the financial planning process. The base budget is the level of funding that would be required to maintain all services at the currently authorized levels, including adjustments for inflation, salary increases, and other predictable cost changes. There is a built-in process that assumes that the current level of service has already been justified and that the starting point for any changes in the budget should be the cost of providing the same services next year.

Budgets can increase or decrease from the base level. Proposed increases in spending to provide additional services are classified as **supplemental budget** requests. When budgets have to be reduced, the change is calculated as a percentage of the base budget. A 10% budget cut means that only 90% of the base budget expenditures for the current year will be approved for the next fiscal year. Savings equivalent to 10% of the budget will have to be proposed.

Significant increases in the fire department budget require early notification and support of the elected officials, well before the municipality's proposed budget is prepared by the budget director. A new program that involves additional employees or a major capital expenditure would require funds that would have to be found by increasing revenues or decreasing expenditures from some other part of the budget. These decisions are made by the elected officials, and, in order to present a balanced proposed budget, the budget director has to have a good idea of what the elected officials are likely to approve.

Local elected officials are both advocates and gatekeepers in developing the local budget. One of their primary roles is to allocate local government resources in a way that best serves the public. They want to keep taxes down and, at the same time, deliver the services that the voters expect. They do this through direct involvement in the budget preparation process and participation in public hearings. Elected officials often request private face-to-face meetings with fire department leadership to discuss issues that are expected to be controversial.

Revenue Sources

The revenue stream that supplies a fire department depends on the type of organization that operates the fire department and the formal relationship between the organization and the local community. There is a wide diversity in organizations and in revenue stream sources. Each type of organization has a different process for obtaining revenue and authorizing expenditures.

Most municipal fire departments operate as components of local governments, such as towns, cities, and counties. A fire district is a separate local government unit that is specifically organized to collect taxes in order to provide fire protection. In some areas, fire departments are operated as regional authorities or an equivalent structure.

Local Government Sources

The mix of revenues available to local governments varies considerably from state to state because state governments set the rules for local governments. There are many variations of the rules, and the fire officer needs to know the rules that apply to his or her jurisdiction. Local governments can collect taxes and spend the revenues for specific purposes. General tax revenues can be spent for any purpose that is within the authorized powers of the local government.

Some funds are restricted and can only be used for certain purposes. A **fire tax district** is created to provide fire protection within a designated area. A special fire protection tax is charged to properties within the service area, in addition to any other county or municipal taxes. For example, the fire tax could be $0.05 per $100 of value on property within the defined area. All of the revenue from the fire tax is dedicated to pay for the provision of fire protection services.

Revenues can also be restricted to many other purposes. In many areas, gasoline tax can be used only to build or maintain roads. Taxes that are levied to repay capital improvement bonds can be used only for that purpose.

Table 13-1 Fiscal Year 2008 Timeline

August–September 2006	Fire station commanders, section leaders and program managers submit their FY08 requests to the fire chief. Their concentration is on proposed new programs, expensive new or replacement capital equipment, and physical plant repairs. Depending on how the fire department is organized, this is where a station commander would request funds for a new storage shed or a replacement dishwasher for the fire station. Documentation to support replacing a fire truck or purchasing new equipment, such as an infrared camera or a hydraulic rescue tool, would be submitted, and a proposal for a special training program would be prepared. In most fire departments, these requests are prepared by fire officers throughout the organization and submitted to an individual who is responsible for assembling the budget proposals.
September 2006	The fire chief reviews the budget requests from fire stations and program proposals from staff to develop a prioritized budget wish list for FY08. This is the time when larger department-wide initiatives or new program proposals are developed. For example, purchasing a new radio system or establishing a decontamination/weapons of mass destruction unit would be part of the fire chief's wish list.
October 2006	The city's budget office distributes FY08 budget preparation packages to all agency or division heads. Included in the package are the application forms for personnel, operating, and capital budget requests. The budget director includes specific instructions on any changes to the budget submission procedure from previous years. The senior local administrative official (mayor, city manager, county executive) also provides specific submission guidelines. For example, if the economic indicators predict that tax revenues will not increase during FY08, the guidelines could restrict increases in the operating budget to 0.5% or freeze the number of full-time equivalent positions.
November 2006	Deadline for agency heads to submit their FY08 budget requests to the budget director.
December 2006	The budget director assembles the proposals from each agency or division head within the local government structure. Each submission is checked to ensure that it complies with the general directives provided by the senior local administrative official. Any variances from the guidelines must be the result of a legally binding agreement, settlement, or requirement or have the support of elected officials.
January 2007	Local elected officials receive the preliminary proposed budget from the budget director. On some budget items or programs, there may be two or three alternative proposals that require a decision from the local elected officials. Once the elected officials make those decisions, the initial budget proposal is completed. This is the "Proposed Fiscal Year 2008 Budget."
February or March 2007	The proposed budget is made available to the public for comment. Many cities make the proposed budget available on an Internet site, inviting public comment to the elected officials. In smaller communities, the proposed budget may show up in the local newspaper.
March or April 2007	Local elected officials conduct a public hearing or town meeting to receive input from the public on the FY08 proposed budget. Based on the hearings and on additional information from staff and local government employees, the elected officials debate, revise, and amend the budget. During this process, the fire chief may be called on to make a presentation to explain the department's budget requests, particularly if large expenditures have been proposed. The department may be required to provide additional information or submit alternatives as result of the public budget review process.
May 2007	Local leaders approve the amended budget. This becomes the "Approved" or "Adopted" FY 2008 budget for the municipality.
July 1, 2007	FY08 begins. Some fire departments immediately begin the process of ordering expensive and durable capital equipment, particularly items that have long lead times for delivery. Many request for proposals are issued in the early months of the fiscal year.
October 2007	Informal 1st-quarter budget review. This review is looking for trends of expenditures to identify any problems. For example, if the cost of diesel fuel has increased and 60% of the motor fuel budget has been spent by the end of September, there will be no funds remaining to buy fuel in January. This is the time to identify the problem and plan adjustments to the budget. Amounts that have been approved for one purpose may have to be diverted to a higher-priority account.
January 2008	Formal midyear budget review. The approved budget may be revised to cover unplanned expenses or shortages. In some cases, this is in response to a decrease in revenue, such as an unanticipated decrease in sales tax revenue. If the money is not coming in, expenditures for the remainder of the year might have to be reduced. Sometimes, revenues exceed expectations and funds become available for an expenditure that was not approved at the beginning of the year.
April 2008	Informal 4th-quarter review. Final adjustments in the budget are considered after 9 months of experience. Year-end figures can be projected with a high degree of accuracy. Some projects and activities may have to stop if they have exceeded their budget, unless unexpended funds from another account can be reallocated.
Mid-May 2008	The finance office begins closing out the budget year and reconciling the accounts. No additional purchases are allowed from the FY08 budget. Purchases are deferred until the beginning of FY09.

The United States Census summarizes the sources of state and local government tax revenues in nine general areas. The top three local tax revenue sources represent about two thirds of the total revenues collected by local governments. These sources are:

- General sales and gross receipts taxes
- Property taxes
- Individual income taxes

The other six listed sources of revenues are:

- Corporation net income taxes
- Motor fuel sales taxes
- Motor vehicle and operators' licenses
- Tobacco product sales taxes
- Alcoholic beverage sales taxes
- Other fees and taxes (e.g., service fees, motel occupancy fees)

Many fire departments also obtain revenue through direct fees for service. Fire departments that operate ambulances often obtain sufficient revenue from charges for patient treatment and transportation to offset the operating costs of the service. Direct revenue might also come from fees for fire prevention permits, special inspections, and other services that are provided to individual property owners or contractors.

Volunteer Fire Departments

There are as many different ways to fund volunteer fire departments as there are paint schemes for fire apparatus. Some volunteer departments are operated by municipal governments and are completely supported by local tax revenues, with all capital and operating expenses being handled by the jurisdiction. The budget process for the jurisdiction would treat the volunteer fire department as a local government agency.

Many volunteer fire departments are organized as independent 501 (c) nonprofit corporations. Some of these volunteer organizations raise their own operating funds through public donations or subscriptions and are entirely independent of local government. In other areas, local tax revenues are allocated to the volunteer corporations. Their relationship with a local jurisdiction may be through a memorandum of understanding, a contract for services, or a regional association of independent volunteer fire departments. Some jurisdictions pass local legislation or approve a charter that details the relationship, duties, services, and compensation that the volunteer fire department will provide to the community in return for the tax revenues.

Tax revenues support the entire operation of some volunteer fire departments, whereas others supplement their tax allocation with fundraising activities. Sometimes, the volunteer corporations own and operate the fire stations and apparatus, whereas other volunteer organizations occupy buildings and operate vehicles that are owned by the local government.

Company Tips

Some volunteer fire departments operate very successful direct mail campaigns to raise funds. Sophisticated marketing techniques can be used to customize the mail campaign to increase the size and percentage of returns they receive on their mail-outs. This is often a more efficient use of volunteer staff than running a bake sale or soliciting door-to-door.

A recent revenue trend involves the installation of cellular telephone equipment at fire department facilities. Telecommunications companies often rent space from a volunteer organization to place their equipment on an existing fire department radio tower. Other volunteer departments rent a small piece of land on the fire station property to a telecommunications company to construct a monopole.

Volunteer organizations use a wide variety of fundraising methods. Direct fundraising often includes door-to-door solicitations or direct mail campaigns. Many volunteer corporations sponsor activities to generate revenue, such as dinners, bake sales, car washes, raffles, and weekly bingo nights. Some volunteer departments generate revenue by renting out social halls or meeting rooms for private events.

Bingo and Other Gaming Activities

One of the traditional fundraising methods for volunteer fire department is to sponsor bingo or other gaming activities. These activities are usually state or regionally regulated and require specific procedures for conducting the games and handling the money. Volunteer fire departments have a revenue advantage over other operators because they do not have to pay their members to run the games.

Some departments hold an annual carnival or other entertainment event that provides revenue to the department. Holding the event on fire department property and using unpaid staff allows the fire department to minimize operating costs. Many of these fundraisers are major events in the community, creating good will and strong citizen awareness and support.

Grants

Municipal and volunteer fire departments can also obtain funds by applying for grants. One of the larger sources of grant funds is the Assistance to Firefighters Grant Program, which distributes federal funds to local jurisdictions. Since the 2001 terrorist attacks, Congress has appropriated increased funding under this act and other grant programs to meet local homeland security needs.

Although the Assistance to Firefighters Grant Program is the largest grant program, there are hundreds of smaller programs offered by the federal government, state governments, private institutions, nonprofit organizations, and for-profit companies. Grants are competitive and require the local fire department to meet specific eligibility and documentation requirements. Many of the grant programs also require the local jurisdiction to provide a percentage of matching funds.

Some state agencies offer grants to local responders from specially designated funds, such as a fee on motor vehicle permits, a surcharge on traffic fines, or a tax on fire insurance premiums.

The U.S. Fire Administration provides a four-step method to develop a competitive grant proposal:

- Conduct a community and fire department needs assessment
- Compare weaknesses to the priorities of the grant program
- Decide what to apply for
- Complete the application

The process of following the four steps in developing a grant proposal requires painstaking compliance with the specific grant instructions. To qualify, applicants have to follow a specific set of procedures and provide a detailed level of financial and operational information. The requirements may seem excessive and intrusive; however, the grant administrators have to ensure that the money is spent wisely and within legislative guidelines.

The Assistance to Firefighters Grant Program applications are evaluated by a committee that ranks them on the basis of pre-established criteria. To be approved, the application must have a request that matches the goal of the grant program, show that it is cost efficient, and demonstrate a strong community or financial need.

The first step is to determine whether a grant program exists that covers your problem area. Once you have identified the program, determine whether your organization is eligible under the rules of the grant program, as well as the process for applying for the grant.

A grant application has to describe how the department's needs fit the priorities of the grant program. The grant applicant needs to analyze the community, conduct a risk assessment, evaluate the existing capabilities of the fire department, and identify the department's needs.

There is strong competition for a limited amount of grant funds, so the applicant must provide a sound justification. Use the information gathered in the needs assessment in a narrative to describe the problem in a concise but detailed manner. For example, consider a request to purchase multiple gas meters to replace the flammable vapor detectors that are currently assigned to ladder and rescue companies. The old detectors measure only the concentration of a flammable vapor in relation to the lower explosive limit, whereas the newer detectors also measure the percentage of oxygen and carbon monoxide in the atmosphere. A part of the needs assessment would identify the increase

in service calls to check on carbon monoxide alarm activations since a local ordinance has been in place.

In addition to the needs assessment, the grant application must demonstrate the department's financial need and why it needs outside assistance to acquire these resources. The application should show that efforts have been made to obtain the funds from other sources. In describing the current financial situation, include factors such as an eroding tax base, expanding community growth, local/state legislation that restricts taxes, and any significant local economic problem, such as the loss of a major employer.

The application also needs to show that this solution provides a benefit at the lowest possible amount of funding. Factors include collaborating with other organizations to share expensive equipment.

Nontraditional Revenue Sources

Many fire departments have solicited donations from a targeted business or industry for a specific purpose. A volunteer fire department serving a resort community solicited donations from the hotel-motel association to purchase an additional aerial platform truck that serves the hotel high-rise district. Equipment for a hazardous materials team could be donated by a company in the community that produces hazardous materials. Some public safety organizations are looking for revenue from even more nontraditional sources. One fire department offered to sell advertising on their vehicles.

Cost Recovery

Local or state regulations often allow a fire department to recover the extraordinary expenses incurred in responding to a hazardous materials incident, particularly decontamination, spill clean-up, and recovery costs. The department can invoice a hazardous materials carrier for extraordinary charges, such as overtime and replacement of expendable supplies (e.g., absorbents and neutralizing agents). In addition, the carrier may be responsible for the expense of repair or replacement of durable equipment, such as contaminated monitoring equipment, chemical protective suits, fire fighter protective clothing, and fire suppression tools.

Cost recovery also applies to some special services supplied by fire departments. For example, if a production company is shooting a film that involves pyrotechnics, the fire department might provide an engine company to stand by for several hours. The cost of overtime to staff the unit and an hourly rate for the use of the apparatus could be charged to the production company.

Expenditures

The annual budget for the fire department describes how the funds that are available are spent during the year. In the case of a municipality, the fire department budget is usually a part of the overall budget for that governmental entity. The format of the budget should comply with recommendations of the **Governmental Accounting Standards Board (GASB)**, which is the standard-setting agency for governmental accounting.

The local government budget typically includes a system of accounts that classifies all expenditures within certain categories and complies with the generally accepted accounting principles as outlined by the GASB. The accounting system normally uses a database system to support a line-item budget. A **line-item budget** is a format in which expenditures are identified in a categorized line-by-line format. The accounting system may have a complex numbering system, such as 02-12301-876543, that includes a category for every type of expenditure. In this example, the line item describes the following:

- "02"—The first two numbers identify the fund. The 02 indicates that this money is coming from the general fund.
- "12301"—The second part (the five-digit number) identifies the department and the division or subdivision where the money is being spent. Called the "object code" in some systems, this one says that the money allocated to the Fire Department (12) and to the Suppression Division (301).
- "876543"—The third part (the six-digit number) provides a detailed description of what the money is being used for. Called the "subobject code" in some systems, this one says that the money is to be used for replacement fire hose.

The accounting system allows budget analysts and financial managers to keep track of expenditures throughout a municipal government. For example, law enforcement, public works, schools, and fire departments all provide employees with work uniforms. The database system would allow a budget analyst to determine how much the municipality spent on all uniforms, as well as breaking expenditures down by department and type of uniform. This could be valuable information when one is considering the savings that could be obtained by purchasing all uniforms from one vendor.

Fire department expenditures are generally divided into three general areas: personnel costs, operating costs, and capital expenditures. The accounting system allows budget analysts and managers to keep track of how much money is spent on fire fighter salaries (personnel costs), fuel for vehicles (operating costs), and construction of new fire stations (capital expenditures).

Personnel Expenditures

In career fire departments, more than 90% of the annual budget is usually allocated to personnel expenses. Some departments spend as much as 97% of their total budget on salaries and benefits. The personnel costs include much more than the advertised salaries. This category includes fringe benefits, such as pension fund contributions, worker's compensation, and life insurance. Civil service regulations, local administrative decisions, and labor contracts often determine the cost of these benefits. Fire fighters have one of the highest worker's compensation rates, based on the history of claims and payments.

Fringe benefits also reflect the estimated cost of providing sick and annual leave benefits to the employee. Many municipalities have a sliding scale, in which the amount of leave earned per pay period increases as the employee gains seniority. The base cost of this fringe benefit equals the amount of the employee's accrued leave. In many cases, the actual cost to the fire department also includes overtime to pay for another employee to fill the vacancy.

Operating Expenditures

The operating budget covers the basic expenditures that support the day-to-day delivery of municipal services. Uniforms, protective clothing, telephone charges, electricity for the fire stations, flashlight batteries, fire apparatus maintenance, and toilet paper are all examples of purchases that would be classified as operating expenses. These funds are usually allocated in categories that allow for some flexibility throughout the year. For example, a fire department could be authorized to spend $250,000 on protective clothing during the year. The actual numbers of coats, pants, helmets, boots, and gloves that would be purchased would depend on the needs and the cost per item at the time of purchase. When a budget needs to be trimmed down, operating expenditures are usually the first area to be considered. For example, employee training, travel, and consulting expenses are usually reduced when revenue falls below expectations.

In large organizations, vehicles are often leased to the fire department and other operating departments from a fleet management division. The cost of a vehicle to the fire department is calculated on a per-mile basis that includes fuel, scheduled maintenance, anticipated repairs, and replacement costs. A portion of the amount that is charged to the fire department goes into a replacement vehicle account that is used to purchase replacement vehicles. The amount that goes into this account for each vehicle is based on the anticipated replacement cost and the projected number of years that the vehicle will be used.

The amount that is charged to the fire department budget is generally based on the classification of each vehicle. Maintenance costs for fire department vehicles are usually high because they tend to be heavily loaded with equipment, including emergency lighting and siren systems, two-way radios, and computers. Large fire apparatus, which is expensive to operate and replace and has relatively low annual mileage, may have a rate of several dollars per mile.

VOICES OF EXPERIENCE

❝ A key component to budget management is understanding how government thinks and communicates. ❞

Budgeting is both an art and a science. Budgeting is a critical element of management, but it tends to receive less attention from new fire officers than tactics and human resources issues. Budget management impacts all facets of fire department operations, from the number of personnel on the apparatus, to the quality of the personal protective equipment. Failure to forecast needs and prepare concise and defensible budgets can have a global impact on the fire department and can increase risk on the fireground.

A key component to budget management is to understand how government thinks and communicates. Government officials tend to prefer long range planning, particularly on capital budget items such as apparatus replacement or facilities. For example, you may find government officials more receptive to a five-year plan to replace all SCBA, as opposed to a large one-time purchase. Understanding these preferences will help to ensure that you can present budget requests in a way that government officials understand.

Also, remember to support your requests with hard numbers and facts, rather than emotion. Our fire department asked the local government officials to adopt response performance goals as a way to communicate to us what they view as good service. This includes identifying how quickly a unit should be out the door for a call and how many personnel should be assembled for a fire attack. How does this apply to budgeting? By having these response performance goals established by government officials, we now have a benchmarking target, as well as solid reasoning and justification for requesting funds to meet *their* goals. By establishing response performance goals, we were able to establish a solid link between fire department operations and budget requests.

Eddie Buchanan
Hanover, Virginia Fire and EMS
22 years of service

Ambulances, which often run up to 40,000 miles a year, have high maintenance costs and have to be replaced every 3 or 4 years, also have a high per-mile rate. Marked sedans and sport utility vehicles (SUVs) used for emergency purposes generally have a longer life span and are charged at a lower cost per mile. If the actual fleet management costs exceed the budget projections because of high fuel costs or unanticipated repairs, the additional amount is sometimes charged back to the fire department at the end of the fiscal year and has to be included in the 4th-quarter adjustments.

Replacement uniforms and protective clothing are another expense calculated in the operating budget (▼ Figure 13-1). For example, the fire department might budget a lump sum of $150 per fire fighter to cover the estimated annual cost of cleaning and repairs to protective clothing and up to $400 per fire fighter for uniforms.

Mandated continuing training, such as hazardous materials and cardiopulmonary resuscitation recertification classes, is also factored into the operating costs. That figure includes the cost of the training per fire fighter per class. The cost to pay another fire fighter overtime to cover the position during the mandated training could also be included.

Capital Expenditures

Capital expenditures refer to the purchase of durable items that cost more than a threshold amount and will last for more than one budget year. Local jurisdictions differ on the amount and the time period that these items are supposed to last. Items such as computers, hydraulic rescue tools, washing machines, and self-contained breathing apparatus are examples of equipment purchased as capital items (▶ Figure 13-2). In general, the municipality establishes a unique inventory record that documents the purchase price, source, assignment, and disposal of each capital item.

Sedans, pick-up trucks, and SUVs are usually part of the capital budget. Specialized vehicles, such as ambulances and heavy fire apparatus, are also capital equipment, but they are often included in a special section of the capital budget because of their cost and complexity. In many larger systems, replacement vehicles are purchased from a special set-aside fund, and only additional vehicles are included in the supplemental capital budget.

Capital improvement projects cover the construction, renovation, or expansion of municipal buildings or infrastructure. These are expensive projects that are often funded through long-term loans or bonds issued by the municipality.

Bond Referendums and Capital Projects

In many states, major capital improvement projects, such as fire station construction and major renovations, are funded through bond programs. A **bond** is a certificate of debt issued by a government or corporation; the bond guarantees payment of the original investment plus interest by a specified future date. The same type of funding mechanism is also used to build roads, water distribution systems, sewer systems, parks, libraries, and similar public facilities. The voters have to approve this type of expenditure through a referendum. If the voters approve, the municipality is authorized to borrow money from investors by issuing bonds up to a set

Figure 13-1 Replacement protective clothing is an operating expense calculated in the budget.

Assessment Center Tips

Remember Annual Expenses

When discussing or responding to a budget proposal, remember to focus on the annual expenses in addition to the initial purchase cost. The annual costs fall into two areas: continuity and personnel.

Consider the purchase of semiautomatic cardiac defibrillators to improve first-responder service for three fire companies. If the department is asking for just three defibrillators, what happens when one of the devices breaks or needs service? Continuity considers the need for a spare unit to be used when one of the new devices is broken or unavailable, as well as the cost of maintenance and repairs. Additional costs would include the initial training of fire fighters to use the devices, plus annual recertification classes.

Personnel costs are the largest expense for career fire departments. A new ladder truck could cost $800,000 to purchase; however, the annual personnel cost to provide staffing for the vehicle could easily exceed $1,000,000. That cost could be difficult to justify for a company that will makes less than 100 responses per year.

A

B

C

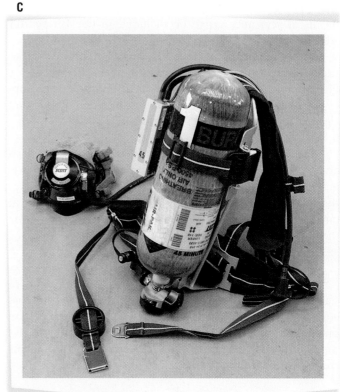

Figure 13-2 Three examples of equipment purchased as capital items. A. Tools. B. Computers. C. SCBA.

amount. The authorization to sell the bonds is usually valid over a period of 5–10 years.

The bonds are repaid over a period of 10–30 years and return a fixed interest rate to investors. Bond funds are classified as a special revenue source and require compliance with a different set of accounting controls and reporting procedures. Annual status reports are required, and there are specific restrictions on the items that can be purchased with the bond revenue. There is a very large market for municipal bonds, with intense competition for lower interest rates. If a city is issuing $100,000,000 in bonds to be repaid over 20 years, a fraction of a percentage difference in the interest rate can have a significant impact to the taxpayers.

Bond funding allows the taxpayers to actually pay for the facility while it is being used, just as a mortgage allows a family to live in a home while they are paying for it. The repayment

period should be at least equal to the anticipated life span of the facility. Bonds are repaid by collecting a special property tax or, if the facility will produce revenue, such as an airport or convention center, the bonds can be repaid from future income.

When bonds are used to build a new fire station, the authorized amount often includes the land, site improvements, building construction, permits, architectural and engineering fees, furniture, and equipment that go into the building, as well as the apparatus and equipment that will be assigned to the new station.

Occasionally, a bond program is used to fund a large fire department apparatus purchase. The St. Louis Fire Department used a bond program to cover the cost of an entire fleet of new apparatus in 1986. That large purchase allowed for the replacement of all existing pumpers and ladders with 30 new quints to implement the "Total Quint Concept." In fiscal year 2000, another bond program was authorized to replace the entire fleet with a new generation of apparatus.

Lower Revenue Means Fewer Resources

There is always a degree of uncertainty in the budget process. Planned expenditures have to be balanced against anticipated revenues a year or more in advance. If the revenues do not meet expectations, adjustments have to be made to reduce spending. If revenues exceed expectations, some extra funds could be available during the year. Local revenue has a history of cyclical behavior, driven by external and internal forces that affect the delivery of services.

Changes in the local economy often result in major changes in the revenue collected by local government. Changes in property values have an impact on property tax revenues; if property values fall, property tax revenue decrease proportionately. Where local governments receive a percentage of sales taxes, a healthy economy provides increasing revenues and a weak economy can result in a sudden drop in revenues. A business that shuts down or reduces employment because of a poor economic trend can have a major impact on the local government budget.

Consider the impact of an empty office, retail, or manufacturing building. When the building is occupied and operating, it provides revenue to local government through property taxes and sales or gross receipts taxes. When the building is vacant, there are no sales or gross receipts to be taxed. In addition, many communities set a lower property tax rate for vacant buildings. The result is that an empty building generates dramatically lower revenues to local government. That is one reason why local officials closely monitor the vacancy rate in office, business, and mercantile space.

If a new business moves into a community, it brings additional revenue to the local government. When a business moves to another jurisdiction, the sales tax revenue also moves to the new jurisdiction. The local jurisdiction could also lose the income tax from people who move with the business. If a property owner goes into bankruptcy, the building could be abandoned and the property taxes might not be paid.

Lower Revenue Options

Sometimes, the results of budget reductions drastically affect fire department operations. Fire departments have to make difficult choices when faced with declining revenues. Five such options are:

- Defer scheduled expenditures, such as apparatus replacement and station maintenance
- Regionalize or consolidate services
- Privatize or contract out elements of the service provided by the department
- Reduce the career work force
- Reduce the size of the department

Fire Marks

Some of the most challenging economic situations in recent history have occurred in California. During the 1970s, property valuations and property tax rates in California increased to the point that a taxpayer revolt occurred. Proposition 13, a measure that restricted the rate that municipalities could raise property tax, was passed by California voters in June 1978. This measure stopped the trend of dramatic annual increases in property taxes in order to pay for increasing municipal expenditures. Several other states, as far away as Massachusetts, adopted similar measures in the subsequent years.

Although Proposition 13 did not have the predicted effect of devastating local government services in California, it significantly restricted future municipal government growth. Many municipalities were unable to expand services and struggled to maintain existing services in the face of increasing costs. The 4700 special districts that provided a variety of services in California felt the most severe impact. These organizations, which included many fire districts, relied solely on property tax revenue and lacked the flexibility that local municipalities had to find other sources of revenue. The freezing of property tax revenue meant that some special districts could not keep up with inflation of their operating costs or provide cost-of-living raises for their employees. Many of the smaller fire protection districts had to look at consolidation or contracting out in order to maintain adequate services. Several regional fire organizations were created to replace smaller independent fire departments.

Sometimes, the local situation becomes catastrophic. In 1994, Orange County, California, announced that its investment pool had suffered a loss of $1.6 billion. The county declared bankruptcy on December 6, initiating a reorganization of local government that took 2 years. Salaries were frozen, and the county cancelled unfilled purchase orders. Any position that was vacant, due to retirement or promotion, was eliminated. For 2 years, the county provided only the services that were legally mandated. Many maintenance and service activities were postponed. Equipment that was scheduled for replacement remained in service. All agencies were required to reduce staff and restructure operations. In the restructuring, the county fire department was eliminated, and the Orange County Fire Authority was created. The new authority operates under finance rules that are different from those that apply to a fire department operated by a county.

Defer Scheduled Expenditures

Deferring scheduled expenditures means delaying the purchase of replacement fire apparatus or other expensive equipment. For example, a department that normally replaces pumpers on a 15-year schedule may decide to stretch the life of the rigs to 18 or 20 years. During the recession of the 1980s, Los Angeles and Chicago could not afford to replace their aging ladder trucks. Instead, they obtained limited funding to rehabilitate some of the existing rigs. The rigs were updated, rewired, repainted, and recertified as aerials. The Los Angeles Service Life Extension Program also replaced older gasoline motors and manual transmissions with diesel engines and automatic transmissions. This provided another 5-7 years of service for the rehabilitated aerials at less than half the cost of new apparatus.

Privatize or Contract Out Elements of the Service Provided

Privatization means replacing municipal employees with contract employees. The concept is that the cost to the municipality to provide the service should be lower because a private company can operate more efficiently than a local government agency. Trash pickup, vehicle fleet maintenance, and school bus operations are three examples of services that are often contracted out by local government. The 1990s saw large, national emergency ambulance providers, such as American Medical Response and Rural-Metro, competing to provide 911 ambulance transportation in place of fire department–based services in many areas.

Some fire departments privatize or contract out services that are so specialized that a private company may be able to provide a better level of service at a lower cost. Paramedic training, special operations classes, hazardous material site clean-up, ambulance billing, apparatus maintenance, and communications system maintenance could all be candidates for privatization.

Regionalize or Consolidate Services

Regional or consolidated fire departments are sometimes established to increase efficiency by reducing duplication in staff and services. Miami-Dade Fire Rescue Department in Florida started in 1935 as a fire patrol with one employee and one truck reporting to the Agriculture Department. The organization grew through numerous mergers with municipal fire departments and fire districts, as well as the expanding population in Dade County. In 1965, it became the Metropolitan Dade County Fire Department, and in 1997, it became the Miami-Dade Fire Rescue Department.

Today, the Miami-Dade Fire Rescue Department provides fire suppression, emergency medical services, and other specialized operations services to a 1924-square-mile area that includes the unincorporated portions of Dade County and 25 municipalities. The municipalities chose to have the county provide fire-rescue service instead of providing their own municipal fire-rescue department. The department has more than 1500 employees operating out of 55 fire stations and responds to an average of 525 calls per day.

Reducing the Career Workforce

When revenues decrease, cities sometimes have to reduce their workforces. Many fire departments try to protect staffing on emergency response vehicles by reducing positions in administrative and support areas. This often has limited effectiveness because these positions usually represent a very small proportion of the entire work force, and the indi-

viduals in these positions often perform important tasks that enable the emergency responders to work efficiently.

If it becomes unavoidable to reduce the number of emergency responders, one of the basic options is to maintain the number of companies in service and reduce the staffing per vehicle. The other option is to limit the number of units in service. Neither option is popular nor attractive; however, budget realities sometimes leave no other choices. In a municipal budget, the elected officials often have to decide where to make cuts, weighing the value of every municipal cost and service in the eyes of the taxpayer.

Some fire departments reduce overtime by temporarily closing selected fire companies for one shift at a time, reassigning the on-duty fire fighters to fill vacancies in other fire stations. This practice creates a Russian-roulette situation. This often has huge political repercussions, particularly if a civilian dies or suffers serious injury in a structure fire where the nearest fire company was closed for the day due to budget restrictions.

New York City laid off more than 40,000 employees, including 1600 FDNY fire fighters, in July 1975 at the start of fiscal year 1976. Although the city hired back 700 of the fire fighters within three days, 900 others lost their permanent fire fighter jobs. It would take 2 years before the city could rehire all the laid-off fire fighters who reapplied to the FDNY. During the time they were laid off, some of them were hired as temporary employees working under a federal Housing and Urban Development grant to board up roofs and windows of fire-damaged buildings, preserving the city housing stock. All of the Housing and Urban Development–funded temporary employees were laid-off FDNY fire fighters.

Reduce the Size of the Fire Department

In the 1990s and early 2000s, several municipal departments were again asked to significantly reduce their career work forces in response to reduced local revenues. Many fire departments had removed companies from multiple-company firehouses. In some cases, an engine company and a ladder company were replaced with a quint.

Safety Zone

Verify Your Sources

One of the best ways to ensure approval of a budget proposal is to clearly identify a federal or state regulation, a local ordinance, or a consensus standard that would mandate the program or require the equipment replacement. The budget officials want to comply with all legally binding requirements but are not motivated to spend limited resources on recommended practices or suggested schedules. If state regulations require replacement of Level A chemical suits after 5 years, the budget request should identify the specific regulation.

Baltimore disbanded 15 engine and ladder companies between 1990 and 2001, reducing the fire fighter work force by 20%. The last series of closings in Baltimore, seven fire stations in 2000, did not reduce the work force, but shifted the staff to reduce fire fighter overtime and create additional EMT transport units.

In May 2003, New York City Mayor Michael Bloomberg ordered the closing of six city fire stations. This was a compromise from his original plan to close nine fire stations and disband up to 40 fire companies. In 2004, Philadelphia proposed shutting down four engine companies and four ladder companies, while activating eight additional medic units. This strategy was intended to reduce costs as well as reallocate resources as a result of a significant decrease in fires and a continuing increase in emergency medical incidents.

Closing a fire station or eliminating a fire company often creates a significant and very high-profile political issue that mobilizes the citizens as well as the fire fighters. No community wants to lose their neighborhood fire company.

Learning the Budget Language

The budget is the method the fire department has to obtain resources. Fire officers working in local fire stations are often in the best position to identify the needs of the department. To participate in the budget process, the fire officer must understand the process, the sources of revenue, and the local budget submission and administrative procedures. This requires the fire officer to learn new terms and a different way of viewing fire department operations.

The Purchasing Process

Most agencies will have a standardized method of making purchases. The fire officer must understand the policies and procedures of the organization. Failure to do so can have significant consequences for the organization. Since most fire departments are either political entities or non-profit organizations, they are accountable to the public for the wise use of funds. Since most fire departments are required to be audited, purchasing violations are found sometimes during the auditing process.

Petty Cash

The policies and procedures for purchasing typically vary by the amount of cost of the item. For example, the fire department may allow for any items under $100 to be purchased directly in a non-competitive manner. Often this is done through a petty cash system. The petty cash system allows for a member of the fire department to be the custodian of an amount of cash that is provided by the organization. This individual is responsible for the cash that they are provided. As members of the fire department purchase small dollar value

items, the petty cash custodian will reimburse them for the expense in exchange for the sales receipt.

Once the petty cash custodian is low on cash, the accumulated receipts are turned over to the finance department in exchange for a like amount of cash. In small organizations, the receipts are turned in to the governing body, such as the fire board that approves the purchases and then writes a check to the petty cash custodian. The check is then cashed and the funds are placed back in the petty cash box. The petty cash custodian will always have the authorized amount in either cash or receipts. This account is regularly audited.

Some personnel tend to think of the petty cash account as an endless fund, which is not true. On each receipt, the appropriate fund/object/fund object codes are assigned to the purchases. Petty cash allows for small purchases to be made without having to obtain purchase orders and charge the items to be paid by check. In effect, there is enough cash available to meet the unspent portion of the budget and each time petty cash is spent it lowers the remaining account balances.

Purchase Orders

Many items are purchased that exceed the petty cash limit that has been set by the organization. When this occurs, a purchase order system is typically used. A purchase order is a method of ensuring that there are sufficient funds available in a budget account to cover a purchase. Like petty cash, most organizations have a limit on the maximum amount of a purchase order. For example, an organization may allow purchase orders to be used for purchases up to $2,000.

When a decision to purchase is made, the fire officer's role is to acquire the item at the most reasonable cost to the organization. To do this, most organizations allow phone bids for these items. For example, if a fire officer needs to purchase a pair of tires, he first determines what size is needed and then calls at least three places to get prices on the tires. He then takes the lowest price and completes a purchase order, which is commonly called a "PO." The purchase order typically requires an authorizing signature by an officer who has control over the budget area from which the funds are paid. In this example, it might be the Chief. The purchase order is then entered into the purchasing system which debits the maintenance account.

The fire officer then takes the purchase order to the location that has the tires and the sale is transacted. A copy of the purchase order is given to the vendor who attaches the sales receipt to it and sends it to the fire department's finance department for reimbursement. In effect, the vendor allows the fire department to charge the merchandise until the purchase order is paid. Additionally, with the purchase order the fire department indicates to the vendor that there are funds available for the purchase and the purchase has been approved. Some vendors do not accept purchase orders.

Requisitions

For purchases that exceed a predetermined amount, such as $2,000, a requisition is required rather than a purchase order. A requisition has even more stringent requirements than a purchase order. When a fire department decides to make a large purchase, such as a new hydraulic extrication tool, a requisition is used.

The requisition differs from a purchase order in that the exact amount of the purchase is not known. Instead the department requests that a specific amount of funds be encumbered, or set aside from the budget account, which will more than cover the cost of the purchase. In the case of the rescue tool, the department might set aside $15,000. A bidding process will be followed and once complete, the requisition will allow for payment of the item.

The Bidding Process

Once funds have been encumbered, the fire department must go about the process of making the purchase. This is done in one of two manners. For smaller items, the fire department may develop specifications for bids. The second method, which is often used for very large or complex purchases, is a request for proposal.

To purchase based upon specifications, or commonly known as "specs," the fire department writes up exactly what they desire in the product. For example, the power unit must not exceed 18″ × 14″ × 21″. With a spec, every requirement must be met for the vendor to be considered. In the example, if a vendor wanted to bid but the power unit was 19″ × 13″ × 20″ their bid would be rejected. This may be important if the compartment that the power unit will be placed in will only accommodate the specified size; however, some fire departments use this method in an attempt so only one vendor or product can meet the specifications. This is an illegal and dangerous practice and the fire department may find itself in a lawsuit with an unsuccessful vendor. The public may be upset as well for wasting taxpayer's money. Specs should be used to specify what must be met for the purchase to be effective for the fire department.

Once developed, a bid sheet listing the specs is sent to vendors for pricing. Most organizations have a bidding list of vendors that wish to do business with them. The vendors will place their name and address on the list so each time a bid comes up, a copy is sent to them for consideration. A notice of bid may also be required to be placed in the local paper. Other organizations place notice of bids in trade magazines and papers. The Internet is becoming a method of getting information out to the public about upcoming bid processes.

Often, the use of specifications eliminates virtually every potential vendor or may greatly increase the price in order to meet the specifications. Sometimes, a bid specification limits the bidders. For example, the purchase of a mobile data computer (MDC) system for use in the fire apparatus would likely use the request for proposal method rather than the bid specification method because of the complexity of the technology.

In a request for proposal, often called an RFP, the fire department gives the general information about what is desired and allows the vendor to determine how they will meet the need. For example, the MDC RFP might require that the system provide coverage to 99% of the city. One vendor might opt to place 10 antenna towers to obtain the coverage while another might opt for only 7 antenna towers but place them at a higher elevation. Both ideas meet the need of the fire department. The fire department also determines how the proposals will be evaluated. For example, the price might be worth 50% of the overall consideration, while delivery time might be worth 5%. Once the RFP is developed, it is sent out to potential vendors for bidding.

Whether the process uses an RFP or bid specifications, vendors are given a specific amount of time to reply. Their replies are placed in a sealed envelope indicating the bid number on the outside of the envelope. At a time that was outlined in the bid specifications or RFP, all bids or proposals are opened in public view. The bids are then awarded to the lowest bidder and the amount is attached to the requisition for payment when delivery is made.

When an RFP is used, each proposal is evaluated to determine how they each meet the proposal. They are typically assessed based on performance as well as price. Once each performance requirement has been addressed and the price is considered, the RFP is awarded to the highest score. The price is then added to the requisition for payment when the item is delivered.

In recent years, there has been a trend to develop intergovernmental or interagency cooperation to streamline purchases and leverage greater discounts. The federal government issues many purchasing contracts for a wide range of equipment. Most vendors honor these rates to other governmental agencies. This streamlines the process because the bidding process has already occurred. In other cases, local organizations bind together to purchase common items. For example, a group of fire departments may jointly develop an RFP to purchase SCBA. This ensures a lower bid because they are purchasing in greater quantity and it allows all agencies to use like equipment.

Navigating the Budgetary Process

Using the example presented in the opening case study, let's navigate through the budgetary process. In the opening case study, the new fire officer suggested the creation of a two-person Alarm Unit 4 that would be the primary responder to most of the activated smoke detector and manual fire alarm systems in the district. Now the fire officer has to find the funds for this unit.

Developing a Budget Proposal

Because this is a new unit providing a new service, the first step in the budget proposal process is to describe what the new unit will be doing and the impact if the unit is not funded.

Overview of Faulty Alarm Enforcement Unit (Alarm Unit 4)

The following is the narrative that is part of the budget submission. It describes what the Faulty Alarm Enforcement Unit (FAEU) will do, the impact of the FAEU on current operations, and the consequences if the new program is not approved.

The FAEU is a two-person response unit that responds, investigates, and mitigates monitored central-station fire and smoke alarm activations in the high-rise district. In FY06, the fire department responded to 9800 activated central station alarms in the high-rise district. In an average 24-hour period, Fire Station 4 responds to 27 reports of activated fire or smoke alarms. Only one in 190 such alarms reports an actual fire or hazardous condition; the remaining 189 are false reports or result from faulty detection equipment.

The FAEU will be the primary fire department unit that will respond to fire and smoke alarm activations that are not accompanied by a report of an actual fire, smoke, or a hazardous condition from a building occupant. The FAEU will handle about 7000 of the 9800 monitored fire alarm activations in the high-rise district in FY08.

In addition to responding to activated fire alarm events, the fire officer assigned to the FAEU will be authorized to enforce the "Faulty Fire Alarm Ordinance" passed by the city in 2004. The fire department has been unable to enforce the ordinance because the staff and reporting system outlined in the ordinance have not been funded. The department submitted funding requests for the faulty alarm ordinance enforcement team in FY05, FY06, and FY07. The goal of the Faulty Fire Alarm Ordinance is to reduce the frequency of false and faulty alarm activations through a focused inspection and code enforcement.

This program does not meet the requirements of the federal Assistance to Firefighters Grant Program. It is not eligible for grant funding from any other source.

Failure to fund the FAEU will have three impacts. The first is the continuing accelerated wear and tear on the fire apparatus assigned to the high-rise district. The pumper assigned to Fire Station 4 required replacement after 8 years instead of the normal 12-year cycle. The accelerated pumper replacement represents an additional capital cost of $13,625 per year. The aerial ladder assigned to Fire Station 4 required replacement after 10 years instead of the normal 17-year cycle. The accelerated aerial replace-

ment represents an additional capital cost of $35,000 per year. Not funding the FAEU means that Engine and Tower 4 will continue to be the primary responders to activated faulty and false fire alarms and will require the department to continue to include an additional $48,625 per budget year to continue to meet the accelerated replacement schedule due to the excessive response workload.

The second impact will be a continuing reduction in unit availability to respond to more urgent situations. As the rate of responses has increased, the availability of Engine and Tower 4 has been decreasing. The FY07 budget request included establishing Quint 4 as a new, full-time fire company to handle the increased workload in the high-rise district and to preserve the department's response time goal. The FAEU proposal is in response to the city council's direction to find an alternative way to address the increased workload. Quint 4 would increase the department's budget by 19 full-time equivalent positions, representing an annual personnel cost of $877,000. The anticipated reduction in responses would eliminate the need for this company.

The third impact will be a reduction in the rate of unnecessary alarms. When the city passed the Faulty Fire Alarm Ordinance, Fire Station 4 was responding to 19 false/faulty fire alarm activations a day. The workload has increased by 40% in the past 4 years, to 27 false/faulty alarm activations a day. Fire departments that have enforced a faulty fire alarm ordinance, similar to the one passed by this city, have documented a 25%–50% reduction after 3 years.

FAEU Annual Personnel and Operating Expenditures

This part of the budget submissions shows the annual expenses to operate the FAEU. There are personnel and operating expenses.

Personnel Expenditures

The FAEU will have a captain, three lieutenants, and four technicians assigned to the unit. In this example, the personnel cost is described in two areas. The first area is the direct salary that is paid for each position. Most civil service pay classification systems use a pay level that includes steps tied to seniority. Budget protocol requires that new positions reflect the average seniority step that the incumbents occupy. That means that the captain is at seniority step 7, the lieutenant is at seniority step 5, and the technicians are at seniority step 3.

The second area is the calculation of the fringe benefits for each employee position. It is calculated as a percentage of the employee's salary. For the FAEU example, fire fighter fringe benefits account for 28.7% of the salary. That means that every $100 of salary costs the city $128.70.

Operating Expenses

The cost of operating the FAEU unit entails three general areas: vehicle cost, office operations, and personnel training. City fleet maintenance has determined the cost to operate and replace a small, all-wheel-drive emergency services SUV to be $0.64 a mile. This covers the fuel and scheduled maintenance of the vehicle. This also includes a per-mile charge that is designed to cover the replacement of the vehicle when it reaches the end of its scheduled service life. Vehicle service life is determined three ways: years of service, miles accumulated, or specific vehicle cost. It is estimated that the unit will accumulate 23,000 miles a year. The city replaces small emergency service vehicles at 80,000 miles or after 10 years. The FAEU will need replacement after 3.5 years.

Office operations include rent, telecommunication charges, and equipment maintenance. In this example, the FAEU is operating out of a fire station, so rent is zero. The office has two hardwire lines to provide service for the phones, Internet access, and fax machine. The city Information Technology office handles this service and has an annual fee per hardwire line that covers the telephones and Internet broadband access. Information Technology also handles computer maintenance and repairs. Like the with the vehicle, the annual maintenance and operating charge covers the anticipated annual costs of maintenance, software upgrades, and repairs. It also includes a cost that will allow replacement of the computers on a 5-year cycle. The FAEU has two desktop computer work stations and two handheld code enforcement computers.

Although the computers have reduced the amount of paper used, each position generates an annual cost in paper, office supplies, and inspection forms. The Fire Prevention Bureau has determined that each inspector position generates $650 a year in office supply expenses.

Operating expenses also include the expenses of mandated annual fire fighter and inspector training. The cost reflects the per-person cost of training, both for the tuition and the position replacement overtime to cover the FAEU person who is receiving the training. This is training required by federal or state regulations, such as blood-borne pathogen and hazardous materials refresher training.

FAEU Capital Budget

The capital budget request would pay for the durable equipment needed to start the FAEU project. The assumption is that there is no equipment available to borrow. Many of the required items would be purchased from existing federal, state, or regional contracts. The budget summary shows the entire first-year capital expenditures to start the program, including the response vehicle, the administrative office, the code enforcement equipment, and the impact of eight new fire department positions.

FAEU Vehicle

During the planning process with the command staff, you need a decision on what type of vehicle would be appropriate for the FAEU. The state, city, or region establishes annual contracts for the purchases of common vehicles. For example, the state offers a range of vehicles for purchase. Here are the ones considered for the FAEU:

- Compact four-door sedan, $11,733
- Midsized four-door sedan, $14,350
- Police pursuit sedan, full-sized, $20,983
- Small all-wheel-drive SUV, emergency services, $22,976
- Large four-wheel-drive SUV, emergency services, $29,290

The operations chief wants the vehicle to be capable of emergency response and to be identical to the small emergency services SUV used for command and staff. In addition to the vehicle, equipment is needed to convert a small SUV into a fire department response vehicle, including emergency lights/siren, mobile fire radio and computer dispatch terminal, two portable fire radios, two self-contained breathing apparatus, and firefighting equipment. Depending on the local budget process, these items may need to be identified in a line-item format consistent with budget documentation protocol (▶ **Figure 13-3**). More than half of the FAEU vehicle cost is for the specialized equipment that goes on or into the vehicle.

FAEU Administrative Office

Because the FAEU will be providing code enforcement and extensive documentation, the budget submission includes the cost of creating a new administrative office with the tools appropriate for a two-person inspection team. The administrative office can go into an available space at Fire Station 4. The capital equipment represents all of the durable items needed to establish the office: desks, chairs, computers, fax machine, and printer. In addition, the capital equipment includes the equipment needed to perform the code enforcement duties, such as two handheld computers, a digital camera, and a code enforcement reference library.

FAEU Uniforms and Protective Clothing

The expense of fire fighter protective clothing and the initial uniform issue are included in the initial year capital expenditures. Although the individuals who will staff the FAEU are already in the city, the creation of eight additional employee positions does necessitate the purchase of protective clothing and uniforms. Because the FAEU program creates the new positions, the expense shows up here.

Ask for Everything You Need

Fire departments try to do the most with the least resources. When submitting a proposal for a new project or service, assume that there are no existing resources and ask for everything. This includes asking for enough tools to ensure continuing operation. For example, the code enforcement

FY 08 Budget
Fire Department - Firefighting Division and Prevention Division
New Program: **Fire Alarm Enforcement Unit**

Initial Year: Capital Expenditures	Unit	Total	Source or comments
Alarm Unit Vehicle:			
1 Small AWD SUV, emergency services	$22,976	$22,976	State vehicle contract
2 Self-contained breathing apparatus	$432	$864	Apparatus section quote
1 Mobile 800 mHz radio and computer terminal	$15,526	$15,526	Communications division quote
1 Fire emergency lighting system	$2,765	$2,765	Apparatus section quote
1 Infared camera	$1,211	$1,211	Firefighting division quote
2 Fire hand tools and hand lights	$272	$544	Firefighting division quote
2 Portable 800 mHz radios	$3,135	$6,270	Communications division quote
1 Safety striping and FD lettering	$411	$411	Apparatus section quote
Total Alarm Unit vehicle cost:		**$50,567**	

Alarm Unit administrative office	Unit	Total	Source or comments
2 Task desk	$520	$1,040	Facilities
2 Task chair	$227	$454	Facilities
2 File cabinet	$200	$400	Facilities
1 Large book case	$175	$175	Facilities
2 Digital documentation camera systems	$836	$1,672	Prevention Division quote
1 Code enforcement reference library	$2,636	$2,636	Prevention Division quote
2 Computer workstation - Model B	$3,200	$6,400	Information Technology
2 Handheld computer - code enforcement package	$3,176	$6,352	Information Technology
1 Fax machine	$344	$344	Information Technology
1 Workstation laser printer - Model B	$1,325	$1,325	Information Technology
Total Alarm Unit administrative office cost:		**$20,798**	

New Uniformed Firefighter Positions - maintaining a two person team on duty every hour of every day	Unit	Total	Source or comments
8 Firefighter protective clothing ensembles	$1,673	$13,384	Firefighting division
4 Fire officer initial uniform issue	$865	$3,460	Firefighting division
4 Firefighter initial uniform issue	$711	$2,844	Firefighting division
New Uniformed Firefighter positions:		**$19,688**	

Initial year Capital Expenditures: $91,053

Annual Operating Expenses	Unit	Total	Source or comments
Personnel Expenditures:			
1 Captain at Pay Step 28-7	$54,433	$54,433	Salary
3 Lieutenant at Pay Step 26-5	$44,654	$133,962	Salary
4 Technicians at Pay Step 22-3	$37,365	$149,460	Salary
Fringe benefits		$96,964	28.7% of firefighter salary
Total annual personnel expenditures:		**$434,819**	

Operating Expenses:	Unit	Total	Source or comments
1 Fire Alarm Unit maintenance and operating	$14,720	$14,720	$.64 per mile, estimate 23,000 miles/year
1 Office rental	$0	$0	Existing Fire Station 4 space
2 Telecomunications charges per hard wire line	$1,884	$3,768	Internet and phone line charges
4 Computer maintenance and operating	$322	$1,288	Information Technology
2 Paper, office supplies and inspection forms	$650	$1,300	Prevention Division, per position cost
8 Mandated fire inspector continuing training	$375	$3,000	Prevention Division, per position cost
4 Mandated fire officer continuing training	$221	$884	Firefighting Division, per position cost
8 Mandated firefighter continuing training	$176	$1,408	Firefighting Division, per position cost
Total annual operating expenses:		**$26,368**	

Annual Personnel and Operating Expenditures: $461,187

Annual Direct Cost Recovery:
Fire Alarm System inspection fee   **$160,600**
System inspections required by ordinance after third false or faulty alarm activation
Estimate four inspection hours a day, two inspector team. $55 per inspector per hour

Annual Indirect Cost Reduction

Elimination of accelerated pumper replacement cost	**$13,625**	Engine 4 - see narrative
Elimination of accelerated aerial replacement cost	**$35,000**	Tower Ladder 4 - see narrative

Net annual cost of program: $251,962

Figure 13-3) Summary budget submission sheet for FAEU.

handheld computer costs $2,600. This is how the city documents code inspections and enforcement—by replacing the old three-part paper inspection forms. It is a critical resource for enforcement activity. Although the FAEU could get by with one, the budget submission is for two computers for the two-person team. This provides the redundancy needed to ensure continued operations.

When submitting for a new project or service, consider the impact of critical resources. Where would the FAEU go if its single handheld code enforcement computer breaks? You do not get what you do not ask for.

Cost Recovery and Reduction

As configured, the annual cost to operate the FAEU is $461,187 a year. The expense impact is reduced through direct and indirect cost recovery and reduction. The FAEU staff are code enforcement officials when handling the 27 false/faulty alarm activations a day in the high-rise district. The local ordinance requires inspection of the fire alarm system by fire prevention personnel after the third false or faulty alarm activation. The ordinance charges $55 per inspector per hour when conducting inspections. It is estimated that the FAEU will perform 4 hours of two-person code enforcement inspections a day, generating $160,000 of inspection fees a year.

In addition, include the indirect cost reduction of replacing Engine and Tower Ladder 4. Eliminating the accelerated apparatus replacement cost represents an annual indirect cost saving of $48,625. The combination of direct and indirect cost reduction means that the net cost of providing the FAEU is $251,962, a 45% reduction of the expense.

You Are the Fire Officer: Conclusion

With the help of the fire operations chief, you write a budget proposal and submit it. Once the budget proposal is completed and submitted, you must anticipate receiving questions from fire department senior staff and the budget office. Every budget cycle receives more proposals that the revenue can support. The time between submitting the proposal and the publication of the advertised budget involves a give and take between the agency head, the budget office, and the local elected officials. It is rare for a new budget proposal to get to the advertised budget without some tweaking or political maneuvering. The FAEU may be approved if the chief withdraws the proposal for an early replacement of Tower Ladder 4. The FAEU may be approved if the staffing is reduced from eight positions working around the clock to two people working a 40-hour weekday schedule.

Wrap-Up

Ready for Review

- A four-step method can be used to apply for a grant:
    - Conduct a community and fire department needs assessment
    - Compare weakness to the priorities of the grant program
    - Decide what to apply for
    - Complete the application
- Annual budget describes how the local government will spend revenue in operating, personnel, and capital expenditures.
- Local governments operate in a fiscal year that generally runs from July 1 to June 30.
- A proposed budget is submitted by the agency heads to the budget director. After negotiations, the proposed budget is released for public review, comment, and hearings.
- After the public comment period, elected officials revise the budget and issue the "Approved" or "Adopted" budget that is the financial document that will guide the jurisdiction.
- During the budget year, informal budget reviews occur during the third and ninth months. Minor revisions may occur.
- A more formal midyear budget review occurs at the 6-month period. The elected officials may reassign resources and revise objectives on the basis of tax revenues and local government issues.
- Lower revenue options:
    - Defer scheduled expenses
    - Regionalize or consolidate services
    - Reduce career work force
    - Privatize or contract out elements of the service provided by the department
    - Reduce the size of the department

Hot Terms

Base budget The level of funding that would be required to maintain all services at the currently authorized levels, including adjustments for inflation, salary increases, and other predictable cost changes.

Bond A certificate of debt issued by a government or corporation; a bond guarantees payment of the original investment plus interest by a specified future date.

Budget An itemized summary of estimated or intended expenditures for a given period, along with financing proposals for financing them.

Expenditures The act of spending money for goods or services.

Fire tax district Special service district created to finance the fire protection of a designated district.

Fiscal year Twelve-month period for which an organization plans to use its funds. Local governments' fiscal years are from July 1 to June 30.

Governmental Accounting Standards Board (GASB) The mission of the Governmental Accounting Standards Board is to establish and improve the standards of state and local governmental accounting and financial reporting that will result in useful information for users of financial reports and to guide and educate the public, including issuers, auditors, and users of those financial reports.

Line-item budget Budget format where expenditures are identified in a categorized line-by-line format.

Revenues The income of a government from all sources appropriated for the payment of the public expenses.

Supplemental budget Proposed increases in spending to provide additional services.

It is September and as fire officer in your agency, one of your job duties is to assist in developing the department's annual budget. In past years the department was able to collect 106% of the previous year's property taxes. However, a new law is limiting this year's collection to 101% of the previous year's property taxes. This means that the budget will not be able to compensate for inflation and cuts will need to be made.

1. What is one option for reducing a fire department budget?

   A. Disband

   B. Regionalize or consolidate

   C. Close facilities

   D. Increase department fundraising activities

2. What is one economic impact that can affect a department's budget?

   A. Property values fall

   B. War

   C. Change in work force

   D. Stocks

3. When reviewing a department's expenditures, under what section would you most likely find how much the department spent on training?

   A. Personnel expenditures

   B. Capital expenditures

   C. Operating expenditures

   D. Facilities expenditures

4. When reviewing personnel expenditures, personnel costs include more than just the salary that an employee receives. Employee benefits make up a substantial amount that significantly impacts personnel costs. What is one employee benefit that is included under personnel expenditures?

   A. Personal Protective Equipment

   B. Training expenses

   C. Pension fund

   D. Union dues

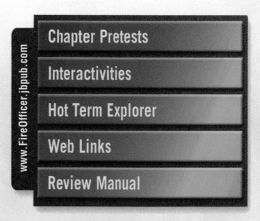

Chapter Pretests

Interactivities

Hot Term Explorer

Web Links

Review Manual

www.FireOfficer.jbpub.com

Fire Officer Communications

www.FireOfficer.jbpub.com

Technology Resources

Chapter Pretests

Interactivities

Hot Term Explorer

Web Links

Review Manual

Chapter Features

Voices of Experience

Getting It Done

Assessment Center Tips

Fire Marks

Company Tips

Safety Zone

Hot Terms

Wrap-Up

NFPA 1021 Standard

Fire Officer I

4.1.2 **General Prerequisite Skills.** The ability to effectively communicate in writing utilizing technology provided by the AHJ; write reports, letters, and memos utilizing word processing and spreadsheet programs; operate in an information management system; and effectively operate at all levels in the incident management system utilized by the AHJ.

4.2.1 Assign tasks or responsibilities to unit members, given an assignment at an emergency operation, so that the instructions are complete, clear, and concise; safety considerations are addressed; and the desired outcomes are conveyed.

(A) Requisite Knowledge. Verbal communications during emergency situations, techniques used to make assignments under stressful situations, and methods of confirming understanding.

(B) Requisite Skills. The ability to condense instructions for frequently assigned unit tasks based on training and standard operating procedures.

4.2.2 Assign tasks or responsibilities to unit members, given an assignment under nonemergency conditions at a station or other work location, so that the instructions are complete, clear, and concise; safety considerations are addressed; and the desired outcomes are conveyed.

(A) Requisite Knowledge. Verbal communications under nonemergency situations, techniques used to make assignments under routine situations, and methods of confirming understanding.

(B) Requisite Skills. The ability to issue instructions for frequently assigned unit tasks based on department policy.

Fire Officer II

5.4.4 Prepare a news release, given an event or topic, so that the information is accurate and formatted correctly.

(A) Requisite Knowledge. Policies and procedures and the format used for news releases.

(B) Requisite Skills. The ability to communicate orally and in writing.

5.4.5 Prepare a concise report for transmittal to a supervisor, given fire department record or records and a specific request for details, such as trends, variances, or other related topics.

(A) Requisite Knowledge. The data processing system.

(B) Requisite Skills. The ability to communicate in writing and to interpret data.

Knowledge Objectives

After studying this chapter you will be able to:

- Discuss the communication cycle.
- Identify ways to improve listening skills.
- Understand the fire officer's responsibilities as an active listener to both subordinates and superiors.
- Describe the ways to counteract environmental noise.
- Identify the conditions that interfere with verbal communication.
- Explain the difference between formal and informal communications.
- Outline fire officer responsibilities for routine and infrequent reports.
- Explain what should be included in a news release.

Skills Objectives

After studying this chapter you will be able to:

- Write a recommendation report using information about an issue or problem.
- Write a news release.

You Are the Fire Officer

At 03:15 hours, your engine company is dispatched along with a truck company to investigate a report of a smoke alarm sounding in a private dwelling. You have been on many similar calls in the past, often caused by the chirping sound indicating a low battery in the smoke alarm. While en route, you tell a member of your crew to bring in an extra battery when you arrive.

As you turn the corner, you are surprised to see smoke and an orange glow from the rear of the house next door to the address where you were dispatched. You quickly radio to dispatch announcing your arrival on the scene "with fire showing" and tell them to "fill out the box." You then order your crew to pull a preconnected hose line to the front door of the burning house to begin an offensive attack. Two members of the truck company move in to perform a primary search, while the other two are designated as the rapid intervention crew.

Three minutes later, you are standing next to the front door, waiting for a report from the attack team and hoping they are going to tell you that the fire is knocked down. Your engineer advises that you are down to a half tank of water and you think to yourself that the second due engine should be arriving any minute now. You radio to dispatch to advise the next arriving engine to lay in a supply line to your engine.

A sudden sinking feeling envelopes you when the dispatcher replies that you are the only engine company assigned to the incident and politely asks if you need assistance. You respond that you have a working fire in a two-story, wood-frame structure; an initial attack is underway, and the primary search is in progress. You also advise dispatch to send the balance of the first alarm assignment plus an EMS unit. The dispatcher repeats your report and request for additional units to back you up.

1. Where was the breakdown in the initial arrival report?
2. Why is it important to complete the entire communications process when you are in an emergency situation?

Introduction to Fire Officer Communication

Many fire officers wear rank insignia that feature bugles, representing the fire officer's speaking trumpet that was used for many years to shout orders on the fireground. This symbol emphasizes the fire officer's important responsibility to communicate. Although the technology has advanced considerably, communication skills are still very important for anyone who serves as a fire officer. An officer must be able to communicate effectively in many different situations and contexts.

A fire officer must be able to process several different types of information to effectively supervise and support the fire company members. One characteristic of public safety services in the 21st century is the rapid expansion of information, particularly regarding newly identified hazards. A fire officer should regularly read at least one fire service trade magazine and visit websites that provide current information in order to stay abreast of new developments. The officer

should make a point of sharing significant information with the company members to keep them well informed.

Effective communication skills are essential for a fire officer. These skills are required to provide direction to the crew members, review new policies and procedures, and simply exchange information in a wide range of situations. A special set of skills is required to effectively transmit radio reports. Communication skills are equally important when working with citizens, conducting tours, releasing public information, and preparing reports.

The Communication Process

Communication is actually a circular process that occurs in repetitive cycles. Successful communication occurs when two people exchange information and develop a level of mutual understanding. Understanding is very important to the process. When information flows from one person to another, the process is effective only when the person receiving the information is able to understand what the other person intended to transmit.

Effective communication does not occur unless the intended message has been received and understood. The message must make sense in the recipient's own terms, and it must convey the thought that the sender intended to communicate. The sender cannot be sure that this has occurred unless the recipient sends some confirmation that the message arrived and was correctly interpreted.

The Communication Cycle

The communication cycle contains five parts:
1. Message
2. Sender
3. Medium (with noise)
4. Receiver
5. Feedback

The Message

The message represents the text of the communication—the part that contains the information. In its purest form, the message contains only the information to be conveyed. Messages do not have to be in the form of written or spoken words. For example, a stern facial expression with purposeful eye contact can convey a very clear message of disapproval, whereas an innocent smile can convey approval and contentment. The messages are clear, although no words accompany them.

The Sender

The sender is the person or entity who is sending the message. We think of the sender as a person, but it could be a sign, a sound, a photograph, or almost anything that is meant to send a message. As a sender, you must ensure that the message is properly targeted to the right person and formulated so that the receiver will understand the meaning. The sender is responsible for making sure that the message is understandable to the receiver, that the receiver properly receives the message, and that the receiver understands the message properly.

A fire officer must be particularly skilled as a sender of verbal information in order to transmit information and instructions to subordinates and coworkers. The tone of voice or the look that accompanies a spoken message can profoundly influence the receiver's interpretation of the meaning. It is difficult, if not impossible, to not send messages to others through body language, mannerisms, and other means. Any behavior is up for interpretation, so we must be careful to send only the intended message and attempt to minimize misinterpretation.

Senders often convey messages that are not intended, not directed to anyone in particular, or not even meant to be messages. A person can outwardly express disappointment at himself or herself; however, others may interpret the look as a message of disappointment or disapproval aimed at them.

The Medium

The medium refers to the method that is used to convey the information from the sender to the receiver. There are many different choices of medium. The medium can be words that are spoken by the sender and heard directly by the receiver. Spoken words or sounds can also be transmitted as electromagnetic waves through a radio system. Written words, pictures, symbols, and gestures are all examples of messages transmitted via a visual medium. The sender normally chooses the medium by which the message is sent. The sender should choose the best medium for the situation, depending on the circumstances, the nature of the message, and the available methods of sending it.

The medium often has a significant influence over how a message is interpreted. When information has to be transmitted to subordinates, a fire officer might have the options of posting a notice on a bulletin board, announcing it at a formal lineup, or mentioning it during a meal at the kitchen table. The medium that is chosen influences the importance that is attached to the message by the receivers. If it is really important, the fire officer might announce the key points at lineup, then direct all members to read a written document and sign a sheet to acknowledge that they have read and understood it. How the message is conveyed can be as important as the message itself.

The Receiver

The receiver is the person who receives and interprets the message. There are many opportunities for error in the reception of a message. Although it is up to the sender to formulate and transmit the message in a form that should be clearly understandable to the receiver, it is the receiver's responsibility to capture and interpret the information. The same words can convey different meanings to different individuals, so the receiver does not automatically interpret a message as it was intended. In the fire service, the accuracy of the information that is received can be vital, so both the sender and the receiver have responsibilities to ensure that messages are properly expressed and interpreted. Many messages are directed to more than one person, so there can be multiple interpretations of the same message.

Feedback

The sender should never assume that information has been successfully transferred unless there is some confirmation that the message was received and understood. When it is important to ensure that a message was received and properly interpreted, the receiver should seek a confirmation from the receiver. Feedback completes the communication cycle, by confirming receipt and verifying the receiver's interpretation of the message. The importance that is attached to feedback depends on the nature and the importance of the message.

When relaying critical information during a stressful event, the sender should have the receiver repeat back the key points of the message in his or her own words. Repeating back the key points is a much more reliable form of confirmation than a simple nod of the head or a single word or phrase, such as "understood" or "10-4." Without feedback, the sender cannot be confident that the message reached the receiver and was properly interpreted.

The communication cycle lies at the core of a fire officer's daily duties. Remember, effective communication should contain all five components of the communications process; if one or more is missing, communication does not occur.

The Basics of Effective Communication Skills

In order to be effective as a fire officer, an individual must develop good communication skills. Almost every task that an officer is expected to perform depends on the ability to communicate effectively. An officer must be effective as both a sender and a receiver of information and, in many cases, as a processor of information that has to flow within the organization. Both written and spoken communication skills are essential for most fire officer positions.

Active Listening

Success as a supervisor depends on how freely your subordinates talk to you, keep you informed, and tell you what is bothering them. Success also depends on your effectiveness in communicating with your superiors and keeping them informed. An individual who lacks good communication skills will have a difficult time in a supervisory role.

One of the first priorities for any officer should be to develop good listening skills. The ability to listen is important before one becomes an officer; however, this basic skill becomes even more important as one advances within the organization. A fire fighter, who is at a relatively low level within the fire department, must listen effectively to information that is coming from a higher level. In order to function effectively as a supervisor, a fire officer must also be able to listen to company members or other subordinates and accurately interpret their comments, concerns, and questions.

Like physical fitness training, listening is a skill that must be continually practiced to maintain proficiency. A typical listening situation for a fire officer could be a meeting with a fire fighter who is expressing a problem or a concern. Listening, in a face-to-face situation, is an active process that requires good eye contact, alert body posture, and frequent use of verbal engagement (▶ **Figure 14-1**). The purpose of active listening is to help the fire officer understand the fire fighter's viewpoint in order to solve an issue or problem.

The techniques below may help to improve your listening skills:

- **Do not assume anything.** Do not anticipate what someone will say.

Figure 14-1 Listening is an active process that requires good eye contact, alert body posture, and frequent use of verbal engagement.

- **Do not interrupt.** Let the individual who is trying to express a point or position have a full say.
- **Try to understand the need.** Often, the initial complaint or problem is a symptom of the real underlying issue. Look for the real reason the person wants your attention.
- **Do not react too quickly.** Try not to jump to conclusions. Avoid becoming upset if the situation is poorly explained or if an inappropriate word is used. The goal is to understand the other person's viewpoint.

Stay Focused

When discussing an issue with a subordinate, it is important to stay focused. In the fire station setting, it is easy to get sidetracked and bring additional unrelated issues into a conversation. Directed questioning is a good method to keep a conversation on the topic at hand. If the speaker starts to ramble, ask a specific question that moves the conversation back to the appropriate subject. For example, if you are attempting to find out why a fire fighter failed to wear a dress uniform shirt to roll call and he starts talking about how he does not like the new collar brass on the dress uniform, you could ask, "How does the collar brass affect whether or not you wear your dress uniform shirt to roll call?"

Ensure Accuracy

Another important attribute for a fire officer is to always ensure accuracy on issues that are being communicated or discussed. A fire officer needs to have up-to-date information on the fire department's standard operating guides, policies, and practices. An officer should also be familiar with the jurisdiction's personnel regulations, the approved fire department budget, and, if applicable, the current union contract. If a fire fighter is misinterpreting a factual point or a departmental policy, the fire officer is obligated to clarify or correct the information. Ignoring an inaccurate statement may only foster erroneous information on the grapevine.

If the fire officer is unsure of the information, it is the fire officer's responsibility to determine the correct information and provide it to the fire fighter. If necessary, it is the fire officer's ability to obtain the accurate information from a chief officer or headquarters staff.

A fire officer sometimes has to exercise control over what is discussed in the work environment. Fire station discussions can easily encroach on subjects that are intensely personal, such as politics, religion, or social values, and can quickly escalate into hurt feelings. To proactively address potential problems, the fire officer needs to establish some ground rules about issues that are allowed to be discussed and to what extent they can be discussed.

Keeping Your Supervisor Informed

Your supervisor also depends on you to share information about what is happening at the fire station or in your work environment. Three areas where a fire officer needs to keep the chief officer informed are:

- **Progress toward performance goals and project objectives**. A fire officer needs to keep his or her chief appraised of work performance progress, such as training, inspections, smoke detector surveys, and target hazard documentation. It is especially important to let the chief know about anticipated problems early enough to get help and to keep the projects running on time.
- **Matters that may cause controversy**. The chief should be informed about conflicts with other fire officers or between shifts, or any potential conflict that extends outside the organization. The chief also needs to know about any disciplinary issues or a controversial application of a departmental policy.
- **Attitudes and morale**. A fire officer spends the workday with a group of fire fighters at a single fire station, whereas the command officer spends much of the workday in meetings or riding alone in a command vehicle. The fire officer should communicate regularly with the chief about the general level of morale and fire fighter response to specific issues.

The Grapevine

In addition to the official communications process, every organization has an informal communication system, often known as the "grapevine." The flow of informal and unofficial communications is inevitable in any organization that involves people; however, the grapevine flourishes in the vacuum that is created when the official organization does not provide the work force with timely and accurate information about work-related issues. Unfortunately, much of the grapevine information is based on incomplete data, partial truths, and sometimes outright lies.

A fire officer can often get clues about what is going on but should never assume that grapevine information is accurate and should never use the grapevine to leak information or stir controversy. In addition, a fire officer needs to deal with grapevine rumors that are creating stress among the fire fighters, by determining accurate information and sharing it with subordinates. The fire fighters appreciate a fire officer's efforts to provide accurate and truthful information.

Unofficial websites, Internet discussion forums, and instant wireless messaging have become the grapevine tools of the 21st century. The fire officer should carefully consider whether to allow fire fighters to participate in online Internet discussion forums while on duty and when using fire department computers. Some of these systems allow users to hide behind screen names, make personal attacks, spread rumors, and make inaccurate observations about department operations. This type of activity is unhealthy for the organization and can create a hostile work environment. Local media have used information obtained from online discussion forums to write stories about fire departments, effectively making the most angry, disgruntled, and irrational members into departmental spokespersons.

Overcoming Environmental Noise

<u>Environmental noise</u> is a physical or sociological condition that interferes with the message. Within this definition, "noise" includes anything that can clog or interfere with the medium that is delivering the message.

Physical noise can take different forms, such as background conversations, outside noises, or distracting sounds that make it difficult to hear. One obvious example is siren noise that can make it difficult to conduct a conversation inside a vehicle or communicate over the radio. Poor reception on a radio frequency or cellular phone can make communications difficult or impossible. A rapid flow of incoming messages can overload the receiver's ability to deal with the information, even if the words come through clearly (► **Figure 14-2**). Just as sounds can interfere with hearing, darkness or bright flashing lights can make it difficult to see clearly and interpret a visual message.

Another form of noise interference occurs when the receiver is distracted or thinking about something else and essentially blocks out most or all of an incoming message. The human brain has difficulty processing more than one input source at a time. For example, you could be having a conversation on the telephone with one individual when someone else calls your name; your attention is diverted momentarily while you try to determine who called you and why. During that moment, you are likely to miss part of the telephone conversation. A similar situation occurs if the sender or receiver is unable to

Figure 14-2 Trying to talk on a cellular phone with poor reception is one example of physical noise interfering with communication.

concentrate on or respond to the message due to fatigue, boredom, or fear.

Sociological environmental noise is more a subtle and difficult problem. Prejudice and bias are examples of sociological environmental noise. If the receiver does not believe that the sender is credible, the message is ignored or results in an inadequate response. A fire officer should be aware of this type of environmental noise when receiving messages. If sociological environmental noise is affecting an officer's ability to communicate effectively when sending messages, it is up to the officer to change that situation. The following list provides seven suggestions to improve communication:

- **Do not struggle for power.** Focus attention on the message, not on whether the sender has the authority to deliver it to the receiver. The situation, not the people involved, should drive the communication and the desired action.
- **Avoid an offhand manner.** If you want your information to be taken seriously, you must deliver it that way. Be clear and firm about matters that are important.

- **Words have meaning.** Select words that clearly convey your thoughts, and be mindful of the impact of your tone of voice.
- **Do not assume that the receiver understands the message.** Encourage the receiver to ask questions and seek clarification if the intent of the message is unclear. A good technique of confirming understanding is to repeat the key points of the message back to the sender.
- **Immediately seek feedback.** If the receiver identifies an error or has a concern about the message, encourage that individual to make the statement sooner rather than later. It is always better to solve a problem or identify resistance while there still is time to make a change.
- **Provide an appropriate level of detail.** Think about the person who is receiving the message and how much information that individual needs. Consider a fire officer telling a fire fighter to set up the annual hose pressure test. The information needs are different for a fire fighter who has 22 years of experience as a pump operator versus a rookie who has 22 days on the job.
- **Watch out for conflicting orders.** Check to make sure that your message is consistent with information coming from other sources. A fire officer should develop a network of peers to consult with when dealing with unfamiliar or confusing situations.

These seven suggestions are all intended to improve communications dealing with administration and supervisory activities. The fireground or emergency scene requires different communications practices.

Emergency Communications

One of the most important communication skills for a fire officer is the ability to communicate effectively and manage communications during emergency incidents. Unlike most personal and business communications, emergency communications require a direct approach. There is no time to be elegant when communicating in the emergency environment. The direct approach entails asking precise questions, providing timely and accurate information, and giving clear and specific orders. We do not have the luxury of taking the time to choose words carefully to avoid hurting someone's feelings; at the same time, we must not be so insensitive that we cause negative reactions that interfere with effective fire fighting.

Under the time pressure of an emergency incident, management of the communications process can be as important as communicating effectively. The incident commander has to manage the exchange of information, so that the most important messages go through and lower-priority or unnecessary communications do not get in the way. This is most important when managing radio communications at an incident, where several individuals may be trying to simultaneously communicate information that they each perceive is important. The incident commander or the fire officer assigned to manage communications must direct the radio traffic.

The key points for emergency communications are:

- Be direct.
- Speak clearly.
- Use a normal tone of voice.
- If you are using a radio, hold the microphone about 2″ from your mouth.
- If you are using a repeater system, allow for this time delay after keying the microphone.
- Use plain English rather than 10 codes.
- Try to avoid being in the proximity of other noise sources, such as running engines (▶ **Figure 14-3**).

"Unit calling, repeat . . ."

Verbal skills are a core competency for the fire officer. Radio messages must be accurate, brief, and clear. An officer should be as consistent as possible when sending verbal messages over the radio. The performance goal should be to sound the same and communicate just as effectively when reporting a minor incident or when communicating under intense stress.

Recordings of radio messages transmitted during emergency incidents are an effective training tool. Listening to others allows a fire officer to identify and emulate techniques that are clear and precise, leaving no doubt about the intent of the message. Listening to recordings of your own radio communications provides valuable feedback, allowing you to compare the output with your thought process at that moment.

Initial On-Scene Size-Up

When arriving on scene of an incident, there are three components that should be communicated to the dispatch center and to other responding units. Each fire department should have standard operating procedures that define exactly what information should be reported and how it should be communicated.

The initial report should describe what you have, state what you are doing, and provide direction for other units that will be arriving. Routinely using the same terminology and format helps ensure that nothing is missed.

An example might sound like this:

Engine 1: Dispatch, this is Engine 1.

Dispatch: Go ahead Engine 1, this is dispatch.

Engine 1: Engine 1 has arrived at 2345 Central Avenue. We have a two-story, wood-frame, single-family dwelling with heavy smoke showing on side Charlie and fire showing from the first floor on side delta. Engine 1 has a supply line and will be making an offensive attack through side alpha. Engine 1 is establishing Central Avenue Command on side alpha.

Figure 14-3 Emergency communications require a direct approach.

Dispatch: Copy Engine 1 arrived at 2345 Central Avenue with a two-story, wood-frame, single-family dwelling—heavy smoke showing on side Charlie and fire showing on side delta first floor. Engine 1 is making an offensive attack through side alpha and establishing Central Avenue Command on side alpha.

Through this simple exchange of information, everyone who needs to know is informed of the situation and the action being taken by the first-arriving unit, and the stage is set for all other units to take action based on standard operating procedures. The dispatcher's repeat of the key information provides solid confirmation that it was received and understood.

Using the Order Model

The order model is a standard method of transmitting an order to a unit or company at the incident scene. It is designed to ensure that the message is clearly stated, heard by the proper receiver, and properly understood. It also confirms that the receiver is complying with the instruction.

Incident Commander: Command to Ladder 2.

Ladder 2: This is Ladder 2—go ahead Command.

Incident Commander: Ladder 2—come in on side Charlie and conduct a primary search on the second floor. Also advise if there is any fire extension to that level.

Ladder 2: Ladder 2 copies—going to side Charlie to do a primary search the second floor and check for fire extension.

Incident Commander: That's correct, Ladder 2.

Radio Reports

Most fire officers, particularly company officers, frequently provide reports relating to emergency incidents over the radio. Radio communications are essential for emergency operations because they provide an instantaneous connection and can link all of the individuals involved in the incident to share important information. Although different radio systems have technological strengths and weaknesses, the value of radio communications for managing emergency incidents is unquestionable.

During an emergency incident, both the sender and the receiver should strive to make their radio messages accurate, clear, and as brief as possible. Conditions are often stressful, with several parties competing for air time and the receiver's attention. There may be only a limited time to transmit an important message and ensure that it is received and understood.

The individual who is transmitting a radio report often feels an intense pressure to get the message out as quickly as possible. The realization that a large audience could be listening often adds to the sender's anxiety level.

When given the option, many fire officers prefer to use a telephone or face-to-face communications instead of a radio to transmit complicated or sensitive information. These options allow a more private exchange of information between two individuals, including the ability to discuss and clarify the information.

Written Communications

Informal Communications

Informal communications include internal memos, e-mails, instant messages, and messages transmitted via mobile data terminals. Generally, informal reports have a short life and are not archived as permanent records. They are used primarily to record or transmit information that will not be needed for reference in the future. This does not mean that these reports are unimportant.

Some informal reports are retained for a period of time and may become a part of a formal report or investigation. For example, a written memo could document an informal conversation between a fire officer and a fire fighter regarding a performance issue. The memo might note that the employee was warned that corrective action is required by a particular date. If the performance is corrected, the warning memo is discarded after 12 months. If the performance remains uncorrected, the memo becomes part of the formal documentation of progressive discipline.

Formal Communications

A **formal communication** is an official fire department document and is usually printed on business stationery with the fire department letterhead. If it is a letter or report intended

Assessment Center Tips

Write a Letter for the Fire Chief

A popular assessment center exercise requires the candidate to prepare a letter for the fire chief's signature. This represents a common assignment for a junior officer in many fire departments. In the scenario, the candidate is provided with all of the necessary information and must demonstrate the ability to prepare a properly formatted and worded letter. The letter is typically a response to a complaint or inquiry by a citizen or community interest group.

This exercise has two primary goals:

1. Assess the candidate's writing skills.
2. Assess the candidate's decision-making skills.

Preparing for this part of the assessment center requires the candidate to learn the proper format for an official letter or formal report. Generally, the assessment center exercise requires a business letter format. One method of preparing for this exercise is to contact the fire chief's administrative assistant and ask how this type of letter is prepared for the Fire Chief. Another method is to go to the local public library or the fire department's library. The basic concepts of proper letter writing can be found in many reference books.

for someone outside the fire department, it is usually be signed by the fire chief or a designated staff officer to establish that it is an official communication. Subordinates often prepare these documents and submit them for an administrative review to check the grammar and clarity. A fire chief or the staff officer who is designated to sign the document performs a final review before it is transmitted.

The fire department maintains a permanent copy of all formal reports and official correspondence from the fire chief and senior staff officers. Formal reports are usually archived for a predetermined length of time.

Standard Operating Procedures

Standard operating procedures (SOPs) are written organizational directives that establish or prescribe specific operational or administrative methods to be followed routinely for the performance of designated operations or actions. SOPs are intended to provide a standard and consistent response to emergency incidents as well as personnel supervisory actions and administrative tasks. Officers are expected to be familiar with the contents of all SOPs that apply to their position and responsibilities, so SOPs are a prime reference source for promotional exams and departmental training.

SOPs are formal, permanent documents that are usually printed in a standard format, signed by the fire chief, and widely distributed, and they remain in effect permanently or until they are rescinded or amended. Many fire departments conduct a periodic review of all SOPs so that they can revise, update, or eliminate any that are outdated or are no longer applicable. Any changes in the SOPs must also be approved by the fire chief.

General Orders

General orders are formal documents that address a specific subject, policy, condition, or situation. They are usually signed by the fire chief and can be in effect for various periods from a few days to permanently. Many departments use general orders to announce promotions and personnel transfers. Copies of the general orders should be available for reference at all fire stations.

Announcements

The formal organization may use announcements, information bulletins, newsletters, websites, or other methods to share additional information with fire department members. These methods are generally used to distribute short-term and nonessential information that is of interest.

Legal Correspondence

Fire departments are often called on to produce copies of documents or reports for legal purposes. Sometimes, the fire department is directly involved in the legal action, whereas in other cases, the action involves other parties but relates to a situation in which the fire department responded or had some involvement. It is not unusual for a fire department to receive a legal order demanding all of the documentation that is on file pertaining to a particular incident. This is a task that can require extensive time and effort. In any situation for which reports or documentation is requested, the fire department's legal counsel should be consulted.

The fire officer who prepared a report may be called on, sometimes years later, to sign an affidavit or appear in court to testify that the information provided in a report is complete and accurate. The fire officer might also have to respond to an **interrogatory**, which is a series of written questions asked by someone from an opposing party. The fire department must provide written answers to the interrogatories, under oath, and produce any associated documentation. If the initial incident report or other documents provide an accurate, factual, objective, complete, and clear presentation of the facts, the response to the interrogatory is often a simple affirmation of the written records. The task can become much more complex and embarrassing if the original documents are vague, incomplete, false, or missing.

Reporting

Report preparation is on the list of fundamental job responsibilities for almost every fire officer position. Like most other government agencies, fire departments are required to create and collect a wide variety of reports for both internal and external use. Every fire officer has to prepare and deliver a variety of reports that provide different types of information, depending on the circumstances. Some reports are prepared on a regular schedule, such as daily, weekly, monthly, or yearly. Other types of reports are prepared only in response to specific occurrences or when requested.

The most important factor to keep in mind in any reporting situation is that a report is intended to provide information. Every report is destined for someone who needs or at least desires the information. To create a useful report, the fire officer must understand the specific information that is needed and provide it in a manner that is easily interpreted.

Verbal Reports

The most common form of reporting is verbal communication from one individual to another, either face-to-face or via a telephone or radio. Fire officers continually report information to higher-ranking officers on various subjects and continually receive information from their subordinates. In many cases, this transfer of information is so routine and informal that it is overlooked as a form of reporting. In order to be effective, the transfer of information must be clear and concise, using terminology that is appropriate for the receiver.

Face-to-face conversation is the most effective means of conveying many types of information. In this mode, the sender and the receiver can engage in a two-way exchange of information that incorporates body language, facial expressions, tone of voice, and inflection along with the words themselves. When verbal communication must be conducted over radio or telephone, the body language, facial expressions, and similar forms of supplementary expression are sacrificed.

Types of Reports

A fire officer should be familiar with several different types of reports. We usually think of reports as written documents; however, some reports are presented orally, and others are prepared electronically and are entered into computer systems. Some reports are formal, whereas others are informal.

Reporting is an important and special type of communication skill that requires preparation and practice. This chapter covers some of the common elements of reporting and provides examples of the most common types of reports that a fire officer might have to prepare. For a new fire officer, there are usually many examples of reports that have been prepared by other officers in the same or similar circumstances that can be used as templates. More experienced fire officers can also provide assistance in mastering the reporting requirements of a new position.

VOICES OF EXPERIENCE

❝ When you use the communication process correctly, it actually works. ❞

Having been a fire officer for a number of years, I have found that many new fire officers fail in some basic points of communication. First, to ensure good communication, do not let yourself get too excited. At times, your emotions may run high and the stress of the situation may make your mind race, but to be effective you MUST control your emotions. A calm mind allows you to communicate your points both clearly and concisely.

Second, actively listen. Active listening requires you to think about what the other person is trying to communicate. If you are not careful, it is easy to pay only partial attention because you are thinking about what you want to say next rather than what the person is actually saying. This can cause your subordinates to feel you do not care about what they have to say.

Lastly, always give feedback. Feedback ensures that the message is interpreted correctly. In emergency scenes, you may miss a critical difference between what you think was said and what was actually said. This is why it is vital to repeat the critical points of the transmission and to acknowledge transmissions. By regularly following the communications process, you will find that it becomes second nature and greatly increases your communication results. When you use the communication process correctly, it actually works.

John House
Springfield, Missouri Fire Department
21 years of experience

Routine Reports

Routine reports generally provide information that is related to fire department personnel, programs, equipment, and facilities. Most fire departments also require company officers to maintain a **company journal** or log book (▶ **Figure 14-4**). A fixture in fire stations since the 19th century, the company journal provides an extemporaneous record of all of the emergency, routine, and special activities that occur at the fire station. Some fire departments have advanced from keeping the journal in a handwritten, hardcovered book to maintaining it as a database on the station computer.

The company journal serves as a permanent reference that can be consulted to determine what happened at that fire station on any particular time and date, as well as who was involved. The general orders usually list all of the different types of information that must be entered into the company journal. The company journal is the place to enter a record of any fire fighter injury, liability-creating event, and special visitors to the station.

Morning Report to the Battalion Chief

Most career fire officers are required to provide some type of morning report to their on-duty battalion chief or superior officer. In many cases, this report is made by telephone or on a simple form that is transmitted by fax or e-mail.

One purpose of the morning report is to identify any personnel or resource shortages as soon as possible after the on duty platoon reports for duty. For example, if the driver/operator calls in sick, the driver/operator from the off-going shift might have to remain at work to keep the engine company in service. The chief needs to know that the company is in an overtime situation until a replacement driver/operator is assigned and arrives.

Monthly Activity and Training Report

The monthly activity report documents the company's activity during the preceding month (▶ **Figure 14-5**). These reports typically include the number of emergency responses, training activities, inspections, public education events, and station visits that were conducted during the previous month. Some monthly reports include details such as a list of the number of feet of hose used and the number of ladders deployed during the month.

In many cases, the officer delegates the preparation of routine reports that do not involve personnel actions or supervisory responsibilities to subordinates; however, the officer is responsible for checking and signing the report before it is submitted.

Two other routine reports are the annual fire fighter performance appraisals and the fire safety inspection. These reports are covered in detail in Chapter 8, *Evaluation and Discipline,* and Chapter 12, *Preincident Planning and Code Enforcement.*

Some fire companies or municipal agencies post a version of a monthly report on their public website. The reports often include digital pictures of the incidents and the people involved. Special events or unusual situations that occurred during the month might also be included; for example, the monthly report might note that the company provided standby coverage for a presidential visit or participated in a local parade.

Infrequent Reports

Infrequent reports usually require a fire officer's personal attention to ensure that the report's information is complete and concise. The special reports that a fire officer may be responsible for include:

- Fire fighter injury report
- Citizen complaint
- Property damage or liability-event report
- Vehicle accident report
- New equipment or procedure evaluation
- Suggestions to improve fire department operation
- Response to a grievance or complaint
- Fire fighter work improvement plans
- Research report

Some situations require two or more different report forms to cover the same event. For example, if a fire engine collides with a private vehicle at an intersection while responding to an emergency call, the fire officer usually spends the rest of the day completing reports, even if there are no injuries. The fire officer must fill out a supervisor's accident report and make sure that the driver completes the required accident report. Another form has to be completed to identify the damage to the fire apparatus and any repairs that are needed. The local jurisdiction's risk management division may also require a report. The insurance company that covers the fire apparatus usually requires the completion of another form.

A **supervisor's report** is required by most state worker's compensation agencies whenever an employee is injured.

Safety Zone

A fire officer is responsible for the timely and proper completion of property damage and injury reports. That does not mean that the fire officer must personally fill out every form. All fire fighters should know how to fill out the required administrative paperwork for a minor injury, blood-borne pathogen exposure, or damage to fire department property.

The late notification of the duty officer on infectious disease encounters and the tardy or incomplete documentation of injuries or toxic exposures are continuing challenges to fire fighter health and safety. All fire fighters should know the reporting procedures to protect them throughout their careers.

out	in	inc #	address/comments				sign
			Fire Station 46 Company Journal				
	0700		**Thursday, May 29 - B shift on Capt Smyth OIC**				
			Engine	**Rescue**	**Medic**	**Chief 9**	
			Lt. Seth (C/B)	Cpt Smyth	Lt. Willow	BC Grave	
			Tech Ayres	Tech Cegar	FF Villani		
			FF Robinson	FF Chase			
			FF Bartlow	FF Adams			
			Not here: Reserve Engine 446 running as Engine 35				
			Reserve Medic 446 running as Medic 11				
			Lt Johnson on Official Representation (at FDIC)				
			Firefighter Tolliver on Annual Leave				
			Tech Madison detailed to Medic 10				
				Engine	**Rescue**	**Medic**	
			Radio	(2) Ayres	(4) Cegar	Willow	
			Keys	Ayres	Cegar	Willow	
			Drugs	Robinson		Willow	
	0745		Medic 46 disinfected				Vil
1030	0800		Rescue 46 to Station 22 to teach hazmat drill				Ceg
1011	1120	0345	7111 Hogarth St - injury		Medic		Vil
1044	1101	0375	8204 Green Heron Way #304 - CPR		Engine		Sth
	1145		Defibrillator #1456 malfunction on event 0375				Sth
			Logistics notified, EMS 9 delivered loaner				
1217	1244	0500	9103 Cross Chase Court - ALS		Medic		Vil
1600	1315		Facilities working on furnace				
1302	1430	0711	4110 Green Springs Dr #101 - sick		Medic		Vil
1533	1551	0987	I-95 N at Newington - Crash		Rescue		Ceg
1533	1607	0987	I-95 N at Newington - Crash		Engine		Sth
1533	1635	0987	I-95 N at Newington - Crash		Medic		Vil
	1700		Reserve Medic 446 returned from Station 11				Ceg
1930	1830		Fire Chief in quarters				
1836	1900	1135	9112 Chapel Oaks Way - auto fire		Engine		Sth
1845	1911	1270	8902 Harrivan Lane - chest pain		Medic		Wil
	2015		FF Sorce on sick leave for tomorrow				
2044	2340	1350	7401 Eastmoreland Ave - 2nd alarm		Rescue		Ceg
2059	2253	1350	7401 Eastmoreland Ave - 3rd alarm		Engine		Ceg
2315	2120		Engine 7 filling in at Station 46				MJ
			Friday, May 30				
0112	0201	0099	10614 Hampton Rd - dyspnea		Medic		Vil
0112	0144	0099	10614 Hampton Rd		Engine		Sth
0249	0305	0143	6600 Springfield Mall - Alarm bells		Engine		Sth

Figure 14-4 Example of a company journal.

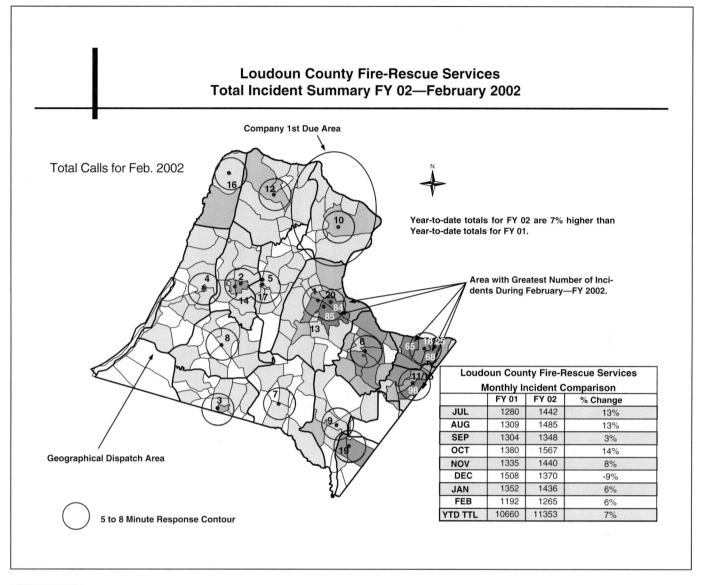

Loudoun County Fire-Rescue Services
Total Incident Summary FY 02—February 2002

Total Calls for Feb. 2002

Company 1st Due Area

Year-to-date totals for FY 02 are 7% higher than Year-to-date totals for FY 01.

Area with Greatest Number of Incidents During February—FY 2002.

Geographical Dispatch Area

5 to 8 Minute Response Contour

Loudoun County Fire-Rescue Services Monthly Incident Comparison			
	FY 01	FY 02	% Change
JUL	1280	1442	13%
AUG	1309	1485	13%
SEP	1304	1348	3%
OCT	1380	1567	14%
NOV	1335	1440	8%
DEC	1508	1370	-9%
JAN	1352	1436	6%
FEB	1192	1265	6%
YTD TTL	10660	11353	7%

Figure 14-5 Sample monthly report.

Most of these reports are fill-in-the-blank forms that duplicate much of the information required by the local reports. This report is the control document that starts the state file relating to an injury or a disability claim (▶ **Figure 14-6**).

The supervisor's first report must usually be submitted within 24 to 72 hours of the incident. The supervisor completes this report after performing an investigation into the facts. The report often includes who was injured, the nature of the injury, how the incident occurred, what practices or conditions contributed to the incident, any potential loss, the typical frequency of occurrence, and what actions will be taken to prevent the same accident from occurring again.

Special Reports

Fire officers most often have to present one of two different types of reports. A **chronological statement of events** is a detailed account of activities, such as a narrative report of the actions taken at an incident or accident. A **recommendation report** is a document that suggests a particular action or decision.

Some reports incorporate both a chronological section and a recommendation section. For example, the fire fighter line-of-duty death investigations produced by the National Institute for Occupational Safety and Health include a chronological report

SUPERVISOR'S ACCIDENT REPORT
WORKERS' COMPENSATION CLAIMS

Claims Management Inc.
PO Box 342
Sacramento, CA 95812-3042
(916) 631-1250
FAX (916) 635-6288

DATE & TIME
REPORTED:

OSHA CASE NO:

COMPANY Scotts Valley Fire Protection District	LOCATION 7 Erba Lane, Scotts Valley, CA, 95066	LOCATION CODE NO: 1100

A. EMPLOYEE

NAME

JOB TITLE

DEPARTMENT
Scotts Valley Fire Protection District 1100

☒☒ LOST TIME
☒☒ NO LOST TIME

☒☒ FIRST AID

B. TIME AND PLACE OF ACCIDENT

DATE	HOUR	DEPARTMENT 1100	IMMEDIATE SUPERVISOR

IDENTIFY EXACT LOCATION WHERE ACCIDENT OCCURRED *(Be specific)*

JOB OR ACTIVITY AT TIME OF ACCIDENT *(Be specific)*

C. WITNESSES - *List of Names and Addresses*

Name	Address

D. DESCRIBE THE ACCIDENT/ACCIDENT CAUSE - *Please be specific*

E. UNSAFE ACT/CORRECTIVE ACTION TAKEN - *Include both employee and supervisor corrective actions to prevent future occurrences.*

EMERGENCY - WENT TO THE DOCTOR
☒☒ YES

☒☒ NO

If yes, please fill out the following information:

Name of Doctor: _____

Address of Doctor: _____

☒☒ NON-EMERGENCY, BUT PLAN ON SEEING A PHYSICIAN
Physician's Name: _____

Figure 14-6 Supervisor's accident report.

of the event, followed by a series of recommendations that could prevent the occurrence of a similar situation.

The goal of a decision document is to provide enough information and persuasion to make the intended individual or body comfortable with accepting your recommendation. This type of report could include recommendations for employee recognition or formal discipline, for a new or improved procedure, or for adoption of a new device. A decision document usually includes:

- **Statement of problem or issue.** One- or two-sentence statement of the problem or issue.
- **Background.** Brief description of how this became a problem or an issue.
- **Restrictions.** Outline of the restrictions affecting the decision. Factors such as a federal law, state regulations, local ordinances, budget or staff restrictions, or union contract are all restrictions that could affect a decision.
- **Options.** Where appropriate, provide more than one option and the rationale behind each option. In most cases, one option is to do nothing.
- **Recommendation.** Explain why the recommended option is the best decision. The recommendation should be based on considerations that would make sense to the decision maker. If the recommendation is going to a political body and involves a budget decision, it should be expressed in terms of lower cost, higher level of service, or reduced liability. The impact of the decision should be quantified as accurately as possible, with an explanation of how much money it will save, what new levels of service will be provided, or how much the liability will be reduced.
- **Next action.** The report should clearly state the action that should be taken to implement the recommendation. Some recommendations may require changes in departmental policy or budget. Others could involve an application for grant funds or a request for a change in state or federal legislation.

Incident Reports

Some type of incident report is required for every type of emergency response. The nature and the complexity of the report depend on the situation: minor incidents generally require simple reports, whereas major incidents require extensive reports.

The **National Fire Incident Reporting System (NFIRS)** is a nationwide database managed by the U.S. Fire Administration that collects data related to fires and some other types of incidents. Many fire departments send their incident information to the NFIRS for inclusion in national reports and statistics. Typically, the first-arriving officer completes an NFIRS data report for each response, including a narrative

description of the situation and the action that was taken. The full incident report often includes a supplementary report from the officer in charge of each additional unit that responded (▶ **Figure 14-7**).

Many departments use a preferred narrative format when reporting on incidents (▶ **Figure 14-8**).

Some incidents require an **expanded incident report narrative**, in which all company members must submit a narrative description of their observations and activities during an incident. The fire officer should anticipate the need to provide expanded incident reports on incidents in which:

- The fire company was one of the first-arriving units at a fire with a civilian fatality.
- The fire company was one of the first-arriving units at an incident that has become a crime scene or an arson investigation.
- The fire company participated in an occupant rescue or other emergency scene activity that would qualify for official recognition, award, or bravery citation.
- The fire company was involved in an unusual, difficult, or high-profile activity that requires review by the fire chief or designated authority.
- Fire company activity occurred that may have contributed to a death or serious injury.
- Fire company activity occurred that may have created a liability.
- Fire company activity occurred that has initiated an internal investigation.

The author of a narrative should describe any observations that were made en route or on the scene and fully document his or her actions related to the incident. The narrative should provide a clear mental image of the situation and the actions that were taken.

Writing a Report

There are many reasons why a fire officer may be required to prepare a report. The fire officer must clearly understand the purpose in order to write a report that is accurate and presents the necessary information in an understandable format. The report could be intended to simply inform the reader of some new information, or it could be designed to persuade the reader to agree with a particular position. It could be to simply brief the reader, or it might be to provide a complete and systematic research of an issue. If the fire chief wants only to be briefed on a new piece of equipment and instead the assigned fire officer delivers a full research presentation, both parties would be frustrated.

In addition to the formal goal of the report, the fire officer must consider the potential reader. The fire officer must first consider the intended audience for the report. For example, a fire officer may be assigned to prepare a study that proposes closing three companies and using quints that can respond as either engine or ladder companies. If the report is

On Monday, May 12 at 1247 hours, Station 2 was alerted to respond with the Rescue Squad and the Tanker on the second alarm of a three-alarm building fire in Charles County. Chief 1 from the LaPlata VFD arrived on the scene at 8190 Port Tobacco Road, the Charles County Community Services Building, with heavy smoke showing from the rear.

While units from Station 2 were en route to the scene, the evacuation tones were sounded and all civilian county employees operating on the scene were forced to exit the building. Upon arrival, crews began cutting ventilation holes in the roof. But shortly after, they were pulled off due to the poor structural integrity of the building.

While on the scene, the crew of Rescue Squad 2 acted as RIT team number 1. Some crew members also assisted with the water supply operation. Tanker 2 was utilized in the tanker shuttle operation. All together, Tanker 2 hauled 23 loads of water to the scene which totalled 69,000 gallons by the time the operation was complete. Rescue Squad 2 was then assigned to take a transfer at LaPlata Station 1 and remained there for the remainder of the night. Both units were back in quarters at 2200 hrs.

Figure 14-7 Supplementary report.

intended for internal use, the technical information would use normal fire department terminology. However, if the report is being prepared for the fire chief to deliver to the city council, with copies going to the news media, many of the terms and concepts would have to be explained in detail for the general public.

The fire officer who is preparing a report must also determine the appropriate mode and format for the report. This can range from an internal memo format with one or two pages of text to a lengthy formal report that includes photographs, charts, diagrams, and other supporting information.

Because some reports may require the fire officer to make a recommendation, the fire officer must also be able to analyze and interpret data. An officer who is assigned to prepare a report is often required to consult information that is maintained in a database or a records management system. The fire department's records management system typically contains NFIRS run reports, training records, occupancy inspection records, preincident plans, hydrant testing records, staffing data, and activity records. A computer-aided dispatch system uses a combination of databases to verify addresses and determine the units that should be dispatched and maintains information that can be retrieved relating to individual addresses. All of the transactions performed by a computer-aided dispatch system are recorded in another database. This information can be retrieved to examine the history of a particular incident or to identify all of the incidents that have occurred at a particular location.

The fire officer also needs to understand how to use and analyze the data that can be retrieved from the various sources. The fire officer could be trying to identify trends such as an increase or a decrease in the number of fire deaths

Assessment Center Tips

A promotional candidate should be prepared for two different types of verbal presentations. The first type is a meeting with a citizen or community group to explain a fire department practice or issue. This is equivalent to a verbal presentation of a decision document.

The second type of presentation could follow an emergency incident scenario. After the incident management portion of the exercise is completed, the candidate could be required to make a presentation of the incident as if he or she is speaking to the press or reporting to a senior command officer.

Batesville Fire Department Incident Update

Date: 09-16-2002
Time: 2349
First Company: B17
Incident Number: 101539
Address: 710 Werkcastle Lane
Incident Commander: Chief Margaret Hinkler
Dollar Loss: Still being determined
Occupancy: Residential
Displaced Occupants: 5
Adults Injured: 1

Minors Injured: 0
Civilian Fatalities: 0
Civilians Rescued: 1
Fire Fighter Injuries: 0
Fire Fighter Fatalities: 0
Total Fire Fighters on Scene: 25
Total Ambulances: 1
Time to Suppress Incident: 55 Min
Incident Duration: 7 HRS 10 MIN
Total Companies: 5

Incident Description:

On Saturday, September 16, 2002 at 2349 Hours (11:49 PM EST), five fire companies and one ambulance company responded to a residential fire at 710 Werkcastle Lane in Batesville. All companies were under the command of Chief Margaret Hinkler. B17 was the first company to arrive on the scene. B17 reported heavy smoke pouring from the downstairs right corner window of a two-story brick residence. Three minors and an adult female were standing outside of the residence. The adult female reported that her husband was still inside. A four-fire fighter rescue team entered the structure with a thermal imaging device. An adult male was found on his knees, coughing in the front hallway and was extracted from the residence. The fire was confined to the kitchen and family room and suppressed in 55 minutes. A dollar loss is still being determined. A grease fire on the stove was the cause of the fire.

Figure 14-8 Sample incident report.

---

in a downtown district over a period of 10 years. There are a variety of statistical methods that can be used for data analysis. The fire officer should consider taking a specialized course in this subject.

The fire officer might also be looking at variances, such as differences between the projected budget and actual expenses in different accounts in a given year. When a variance is encountered, the analysis should consider both the cause and the effect. If the budget included an allocation of $10,000 for training and the expenditure records indicate that only $500 was spent, there would be a variance of $9,500 that could be reallocated for some other purpose.

This variance could also indicate that very little training was actually conducted.

Presenting a Report

A fire officer should be prepared to make an oral presentation of a decision document or to speak to a community group on a fire department issue. Using the written report as a guide, a verbal presentation should consist of four parts:

1. **Get their attention.** Have an opening that entices the audience to pay attention to your message.
2. **Interest statement.** Immediately and briefly explain why listeners should be interested in this topic.
3. **Details.** Organize the facts in a logical and systematic way that informs the listeners and supports the recommended decision.
4. **Action.** At the close of your presentation, ask the audience to take some specific action. The most effective closing statements are actions that relate directly to the interest statement at the beginning of the presentation.

Preparing a News Release

Fire organizations need to be able to communicate effectively through the local mass media. A news release allows a fire department to reach a large audience at virtually no cost. The release could be structured to promote a public education message, draw attention to the community's risk problems, announce a new program or the opening of a new facility, or simply develop good public relations. A well-prepared news release encourages the media to cover your story.

The first step in preparing a news release is to formulate a plan. Consider what the release is to accomplish. Is it an urgent fire safety message or an invitation to the fire department picnic? Who is the target audience? Most messages are directed toward some particular group or audience. A safety message about a smoke alarm program is directed toward those residents who do not have smoke alarms.

During this first step, you should also identify what makes the story interesting and worthy of the time and energy the media source must expend to present it. The media may have hundreds of releases that they must prioritize in order to decide which ones to give attention. If the news release is about smoke alarms, the release should emphasize why this subject is important.

The second step is to develop the concept and write the release. The format should be clear, concise, and well organized. The first paragraph or two should cover the "who, what, when, where, and why" of the story. Successive paragraphs should provide further information, progressing from the most important information to the least important. Do not be wordy in a press release, and stick to the facts. The idea is to give the media enough information to convince them that the story

warrants coverage. Each media outlet considers the release in the context of whether it would be of interest to their audience. If the release is on target, a reporter will follow up.

Make your release unique, with an interesting perspective. Consider using quotes that would appeal to the human side of the reader. For example, "The fire spread though the house so rapidly that the elderly couple was unable to escape, and they perished. Fire Department Investigator Jones indicated that they died because the battery had been removed from their smoke alarm."

The use of the department's letterhead helps draw the attention of the news agency. Formatting can also help bring attention to the release. Use at least 1-inch margins, double space the text, and make the news release fit on one page. Draw attention by keeping it neat and clearly organized, and make sure that there are no typographical or grammatical errors.

At the top of the page, print "NEWS RELEASE" in all caps. The time and date should appear on the following line. Separate this heading from the text by a series of five to seven pound signs (#######). Pound signs should also be used to separate the text from the ending, which should include a contact name and number for further information.

The last step is to get the news release out to the media. It can be distributed through fax, e-mail, or the postal service if there is sufficient time. It is important to distribute the release to all media outlets equally. Giving exclusive stories to a favorite news outlet can damage a fire department's credibility.

Using Computers

Just a few years ago, most fire officers would have scoffed at the notion that there was a place in the fire station for a computer. That perception has drastically changed, and today's fire officer must be as proficient with a computer as with a portable radio. In most fire stations, the officer spends more time at the computer than talking on the radio.

Although many officers understand the basics of computers, it is important for any prospective officer to understand the general principles. Computers are made up of physical components called hardware. Hardware includes the keyboard, monitor, printer, and central processing unit, or CPU. The CPU houses the internal hardware components, such as the processor, memory, hard drive, and motherboard. The motherboard is a circuit board that provides the connections to all of the components of the computer. The prossessor is the component of the computer that actually does the calculations and processes the information.

The memory is also attached to the motherboard. Memory is like the in-box on your desk. It holds all the items that you are currently using and keeps them readily at hand. The hard drive is also attached to the motherboard. The hard drive is a device that stores information in a more permanent fashion, like the filing cabinet in your office. You place files in it that you want to save long term. A greater quantity of information can be stored in the hard drive, but it cannot be accessed as quickly as the files in your in-box. Like your in-box and filing cabinet, when the memory and hard drive are nearly full, it takes significantly longer to find the files.

Computer hardware alone is of little value. Software is needed to instruct the computer on what to do. Software contains all of the instructions that tell the computer how to function. There are many different types of software, and each tells the computer how to perform specific actions. The most common types of software include operating systems; programs for word processing, spreadsheets, presentations, Internet access, and communication; and database packages.

Operating System

Operating system software enables the computer to perform its basic functions. Every computer must initially be loaded with an operating system to manage all other programs, which are called application programs. The operating system performs basic tasks, such as recognizing input from the keyboard, sending visual output to the monitor, keeping track of files, and controlling external devices, such as printers. The application programs interface with the operating system to allow them to perform their tasks. Some common operating systems are Linux®, Mac OS®, and Windows®, including Windows 2000, XP, and ME.

Word Processing

Word processing software is used to create documents, similar to a highly advanced typewriter. Word processing programs allow a user to type letters, memos, and reports. Most programs also have templates, which are pre-made layouts of the most common letters, memos, and reports, as well as fax cover sheets. The most common word processing programs are Microsoft Word® and Corel WordPerfect®.

Spreadsheets

Spreadsheet software is used to tabulate numbers and information in columns and rows. It is often used to analyze data and to develop charts. A spreadsheet resembles a ledger sheet, with the columns represented by letters and the rows are identified by numbers. The intersection of a column and a row is called a cell. A cell can contain numbers, letters, or a formula.

A formula is used to make calculations. For example, a formula in cell A2 might read "=sum(C3+D4)," which means that it adds the number in cells C3 and D4 and places the total in cell A2.

Spreadsheets have immense calculation capabilities. Each brand of software uses different methods of inputting formulas, so the user needs to learn the specific program being used. Some of the more common spreadsheet programs are Microsoft Excel®, Microsoft Works®, Corel Quattro Pro®, and Lotus 1-2-3®.

Presentation Software

Audiovisual presentations are often used to communicate effectively with an audience. Presentation software is used to place information in a format that allows the audience to visually understand the message being presented. Presentation software has taken over the functions that used to be performed with slide shows, overhead transparencies, and video recorders. Today, a computer and a projection system allow the user to create slides and videos that can be projected onto a screen, produce handouts, and manage instructor notes. The most common presentation software programs are Microsoft PowerPoint®, Corel Presentation®, and Lotus Freelance Graphics®. These programs allow a fire officer to quickly develop a high-quality presentation that includes pictures and graphics.

Internet Access

The Internet is a network of global computers, all linked together by high-speed, high-capacity connections. Special web browser software is needed in order to access the Internet. The most common web browsers are Microsoft Internet Explorer® and Netscape Navigator.

Many fire departments provide Internet access for fire station computers and allow employees to use them for work-related purposes. The computer must have a connection to the Internet, through a telephone line or data cable, as well as the browser software. Most connections are through an Internet service provider, or ISP, who charges a monthly fee for access to the Internet.

Once a computer is connected to the Internet, a user can access the World Wide Web, which is a series of computers loaded with web pages on every imaginable topic. In essence, any computer that is connected to the Internet can communicate with any other connected computer, anywhere in the world. The scope of information that can be accessed is beyond estimation.

Communication

Communication software is required for the computer to talk to other electronic devices. The most common form of communication is e-mail, which is equivalent to a memo or letter transmitted via the Internet. E-mail programs allow the user to type messages, attach files, and send them instantaneously to another e-mail address. The most common e-mail programs are Microsoft Outlook®, Netscape Messenger®, and Eudora®. For a communication program to send the message, the computer must be connected to the Internet. Once the connection is established, the message is sent to the appropriate e-mail address, where it can be retrieved by the receiving party.

Database

Database software is a special type of program that allows the user to store and manage data. Databases store information in tables that are similar to spreadsheets. A database is composed of a series of interrelated tables. For example, the department might create a table that lists all of the addresses of every household in town, another table that lists all the addresses where fires have occurred, another that tracks the addresses where every smoke alarm has been installed, and yet another that lists the public education events by address. Using database software, reports can be generated to pull data from any of the various tables. For example, if the department wanted to know whether there has been a reduction in the number of fires in areas where public education has been delivered, this information could be extracted from the database very rapidly. Without a database, it would be very time consuming to search for this information. The most common database programs are Microsoft Access®, Oracle®, and FileMaker Pro®.

Information Management

It is readily apparent that huge amounts of information are stored electronically. Many businesses and organizations are working toward paperless environments in which everything is stored and processed electronically. Like books in a library, this requires a highly advanced management system to allow documents and information to be filed appropriately, so that they can be retrieved when they are needed

in the future. A lost electronic file is no different than a lost paper file; it is the same as no file at all.

Many fire departments have established local networks that link every computer within the organization. Each computer is connected to a server, where information can be stored, shared, and properly protected from loss. Typically, each user is given a private storage location on the server that no one else can access. There may also be a location on the server that allows only certain users to have access to certain types of information. For example, the Fire Prevention Division may have a location on the server that holds all of the permit files. Only employees in this division can access these files, but each person within the division can access any file.

Most organizations also have a department-wide location on the server that anyone on the system can access. This allows all employees to share files. This can also generate a huge amount of files, so specific naming conventions are established to allow quick access to the relevant files.

The advantage of having all data stored on the network is that it allows all of the data to be regularly archived in a central location. If data are stored on an individual computer's hard drive, that information is usually not archived regularly by the user. This creates the potential for the permanent loss of files.

Summary

Effective communication is a core success skill for the fire officer. The communication skill set becomes more important as the individual rises higher in the fire department rank structure. The 19th century fire officer needed only the speaking trumpet to direct fire fighters at emergencies. Today's fire officer uses computers and wireless communication to function in the 21st-century environment.

You Are the Fire Officer: Conclusion

When you hastily advised dispatch of your arrival and told them "to fill the box," you failed to confirm that the message was received and understood. You made the false and dangerous assumption that you had effectively communicated your message. The dispatcher might not have heard you or might not have understood the terminology that you used.

When you realized that your help was not coming, you communicated your message to the dispatcher in a clear and concise manner and confirmed that the dispatcher understood and could act on it. Good communication skills are essential in emergency situations. You must ensure that the message is not only received but also understood.

Wrap-Up

Ready for Review

- There are five parts to the communication cycle: message, sender, medium, receiver, and feedback.
- To improve your listening skill: do not assume, do not interrupt, try to understand the need, and do not react too quickly.
- Fire officers who are active listeners are also up to date on the department's guides, policies, and procedures.
- Fire officers keep their bosses informed on progress toward goals and projects, potential controversial issues, fire fighter attitude, and morale.
- Environmental noise can be counteracted by avoiding a power struggle, communicating clearly and firmly, carefully choosing words, confirming that the receiver understands the message, and providing consistent and appropriately detailed messages.
- Informal reports are not considered official fire department permanent records.
- The company journal is the extemporaneous record of fire station activity.
- A written report will have formal, informal, and unanticipated readers.
- A generic decision document has a statement of problem/issue, background, restrictions, options, recommendation, and next action.

Hot Terms

Chronological statement of events A detailed account of the fire company activities as related to an incident or accident.

Company journal Log book at the fire station that creates an extemporaneous record of the emergency, routine activities, and special activities that occurred at the fire station. The company journal also records any fire fighter injury, liability-creating event, and special visitors to the fire station.

Environmental noise A physical or sociological condition that interferes with the message in the communication process.

Expanded incident report narrative A report in which all company members submit a narrative on what they observed and activities they performed during an incident.

Formal communication Represents an official fire department communication. The letter or report is presented on stationery with the fire department letterhead and generally is signed by a chief officer or headquarters staffperson.

General orders Short-term documents signed by the fire chief and lasting for a period of days to 1 year.

Informal communications Internal memos, e-mails, instant messages, and computer-aided dispatch/mobile data terminal messages. Informal reports have a short life and are not archived as permanent records.

Interrogatory A series of formal written questions sent to the opposing side. The opposition must provide written answers under oath.

National Fire Incident Reporting System (NFIRS) A nationwide database at the National Fire Data Center under the U.S. Fire Administration that collects fire-related data in order to provide information on the national fire problem.

Recommendation report A decision document prepared by the fire officer for the senior staff. The goal is support for a decision or action.

Standard operating procedures (SOPs) A written organizational directive that establishes or prescribes specific operational or administrative methods to be followed routinely for the performance of designated operations or actions.

Supervisor's report A form that is required by most state worker's compensation agencies that is completed by the immediate supervisor after an injury or property damage accident.

Fire Officer in Action

In the morning during apparatus checks, you notice that there is some tension among the shift personnel. They are very terse with you, and there have been some vocal disagreements between personnel as well. You are not exactly sure what is going on, so you call the acting engine lieutenant into your office. Acting Lieutenant Winkelman has worked with you for several years, and you feel like you can talk openly with him. You tell Winkelman that you've noticed some tension among the shift this morning and wonder if he may know the cause. Winkelman advises you that a memo you put out last shift regarding use of the Internet and television is the cause. The personnel are upset because the memo was placed in their mailboxes but not discussed at the shift briefing.

1. What is one basic technique for effective communications?

    A. Do not react too quickly

    B. Do not listen to your shift

    C. Effectively use time by dealing with multiple issues concurrently

    D. Do not confront

2. If a fire fighter is rambling on and not being very clear in his or her question, what can you do to solve the problem effectively?

    A. Yell at him or her to stop rambling

    B. Walk away

    C. Use direct questioning

    D. Have a counseling session

3. What is one area that you need to keep your supervisor informed about?

    A. Matters that may cause controversy

    B. Strengths and weaknesses

    C. Your educational plan

    D. Staffing

4. Regarding the grapevine, what is something the fire officer should do?

    A. Believe a rumor

    B. Start a rumor

    C. Leak information

    D. Squash rumors that create stress

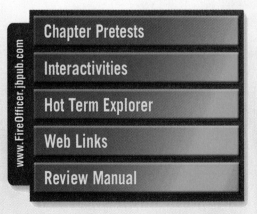

www.FireOfficer.jbpub.com

- **Chapter Pretests**
- **Interactivities**
- **Hot Term Explorer**
- **Web Links**
- **Review Manual**

Managing Incidents

NFPA 1021 Standard

Fire Officer I

4.6.3* Implement an action plan at an emergency operation, given assigned resources, the type of incident, and a preliminary plan, so that resources are deployed to mitigate the situation.

(A) Requisite Knowledge. Standard operating procedures, resources available for the mitigation of fire and other emergency incidents, an incident management system, scene safety, and a personnel accountability system.

(B) Requisite Skills. The ability to implement an incident management system, to communicate orally, to manage scene safety, and to supervise and account for assigned personnel law-making process at the local, state/provincial, and federal levels; and the functions of other bureaus, divisions, agencies, and organizations and their roles and responsibilities that relate to the fire service.

4.6.4 Develop and conduct a postincident analysis, given a single-unit incident and postincident analysis policies, procedures, and forms, so that all required critical elements are identified and communicated, and the approved forms are completed and processed in accordance with policies and procedures.

(A) Requisite Knowledge. Elements of a postincident analysis, basic building construction, basic fire protection systems and features, basic water supply, basic fuel loading, fire growth and development, and departmental procedures relating to dispatch response tactics and operations and customer service.

(B) Requisite Skills. The ability to write reports, to communicate orally, and to evaluate skills.

Fire Officer II

5.6.1 Produce operational plans, given an emergency incident requiring multiunit operations, so that required resources and their assignments are obtained and plans are carried out in compliance with approved safety procedures, resulting in the mitigation of the incident.

(A) Requisite Knowledge. Standard operating procedures; national, state/provincial, and local information resources available for the mitigation of emergency incidents; an incident management system; and a personnel accountability system.

(B) Requisite Skills. The ability to implement an incident management system, to communicate orally, to supervise and account for assigned personnel under emergency conditions; and to serve in command staff and unit supervision positions within the incident management system.

5.6.2 Develop and conduct a postincident analysis, given multiunit incident and postincident analysis policies, procedures, and forms, so that all required critical elements are identified and communicated and the approved forms are completed and processed.

(A) Requisite Knowledge. Elements of a postincident analysis, basic building construction, basic fire protection systems and features, basic water supply, basic fuel loading, fire growth and development, and departmental procedures relating to dispatch response, strategy tactics and operations, and customer service, the ability to write reports, to communicate orally, and to evaluate skills.

(B) Requisite Skills. The ability to write reports, to communicate orally, and to evaluate skills.

Additional NFPA Standards

NFPA 1500, *Standard on Fire Department Occupational Safety and Health Program*

NFPA 1521, *Standard for Fire Department Safety Officer*

NFPA 1561, *Standard on a Fire Department Incident Management System*

Knowledge Objectives

After studying this chapter, you will be able to:

- Explain how the Incident Management System was created.
- Describe the National Incident Management System.
- Describe the role and elements of the Federal Response Plan.
- Explain the importance of fire fighter safety and accountability within the Incident Management System.
- Describe the "two-in-two-out" rule.
- Explain how to effectively accomplish a transfer of command.
- Describe the fire officer's role in incident management.
- Explain the use of divisions and groups within the Incident Management System.
- Describe the task level of incident management.
- Describe the postincident review.

Skills Objectives

After studying this chapter, you will be able to:

- Demonstrate making an initial radio report.
- Demonstrate conducting a face-to-face transfer of command.

You find yourself responding to your first fire as the officer in charge of the first-due engine company. You have practiced this over and over in your head, but as you get closer to the column of smoke rising above the trees, your thoughts go into overdrive. The call is in an area that is transitioning from rural to suburban, with many new houses under construction. Your heart races as you turn the corner and see what appears to be a vacant barn at the end of the street, with fire venting out of all the windows and the roof. You know that you will be first on the scene, at least 3 minutes ahead of the next-arriving company.

1. As the first-arriving officer, what is your responsibility for an incident management system?
2. What mode of command will you assume on this fire?
3. What components of the incident management system would you use?

Introduction to Managing Incidents

Every fire officer should be prepared to function in a variety of roles in the Incident Management System. A fire officer must be prepared to perform the duties of a first-arriving officer at any incident, including assuming initial command of the incident, establishing the basic management structure, and following standard operating procedures. A fire officer must also be fully competent at working within the incident management system at every incident. This chapter introduces the model procedures for incident management. Chapter 16, *Fire Attack* focuses on the fire attack and on applying the incident management procedures to this type of situation.

How Incident Command Developed

Almost all fire departments today are organized in a paramilitary structure, with fire officers holding ranks similar to those of military officers. The choice of this organizational model is appropriate when the similarities between military combat and fighting fires are considered. Just like the army, in fire departments, low-level officers are in charge of companies, midlevel officers are in charge of battalions, higher-ranking officers are in charge of divisions, and the highest-raking officers manage the overall organization. In many cities, the roots of this system evolved with the transition from independent volunteer fire companies to a city fire department; this change occurred under the leadership of experienced military officers.

Hierarchical command and control is one of the foundations of the Incident Management System. Fire departments have used different variations of hierarchical command structures throughout history; however, there were numerous variations of the systems and procedures based on the preferences of different fire departments. Standardized command systems, which are based on model structures and procedures that can be applied and understood by every fire department, are a relatively recent development.

Incident management systems have evolved over the past 30 years. The evolution of modern incident management systems can be traced back to the Incident Command System (ICS), which was developed under the FIRESCOPE organization in California, and the Fireground Command System (FGC), which was developed by the Phoenix Fire Department. These two systems, which have many basic similarities, were adopted by fire departments across the United States and Canada. During the 1990s, the National Fire Service Incident Management Consortium was established to merge these two systems, creating the **Incident Management System** (IMS) and expanding its application to a wider variety of emergency situations.

The federal government, which has always had a major role in the response to natural disasters and forest fires, also played an important role in the development of modern incident management systems. The Federal Emergency Management Agency (FEMA) became even more directly involved in emergency incident management systems as a result of the 1995 Oklahoma City bombing and the September 11, 2001 terrorist attacks. Each of these incidents involved numerous agencies working together in highly complex and dangerous situations. In March 2004, the Department of Homeland Security announced that all emergency responders involved in operations under federal government command and control would be mandated to use the National Incident Management System (NIMS). The NIMS model applies the same basic principles as ICS, FGC, and IMS, but on an even larger scale.

One of the essential attributes of modern incident management systems is that they can be implemented at any scale and with any required scope. Every incident, even a single-company response, can use the Incident Management System. If the situation expands, the Incident Management System can expand incrementally along with it. Through the use of well-known models and common terminology, resources from different agencies should be able to integrate and coordinate their efforts, even if they have never previously worked together.

The recognition that fire fighter accountability is an essential safety issue also influenced the development and application of the Incident Management System. The use of a standard Incident Management System allows the incident commander to coordinate and keep track of the location and assignment of every resource that is committed to an incident. When everyone is integrated into a standard system and understands that system, it becomes feasible to keep track of everyone involved in an incident.

FIRESCOPE and ICS

FIRESCOPE (**FI**refighting **RES**ources of **C**alifornia **O**rganized for **P**otential **E**mergencies) was created in the wake of several massive wildfires in southern California during the 1960s and 1970s. These large and fast-moving fires spread into residential areas and burned hundreds of homes, creating tremendous challenges for California's fire protection agencies. Resources from dozens of agencies were involved in fighting the fires, which crossed numerous boundary lines and often burned within multiple jurisdictions at the same time. Within each jurisdiction, the local fire department was responsible for firefighting efforts, including requesting and managing mutual aid resources from dozens of other agencies. The established command and control systems proved to be ineffective at managing operations of this scale and complexity.

Several serious problems were encountered in these major fire situations. Resources were often deployed to combat a fire in one jurisdiction that could have been better used in another jurisdiction. Resources from different agencies had tremendous problems communicating and coordinating their actions. Deficiencies in radio system compatibility, standardized terminology, and equipment compatibility all led to resources being used in a disconnected manner. The organizations lacked the interoperability that was needed to provide a seamless operation.

The FIRESCOPE organization was created by the major fire agencies in southern California to develop a more effective system to deal with major incidents involving multiple agencies, such as wildland fires, earthquakes, or other large-scale emergencies (▶ **Figure 15-1**). FIRESCOPE set out to resolve the jurisdictional, interoperability, and standardization issues that are often encountered in this type of operation. The objective was to create a system that could coordinate efforts among multiple jurisdictions and be able to use all fire service resources as interchangeable units within an overall structure. The FIRESCOPE organization developed a standardized method of setting up an incident management structure, coordinating strategy and tactics, and disseminating information.

By 1980, this model, known as the Incident Command System, was implemented by most of the fire agencies in California. In 1982, it was adopted as a cornerstone of the National

Figure 15-1 FIRESCOPE was developed to coordinate efforts during large-scale wildland fires.

Interagency Incident Management System (NIIMS), and a year later, the National Fire Academy began teaching ICS as the model system for emergency management.

Fire Ground Command

As California was developing the FIRESCOPE system, the Fire Ground Command concept was created in Arizona. The FGC model was developed within the Phoenix Fire Department as a system to manage the everyday incidents that were faced by fire departments, from one-alarm fires to multiple-alarm incidents. Many of the principles of FGC were similar to those of ICS, although there were a few differences in terminology. The major difference was that FGC was designed for implementation in routine situations and could be expanded for larger-scale situations, whereas ICS was designed for large-scale operations and could be implemented on a reduced scale for smaller incidents.

Incident Management System

Several federal regulations and consensus standards, including NFPA 1500, *Standard on Fire Department Occupational Safety and Health Program*, were adopted during the 1980s. These standards mandated the use of an incident command or IMS at emergency incidents. NFPA 1561, *Standard on a Fire Department Incident Management System*, released in 1990, identified the key components of an effective system and the importance of using such a system at all emergency incidents. Fire departments could meet the requirements of NFPA 1561 by adopting either ICS or FGC.

In the following years, the users of different systems from across the country formed the *National Fire Service Incident*

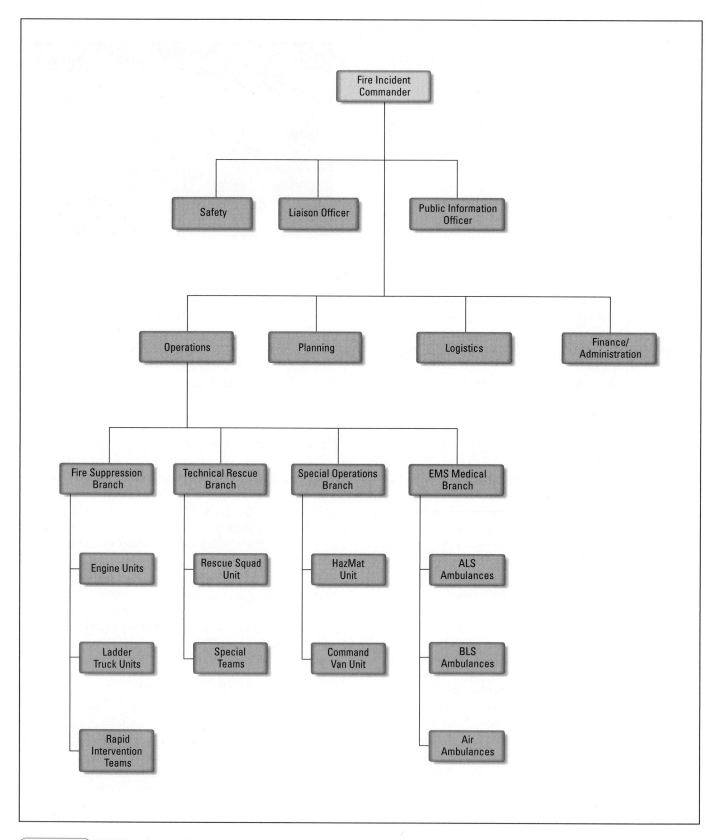

Figure 15-2 NIMS Organizational Chart for fire departments.

Management System Consortium to develop "model procedure guides" for implementing effective incident management systems at various types of incidents. The resulting system, which blended the best aspects of both ICS and FGC, is now known formally as the Incident Management System. The IMS model can be used at any type or size of emergency incident and by any type or size department or agency. NFPA 1561 is now called *Standard on an Emergency Services Incident Management System.*

National Incident Management System

The **National Incident Management System** (NIMS) is the latest development in the evolution of incident management systems. Since the September 11, 2001 terrorist attacks and the creation of the Department of Homeland Security, the federal government's interest and involvement in incident management have increased significantly. The NIMS is designed to ensure that all of the agencies involved in a federal response to mega-incidents will be able to fully integrate and coordinate their efforts. If an incident escalates to the point that it requires a federal response, all of the participating agencies will be expected to use the NIMS model (◀ **Figure 15-2**).

The federal government cannot require state or local agencies to use NIMS; however, it does require compliance in order to receive federal reimbursement for their expenses and losses. This is similar to the effect of federal transportation funding and the national speed limit. States do not have to comply with a national speed limit, but they do not receive their federal payments for interstate highway development if they do not comply.

The NIMS is, in reality, a higher-level application of the IMS model. Because most fire departments already use IMS, or some variation of ICS or FGC, they should have little difficulty adapting to comply with NIMS. The NIMS mandate requires significant changes for agencies that were not accustomed to using any incident management system. Because of this, it is vitally important for fire departments to review, adopt, and train on this system.

Federal Response Plan

The NIMS incident management system is directly related to the **Federal Response Plan (FRP)**. This plan was issued in 1992 to address the federal government's response to disasters such as hurricanes, floods, fires, explosions, or any similar emergency in which the President determines that federal assistance is needed to support local resources. Although the FRP system is designed for very large-scale incidents, it can be implemented incrementally. Most state disaster plans are based on the same model.

FEMA is responsible for developing and maintaining the FRP. The plan delegates the responsibilities for coordinating different aspects of emergency preparedness, planning, man-

agement, and disaster assistance functions to different federal agencies during a national disaster. (▼ **Figure 15-3**) shows the assignment of responsibilities to specific organizations. The FRP consists of the following sections:

- **Basic Plan**. Policies and concept of operations that guide how the federal government will assist state and local governments; summarizes federal planning assumptions, response, and recovery actions and responsibilities.
- **Emergency Support Function (ESF) annexes**. Each ESF annex defines the mission, policies, concepts of operations, and responsibilities of primary and support agencies for a specific set of functions. The ESF annexes that are most likely to be of interest to a fire officer include:
  - **ESF 4**: Firefighting—the U.S. Forest Service is the primary agency
  - **ESF 6**: Mass Care—the American Red Cross is the primary agency in coordinating mass care operations
  - **ESF 8**: Health and Medical Services—the Department of Health and Human Services is the primary agency
  - **ESF 9**: Urban Search and Rescue—FEMA is the primary agency
  - **ESF 10**: Hazardous Materials—the Environmental Protection Agency is the primary agency

The U.S. Forest Service is assigned the responsibility for firefighting because it is responsible for wildland fires on federal lands. The Forest Service role includes the overall coordination of the Incident Management System for federal disasters. The incident management resources that are used for large-scale forest fires can also be used to provide

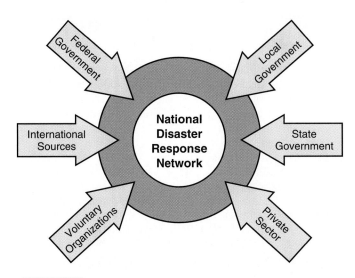

(**Figure 15-3**) Federal Response Plan.

incident management teams, equipment, and supplies to support state and local agencies for other types of incidents.

- **Recovery function annex.** Planning and concepts of operations that are directed toward helping disaster victims and affected local governments to return to normal and minimize future risk. Delivery systems include assistance to individuals, families, businesses, and local or state governments.

- **Support annexes.** Mission and policies that support overall federal disaster operations: including community relations, congressional affairs, donations management, financial management, logistics management, and occupational safety and health.

- **Incident annexes.** These plans document the mission and concept of operations in specific types of situations. The terrorism annex is the first in a series of new incident annexes.

Fire Fighter Accountability

While the evolution of incident management systems was occurring, the issue of fire fighter accountability on emergency scenes also gained prominence. Although the concept had been used in the United Kingdom for several years, it was required by the first edition of NFPA 1500 in 1987, which was adopted by many fire departments in the United States and Canada during the 1990s.

One of the factors that influenced many fire departments to address the accountability issue was a series of incidents and subsequent legal actions involving the failure to keep track of the location and function of all resources operating at an emergency incident. Seattle is a case in point. The Seattle Fire Department lost six fire fighters in a series of incidents between 1987 and 1995. The Washington State Department of Labor and Industry, in the investigation of these incidents, found the Seattle Fire Department negligent in SCBA training and tracking of fire crews at large-scale operations. The primary issue in these findings was not only that a fire fighter died in the line of duty, but that a significant amount of time elapsed before the incident commander was aware that a fire fighter was missing. The City of Seattle received a large fine, and, as part of the corrective action, the fire department purchased personal alert safety system devices and implemented a passport-style accountability system.

As Seattle was dealing with the fines and accompanying negligence lawsuits, the National Institute for Occupational Safety and Health (NIOSH) was also looking at the number of incidents in which fire fighters died as a result of becoming lost, disoriented, or running out of air.

In September 1994, NIOSH released a report *A Request for Assistance in Preventing Injuries and Deaths of Fire Fighters*. This alert bulletin made the following statement.

"A recent NIOSH investigation identified four factors essential to protecting fire fighters from injury and death: (1) following established firefighting policies and procedures, (2) implementing an adequate respirator maintenance program, (3) establishing fire fighter accountability at the fire scene, and (4) using personal alert safety system devices at the fire scene. Deficiencies in any of these factors can create a life-threatening situation for fire fighters."

An additional influence for improved incident management and accountability resulted from a request from the International Association of Fire Fighters (IAFF) for a clarification of the OSHA respiratory protection regulations. This regulation, 29 CFR 1910.134, from the Code of Federal Regulations, was originally written for an industrial environment, when employees were operating in confined spaces, toxic environments, or in oxygen-deficient atmospheres. These work areas are classified as immediately dangerous to life and health (IDLH).

The IAFF question resulted in a 1996 OSHA ruling that affirmed that fire fighters operating in a structure fire were operating in an IDLH atmosphere and that fire departments must comply with specific OSHA regulations while self-contained breathing apparatus is being used. The rules state that at least two fire fighters must enter the IDLH area together and remain in visual or voice contact with one another at all times. In addition, at least two properly equipped and trained fire fighters must:

- Be positioned outside the IDLH atmosphere
- Account for the interior teams
- Remain capable of rescue of the interior team or teams

This interpretation became known as the **two-in-two-out rule** and evolved into the Rapid Intervention Team concept that was incorporated into NFPA 1500.

Safety Zone

Two different federal agencies, which are both involved in fire fighter safety, are often mistaken for the other. NIOSH, which is a part of the Centers for Disease Control and Prevention in the Department of Health and Human Services, is the federal agency responsible for conducting research and making recommendations for the prevention of work-related injury and illness. One of the ongoing NIOSH projects is to investigate fire fighter fatalities in order to identify causal factors for research purposes; however, the agency has no regulatory powers.

OSHA, the Occupational Safety and Health Administration, is part of the Department of Labor. OSHA is a regulatory agency that is responsible for setting and enforcing standards and for providing training and education to improve workplace safety and health. In many cases, OSHA, or the equivalent state agency, has the responsibility to issue correction orders and take enforcement actions when regulations are not followed.

The Fire Officer's Role in Incident Management

Every fire officer should be prepared to function as an initial incident commander, as well as a company-level supervisor within the Incident Management System. As a company-level officer, the incident commander has basic responsibilities that include supervising the work of a group of fire fighters, while also reporting to a higher-level officer and working within a structured plan at the scene of an incident. In addition, a company-level officer must be prepared to assume overall command of an incident and implement the Incident Management System.

Typically, a company-level officer is the highest-ranking individual on the first-arriving vehicle at an emergency incident. That officer has the initial responsibility to establish command and manage the incident, at least until a higher-ranking officer arrives, in addition to supervising the members of his or her company. Managing an incident requires the fire officer to develop strategies and tactics, determine required resources, and decide how those resources will be used. The fire officer needs to set up the basic command and control structure, following the appropriate incident management procedures. The following section describes some of the basic elements of a command structure and organization.

The Incident Management System (IMS) is designed to be flexible and to be implemented incrementally. The command structure for an incident should only be as large as the incident requires. The objective in applying a standard command structure is not to assign someone to perform every role on the organization chart. The real objective is to use the model structure to assign all of the functions that *have* to be performed at that incident. The command structure has to fit the situation, not the organization chart.

One of the primary benefits of the Incident Management System is to maintain a manageable span of control. One person can provide effective supervision over only a limited number of subordinates, whether those subordinates are individual workers or supervisors directing groups of workers. For emergency operations, a recommended span of control is three to five individuals reporting to one supervisor. The span of control is maintained by adding levels of management as an officer's effective span of control is exceeded.

Levels of Command

There are three standard levels in the command organization. A set of responsibilities is assigned to each level, and all of the basic responsibilities must be covered; however, it is not necessary to assign different officers to all three levels at every incident. In many cases, particularly at small-scale incidents or in the early stages of a larger incident, one officer covers all three levels simultaneously. When the incident expands, the management responsibilities are subdivided, and all three levels are likely to be implemented. Depending on the size and complexity of the incident, a fire officer may have a role at any level.

The overall direction and goals for the incident are set at the **strategic level**. The incident commander always functions at the strategic level. A strategic goal, in a defensive situation, could be to stop the extension of a fire to any adjacent structure.

Tactical-level objectives define the actions that are necessary to achieve the strategic goals. A tactical-level supervisor would be assigned to manage a group of resources to accomplish the tactical objective. In medium- to large-scale incidents, the tactical level components would be called divisions, groups, or sectors, and each of these components could include several companies. A company-level officer would be in charge of each company, and all of the company-level officers would report to the tactical level supervisor.

Tactical assignments are usually defined by a geographical area (e.g., one part of a building) or a functional responsibility (e.g., ventilation), or sometimes by a combination of the two (e.g., ventilation in a particular part of the building).

In large incidents, an additional level of management may be added to maintain a reasonable span of control. Branches are established to group tactical components. An officer assigned to a branch would oversee some combination of divisions, groups, or sectors. It is unlikely that a company officer would be required to fill a branch-level position.

Task-level assignments are the actions required to achieve the tactical objectives. This is where the physical work is actually accomplished. Individual fire companies or teams of fire fighters perform task-level activities, such as searching for victims, operating hoselines, or opening ceilings.

Strategic-Level Incident Management

Responsibilities of Command

The **incident commander** is the individual who is responsible for the management of all incident operations. The incident commander is responsible for the completion of three strategic priorities:

1. Life safety
2. Incident stabilization
3. Property conservation

The incident commander is also responsible for:

- Building a command structure that matches the organizational needs of the incident.
- Translating the strategic priorities into tactical objectives.
- Assigning the resources that are required to perform the tactical assignments.

Functions of Command

The Incident Management System (IMS) requires the incident commander to perform a standard set of functions; these functions are neither rank nor personality driven. There are nine functions of command.

- Determining strategy
- Selecting incident tactics
- Setting the action plan
- Developing the IMS organization
- Managing resources
- Coordinating resource activities
- Providing for scene safety
- Releasing information about the incident
- Coordinating with outside agencies

Establishing Command

The first fire officer or fire department member to arrive at the scene is always required to assume command of the incident. In most cases, the officer in charge of the first-arriving company is the initial incident commander. The initial incident commander remains in charge of the incident until command is transferred or the situation is stabilized and terminated. If the incident expands, a higher-ranking officer usually arrives and assumes command, and the initial incident commander's role changes.

Command must be established and the IMS must be used at every event. Many incidents, such as vehicle fires and EMS responses, are handled by a single company under the supervision of a company-level officer. These incidents do not require a highly structured command presence; nevertheless, the basic functions of incident command must still be performed. For single-company incidents, establishing command may only require notification to the dispatcher that the company is on the scene and is handling the situation.

For incidents requiring two or more companies, the first fire department member or company-level officer on the scene must establish command and initiate an incident management structure that is appropriate for the incident.

Activating the command process includes providing an initial radio report and announcing that command has been established. The initial radio report should provide an accurate description of the situation for units that are still en route so that they can visualize what to expect. The report should also identify the actions that arriving units will take. In that report, the company officer should include:

- Identification of the company or unit arriving at the scene
- A brief description of the incident situation. That may include providing the building size, height and occupancy; number of vehicles in an accident, or other similar information.

Getting It Done

Examples of Initial Radio Reports

For a single-company incident: "Engine 2 is on the scene of a fully involved auto fire with no exposures. Engine 2 can handle."

For an EMS incident: "Quint 618 is on the scene of a two-vehicle crash with multiple patients and one trapped. Transmit a special alarm for a heavy rescue company, a second suppression company, and an EMS task force. Quint 618 will be Palisade Command."

For an offensive structure fire: "Engine 2 is on the scene of a two-story motel with a working fire in a second-floor room on side Charlie. Engine 2 has a hydrant and is going in with a hand line to start primary search. This will be an offensive fire attack. Engine 2 will be Cedar Avenue Command."

For a defensive structural fire: "Engine 422 is on the scene of a one-story strip shopping center with an end unit heavily involved. Engine 422 is laying a supply line and will be attacking the fire with a master stream to cover the exposure on side Delta. This is a defensive fire. Engine 422 will be Loisdale Command."

- Obvious conditions, such as a working fire, multiple patients, hazardous materials spill, or a dangerous situation, like a man with a gun
- Brief description of action to be taken. "Engine 2 is advancing an attack line into the first floor."
- Declaration of strategy to be used (offensive or defensive)
- Any obvious safety concerns
- Assumption, identification, and location of command
- Request for additional resources (or release of resources), if required

Command Options

The first-arriving company-level officer has three options when arriving at the incident and assuming command: investigation, fast attack, or command mode. The decision is based on the situation that is presented to the fire officer.

Investigation Mode

On arrival, there may be no visible indicators of a significant event or the appearance of a very minor situation. These situations generally require investigation by the first-arriving company, whereas other responding companies remain uncommitted. The first-arriving company-level officer can usually perform the role of initial incident commander as well as personally investigating the situation along with the company members.

Fast-Attack Mode

Some situations require immediate action by the first-arriving fire company. Every person on the first-arriving company might be needed in order to make a difference in the outcome of the incident. This might be the case when a single company arrives at a fire with occupants in imminent danger at upper-floor windows. In this situation, the company-level officer should work with the company members and perform the initial incident command responsibilities through a portable radio.

The fast-attack mode should not last more than a few minutes and ends when:

1. The situation is stabilized.
2. The situation is not stabilized and the company officer must withdraw to the exterior and establish a command post.
3. Command is transferred to another officer.

Command Mode

Some events are so large, complex, or dangerous that they require the immediate establishment of command by the first-arriving company-level officer. In these cases, the company-level officer's personal involvement in tactical operations is less important than the command responsibility. The company-level officer should establish a command position in a safe and effective location and initiate a **tactical worksheet** (► **Figure 15-4**). The role of the initial incident commander at

this type of situation is to direct incoming units to take effective action. While the initial incident commander remains outside, the rest of the company members should do as follows:

- Initiate fire suppression with one of the members assigned as the acting company officer. The acting company officer must be equipped with a portable radio, and the crew must be capable of performing safely without the initial incident commander.
- The initial incident commander could assign the remaining company members to work under another company officer.
- The crew members could stay with the initial incident commander to perform staff functions that assist command.

Transfer of Command

If the incident is large enough, a chief officer usually assumes command from the first-arriving company-level officer. At a large incident, command could be transferred more than once, depending on the situation and the chain of command, as successively higher-ranking or more qualified officers arrive at the incident scene.

Command should not be transferred automatically simply because a higher-ranking officer has arrived, and it should not be transferred more times than are necessary to effectively manage the incident. Command should be transferred only to improve the quality of the command organization.

Transfer of command is performed in a structured manner:

1. The officer assuming command communicates with the initial incident commander. This can occur over the radio; however, a face-to-face meeting is preferred.
2. The initial incident commander briefs the new incident commander, including:
   - Incident conditions—the location and extent of fire, the number of patients, status of the hazardous materials spill or leak, and so on
   - Tactical worksheet and incident action plan
   - Progress toward completion of the tactical objectives
   - Safety considerations
   - Deployment and assignment of operating companies and personnel
   - Need for additional resources
3. Command is officially transferred only when the new incident commander has been briefed. The new incident commander determines the assignment that is most appropriate for the relieved incident commander.

The initial incident commander should always review the tactical worksheet with the new incident commander. The worksheet outlines the status and location of personnel and resources in a standard form. This is especially important on those events where the first-arriving company initially started in the command mode.

NORTHERN VIRGINIA INCIDENT COMMAND SYSTEM
INITIAL INCIDENT COMMAND WORKSHEET

ADDRESS:

OCCUPANCY TYPE:

MODE OF ATTACK:

SIZE/ CONSTRUCTION:

☐ OFFENSIVE   ☐ TRANSITION   ☐ DEFENSIVE

STAGING LOCATION:

BASE: (highrise)

RIT TEAM

RADIO CHANNELS: TACTICAL/ _____   _____ / _____

COMMAND/ _____   _____ / _____

PAR CHECKS: ☐ 20 MIN   ☐ 40 MIN   ☐ 60 MIN   ☐ 80 MIN

RESOURCES		DIVISION/GROUP ASSIGNMENTS	
1ST ALARM	**ASSIGNMENT**		
ENG		FLR/DIV	
ENG			
ENG			
ENG		FLR/DIV	
TRK			
TRK			
RSC		FLR/DIV	
EMS			
ECPT			
2ND ALARM		FLR/DIV	
ENG			
ENG			
ENG		GROUP	
TRK			
EMS			
BFC		GROUP	
ECPT			
3RD ALARM			
ENG			
ENG			
ENG			
TRK			
EMS			
BFC			
SPECAL CALL			

SIDE CHARLIE

SIDE BAKER   SIDE DAVID

SIDE ADAM

TASKS	YES/NO	TASKS	YES/NO
PRIMARY SEARCH (10 minutes)		REHAB	
SECONDARY SEARCH		UTILITY CONTROL	
ALL CLEAR		FM/INVESTIGATOR	
FIRE CONTROL		GAS COMPANY	
EXPOSURES		ELECTRIC COMPANY	
EXTENSION		TELEPHONE COMPANY	
EXTINGUISHMENT		WATER COMPANY	
VENTILATION		BUILDING INSPECTOR	
SALVAGE		HEALTH INSPECTOR	
OVERHAUL		RED CROSS	
LIGHT/AIR UNIT		HAZ MAT REGULATORY	
MEDICAL AID STATION		POLICE	
MAPS/PREPLANS		VDOT	

Figure 15-4 Tactical Worksheet.

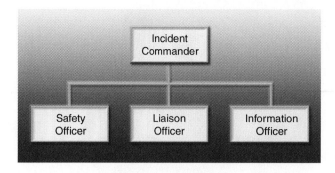

Figure 15-5 The command staff report directly to the incident commander.

After the transfer of command has occurred, the new incident commander determines the most appropriate assignment for the previous incident commander. A company-level officer may be assigned to supervise that company assigned as a division or **group supervisor** or may remain with the new incident commander at the command post. The size and the complexity of the incident determine how much the management structure will need to be expanded.

After the Transfer of Command: Building the Incident Management System

Command Staff

After command has been transferred, the company-level officer may be assigned to the **command staff**. Other fire officers who respond to an incident might also be assigned to perform command staff functions. Individuals on the command staff perform functions that are reported directly to the incident commander. The safety officer, liaison officer, and information officer are always part of the command staff; these duties cannot be delegated to other sections of the incident organization. In addition, aides, assistants, and advisors may be assigned to work directly for the incident commander (▶ **Figure 15-5**).

An aide is a fire fighter (sometimes a fire officer) who serves as a direct assistant to a command officer. In many fire departments, operational-level command officers have regularly assigned aides who drive the command vehicle and perform administrative support duties, as well as functioning as aides at incident scenes. In other departments, aides are assigned only as needed at incident scenes. The incident commander may assign more than one aide to perform support functions at a major incident. Aides can also be assigned to other officers in the command structure.

Safety Officer

The **safety officer** is responsible for ensuring that safety issues are managed effectively at the incident scene. The safety officer is the eyes and ears of the incident commander for identifying and evaluating hazardous conditions, watching out for unsafe practices, and ensuring that safety procedures are followed. Normally, the safety officer is appointed early during an incident. As the incident becomes more complex, additional qualified personnel can be assigned as assistant safety officers to subdivide the responsibilities.

The safety officer is an advisor to the incident commander but has the authority to stop or suspend operations when unsafe situations occur. This authority is clearly stated in national standards, including NFPA 1500, *Standard on Fire Department Occupational Safety and Health Program*; NFPA 1521, *Standard for Fire Department Safety Officer*; and NFPA 1561, *Standard on Emergency Services Incident Management System*. Several state and federal regulations require the assignment of a safety officer at hazardous materials incidents and certain technical rescue incidents.

The safety officer should be a qualified individual who is knowledgeable in fire behavior, building construction and collapse potential, firefighting strategy and tactics, hazardous materials, rescue practices, and departmental safety rules and regulations. A safety officer should also have considerable experience in incident response and specialized training in occupational safety and health. Many fire departments have full-time safety officers who perform administrative functions relating to health and safety when they are

VOICES OF EXPERIENCE

❝ There are three common firefighting misconceptions. ❞

1. **The best officers in the fire service are the "strong silent types", who do not issue too many orders.**

   The most important trait of a fire officer is the ability to listen and communicate while under great stress. As a supervisor you must communicate with your company.

2. **It is best not to overestimate the number of fire fighters required to fight the fire, because you may look foolish and transmit an unnecessary greater alarm.**

   Transmit the alarm or call the mutual aid companies. In this instance it is better to make an error of commission than omission. It is always best to have reserve forces at a fire or emergency.

3. **Fire fighters should be expected to risk their lives.**

   This used to be true, but it is no longer. The safety of fire fighters is considered as high a priority as the safety of citizens. Fire fighters experience risk responding to fires and emergencies and by entering burning buildings. Incident commanders do not expose fire fighters to danger. The priorities of firefighting and emergency work are life safety first, and this includes the lives of fire fighters. Incident stabilization is the second priority of emergency work, and property protection is the third priority.

Vincent Dunn
New York City, New York Fire Department
42 years of service

not responding to emergency incidents. The National Fire Academy's *Incident Safety Officer* and *Advanced Safety Operations and Management* courses are excellent resources for people interested in serving in this capacity.

Liaison Officer

The **liaison officer** is the incident commander's point of contact for representatives from outside agencies and is responsible for exchanging information with representatives from those agencies. During an active incident, the incident commander may not have time to meet directly with everyone who comes to the command post. The liaison officer position takes the incident commander's place, obtaining and providing information, or directing people to the proper location or authority. The liaison area should be adjacent to, but not inside, the command post.

The incident commander can also assign a liaison officer to directly represent the fire department with another agency. At a complex incident involving extensive interaction between the police and fire departments, the fire department could assign a liaison officer to work with the police, or the police commander could assign a liaison officer to the fire department command post.

Public Information Officer

The **public information officer** is responsible for gathering and releasing incident information to the news media and other appropriate agencies. At a major incident, the public wants to know what is being done. The public information officer serves as the contact person for media requests so that the incident commander can concentrate on managing the incident. A media briefing location should be established that is separate from the command post.

General Staff Functions

When an incident is too large or too complex for just one person to manage effectively, the incident commander may appoint officers to oversee major components of the operation. Four standard components are defined in the IMS model. Everything that occurs at an emergency incident can be divided among these four major functional components ▶ **Figure 15-6**):

1. Operations
2. Planning
3. Logistics
4. Finance/administration

The incident commander decides which (if any) of these four components need to be activated, when to activate them, and who should be placed in each position. Remember that the blocks on the IMS organization chart refer to function areas or job descriptions, not to positions that must always be staffed. The positions are assigned only when they are needed and, if a position is not assigned,

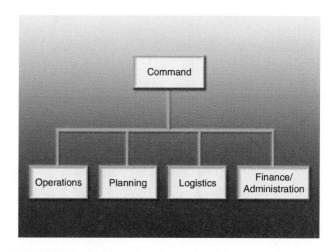

Figure 15-6) Major functional components of IMS.

the incident commander is responsible for managing that function.

The chiefs in charge of the four major sections are known as the **IMS general staff**. The four chiefs on the IMS general staff, when they are assigned, may conduct their operations from the main command post or from a different location. At a large incident, the four functional organizations may operate from different locations, but the general staff chiefs are always in direct contact with the incident commander.

Operations

The **operations section** is responsible for the management of all actions that are directly related to controlling the incident. The operations section fights the fire, rescues any trapped individuals, treats the patients, and does whatever else is necessary to deal with the emergency situation. This is the part of the organization that produces the most visible results.

For most structure fires, the incident commander directly supervises the functions of the operations section. A separate **operations section chief** is used at complex incidents so that the incident commander can focus on the overall situation while the operations section chief focuses on the strategy and tactics that are required to get the job done.

Planning

The **planning section** is responsible for the collection, evaluation, dissemination, and use of information relevant to the incident. The planning section works with status boards and preincident plans, as well as building construction drawings, maps, aerial photographs, diagrams, and reference materials.

The planning section is also responsible for developing and updating the incident action plan (IAP). The planning section is similar to the scheduling department at a company. It plans what needs to be done by whom and what resources are needed. To perform this role, the planning section must be in close and regular contact with all of the other sections.

Fire Marks

The bombing of the Murrah Federal building in Oklahoma City on April 19, 1995 added a chapter to incident management experience. The response to the collapse of the nine-story office building included a host of local, regional, and state agencies and 11 of the 28 federally funded Urban Search and Rescue task forces. The after-action report identified the value of establishing a written IAP to outline the strategic goals, tactical objectives, and support requirements for the incident for a 12-hour period. Although written IAPs had been used for many years in wildland fires, the Oklahoma City bombing emphasized the need to formalize their use in other types of emergency incidents.

The incident commander activates the planning section when information needs to be obtained, managed, and analyzed. The **planning section chief** reports directly to the incident commander. Individuals assigned to planning functions examine the current situation, review available information, predict the probable course of events, and prepare recommendations for strategies and tactics. The planning section also keeps track of resources at large-scale incidents and provides the incident commander with regular situation and resource status reports.

The planning functions may be delegated to subunits. These include: resources unit, situation unit, documentation unit, demobilization unit, and technical specialists.

Incident Action Plan

The **incident action plan (IAP)** is a basic component of IMS; all incidents require an action plan. The IAP outlines the strategic objectives and states how emergency operations will be conducted. At most incidents, the IAP is relatively simple and can be expressed by the incident commander in a few words or phrases. A written IAP is required for large or complex incidents that have an extended duration. The IAP for a large-scale incident can be a lengthy document that is regularly updated and used for daily briefings of the command staff.

Logistics

The **logistics section** is responsible for providing supplies, services, facilities, and materials during the incident. The **logistics section chief** reports directly to the incident commander. Among the responsibilities of this section would be keeping apparatus fueled, providing food and refreshments for the fire fighters, obtaining the foam concentrate needed to fight a flammable liquids fire, and arranging for a bulldozer to remove a large pile of debris. The logistics section is similar to the purchasing and support services departments of a large company. It ensures that adequate resources are always available and functional.

In many fire departments, these logistical functions are routinely performed by personnel assigned to a support services division. These groups work in the background to ensure that the members of the operations section have whatever they need to get the job done. Resource-intensive or long-duration situations may require assignment of a logistics section chief because service and support requirements are so complex or extensive that they need their own management component.

The logistics section may use subunits to provide the necessary support for large incidents. These units may include: a supply unit, facilities unit, ground support unit, communications unit, food unit, and medical unit.

Finance/Administration

The **finance/administration section** is the fourth major IMS component under the incident commander. This section is responsible for the administrative, accounting, and financial aspects of an incident, as well as any legal issues that may arise. This function is not activated at most incidents, because cost and accounting issues are usually addressed before or after the incident. A finance/administration section may be needed at large-scale and long-term incidents that require immediate fiscal management, particularly when outside resources must be procured quickly.

A finance/administration section is usually established during a natural disaster, when state or federal expense reimbursements are expected, or during a hazardous materials incident in which reimbursement may come from the shipper, carrier, chemical manufacturer, or insurance company. The finance section is equivalent to the finance department in a company; it accounts for all activities of the company, pays the bills, ensures that there is enough money to keep the company running, and keeps track of the costs.

The finance/administration section may incorporate subunits to efficiently handle its duties. These include a time

Getting It Done

Using an IAP in a Nonfire Incident

Fairfax County, Virginia, used the FEMA IAP process to coordinate operations during a 1996 blizzard. Operating out of the emergency operations center, the Operations Chief faxed out a copy of the incident action plan for every 12-hour period to every work location. In addition to clearly stating the work objectives for that period, the chief could also describe the resources committed to snow removal, dialysis patient transports, and emergency response. Every company officer had a copy of the plan. This procedure provided consistent and clear communications from the fire chief to the newest fire fighter.

unit, procurement unit, compensation and claims unit, and cost unit.

Tactical Level of Incident Management
Divisions, Groups, and Sectors

Divisions, groups, and sectors are tactical-level management elements that are used to assemble companies and resources for a common purpose. The flexibility of the IMS enables organizational units to be created as needed, depending on the size and the scope of the incident. In the early stages of an incident, individual companies are often assigned to work in different areas or perform different tasks. As the incident grows and more companies are assigned to areas or functions, the incident commander can establish tactical-level units to place multiple resources under one supervisor. This keeps the incident commander's span of control within desired limits and provides direct coordination of common efforts.

Divisions represent geographical operations, such as one floor or one side of a building. **Groups** represent functional operations, such as the ventilation group. The term **sector** is a generic term that can be applied to either a geographical or a functional component. An officer is assigned to supervise each division, group, or sector as it is created.

These organizational units are particularly useful when several resources are working near each other, such as on the same floor inside a building, or are doing similar tasks, such as providing ventilation. The assigned supervisor can directly observe and coordinate the actions of several crews.

Divisions

A division is composed of the resources responsible for operations within a defined geographical area. This area could be a floor inside a building, the rear of the fire building, or a geographical area at a brush fire. Divisions are most often used during routine fire department emergency operations. By assigning all of the units in one area to one supervisor, the incident commander has to communicate with only that one individual. The assigned supervisor coordinates the activities of all resources that are working within that area, in accordance with the incident commander's strategic plan.

The division structure provides effective coordination of the tactics being used by different companies working in the same area. For example, the **division supervisor** would coordinate the actions of a crew that is advancing a hoseline into the fire area, a crew that is conducting search-and-rescue operations in the same area, and a crew that is performing horizontal ventilation.

Groups

An alternative way of organizing resources is by function rather than by location. A group is composed of resources assigned to a specific function, such as ventilation, search

and rescue, or water supply. The officer assigned to supervise the ventilation group uses the radio designation "ventilation group" and is required to coordinate activities with other division, group, or sector supervisors.

Groups are responsible for performing an assignment, wherever it may be required, and often work within more than one division. Sometimes, groups are established with both functional and geographical designations, such as a west wing search-and-rescue group and an east wing search-and-rescue group.

Division/Group/Sector Supervisor Responsibilities

In many cases, a company-level officer is assigned by the incident commander to function as a division/group/sector supervisor. When an incident is rapidly escalating, the company-level officers on the first alarm are often assigned to geographical or functional areas and designated as the division/group/sector supervisor. For example, the company officer in charge of the first-arriving ladder company could be assigned as the "ventilation group" supervisor. Additional companies assigned to ventilation would become part of this group. When additional command officers arrive, the ventilation group supervisor might be relieved by a chief officer and revert to functioning as a company-level officer.

As a division/group/sector supervisor, the company-level officer directly supervises and monitors the operations of assigned resources and provides updates to command. Duties of a division/group/sector supervisor include:

- Using an appropriate radio designation, such as roof division, division A, or rescue group
- Completing the objectives assigned by command
- Accounting for all assigned companies and personnel
- Ensuring that operations are conducted safely
- Monitoring work progress
- Redirecting activities as necessary
- Coordinating actions with related activities and adjacent division/group/sector supervisors
- Monitoring welfare of assigned personnel
- Requesting additional resources as needed
- Providing the incident commander with essential and frequent progress reports
- Reallocating or releasing assigned resources

Division, group, and sector supervisors all function at the same level within IMS, regardless of the rank of the individual who is assigned. Divisions do not report to groups, and groups do not report to divisions; the supervisors are required to coordinate their actions and activities with each other. A division supervisor must be aware of everything that is happening within the assigned geographical area, so a group supervisor must coordinate with the division supervisor when the group enters the division's geographical area, particularly if the group's assignment will affect the division's personnel, operations, or safety. Effective communication among divisions and groups is critical during emergency operations.

Branches

A <u>branch</u> is a supervisory level established in either the operations or logistics function to provide a span of control. At a major incident, several different activities may occur in separate geographical areas or several distinct, but related functions may be performed. The span of control might still be a problem, even after the incident commander has established divisions/groups/sectors. In these situations, the incident commander can establish branches to place a <u>branch director</u> in charge of a number of divisions/groups/sectors. For example, after a tornado, the incident commander could establish two branches, if the destruction occurred in two separate areas across town. If there are numerous injuries, the incident commander could establish a medical branch to coordinate all aspects of caring for the victims, including triage, treatment, and transportation.

Location Designators

IMS uses a standard system to identify the different parts of a building or a fire scene. Every fire fighter must be familiar with this terminology.

The exterior sides of a building are generally known as sides A (Alpha), B (Baker), C (Charlie), and D (Delta). The front of the building is side A, with sides B, C, and D following in a clockwise direction around the building. The companies working in front of the building are assigned to division A, and the radio designation for their supervisor is "Division A." Similar terminology is used for the sides and rear of the building.

The areas adjacent to a burning building are called exposures. Exposures take the same letter as the adjacent side of the building. A fire fighter facing side A can see the adjacent building on the left (Exposure B) and the building to the right (Exposure D). If the burning building is in a row of buildings, the buildings to the left are called Exposures B, B1, B2 and so on. The buildings to the right are Exposures D, D1, D2 and so on.

Within a building, divisions commonly take the number of the floor on which they are working. For example, fire fighters working on the fifth floor would be in division 5, and the radio designation for the chief assigned to that area would be "Division 5." Crews performing different tasks on the fifth floor would all be part of this division. Common simple terminology should be used for designations whenever possible, such as "roof division," and "basement division."

Fire Officer Greater Alarm Responsibilities

The incident commander calls for additional resources or greater alarms when more resources are needed at an incident. In many cases, the incident commander calls for a greater alarm to have stand-by resources immediately available, in case they are needed or in anticipation of a future need. There are many duties that a company-level officer may be called on to perform when responding as an additional resource. These include:

- Reinforcing a fire attack strategy by adding resources.
- Relieving an exhausted crew from the first alarm and performing the fireground task that they were performing.
- Performing support activities, such as salvage, overhaul, moving equipment to a forward staging area, going door-to-door to evacuate a neighborhood, assisting occupants in recovering personal items, or providing runners to support a face-to-face communications system.
- Maintaining a ready reserve in a staging area.
- Performing additional related duties. This catchall phrase means that the officer and/or company may be performing tasks that are not traditional fire company tasks, but are time or mission critical. For instance, they could be assigned to an evacuation center to provide assurance and assessment of the occupants who fled from an apartment fire.

Sometimes, the additional resources are dispatched as strike teams or task forces, as opposed to individual companies. This means that the companies are directed to combine forces to respond and operate as a group. They could be assigned to meet at a designated location and travel to the scene together or to assemble at the staging area.

Staging

<u>Staging</u> refers to a standard procedure to manage uncommitted resources at the scene of an incident. Instead of driving up to the incident scene and committing apparatus and crews without direction from the incident commander, staged resources stand by uncommitted and wait for instructions. These resources are immediately available for use

---

Assessment Center Tips

Rapidly Organize Your Greater Alarm

You will probably need to call for a special, additional, or greater alarm when handling a simulated emergency incident during a promotional examination. Practice identifying the staging location as you call for additional help. For example: "5th Street Command to Communications, transmit a second alarm. Staging for the second alarm units will be at the shopping center parking lot at 5th Street and Avenue D. Designate the officer on the first-arriving engine to assume the Staging Officer assignment."

Make sure that the terms you use and the assignments you make comply with the IMS used in your jurisdiction. In an assessment center, you are evaluated on how well you implement the local system.

where and when the incident commander needs them. This allows the incident commander to determine the most appropriate assignment for each unit.

Many fire departments have policies that automatically commit only one or two first-arriving companies and direct all others to level I staging. In level I staging, only the pre-designated units respond directly to the scene, and later-arriving units remain uncommitted and wait for instructions. For example, the second-arriving aerial ladder would stop about a block short of the fire building, where it would have the option of going to whichever side it is needed. Standard operating procedures guide when units stage, where they stage, and how they communicate with the incident commander to indicate they are available for an assignment.

Level II staging, which is generally used for greater alarm incidents, directs responding companies to a designated stand-by location away from the immediate incident scene. A designated officer supervises the level II staging area with the "staging" radio identification and assigns units from the staging area as they are requested by the incident commander. The incident commander might direct the staging officer to maintain a minimum level of resources in the staging area.

Task Level of Incident Management

Individual companies, sometimes referred to as single resources, operate at the task level. A company is considered as a single resource, because it is the basic work unit assigned to perform most emergency scene tasks. Certain types of tasks are routinely assigned to engine companies, ladder companies, rescue companies, and other types of companies. The normal role of a company-level officer is to supervise a group of fire fighters who are operating as a company at the task level.

At a simple incident, the company-level officers report directly to the incident commander. When divisions/groups/sectors are established, company-level officers report to their assigned division/group/sector supervisor and continue to supervise the operations of their respective companies in performing assigned tasks. Sometimes, the assigned supervisor is one of the company-level officers. Company-level officers advise the division/group/sector supervisor of work progress, preferably face-to-face. All company-level officer requests for additional resources or assistance must go through the division/group/sector supervisor.

Task Forces and Strike Teams

Task forces and strike teams are groups of single resources that have been assigned to work together for a specific purpose or for a period of time. Grouping resources reduces the span of control by placing several units under a single supervisor.

A **task force** includes two to five single resources that are assembled to accomplish a specific task. For example, a task force could be composed of two engines and one truck company, or two engines and two brush units, or one rescue company and four ambulances. A task force operates under the supervision of a **task force leader**. In many cases, the designated task force leader is one of the company-level officers. All communications for the separate units within the task force are directed to the task force leader.

Task forces are often part of a fire department's standard dispatch philosophy. In the Los Angeles City Fire Department, a task force consists of two engines and one ladder truck, staffed by a total of 10 fire fighters. Some departments create task forces of one engine and one brush unit for response during wildland fire season. The brush unit responds with the engine company wherever it goes.

A **strike team** is five units of the same type with an assigned leader. A strike team could be five engines (engine strike team), five trucks (truck strike team), or five ambulances (EMS strike team) (▼ **Figure 15-7**). A strike team operates under the supervision of a **strike team leader**. In most cases, the strike team leader is a command officer.

Strike teams are commonly used for wildland fires, where dozens or hundreds of companies may respond. During wildland fire seasons, many departments establish strike teams of five engine companies that are dispatched and work together on major wildland fires. The assigned companies are assigned to rendezvous at a designated location and then respond to the scene together. Each engine has an officer and a crew of fire fighters, and one officer, typically a battalion chief, is designated as the strike team leader. All communications for the strike team are directed to the strike team leader.

EMS strike teams, consisting of five ambulances and a supervisor, are often organized to respond to multicasualty incidents or disasters. Rather than requesting 15 ambulances and establishing an organizational structure to supervise 15 single resources, the incident commander can request three EMS strike teams and coordinate their operations through the three strike team leaders.

Figure 15-7 A strike team is five units of the same type under one leader.

Greater Alarm Infrastructure

Small fire departments also tend to have limited infrastructure support. When they are working at a major incident, there could be a considerable delay before a rehabilitation sector is established with rest and recovery supplies. Fire officers should prepare for this by making sure that their apparatus carries enough water and food to support the fire company for a reasonable period of time. In wildland areas, many companies plan for a 2-day excursion, carrying 5 gallons of water and six energy bars per fire fighter. In the summer, 10 gallons of drinking water per fire fighter per day could be required.

Postincident Review

One of the important responsibilities of a fire officer is to complete a postincident review of significant operations. Some form of review should be conducted at the company level after every call in which the company performs emergency operations. In most cases, this can be an informal discussion conducted by the company officer to review the incident, discuss the situation, and evaluate team performance. Every situation should be viewed as a potential learning experience, and the company officer should provide feedback to the crew members to reinforce positive performance and identify areas where there is room for improvement. When a deficiency is noted, a plan for addressing the problem should be developed at the same time.

Situations that involve multiple company operations should be reviewed from the perspective of how well the overall team performed, in addition to individual fire officers conducting company-level reviews. The format of the multi-company review necessarily depends on the nature and the magnitude of the incident, whether it was a one-room house fire involving three or four companies or a five-alarm blaze that involved dozens of companies and activated every component of the fire department. It is relatively easy to gather the companies that were involved in a one-alarm fire, soon after the event, to conduct a basic review of the incident. The critique of a large-scale incident can be a major event involving extensive planning, scheduling, and preparations.

Some organizations conduct critiques only for multiple-alarm or unusually large-scale incidents or for situations where something obviously went wrong. This creates the impression that there is nothing to be learned from the more ordinary-sized incidents, unless someone made a mistake. Whether it is conducted on a large or small scale, the primary purpose of a review process should always be to serve as an educational and training tool, not to place blame for improper or deficient actions. It is imperative for the officer to approach every postincident review in a nonthreatening manner and to provide an open format for discussion as an aid to future training.

Company Tips

Reviews should always focus on lessons and positive experiences; however, there will always be situations where errors were made and cannot be ignored. If a company performed in a seriously deficient manner, the serious discussion between the fire officer and his or her immediate supervisor should occur in private, not in front of a room full of observers. At the company level, poor performance by one particular crew member should be discussed one-on-one in the office, not at the kitchen table. Areas where the whole crew needs to make improvements should be discussed with the entire crew.

Whether it is a single-company review or a formal critique of a major incident, the review process should be conducted routinely and in a systematic manner. The same basic factors should be examined and discussed. Many fire departments have a standard format for conducting a critique and written procedures to guide the officer who is responsible for planning and organizing it.

Preparing Information for an Incident Review

In most cases, the review of a multiple-company incident is conducted by the officer who was the incident commander; however, the preparatory work is often delegated to one of the lower-ranking officers. A battalion chief might assign the officer in charge of the first-due company to prepare the information and make the necessary arrangements for the review of a one-alarm incident. The battalion chief could be responsible for preparing the information for a critique of a multiple-alarm situation that will be presented by a higher-ranking chief officer; however, the work of gathering and assembling the information might still be delegated to a company officer or staff officer.

Before the actual review occurs, a significant amount of information should be gathered and assembled for presentation at the actual critique. This includes as much information that can be obtained about the situation leading up to the incident and whatever occurred before the arrival of the first fire department unit. This information sets the stage for a discussion of the actual event from an operational perspective and often reveals important factors that were unknown before the incident occurred. In many cases, the presentation of this information causes individuals who were involved in the incident to understand what happened and what they observed at the time but did not have the opportunity to interpret.

In the case of a fire involving a building, this process should begin with a review of the building, including its size and arrangement, construction type, date of construction, known modifications or renovations, and any known history that could be significant. All built-in fire protection features, including firewalls, automatic sprinklers, standpipes, and alarm and detection systems should be identified, providing as much detail as possible. Any structural factors that could affect fire spread, such as unsealed openings around pipes, open shafts, and concealed spaces, should also be identified.

Details related to the occupancy should also be obtained. Was the building occupied, unoccupied, vacant, abandoned, or undergoing renovation? If it was a business, what did they do, make, or store on the premises? What was the fire load? Did the fire grow and spread more quickly because of the nature of the contents?

Do we know where, when, and how the fire started? Do we know what was burning initially and how the fire spread? Did the fire involve the contents or the structure or both? Do we know who discovered the fire, what they observed, and when the fire department was called?

If the desired information about the building and occupancy is not readily available, the officer who is gathering this information can refer to several sources of information, such as the inspection file at the fire prevention bureau or the city department that issues building permits and keeps copies of the plans on file. Valuable information can often be obtained from preincident plans, and it is an excellent quality control technique to compare the prefire plan information with the situation that was actually encountered. This is also a good opportunity to determine whether the information on file was useful to the incident commander.

The fire investigator can usually provide information about the fire's cause and origin, and checking with the communications center provides information about the source of the alarm and a complete listing of the resources that responded. This should include a list of apparatus as well as the staffing level on each vehicle, if this is a variable. For volunteer departments, the number of personnel that were on the scene should be noted. A recording of the incident radio communications should be obtained, as should photographs or videos of the incident.

If water supply was a significant factor in the incident, additional information about the water system should be obtained and evaluated. The locations of hydrants and available flow data should be obtained. The main sizes and whether the area is within a distribution grid or at the end of a dead-end main could be important.

All of this information should set the stage for the analysis of the situation that occurred from the time the first unit arrived at the scene. This analysis of the situation before arrival often reveals factors or circumstances that were unknown to the companies that were involved in the incident. The officer who is preparing for the incident review should be prepared to present all of this information in a manner that is factual and informative.

Conducting a Critique

The actual critique should be scheduled as soon as convenient after the event, allowing time to gather and prepare the necessary information. It is best to capture the information while it is still fresh in the minds of the participants. If it was a small-scale incident, it may be possible to have all of the involved companies in attendance so that all of the crew members can be involved in the discussion and learning experience. The attendance at a critique of a larger-scale incident could be limited to company officers and command officers; however, it is a good idea to invite the entire crew of any company that played a significant role or was involved in some unusual occurrence.

The critique should be conducted by the officer who was in command of the incident or by a designated officer who is prepared to discuss the situation that occurred. The officer acting as moderator should begin by establishing ground rules for the review, reinforcing the point that that a critique is not designed to find fault nor to criticize the actions of others. It should instead provide an honest assessment of the operation, noting strengths and weaknesses that can be used to improve future performance.

The critique should start with an overview presentation of the background material and basic information about the incident, including the timeline and the units that were dispatched. The first-arriving officer should then be asked to describe the situation as it was presented on arrival and the actions that were taken. This description should include the first officer's role as the initial incident commander as well as the tactical operations performed by that company.

Each successive company should then take a turn explaining what they saw and what they did. An effective visual method is to draw a plot layout of the area on a dry erase board and to have each company-level officer locate where they positioned their apparatus, where they operated, and what actions they took. This should gradually produce a complete diagram of the incident. Photograph, videos, and radio recordings can be inserted at pertinent points in the discussion. If a large number of companies were involved in the incident, the presentation could be made by division/group/sector officers instead of individual company officers.

During this process, the moderator should keep the analysis directed at key factors, including the initial strategy and any changes in strategy that occurred, how the command structure was developed, how resources were allocated, and what special or unusual problems were encountered. The discussion should also focus on standard operating procedures, including whether or not they were followed and how well they worked in relation to the situation that occurred. The officer acting as moderator should help the participants differentiate between problems that occurred because procedures were not followed and areas where procedures should be changed or updated (▶ **Table 15-1**). The best way to evaluate the effectiveness of procedures is to determine whether following them actually produced the anticipated results. If there was a deviation from standard procedures, the reasons and the impact should be discussed.

Once every company-level officer has had a chance to make a presentation, the officer directing the critique should provide

---

Table 15-1 Incident Review Questions

- Did the preincident plan provide accurate and useful information?
- Are there factors that could have or should have been addressed by fire prevention before the incident?
- Were the appropriate units dispatched based on procedures and the information that was received?
- Were the units dispatched in a timely manner?
- Was the appropriate information obtained and transmitted to the responding units?
- What was the situation on arrival?
- What was the initial strategy as determined by the initial incident commander?
- How did the strategy change during the incident?
- How was the incident command structure developed?
- Were there adequate resources for the situation?
- How were the resources allocated and assigned?
- Were standard operating procedures followed?
- Do any standard operating procedures need to be changed?
- What unusual circumstances were encountered, and how were they addressed?
- Is additional training needed?
- Did all support systems function effectively?

---

his or her perspective and assessment of the operation, including both the positive and the negative factors. The critique officer should make a point of evaluating the outcome in terms of where and when the fire was stopped and the extent of damage that occurred. If the outcome was positive and everything met expectations, praise should be widely distributed. Where there is room for improvement, it should be noted in terms of valuable lessons learned.

It is a good technique to have the assembled group participate in identifying the positive and negative points and listing the actions that should be taken to improve future performance. These points should be listed on the board for everyone to see as the final product of the critique.

Documentation and Follow-up

The last step in conducting an incident review is to write a summary of the incident for departmental records. The results of this review should be documented in a standard format; many fire departments use a form to document the essential findings. Use of a form allows the information to be stored in a consistent manner and ensures that all relevant information is collected. The form should also list recommendations for changes in procedures, if necessary.

The completed package should then be forwarded to the appropriate section or sections within the department for review and follow-up. The training division should receive a copy of every incident review to determine training deficiencies and needs. Recommendations for policy changes should also be forwarded to the appropriate personnel for consideration.

The completed report should include all of the basic incident report information included from the National Fire Incident Reporting System as well as the operational analysis information.

You Are the Fire Officer: Conclusion

As the first-arriving fire officer, it is your responsibility to immediately assume the role of the incident commander and to begin developing a strategic plan, identifying tactical objectives, and assigning resources. The mode of command also needs to be determined. Because it appears that this will be a defensive operation, the most appropriate choice would be a command mode. You can assign your company to a defensive position and prepare to direct the actions of the subsequent-arriving companies.

This incident most likely does not need any of the general staff positions to be filled. You should assign the command staff position of safety officer and establish a division to cover each of the exposure sides of the involved barn. One or two single resources will probably be assigned to each division.

Wrap-Up

Ready for Review

- The incident management system has evolved from the southern California FIRESCOPE and the Phoenix Fire ground Command System.
- The Federal Response Plan outlines the federal government response to natural catastrophe, fire, flood, explosion, or any instance in which the President determines that federal assistance is needed on a local operation.
- Negligence lawsuits in Seattle and an OSHA ruling on respiratory protection drove the emphasis on fire fighter safety and accountability at an emergency incident.
- The OSHA ruling on the OSHA *Respiratory Protection* regulations applied an industrial standard to structural firefighting for operations in confined space or in toxic or oxygen-deficient atmospheres that became the two-in-two-out rule.
- The first-arriving fire officer is responsible for taking command of an incident.
- Establishing command includes providing an initial radio report.
- Strategic-level command covers the overall direction and goals of the incident.
- Tactical-level command objectives achieve the strategic-level command goals through the resources assigned to the division or group supervisor.
- Task-level command is where the tactical level objectives are achieved by the work performed by individual companies or individuals. This is the work that is actually performed at an incident.
- The first-arriving fire officer has three options when arriving at the incident: investigation, fast attack, or command.
- If the first-arriving fire officer selects the command mode, the officer needs to remain outside and initiate a tactical worksheet.
- Transfer of command is generally a face-to-face meeting in which the tactical worksheet is reviewed.
- Divisions represent geographical operations, whereas groups represent functional operations. Sector is a generic term that can be applied to either a geographical or a functional component.
- One of the most important responsibilities of a fire officer is to complete a postincident review of significant operations.
- Situations that involve multiple-company operations should be reviewed from the perspective of how well the overall team performed and how individual companies performed.

Hot Terms

Branch A supervisory level established in either the operations or logistics function to provide a span of control.

Branch director A supervisory position in charge of a number of divisions and/or groups. This position reports to a section chief or the incident commander.

Command staff Positions that are established to assume responsibility for key activities in the Incident Management System that are not a part of the line organization that include safety officer, public information officer, and liaison officer.

Divisions A supervisory level established to divide an incident into geographical areas of operations.

Division supervisor A supervisory position in charge of a geographical operation at the tactical level.

Federal Response Plan (FRP) The Federal Response Plan (for Public Law 93-288, as amended) describes the basic mechanisms and structures by which the federal government will mobilize resources and conduct activities to augment state and local disaster and emergency response efforts.

Finance/administration section A section of the Incident Management System that is responsible for the accounting and financial aspects of an incident, as well as any legal issues that may arise.

Group A supervisory level established to divide the incident into functional areas of operation.

Group supervisor A supervisory position in charge of a functional operation at the tactical level. This position reports to a branch director, the operations section chief, or the incident commander.

IMS general staff The group of incident managers composed of the operations section chief, planning section chief, logistics section chief, and finance/administration section chief.

Incident action plan (IAP) The objectives reflecting the overall incident strategy, tactics, risk management, and member safety that are developed by the incident commander. Incident action plans are updated throughout the incident.

Incident commander The person who is responsible for all decisions relating to the management of the incident and is in charge of the incident site.

Incident Management System (IMS) A system that defines the roles and responsibilities to be assumed by personnel and the operating procedures to be used in the management and direction of emergency operations; the system is also referred to as an incident command system (ICS).

Liaison officer The incident commander's representative or a point of contact for representatives from outside agencies.

Logistics Section Responsible for providing facilities, services, and materials for the incident. Includes the communications, medical, and food units within the service branch, as well as the supply, facilities, and ground support units within the support branch. The logistics section chief is part of the general staff.

Logistics section chief A supervisory position that is responsible for providing supplies, services, facilities, and materials during the incident. The person in this position reports directly to the incident commander.

National Incident Management System (NIMS) A system that provides a consistent nationwide approach for federal, state, local, and tribal governments to work effectively and efficiently together to prepare for, prevent, respond to, and recover from domestic incidents, regardless of cause, size, or complexity.

Operations section Responsible for all tactical operations at the incident. In the national model, the operations section can be as large as five branches, 25 divisions/groups or sectors, or 125 single resources, task forces, or strike teams.

Operations section chief A supervisory position that is responsible for the management of all actions that are directly related to controlling the incident. This position reports directly to the incident commander.

Planning Section Responsible for the collection, evaluation, dissemination, and use of information about the development of the incident and the status of resources. Includes the situation status, resource status, and documentation units as well as technical specialists. The planning section chief is part of the general staff.

Planning section chief A supervisory position that is responsible for the collection, evaluation, dissemination, and use of information relevant to the incident. This position reports directly to the incident commander.

Public information officer A command staff position that is responsible for gathering and releasing incident information to the news media and other appropriate agencies. This position reports directly to the incident commander.

Safety officer The person who is responsible for all decisions relating to the management of the incident and is in charge of the incident site.

Sector Either a geographical or a functional assignment.

Staging A specific function in which resources are assembled in an area at or near the incident scene to await instructions or assignments.

Strategic level Command level that entails the overall direction and goals of the incident.

Strike team Specified combinations of the same kind and type of resources, with common communications and a leader.

Strike team leader A supervisory position that is in charge of a group of similar resources.

Tactical-level Command level in which objectives must be achieved to meet the strategic goals. The tactical-level supervisor or officer is responsible for completing assigned objectives.

Tactical worksheet A form that allows the incident commander to ensure all tactical issues are addressed and to diagram an incident with the location of resources on the diagram.

Task force Any combination of single resources assembled for a particular tactical need, with common communications and a leader.

Task force leader A supervisory position that is in charge of a group of dissimilar resources.

Task-level Command level in which specific tasks are assigned to companies; these tasks are geared toward meeting tactical-level requirements.

Two-in-two-out rule OSHA Respiratory Regulation (29 CFR 1910.134) that requires a two-person team to operate within an environment that is immediately dangerous to life and health (IDLH) and a minimum of a two-person team to be available outside the IDLH atmosphere to remain capable of rapid rescue of the interior team.

Lieutenant Sobol and his engine company are dispatched to a working residential structure fire with multiple reports of fire from the second floor of a structure. Additional reports advise that there is no one at the residence, but the fire is threatening two adjacent structures. Based on the additional information, Lieutenant Sobol upgrades the fire to a second alarm to bring three additional engine companies. On arrival, Lieutenant Sobol finds a two-story 2500-square foot wood-framed residential structure with 50% involvement on the second floor. He advises his crews to pull lines, establish a water supply, and prepare for an offensive attack. On completion of a full-scene size-up, Lieutenant Sobol confirms that the fire is threatening two structures on sides B and D that are within 15 feet of the fire structure. He reassigns two of his first-arriving crews to set up exposure hoselines on sides B and D to protect the adjacent structures. Lieutenant Sobol calls dispatch and establishes incident command.

1. What is one of the four tactical responsibilities of the incident commander?

- **A.** Conduct a scene size-up
- **B.** Develop an IAP
- **C.** Stabilize the incident and provide life safety
- **D.** Set up a division

2. What is one of the nine functions of command?

- **A.** Provide tactical objectives
- **B.** Conduct a scene size-up
- **C.** Set up a PIO
- **D.** Establish rehabilitation

3. What should be included in an initial radio report?

- **A.** Time of day
- **B.** Any obvious safety concerns
- **C.** How many personnel you have
- **D.** An update of the fire

4. What are the three methods of attack?

- **A.** Offensive, defensive, switchable
- **B.** Interior, exterior, transitional
- **C.** Offensive, defensive, transitional
- **D.** Interior, defensive, transitional

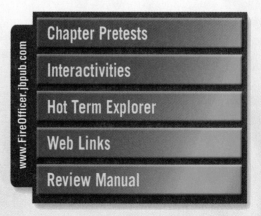

www.FireOfficer.jbpub.com

Chapter Pretests

Interactivities

Hot Term Explorer

Web Links

Review Manual

www.FireOfficer.jbpub.com

Fire Attack

NFPA 1021 Standard

Fire Officer I

4.6* **Emergency Service Delivery.** This duty involves supervising emergency operations, conducting preincident planning, and deploying assigned resources in accordance with the local emergency plan and according to the following job performance requirements.

4.6.2 Develop an initial action plan, given size-up information for an incident and assigned emergency response resources, so that resources are deployed to control the emergency.

> **(A) *Requisite Knowledge.** Elements of a size-up, standard operating procedures for emergency operations, and fire behavior.

> **(B) Requisite Skills.** The ability to analyze emergency scene conditions; to activate the local emergency plan, including localized evacuation procedures; to allocate resources; and to communicate orally.

4.6.3* Implement an action plan at an emergency operation, given assigned resources, type of incident, and a preliminary plan, so that resources are deployed to mitigate the situation.

> **(A) Requisite Knowledge.** Standard operating procedures, resources available for the mitigation of fire and other emergency incidents, an incident management system, scene safety, and a personnel accountability system.

> **(B) Requisite Skills.** The ability to implement an incident management system, to communicate orally, to manage scene safety, and to supervise and account for assigned personnel under emergency conditions.

Fire Officer II

5.6 **Emergency Service Delivery.** This duty involves supervising multiunit emergency operations, conducting preincident planning, and deploying assigned resources, according to the following job requirements.

5.6.1 Produce operational plans, given an emergency incident requiring multiunit operations, so that required resources and their assignments are obtained and plans are carried out in compli-

ance with approved safety procedures, resulting in the mitigation of the incident.

> **(A) Requisite Knowledge.** Standard operating procedures; national, state/provincial, and local information resources available for the mitigation of emergency incidents; an incident management system; and a personnel accountability system.

> **(B) Requisite Skills.** The ability to implement an incident management system, to communicate orally, to supervise and account for assigned personnel under emergency conditions, and to serve in command staff and unit supervision positions within the Incident Management System.

Additional NFPA Standards

NFPA 1561, *Standard on Emergency Services Incident Management System*

NFPA 1710, *Standard for the Organization and Deployment of Fire Suppression Operations, Emergency Medical Operations, and Special Operations for the Public by Career Fire Departments*

Knowledge Objectives

After studying this chapter, you will be able to:

- Describe how to supervise a single company.
- Describe how to supervise multiple companies.
- Describe how to size up the incident.
- Describe Lloyd Layman's five-step size-up process.
- Describe the National Fire Academy size-up process.
- Describe risk-benefit analysis.
- Describe how to determine fire flow.
- Describe how to develop an incident action plan.
- Describe how to assign resources.
- Describe the tactical safety considerations.
- Describe the general structure fire considerations.

Skills Objectives

After studying this chapter, you will be able to:

- Perform a scene size-up.
- Assign resources.

You are the officer in charge of the first due engine company responding to an alarm for a report of a fire in an apartment building. As you turn the corner, you see fire coming from a second-floor window in a five-story brick building. Gray smoke is billowing from the rest of the second-floor windows and several third- and fourth-floor units. Several people are waving out their windows, waiting to be rescued.

1. What information would you consider in order to make an accurate size-up?
2. What are your strategic priorities?
3. What tactics might you use to meet those strategies?

Introduction to Fire Attack

Every fire officer must fully understand the basic elements of firefighting strategy and tactics, as well as the techniques of supervision and incident management. Supervising operations at the scene of a fire is the very essence of the job of a fire officer, even for officers who are normally assigned to perform some other function or position within the fire department (▼ Figure 16-1).

Individuals at the Fire Officer I or Fire Officer II levels most often serve as fire officers and directly supervise a crew of fire fighters who operate at fire and other types of emergency incidents. The fire officer is a vital player with important responsibilities for the safety of fire fighters as well as civilians, as well as ensuring that the goals and objectives of the fire department are achieved.

This chapter explains how to perform a comprehensive scene size-up of a fire situation, how to calculate fire flow requirements, how to determine incident priorities, and how to develop tactical assignments to control a fire. It also covers establishing of command, command transfer procedures, multiple alarm incidents, and special considerations.

Supervising a Single Company

The primary responsibility of a fire officer is to supervise the company while assigned tasks are being performed. The company is a basic tactical unit for emergency operations, and the fire officer is generally a participant in the work as well as the unit supervisor. This means that when the company is assigned to advance an attack line into a structure, the fire officer is normally with the crew members, directing and leading them as well as continually evaluating the environment for hazards such as backdraft, flashover, or structural collapse. In most cases, the fire officer is also a hands-on participant, assisting in pulling and advancing the hose. The officer's personal physical involvement must never compromise his or her supervisory responsibilities.

In addition to leading and participating in company-level operations, the fire officer should also be evaluating their effectiveness. The officer must determine whether the fire stream is making progress or losing ground against the fire. The officer must also judge whether the ventilation opening is of sufficient size to effectively ventilate the products of combustion so that the attack line can advance. When problems are identified, it is up to the officer to figure out how to resolve them. If the hoseline is not making progress, the officer must determine whether an additional line or a larger line is needed or if the attack should be made from a different direction.

The fire officer is always responsible for relaying relevant information to his or her supervisor. In the case of a fire officer at a fire, this would include informing that supervisor when the company's assignment has been completed or when an assignment will be delayed or cannot be completed. This would also include informing the supervisor immediately when problems or hazards are identified, such as signs of collapse. The fire officer serves as eyes and ears for the incident

Figure 16-1 Supervising operations at the scene of a fire is the very essence of the job of a fire officer.

commander or for an assigned supervisor in the incident management structure.

Closeness of Supervision

Supervising a fire company at an emergency incident requires a fire officer to provide closer direct supervision than required during most routine nonemergency duties. The inherent risks associated with the emergency operations demand close supervision at all times. Fire fighters who have been assigned specific tasks to complete, such as advancing a hoseline to the seat of the fire or performing vertical ventilation, often focus on the task and ignore everything else. Aggressive fire fighters are good at getting the job done but are often too involved in the task to pay attention to their surroundings. The role of the fire officer should be to look at the big picture: monitoring progress, coordinating activities with other companies, and looking out for hazards.

The level of supervision should be balanced with the experience of the company members and the nature of the assignment that is being performed. An inexperienced crew performing a high-risk task requires more direct supervision than a highly experienced crew performing a more routine task. The greater the danger, the greater the level of supervision that is needed. A company that is making an interior attack on a fire in a large warehouse requires more direct supervision than a company operating a master stream to protect an exposure. The fire officer must always know where the crew members are located and what they are doing; however, in an interior attack situation, the fire officer should be able to see all crew members and communicate directly with them at all times.

Form of Leadership

Emergency scene supervision also requires a more authoritarian form of leadership than in nonemergency activities. During nonemergency situations, the fire officer should normally use a participative style of leadership in order to promote productivity and group cohesiveness. In this environment, there is time for discussion and feedback before decisions are made, and the consequences seldom involve an imminent risk of injury or death. Emergency operations present a very different environment for leadership and decision making.

During an emergency incident, time is often of the essence. Decisions are needed quickly, and actions must be implemented in a timely manner. These situations do not provide the luxury of time to gather the team, discuss what needs to be done, and reach a group consensus. The fire officer must frequently make decisions with little or no input from subordinates, and every member of the team must be prepared to perform a specific set of tasks with a minimum of instruction.

When the fire officer gives the order to lay a line, the fire fighter who is assigned to catch the hydrant must know exactly what to do. While that fire fighter is dragging the hose to the hydrant, the fire officer is gathering information to develop the next step in the action plan. The fire fighter who will be advancing the attack line can anticipate what comes next and does not need a lengthy explanation from the officer.

The key to using an authoritative style of leadership effectively is to develop the trust and confidence of the subordinates before the incident. Fire fighters must trust the competence of their officer to lead effectively in an authoritative manner. A fire officer who has demonstrated sound knowledge of the technical aspects of firefighting and has developed a trusting relationship with the fire fighters is ready to lead them into action. Fire fighters are less likely to trust the judgment of an officer who does not appear to fully understand the job and has not made it clear that their safety is foremost in his or her mind.

The fire officer must strike a balance with subordinates. Fire fighters must be prepared to follow the officer's directions; however, they should also feel confident giving relevant information to the officer. For example, a fire fighter who has been given orders to perform vertical ventilation should not hesitate to inform the officer of safety conditions that the officer might not know about.

Supervising Multiple Companies

Generally, the first-arriving officer at a fire incident assumes the role of incident commander. When operating as the incident commander, the fire officer has an even greater level of responsibility; the incident commander is responsible for every company on the scene and for management of the overall operation.

As covered in detail in Chapter 15, the initial incident commander's immediate responsibilities include conducting a size-up, developing an action plan to mitigate the situation, assigning the resources to execute the plan, developing a command structure to manage the plan, and ensuring that the plan is completed safely. The fire officer who is functioning as the incident commander is responsible for all of these functions.

The fire officer might also be assigned as a division/group/ sector leader, supervising multiple units, or even as a branch director within the Incident Management System (IMS). In each of these situations, the officer serves as a relay point within the command structure. Direction comes down through the system, from the incident commander, through the intermediate levels, to the individual companies or units. At the same time, information is transmitted upward from subordinates to the officer, who either acts on the information directly or relays the information up to the next level. For example, a division supervisor could be assigned to oversee operations directed toward keeping the fire out of an exposure. The division supervisor would be expected to manage that operation and supervise the assigned companies, giving regular progress reports to the incident commander as long as everything is going well. If one of the companies reports that they can see flames in the windows of the exposure, the division supervisor would immediately relay that information to the incident commander.

Standardized Actions

A fire officer has great latitude in deciding how work is accomplished during nonemergency activities; however, the opposite is true of most emergency scenes. Because of the hazardous nature of the situation and the need for everyone to coordinate efforts and think in the same manner, emergency operations must be conducted in a very structured and consistent manner. This is accomplished by placing a strong emphasis on standard operating procedures (SOPs).

SOPs provide a framework to allow activities at an emergency scene to be completed in an efficient manner and with all components of the organization working in concert with each other. The SOPs are the equivalent of a playbook for a football team; they explain the standard approach that should be followed in a particular situation. This allows the incident commander to develop an incident action plan. The use of SOPs also promotes a standard approach to safety.

The same concepts are behind the use of a standard IMS. The use of an IMS allows for a consistent approach to developing a command structure that allows for efficient management and effective control at emergency scenes. The consistency allows everyone operating at the scene to know precisely to whom they report and from whom they receive their orders. The IMS incorporates a process that identifies a strategic goal for the incident, followed by an incident action plan that is translated into specific objectives for each tactical unit.

Most fire department incidents are relatively small scale and short duration; however, some situations require the use of the jurisdiction's local emergency plan. This is a plan that is designed to be implemented when a significant emergency occurs and requires the response of multiple agencies. The plan predetermines what agencies are responsible for which actions. For example, a tornado would require a significant response from the fire department, the police department, and the Emergency Medical System (EMS). The local emergency plan would specify the primary responsibility of each agency and how the command system would be structured for that type of incident. A fire officer must know the workings of this plan in order to effectively manage the fire department's resources during a crisis.

In major disasters, state and national resources may be required to supplement local resources. Local agencies, such as the law enforcement, EMS, emergency management, water department, sewer department, and public works, are typically contacted and activated directly. For more severe emergencies, state organizations may be needed. Typically, these are coordinated through a state emergency management agency, which can make the appropriate contacts for state aid, such as overhead command teams, the National Guard,

and the statewide mutual aid system for fire, law enforcement, and EMS. In emergencies that require a national response, the Federal Emergency Management Agency (FEMA) is the coordinating agency.

Command Staff Assignments

Every fire officer should also be prepared to serve in a command staff role at an emergency incident. The command staff positions include safety officer, liaison officer, and public information officer. While working in one of these positions, the fire officer reports directly to the incident commander. The incident commander could assign any available and qualified officer to perform one of these roles at an incident scene.

The safety officer is responsible for overseeing the incident from a safety perspective, keeping the incident commander informed of safety concerns and taking preventive action when an immediate hazard is identified. When assigned as liaison officer, the fire officer functions as the link between the incident commander and representatives from various agencies. At a hazardous materials incident, this could be the property manager or a chemical manufacturer. When operating as the public information officer, the fire officer is responsible for gathering information that is to be released to the general public, developing news releases, and giving interviews or press conferences. The person in this position acts as the public spokesperson for the incident commander.

Sizing Up the Incident

Sizing up a fire incident is a critical skill for a fire officer. Size-up is a systematic process of gathering and processing information to evaluate the situation and then translating that information into a plan to deal with the situation. The art of sizing up an incident requires a diverse knowledge about emergency incidents. The fire officer could be faced with a house fire, a hazardous materials release, or a major automobile collision in one shift. Each situation requires the ability to recognize and analyze the important factors, develop a plan of action, and then implement the plan. Each type of incident involves specialized knowledge.

Many new fire officers think that size-up begins when they get their first glimpse of the incident scene and ends when they have decided on a plan of action. In reality, size-up should begin long before arrival and continue until the incident is stabilized. Deciding on an initial plan of action after arriving on the scene is only one step in a continuous process. As the plan is executed, new information must be gathered and processed to determine whether or not it is working as anticipated and to make any adjustments that are required.

The initial size-up at the scene of an incident must often be conducted under intense pressure to "do something." This pressure could be coming from the public, from fire companies that are arriving and anxious to get into action, and from within the fire officer who feels the anxiety of being responsible for deciding what to do quickly. The end result of a good size-up is an **incident action plan** that considers all the pertinent information, defines strategies and tactics, and assigns resources to complete those tactics.

In many cases, a comprehensive evaluation of a situation requires information that is simply not available to the fire officer at the time. A good size-up requires an officer to look carefully and assess what can be seen, make reasonable assumptions about what cannot be seen, and anticipate what is likely to happen (▼ Figure 16-2). An experienced officer should be able to develop a reasonable plan based on that assessment, then refine the information and adjust the plan as more information becomes available.

In some cases, the additional information could result in a major change in the plan; however, this does not necessarily mean that the initial size-up or the initial plan was inappropriate. In many cases the incident commander has to begin with a plan based on the information that is available, while remaining flexible if the additional information points toward a different approach. The incident commander must con-sciously differentiate between what is known, what is assumed, and what is anticipated in processing size-up information.

The first phase of size-up begins long before the incident occurs. Everything the fire officer learns, observes, and experiences goes into his or her memory banks, which are used to make size-up decisions. The fire officer's knowledge of the typical types of building construction used in the area, the available water sources, and the available resources are useful in sizing-up incidents. Whenever a fire officer completes a preincident plan or conducts a familiarization visit, he or she is performing part of the size-up for a possible incident at that location.

Prearrival Information

The specific size-up for an incident begins with the dispatch. The name, location, and reported nature of the incident all help the fire officer begin anticipating what is going on at the scene. Seasoned fire officers know that dispatch reports indicating multiple calls to 9-1-1 help the officer develop a vision of what to expect before he or she actually sees the incident.

On-Scene Observations

The ability to quickly size-up a fire situation requires both a systematic approach and a solid foundation of reference information. Most fire department SOPs list the essential size-up factors that should be considered by a first-arriving officer. These include building size and arrangement, type of construction, occupancy, fire and smoke conditions, and other factors, such as weather and time of day. The SOPs should guide the officer's systematic thinking to ensure that all of the important factors are considered.

A fire officer must understand and recognize basic fire behavior. It is easy enough to identify fire coming out of a window and smoke pushing out from the eaves of a house, but it takes a firm grasp of conduction, convection, and radiation to be able to predict where the fire is burning and where it will spread. An officer who has studied and observed fire behavior is able to process this information very quickly, without having to spend an excessive amount of time looking at the building and thinking about what is going on inside.

One of the most significant factors in size-up is visualization. Every previous situation that the fire officer has experienced or observed, including some that might have been observed only in training sessions, photographs, or videos, is stored in the individual's memory. When a new situation is observed, the mind subconsciously looks for a matching image to create a template for this new observation. Instead of methodically processing the information, the brain can instinctively jump to a similar observation and apply the stored experience to the new set of circumstances. The same process works for other human senses, particularly smell, taste, and feel, fostering rapid recognition and interpretation of many situations. Unfortunately, the process does not work when one is faced with an entirely new and unanticipated situation.

Figure 16-2 Sizing up a fire scene requires an officer to look at each side of the building, assess what can be seen, make assumptions about what cannot be seen, and theorize what will happen next.

Experienced fire officers know that the more churning that is present in the smoke, the hotter the fire and the closer it is to the seat of the fire. Also, darker smoke is generally closer to the seat of the fire than lighter-colored smoke. This helps determine where the seat of the fire is located. By evaluating where the smoke is coming from the building, the fire officer can predict where the fire will be traveling unless there are mitigating circumstances.

The volume and color of the smoke also aid the officer in determining the need for ventilation. Because heat and smoke travel together, large amounts of smoke coming from a structure normally indicate that there is a need to ventilate in order to let the heat out of the building. Dark smoke is also an indication that there are more carbon particles that are suspended within it and the possibility of a lack of oxygen. This could indicate a potential for backdraft requiring vertical ventilation techniques.

Understanding fire behavior is also needed to develop action plans. Fires in the ignition phase could be extinguished with a portable fire extinguisher. Fires in the growth phase may require a small hand line. A fully developed fire normally requires multiple small hand lines, large hand lines, or master streams.

Fuel loading is also an important factor during size-up. A fire in class A combustibles normally dictates a direct attack with water. Those involving Class B fires require the use of foams. When using foams, the officer must determine what type and how much foam will be needed to complete the job before beginning the application.

In addition to departmental SOPs, various systems have been developed to assist in performing a size-up of a fire incident. Each system provides a slightly different list of considerations and procedures, but they are all designed to accomplish the same objective: identifying critical information to develop a plan of action. A fire officer should use the system that is officially adopted by the fire department; however, it is also a good practice to explore other systems and consider their approach. An officer should practice performing size-ups until the process becomes second nature.

Lloyd Layman's Five-Step Size-Up Process

In 1940, Chief Lloyd Layman authored *Fundamentals of Fire Tactics*, which became a standard text for several generations of fire officers. His publication was based on a military model of training, with the emphasis on establishing general principles that many different departments could use.

Layman presented a five-step process for analyzing emergency situations:
1. Facts
2. Probabilities
3. Situation
4. Decision
5. Plan of operation

Facts

Facts are the things that are known about the situation. Chief Layman was an early advocate of preincident planning. This allowed the first-arriving officer to quickly determine the type of construction, whether fire protection systems are in place, fire flow requirements, water supply sources, special hazards, and other significant factors. Additional facts will be known on dispatch, such as time of day, location of incident, weather, and specific resources that the fire department is sending. Visible smoke and fire conditions are additional facts. The more facts the fire officer can secure, the more accurately the situation can be sized up.

Probabilities

Probabilities are things that are likely to happen or can be anticipated based on the known facts. We know the facts, but we have to estimate or predict the probabilities. We may know that the structure has a fire sprinkler system and we can anticipate that it will probably work, but we cannot be *sure* that that it is going to work. It may have been turned off before the fire for maintenance, or it could have been damaged by an explosion. Fire officers consider the most likely possibilities based on the known facts.

The fire officer has to anticipate where the fire is likely to spread. If the fire is burning freely in the attic, what impact is it likely to have on the stability of the building? This skill requires an in-depth knowledge about factors such as building construction, fire behavior, hazardous materials, sprinkler system performance, and human behavior in fires.

Another aspect of considering probabilities is the amount of time it takes to accomplish tasks. When a decision is made, there is a time lag before the task is executed. The fire officer should anticipate that the fire will continue to spread during this period. How far is the fire likely to advance before the hoselines are in place to attack? How long will it take to get ground ladders to Side C? How fast can a crew with a back-up hoseline get to your location? How large will the fire be in 10 minutes?

Situation

The situation assessment involves three dimensions, and all three parts must be considered and factored into the incident action plan. The first consideration is whether the resources on scene and en route will be sufficient to handle the incident. The incident commander should know which companies are responding and should be able to determine whether additional companies will be needed to control the problem. If additional resources are needed, the incident commander needs to have an idea of how long it will take them to respond. The incident commander has to anticipate the time lag between requesting resources and having them on the scene available to use.

Many fire departments have policies that require an incident commander to request additional resources whenever all of the units have been or will be committed, in order to provide an on-scene reserve. Some departments require the first-arriving company to immediately call for a second alarm if any smoke or fire is showing from a high-rise building.

The second consideration is the specific capabilities and limitations of the responding resources in relation to the problem. Consideration should be given to the amount of water carried on the apparatus, pump capacities, length and diameter of hose carried, and length and type of aerial devices. The number of personnel on each vehicle is an important factor. A company staffed with just two fire fighters cannot perform as much work as a company with five fire fighters. The incident commander might have to call for additional resources or adjust performance expectations downward.

The third consideration consists of the capabilities and limitations of the personnel, based on training and experience. As a fire officer, consideration should be given to your personal experience with this particular type of incident, as well as the experience and capabilities of your crew members. If two fire fighters out of a four-person crew are rookies, the company probably has limited capabilities and requires more direct supervision.

Decision

The fourth step in Layman's size-up procedure is making fire attack decisions. This step requires the fire officer to make specific judgment decisions based on the known facts and probabilities, as well as the situation evaluation. This step requires the officer to answer four questions:

1. Are there enough resources responding to and on the scene to extinguish the fire or mitigate the situation?
2. Are sufficient resources available and do conditions allow for an interior attack?
3. What is the most effective assignment of on-scene resources?
4. What is the most effective assignment of responding resources?

The answers to these questions allow the fire officer to develop the overall strategy and the underlying tactics to mitigate the incident.

Plan of Operation

The final step in Layman's decision process is to develop the actual plan that will be used to mitigate the incident. Most other methods of size-up do not include this step as part of the size-up; they refer to the information that is obtained during size-up being used to develop a plan of action.

National Fire Academy Size-Up Process

The National Fire Academy (NFA) developed a size-up system, which also includes three phases. The process has been widely distributed through the Managing Company Tactical Operations series of courses published by the NFA. The three basic phases of size-up are:

- Phase one: preincident information
- Phase two: initial size-up
- Phase three: ongoing size-up

Phase One: Preincident Information

Phase one considers what you know before the incident occurs. This consideration closely mirrors the facts step identified by Layman. Preincident plans often provide valuable information for this phase (▼ **Figure 16-3**). Information about the building and the occupancy, such as the building layout and construction type, built-in protection systems, type of business, and nature of the contents, are all needed for an accurate size-up. Preincident information should also identify water supply sources, including their location, accessibility, and capacity. If preincident plans are not available, the information must be determined through on-scene observation and research. The fire officer's intuition and experience may have to be relied on until this information can be obtained.

Fire officers are preloaded with a large amount of useful information that figures into the size-up process. Environmental information includes factors such as heat and cold extremes, humidity, wind, and snow or ice accumulations. The time of day can also be a critical piece of information; responding to fire in a multifamily residential building at 2:00 A.M. involves different considerations from the same fire at 2:00 P.M. A fire officer also needs to know departmental resources, including physical and human resources, as well as the assistance that is available from mutual aid departments.

Figure 16-3 Preincident plans often provide much of the information the fire officer needs during the preincident information phase.

Phase Two: Initial Size-Up

The second phase of the size-up begins on receipt of an alarm. There are three questions that need to be answered:

1. What do I have?
2. Where is it going?
3. How do I control it?

To determine what you have, start with the information that was established in the first phase and build on it. This phase considers the specific conditions that are present at the incident. This could require a walk around a fire building to observe conditions, such as the location and the size of the fire, as well as the volume, color, movement, and location of the smoke. The size-up should also look at the size and construction of the building, identify any exposures that are present, and look for indications of special concerns, such as the possibly that lightweight construction could be involved.

To determine where the fire is going, the fire officer must consider the current location and stage of development of the fire; in what direction is it likely to spread; and whether there is a risk of flashover or backdraft. The accuracy of these predictions is heavily dependent on the fire officer's knowledge and experience. This stage is closely associated with the probabilities step noted by Layman.

The third question is, "How do I control it?" This is where a fire officer has to consider different alternatives in relation to the available resources, including apparatus, equipment, and personnel. The development of strategy and tactics must be based on the resources that are available and, if additional resources are needed, the time it will take for them to arrive. This stage is closely associated with the "own situation" step noted by Layman.

Assessment Center Tips

WALLACE WAS HOT

As an aid to help officers remember the critical points that should be considered for a size-up, the NFA uses the mnemonic WALLACE WAS HOT.
- **W**ater
- **A**pparatus/personnel
- **L**ife
- **L**ocation/extent
- **A**rea
- **C**onstruction
- **E**xposure
- **W**eather
- **A**uxiliary
- **S**pecial hazards
- **H**eight
- **O**ccupancy
- **T**ime

Phase Three: Ongoing Size-up

The third phase in the NFA size-up system addresses the need to continually size up the situation as it evolves. This includes an ongoing analysis of the situation and an ongoing evaluation of the effectiveness of the plan that is being executed. The incident commander should always be prepared to modify the plan if the situation changes, including switching between offensive and defensive strategy. Additional resources could be required to address unanticipated needs or to replace tired crews. The ongoing evaluation could also indicate that resources can be redeployed from one task to another where they would be more effective.

This ongoing evaluation requires a constant flow of feedback information to the incident commander. Specifically, the incident commander needs to know when:
- An assignment is completed.
- An assignment cannot be completed.
- Additional resources are needed.
- Resources can be released.
- Conditions have changed.
- Additional problems have been identified.
- Emergency conditions exist.

Risk/Benefit Analysis

A risk/benefit analysis, using all of the previous information, must also be incorporated into the size-up process. The risk/benefit analysis is the key factor in selecting the appropriate strategic mode for the incident. The strategic mode regulates the degree of risk to fire fighters that is acceptable in a given situation. All fire department operations involve some inherent and unavoidable risks; however, many risks can and should be avoided. The degree of risk that is acceptable is determined by the realistic benefits that can be anticipated by taking a particular course of action.

The incident commander must always consider the risks to fire fighters in relation to expected benefits that may be gained from those actions. The potential benefits include the possibility of saving lives or preventing injury to persons who are in danger, preventing property damage, and protecting the environment from harm. The major risk factor is the possibility of death or injury to fire fighters. Potential occurrences, such as flashover, backdraft, structural collapse, or lost or disoriented crew members, all have to be evaluated. Each alternative action plan involves a different combination of risks and benefits.

All firefighting operations involve some inherent and unavoidable risks; however, we have the ability to carefully manage these risks, depending on the situation. Although a certain degree of risk has to be recognized in every offensive situation, we take advantage of training, experience, protective clothing and equipment, communications equipment, SOPs, accountability systems, and rapid intervention crews to work with a reasonable level of safety in a hazardous environment.

Fire fighters accept a higher personal risk when it is necessary to save a life; however, there is no reason to expose fire fighters to this type of risk when there are no lives to be saved. When defensive operations are conducted, the risks to fire fighters are reduced to an absolute minimum. There is no justification for risking the lives of fire fighters when no property value can be saved, such as in fires in vacant or abandoned buildings or situations where the damage is already done. Building construction features, such as lightweight trusses, can also cause the risks of offensive operations to outweigh the potential benefits.

The risk analysis determines the appropriate strategy for an incident: offensive, defensive, and transitional. An **offensive attack** is an advance into the fire building by fire fighters with hoselines or other extinguishing agents to overpower the fire. Offensive attack is the activity that drives most fire department training, operations, and organizational structures.

When this mode is selected, the incident commander must consider that the benefits associated with controlling and extinguishing the fire outweigh the risks to the fire fighters. This necessarily requires the resources to mount an attack that is powerful enough to overwhelm the fire within the time that is available to operate safely inside the burning building. The risks of an offensive attack can be justified only when there are realistic benefits. There must be lives or valuable property that can be saved in order to justify taking any risk.

The **defensive attack** is used when the risks outweigh the expected benefits. In this mode, fire fighters are not to be allowed to enter the structure or to operate from positions that involve avoidable risks. Defensive operations are conducted from the exterior, using large streams to contain and, if possible, overwhelm a fire. This attack could be required for various reasons, including the risk of a structural collapse or the lack of adequate resources to conduct an adequate interior attack to control the fire. Defensive strategy is also the appropriate choice in a situation where the building and contents would be a total loss even if an aggressive interior attack could control the fire.

A **transitional attack** refers to a situation in which an operation is changing or preparing to change. This attack particularly applies to a situation where an offensive attack is initiated, with the recognition that it could be unsuccessful or the situation could change quickly and change the risk/benefit analysis. The offensive attack is conducted cautiously and in a manner that allows interior attack crews to be quickly withdrawn. At the same time, back-up resources are positioned in defensive positions in case they are needed. This strategy could also be used to quickly conduct a search and rescue operation ahead of a fire, knowing there is only a limited time for crews to get in and get out.

A transitional attack could also apply to a situation in which an initial attack is made with an exterior master stream to knock down a large body of fire, while crews prepare to conduct an offensive interior attack. The switch to offensive attack should occur only if there are still lives or property to be saved after the heavy streams have been shut down.

Determining Fire Flow

Several methods have been developed to estimate the water flow that will be needed to extinguish a fire in a structure. The most common method of extinguishing a fire is by using water to cool the burning material. In order to extinguish the fire, the volume of water that is applied must be sufficient to absorb the heat that is being released. The estimated fire flow is an approximation of the rate of water application (usually measured in gallons per minute [gpm]) that would be required to control a fire in a particular building or section of a building.

Fire flow is an important consideration in the development of strategy and tactics. The incident commander can use this information to estimate the number of hoselines that will be required, the number of engine companies that will be required to operate those hoselines, and the water supply that will be needed to mount an effective fire attack. If the incident commander does not have the resources to extinguish the fire, the option is to use the available resources in a defensive mode to contain the fire.

The Kimball Rule of Thumb

In 1966, the National Fire Protection Association published *Fire Attack 1: Command Decisions and Company Operations*, by Warren Y. Kimball, the manager of the NFPA Fire Service Department. Like Layman's book a decade earlier, Kimball's book served a wide fire service audience. Kimball provided a rough rule of thumb for estimating required fire flow at 3 gpm per 100 cubic feet.

The volume of a building is determined by length multiplied by width multiplied by height. Using this formula, a three-story building that is 50 feet wide, 100 feet long, and 10 feet high per floor would contain 150,000 cubic feet of interior volume. Multiplying 150,000 square feet by 0.03 (3 gpm for each 100 cubic feet) means that this building would require 4500 gpm.

This figure represents an estimate of the total flow that would be required to extinguish a fire, assuming that the building is ordinary construction, with an average fire load, and is fully involved. This also assumes that the water will be directed into the burning building using large-caliber streams.

Iowa State University Rate-of-Flow Formula

The **Iowa State University Rate-of-Flow Formula** was developed to determine the flow needed to knock down a fire in a fully involved structure or in an enclosed compartment within a structure. This formula is particularly applicable to fires in rooms or enclosed spaces, where it is feasible to apply fog streams directly into the heated atmosphere. The

VOICES OF EXPERIENCE

❝ Fire fighters on a fire ground are similar to basketball players on a court because they are trying to accomplish a goal as a team. ❞

I have always equated the fire officer's communications on the fire ground to the 30 second timeout in a basketball game. Fire fighters on a fire ground are similar to basketball players on a court because they are trying to accomplish a goal as a team. Each player or fire fighter has his or her own role in the team effort, and if each understands clearly what his or her role is, then the task has a much greater chance of being accomplished successfully.

In a basketball game, coaches are given 30 second timeouts to communicate what he or she wants the team to do. Thirty seconds is not much time to communicate anything. What are the keys to making the most out of this quick communication?

First, know the purpose for your timeout. Organize your thoughts prior to calling your fire fighters together. "What do I really want my team to know?" This doesn't mean you need to rehearse a speech, but take a moment and organize your thoughts. The last thing you want is to gather your fire fighters around you and then start wondering what to say. Remember, you called the timeout for a reason. At the end of your message, there should be no doubt what that reason was.

Second, identify the key tasks and convey them directly to the person who is responsible for making the task happen. Don't speak to the group if you expect a certain person to carry out a task. Speak directly to him or her while making eye contact.

As a fire officer, you need to understand that effective communication both on and off the fire ground is one of your most important skills. When you look at effective fire officers that you have worked for, what is it about them that made them so effective? I'll bet that they were great communicators who said what needed to be said briefly and concisely.

Bart Ridings
Air National Guard Fire Department
182nd Airlift Wing, Peoria, Illinois
19 years of service

calculation is based on a flow equivalent to 1 gallon per minute per 100 cubic feet for 30 seconds, which is directly related to the ability of water in the form of fog droplets to absorb heat within an enclosed space.

The knockdown flow is determined by dividing the volume of the compartment or building in cubic feet by 100. For a room that is 50 feet wide, 50 feet long, and 10-foot high, the Iowa rate-of-flow formula would require a flow of 250 gpm for 30 seconds to achieve knockdown.

NFA Fire Flow Formula

The <u>National Fire Academy fire flow formula</u> was developed for use in strategy and tactics classes. The formula is a simplified version of more complex calculations developed by the insurance industry, as well as the Kimball and Iowa approaches. This system can be used to estimate the required flow for a building that is only partially on fire and also accounts for water required for exposure protection. This system uses a four-part calculation and is based on square footage rather than cubic footage:

Example:

Estimate the fire flow for a two-story building that is 30 feet by 40 feet in area, is 25% involved in fire, and has an exposure building 25 feet away on each side:

Step 1—Determine the fire flow requirement per floor:

- For each floor of a building, estimate the length and width of the building and divide by three.

Needed Fire Flow (NFF) = (Length × Width)/3

NFF = (L × W)/3

NFF = (30 × 40)/3

NFF = (1200)/3

NFF = 400 gpm

Step 2—Determine the fire flow requirements for the total building:

- Multiply the NFF per floor times the number of floors.

NFF = 400 gpm × 2 floors

NFF = 800 gpm

Step 3—Determine the NFF based upon the percent of fire involvement for the building:

- Multiply the NFF by the percent involvement.

NFF = [(L × W)/3] × [# of floors] × [% involved]

NFF = 800 gpm × 25%

NFF = 200 gpm

Step 4—Add 25% for each exposure:

- Increase the NFF by 25% for each exposure.
- An exposure is any building within 50 feet (Buildings more than 50 feet but less than 100 feet can also be counted).
- An exposure is also any floor in the building that is above the fire floor but less than 5 floors above the fire floor.

For two exposures add 25% + 25% = 50%

NFF = 300 gpm + 50% (300)

NFF = 450 gpm

Developing an Incident Action Plan

Once an accurate size-up has been performed, the incident commander must develop an incident action plan, based on the incident priorities. There are two major components to the incident action plan:

1. The determination of the appropriate strategy to mitigate an incident
2. The development of tactics to execute the strategy

Strategies are general, whereas tactics are more specific and measurable. In nonfire terms, strategies are equivalent to goals, whereas tactics are the objectives to meet the goals. The incident commander identifies the strategic goal and then identifies the tactical objectives that are required to reach the goal. Resources (companies) are assigned to perform individual tasks that allow the tactical objectives to be achieved.

SOPs are used to provide a consistent structure to the process of establishing strategies, tactics, and tasks. Without SOPs, every situation might be managed differently, depending on the individual who was in command and his or her interpretation of the situation, as well as the decisions made by individual company officers. SOPs guide the decision-making process and ensure consistency between officers and events. On arrival at a house fire with smoke showing, the SOP might require the first-arriving officer to assume command, the first-arriving engine company to begin a fire attack, and the second-arriving engine to provide a supply line for the first engine. The SOPs assist everyone involved in the process by outlining what actions to take when presented with a given situation. In effect, the SOPs are pre-established components of an incident action plan.

Incident Priorities

The three basic priorities for an incident action plan are:

1. Life safety
2. Incident stabilization
3. Property conservation

The priorities are always presented in this order because life safety concerns always takes precedence over stabilizing the incident. Incident stabilization and property conservation are often addressed simultaneously, although property conservation is lower on the priority list. If the incident commander does not have sufficient resources to perform both, property conservation measures might have to be delayed until the incident has been stabilized. In reality, many of the actions that are taken to stabilize an incident also work toward preserving property.

The life-safety priority refers to all people who are at risk due to the incident, including the general public as well as fire department personnel. The incident stabilization priority is directed toward keeping the incident from getting any worse. If one structure is fully involved, protecting the exposures is part of incident stabilization. The burning structure cannot be saved, but the fire will not spread beyond that building.

Property conservation is directed toward preventing any additional damage from occurring. This could include minimizing water damage by covering building contents with salvage covers or removing valuable property from harm's way. Property can also be protected by venting smoke and heat from a building and then covering the openings to keep wind and rain from causing additional damage.

Tactical Priorities

At most fires and other emergency scenes, the incident commander is faced with a number of issues and considerations that require attention. If an abundance of resources all arrived simultaneously, it would be fairly simple to assign them in a manner that would get everything done immediately. Because this is rarely the case, fire officers need a simple method to prioritize what needs to be accomplished and then allocate resources to address the most critical concerns first. Tactical priorities provide a list of considerations that must be addressed and an order of priority for dealing with them. The tactical priorities and the information obtained in the size-up are used to develop the incident action plan.

Lloyd Layman developed a list of seven factors that should be considered to assist in developing an action plan. Although there are other methods to assist fire officers in quickly developing a plan, Layman's method remains one of the most popular, 50 years after it was first published.

RECEO VS is an acronym that covers the critical factors in developing a strategy. The first five factors are in a priority order, whereas the last two may be used at any point to support the first five. The acronym represents:

- Rescue
- Exposures
- Confinement
- Extinguishment
- Overhaul
- Ventilation
- Salvage

Rescue

Because life safety is our highest incident priority, rescue is at the top of the list. Rescue includes citizens and fire fighters, although animals could be added to the list in some circumstances (▶ **Figure 16-4**). The rescue category includes all functions related to searching for potential victims and removing them from danger to a location of safety, as well as any other activity that is necessary to reduce the risk of death or injury. Medical care after the occupants have been removed would be included as a rescue priority.

(**Figure 16-4**) Rescue is the most important tactical priority.

Because rescue is the first priority, there could be a situation in which the first-arriving company is committed to rescue a trapped occupant while the fire is allowed to burn. This does not mean that fire attack is always delayed until rescue has been completed. In many cases, the best method of protecting the occupants from harm is to quickly extinguish the fire. In other cases, the fire has to be attacked to give the rescuers time to enter the structure and search for occupants.

Exposures

Once the rescue factor has been addressed, the next priority is to keep the fire or the problem from getting bigger. Exposure protection is most important in a situation in which the incident commander does not immediately have the resources that would be needed to fully control or extinguish the fire. In most cases, the best method of protecting an exposed property is to make an aggressive attack and extinguish the fire before it can spread. In the same manner that quickly extinguishing the fire reduces the rescue problem, rapid extinguishment also alleviates the exposure problem.

Figure 16-5 Protecting an exposure.

If the fire is beyond the capabilities of an initial attack, the best decision a fire officer can make is to protect the adjacent properties rather than waste resources attempting an inadequate attack on the original fire building. If there are insufficient resources to save the fire building, use the available resources to protect adjacent buildings (▲ **Figure 16-5**).

Confinement

The third priority is to prevent the fire from spreading to the uninvolved areas of the same property. If the fire is in one room, the objective should be to confine the fire to that room of origin. If confining the fire to one room is not possible, then the objective could be to confine it to the floor or area of origin.

Confinement can also be directed toward supporting rescue efforts. Confining the fire to the room of origin can provide time for other crews to perform rescue. Confinement could also be directed toward protecting exit stairways and corridors in order to safely remove occupants.

Company Tips

NFPA 1710

NFPA 1710, *Standard for the Organization and Deployment of Fire Suppression Operations, Emergency Medical Operations, and Special Operations for the Public by Career Fire Departments*, calls for 14 fire fighters to conduct a basic fire attack on a single-family dwelling. The standard states that these resources should be assembled on the scene within 9 minutes after dispatch. The minimum staffing is required to operate two hand lines and provide a search team, a ventilation team, and a rapid intervention crew, plus apparatus operators, an incident commander, and a safety officer. The standard calls for an uninterrupted water supply capable of delivering 400 gpm and a minimum initial flow of 300 gpm through two hand lines.

Extinguishment

The next priority is to extinguish the fire or mitigate the incident. Ultimately, the goal is to extinguish all fires, and in many cases, rapid extinguishment takes care of the rescue, exposure, and confinement priorities. The priority order becomes important when rapid extinguishment is not feasible. The higher priorities must always be considered ahead of extinguishment. There are many tactical options for accomplishing extinguishment, using either offensive or defensive strategies.

Overhaul

The fifth and last of the tactical priorities is overhaul, the activity that makes sure that the fire is completely out. After all of the visible fire has been extinguished, the area must be checked for residual fire, including any fire that could have extended into areas not originally involved, such as walls, ceilings, and attic spaces. Incomplete overhaul may allow the fire to rekindle.

Ventilation

Ventilation may be required at different points in the time sequence because it can support the successful completion of any or all of the other priorities. Ventilation is designed to remove heat, smoke, and the products of combustion from a fire area and allow cool, fresh air to enter. Effective ventilation can improve the survival chances for occupants and also can improve conditions to allow fire fighters to enter and operate inside a building (▶ **Figure 16-6**). Releasing smoke and heat can also reduce the risk of backdraft or fire spread within a building and reduce the damage to contents. Ventilation could be the first action taken by crews on the fire scene, even ahead of rescuing the victim and extinguishing the fire, particularly if there is a potential for backdraft.

Ventilation was not included in Layman's prioritized list because he advocated an indirect attack, applying high-pressure fog stream into a sealed structure. Any ventilation decreased the effectiveness of this attack method. Because few departments today use a pure, indirect attack as outlined by Layman, ventilation is often a high priority.

Salvage

Salvage is the other tactical activity that does not have a specific place on the priority list. Salvage includes protecting or removing property that could be damaged by the fire, smoke, water, or firefighting operations. Salvage also includes securing the building and protecting it from weather. Salvage can be placed wherever it is appropriate in a particular situation and is often performed in parallel with the other priorities, if resources are available. Salvage often involves placing salvage covers to protect property during overhaul. It can also include actions such as only using enough water to extinguish the fire in order to limit water damage.

Figure 16-6 Proper ventilation enables fire fighters to control a fire rapidly. A. Unvented structure. B. Vented structure.

Determining Task Assignments

Once the fire officer has determined the tactical priorities for an incident, the tactical priorities are subdivided into tasks and assigned to companies. Completion of the tasks results in accomplishment of the tactical objectives. Tasks are specific assignments that are typically performed by one company or a small number of companies working together.

Few departments have the luxury of having adequate resources to initially assign every required task. In most cases, at least during the early stages of an incident, there are more tasks to be performed than there are companies available to do the work. The incident commander must make assignments based on tactical priorities and available resources. The incident commander must prioritize assignments and distribute them to companies as they arrive or become available. Companies that have completed one task assignment may have to be reassigned to another.

There are no hard and fast rules to assist an officer in making task assignments. The incident commander should use the tactical priorities to determine the relative importance of each task that needs to be performed, in the context of the specific situation. SOPs often guide these decisions, as do the staffing and standard equipment on each piece of apparatus.

The normal function of the company should also be considered, when possible. In most cases, an engine company would be assigned to attack the fire, while a medic unit provides medical care and a truck company is assigned to ventilate. At times, the incident commander has to assign companies to perform tasks that might not fit their normal role, simply because of the importance of the task and the limited resources that are available. For example, if no medic unit is on scene, the incident commander might have to assign an engine company to provide medical care. If the truck company will be delayed, an engine company could be assigned to perform ventilation.

The rescue priority could be assigned to different companies to perform primary and secondary search of the structure, raise ladders, remove trapped occupants, provide medical care and transport, and establish a rapid intervention crew (▶ **Figure 16-7**).

Task assignments for the exposure priority typically include establishing a water supply and setting up master streams for use on the fire building or on the exterior of the exposure. Placing a hand line on the unburned side of a fire-wall or in the cockloft of an exposure building could also be a task assignment, as could removing combustible materials from the windows of exposures.

Task assignments for fire confinement typically include advancing hand lines to the room of origin, into stairways, and into the attic or the floor above the fire. Ventilation tasks could be performed to support confinement efforts.

Task assignments for extinguishing a fire typically include establishing a water supply, advancing a hand line to the seat of the fire, and applying water or other extinguishing agents. When making assignments for hoselines, the fire officer should indicate the line size as well as placement, unless predetermined by departmental SOPs.

Task assignments for overhauling a fire typically include pulling ceilings and walls in the burned areas, removing door and floor trim where charred, checking the attic and basement for extension, checking the floors above and floor below the fire, removing or wetting all burned material.

Task assignments for ventilating a fire typically include vertical ventilation, horizontal ventilation, positive pressure ventilation, negative pressure ventilation, and natural ventilation. When making assignments for ventilation, the officer should indicate the location of entry points, the potential victims, and location of the fire to ensure that proper techniques are used.

Task assignments for salvage typically include throwing salvage covers over large, valuable items; removing lingering smoke; soaking up water from floors; deactivating sprinklers; and removing important documents or memorabilia.

Assigning Resources

The level of resources that respond to an incident vary greatly from one department to another. One fire department may respond with a single engine and tanker, whereas another might send three engines, a ladder truck, and a rescue

Figure 16-7 Task assignments for a rescue strategy may include performing a primary and secondary search of the structure.

truck to every structure fire. Urban fire departments often have more than 100 companies available for immediate response, if they are needed.

Whether the community is large or small, some situations exceed the capabilities of any organization and require assistance from other agencies or jurisdictions. In most cases, the first level of assistance for fires is mutual aid from surrounding fire departments. Most fire departments also have working relationships with other local agencies, such as law enforcement and public works, to obtain the resources that are likely to be needed in most situations.

When the need for resources exceeds the normal capabilities of the fire department and involves numerous other agencies, a fire officer may have to activate the local emergency plan. This plan defines the responsibilities of each responding agency and outlines the basic steps that must be taken for a particular situation. For example, firefighting is normally the responsibility of the fire department, along with technical rescue and hazardous materials incidents. Traffic control, crowd control, perimeter security, and terrorist incidents are usually the responsibility of the police or sheriff's department. Every fire officer should be familiar with the local plan and the role of the fire department within it.

The most common method a fire officer can use to activate the local emergency plan is to notify the dispatch center, which should have the information and procedures to either activate the local emergency plan or notify the emergency management office. The emergency management office normally is responsible for maintaining and coordinating the plan, as well as facilitating the response to emergency situations.

The local emergency plan usually includes an evacuation component that can be used for any situation where the residents or occupants of an area have to be protected from a dangerous situation. A hazardous material release or major fire event could require the evacuation of a local area. In many cases, the evacuation plan can be activated on a limited scale, without activating the overall local emergency plan.

The police department is often a primary agency for notifying people within an evacuation area of what to do. The use of emergency sirens, the emergency broadcast system, or a reverse 9-1-1 telephone system can greatly aid in the process. For large-scale or long-term evacuations, the Red Cross may need to be contacted to establish emergency shelters for the residents. Depending on the situation, fire crews could be available to assist evacuating residents; however, they might be occupied with fighting a fire or other mitigation actions. Mutual aid companies or other agencies could be needed to assist with evacuation.

When an evacuation area is being established, consideration must be given to the nature of the event. A natural gas leak in a street generally requires short-term evacuation of residents in the immediate area. An apartment building fire could result in several residents being displaced for a long period of time, but it affects only those who live in the building or complex. A major hazardous materials release could result in the evacuation of a large area and thousands of people. The fire officer should consult the North American Emergency Response Guidebook, the local hazardous materials team, or CHEMTREC for guidance in determining evacuation distances, based on the product, quantity, and environmental conditions.

Tactical Safety Considerations

Fighting fires is an inherently dangerous activity that exposes fire fighters to a wide variety of risks and hazards. Many advances have been made in methods to protect fire fighters from different hazards, ranging from improved protective clothing and equipment, to more effective methods of fighting fires, to better procedures for managing fire incidents. Although all of these advances have been made, the dangers still exist and continue to evolve.

Modern protective clothing and self-contained breathing apparatus provide much more protection than was ever available to earlier generations of fire fighters. In order to provide this protection, the full ensemble of personal protective clothing and equipment must be worn whenever fire fighters are exposed to hazardous conditions. The fire officer must enforce an unwavering commitment to the use of personal protective equipment (PPE), personally and by all subordinates.

PPE allows fire fighters to enter deeper into burning buildings and stay longer, often without sensing the level of heat in that environment. PPE has saved many lives and prevented many injuries; however, it also exposes fire fighters to an increased risk of being trapped by a sudden flashover, becoming disoriented inside a burning building, or running out of air in a highly toxic atmosphere.

The fire officer should always consider the risk/benefit ratio when placing personnel in hazardous situations. Normally, this is achieved through a determination of either offensive or defensive strategy. Fire officers must closely watch and supervise activities during interior fire attack operations to keep the risk level within reasonable limits.

The weight and bulk of modern PPE must also be considered, particularly during extreme weather conditions. Although fire fighters are well protected from the fire, they may have to be rotated and given sufficient time for rehabilitation between work periods. Resources must be managed effectively to maintain a safe work environment.

Newer buildings and renovations to older buildings often incorporate lightweight construction components that can collapse quickly and without warning. The contents of most structures include many items that are petroleum based, which burn hotter and faster than wood or cloth materials. Modern buildings are also more airtight to save on heating and cooling bills; however, this tends to prevent the release of heat and products of combustion, hastening the development of flashover conditions. All of these factors increase the level of risk that is likely to be encountered during interior fire attack.

Scene Safety

Many changes have occurred to improve the safety of fire fighters on the fire ground, including rapid intervention crews and improved accountability systems. However, scene safety encompasses much more. A fire officer must consider the safety factors and procedures that apply to every situation.

Many fires occur at night, and the typical fire scene is full of tripping hazards and obstacles. The use of lights to adequately illuminate the incident scene is an important safety consideration. Similarly, during cold weather, the fire officer may need to have abrasive materials spread about the scene to improve traction on ice.

Apparatus are typically positioned on roadways, and many emergency operations are conducted in the street, exposing fire fighters to traffic hazards. Apparatus should be positioned to protect the scene, and the officer should request traffic control to reduce the risk of fire fighters being struck by vehicles.

Hazardous areas at the incident scene, such as holes and potential collapse zones, should be clearly identified. Some fire departments use a special red barrier tape to establish exclusion zones around hazards. Fire fighters should recognize that the red tape is used to warn them of hazards, whereas yellow tape is used to control access to the scene by the general public.

Rapid Intervention Companies

The **rapid intervention crew (RIC)** has become a standard component of emergency operations, whenever fire fighters are operating in hazardous conditions. NFPA 1561, *Standard on Emergency Services Incident Management System*, defines an

RIC as a minimum of two fully equipped personnel on site, in a ready state, for immediate rescue of injured or trapped firefighting personnel. Most fire departments meet the RIC requirement by adding one additional fire company to the first-alarm assignment for a structure fire, either as part of the initial dispatch or as soon as a working fire is reported. Some departments refer to this company as the "safety engine" or the "FAST Truck" (Fire fighter Assistance and Safety Team).

Many departments have also developed an SOP to address RIC training and deployment; the training includes techniques for locating and removing fire fighters who are in need of assistance. The standard radio term "Mayday" is

Getting It Done

Taking Care of Yourself

The fire officer should encourage fire fighters to stash a spare pair of socks, gloves, a towel, and a watch cap on the engine. Getting soaked in a winter fire is a common occupational hazard. Adding a spare t-shirt, shorts, and coveralls means that the fire fighter can have completely dry clothes when soaked in the winter or after a technical decontamination.

Exposure to asbestos building components, drug laboratories, and chemical/biological weapons are three situations in which a fire crew may have to disrobe on the fireground to be decontaminated. A technical decontamination may result in the fire fighters losing their uniforms to a decontamination team. Without a change of clothing, the only option to the fire fighters may be a disposable Tyvek gown.

Due to these same risks, it is more important than ever for the fire fighters to change out of their duty uniforms and clean up before they arrive home, even after a slow duty day. Infectious diseases, asbestos, industrial chemicals, and products of combustion are hazards that should stay at the fire station and not be brought home to expose family members.

The fire officer should also encourage the crew to carry an on-duty wallet. This wallet replaces the regular wallet while the fire fighter is on duty. The on-duty wallet can be lost during decontamination without too much disruption to the fire fighter's personal life. The on-duty wallet includes a duplicate driver's license, copies of any state or department mandated certification cards (like EMT-paramedic), a phone calling card, and, if desired, a credit card. The fire fighter should include enough money to get through the day and should leave the regular wallet locked in the car or the fire department locker.

Make up a set of duty keys. Make duplicates of the keys you need to get in the fire station, your locker, and your car. Like the on-duty wallet, you can lose your duty keys and not have your life too disrupted. Do not forget to put some type of durable identification tag on your duty keys so you can get them back after they are sanitized! Lock up your regular keys, wedding ring and other valuable jewelry before you report for duty. It may be a good idea to use a combination lock to secure your department locker.

used to indicate that a fire fighter is lost, missing, or in life-threatening danger. The incident commander's standard response to a Mayday is to activate the RIC. When this occurs, the RIC operation has priority over all other tactical functions.

The RIC is generally positioned outside the building, near an entrance, ready for immediate action. The SOP generally specifies a list of equipment that will be prepared for use, including forcible entry tools, a thermal imaging camera, a search rope, medical equipment, and a spare SCBA or full replacement cylinder. Every RIC team member should have a portable radio and should be wearing full PPE. While standing by, the RIC members should closely monitor fireground radio communications to keep track of activities and listen for a Mayday or any indication of a problem.

Personnel Accountability Report

A **personnel accountability report (PAR)** is a systematic method of accounting for all personnel at an emergency incident. When the incident commander requests a PAR, each fire officer physically verifies that all assigned members are present and confirms this information to the incident commander. This should occur at regular, predetermined intervals throughout the incident, generally every 10 minutes. A PAR should also be requested at tactical benchmarks, such as a change from offensive strategy to defensive strategy.

The fire officer must be in visual or physical contact with all company members in order to verify their status. The fire officer then communicates with the incident commander by radio to report a PAR. Any time that a fire fighter cannot be accounted for, he or she is considered missing until proven otherwise. A report of a missing fire fighter always becomes the highest priority at the incident scene.

If unusual or unplanned events occur at an incident, a PAR should always be performed. For example, if there is a report of an explosion, a structural collapse, or a fire fighter missing or in need of assistance, a PAR should immediately take place so it can be determined how many personnel might be missing, the missing individuals can be identified, and their last known location and assignment can be established. This last location would be the starting point for any search and rescue crews.

General Structure Fire Considerations

Chapter 12, *Code Enforcement and Preincident Planning* divided buildings into use groups and briefly discussed common firefighting hazards that are associated with some of those groups. This section provides an overview of typical firefighting hazards, listed by traditional fireground classifications. There are dozens of strategy and tactics textbooks, seminars, and classes that can provide much more specific information about these hazards.

Single-Family Dwellings

More civilian fire deaths occur in single-family and two-family dwellings than any other type of occupancy in the United States. Unlike many other structures, these dwellings are not required to undergo a regular safety inspection by the fire department. This classification includes a wide range of structures, from a single-story frame house of less than 1,000 square feet to a mansion with more than 10,000 square feet on a private estate in a nonhydrant area. Balloon frame construction, which was common until the 1930s, incorporated void spaces in the exterior walls, where a fire could spread rapidly from the basement to the attic (▼ **Figure 16-8**). Since the mid-1980s, most single-family dwellings have been constructed with lightweight structural components supporting floors and roofs. In both cases, a fire can be expected to spread rapidly in these buildings. In addition, newer and renovated homes have better insulation. That means that there could be less smoke showing from the outside and a more rapid heat buildup, leading to flashover conditions on the inside.

An almost infinite variety of hazards can be found in many single-family dwellings, including improper storage of gasoline, faulty electrical wiring, and poor housekeeping habits. Each of these conditions increases the probability that a fire will occur at that home. Many Americans operate home-based

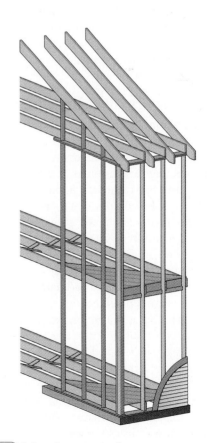

Figure 16-8 Balloon frame construction.

Company Tips

Zero-Clearance Chimneys

A recent problem is the deterioration of zero-clearance chimneys. Constructed decades ago, a zero-clearance chimney has an inside steel flue and an outside flue, separated by 1 inch of air. At some point, the owner has added glass doors to the fireplace, increasing the temperature inside the flue beyond its design parameters. To further complicate this scenario, an artificial log may be used in place of a traditional log. The artificial log plus fireplace doors plus an aging flue cause a failure of the flue that manifests itself by heating up the structural wood components around the flue. In some cases, the homeowner discovers the problem when someone knocks on their front door on a bitter cold winter night to tell them that their attic is on fire.

business, often in service industries, and have a home office loaded with computers and paper files. Other homes contain storage for a retail business in the basement, attic, garage, or spare bedroom. Other home-based businesses, such as vehicle repair businesses or home-cleaning services, store and use chemical products.

A trend in many older communities is to convert buildings that were originally designed as single-family homes into apartments. Another trend in many areas is to have multiple families living together in one crowded house. This means that you might arrive at a fire in a single-family dwelling and encounter a dozen trapped residents.

Low-Rise Multiple-Family Dwellings

Low-rise multiple-family dwellings that have been built since the mid-1980s are often type V (wood frame) construction, with lightweight wood truss components in the floors and roof. Creative building designs often create voids, as well as architectural features like cathedral ceilings that frustrate and limit the ability of fire companies to open up ceilings to check for fire extension. Like newer single-family dwellings, newer or renovated multiple-family dwellings usually have extensive insulation.

Often, the first-arriving fire companies responding to an activated fire alarm in a garden-style apartment often report nothing showing on arrival. When they force the door of the apartment that has the activated smoke detector, they encounter thick smoke boiling out of the apartment. Sometimes, a room has already flashed over, and the fire has died out due to oxygen starvation.

With the extensive amount of insulation, a room-and-contents fire in a type V building that has lightweight wood trusses is also extremely prone to early failure. This is the result of a combination of factors. First, a room-and-contents fire generates greater amounts of heat than in the past be-

cause of the amount of plastics typically found in residences today. The tighter insulation retains more of this heat than ever before. Last, lightweight wood trusses require minimal fire damage to cause collapse, which often occurs suddenly.

Multiple-family dwellings with NFPA compliant R-13 automatic sprinkler systems provide greater protection to the occupants but may increase the degree of difficulty for the fire fighters. The R-13 system controls fires that are in the living space, but it does not protect the attic area or the void spaces. In addition, the authority having jurisdiction over building construction allows taller buildings with longer hallways when a fire sprinkler system is in place. Also creative landscaping that restricts fire apparatus access to all sides of the structure is possible. Given a choice, a fire in a sprinkler-protected property is smaller, creates less damage, and provides the greater good. Your preincident plan should identify the best apparatus access positions and the hose lays required to get to every space in the building. This is especially vital if the buildings are in a gated community. Make sure that the fire apparatus can actually get into the property, and make provisions to allow prompt access to the property every hour of every day.

High-Rise Multiple-Family Dwellings

Although most fires occur in residential structures, only a small number of those occur in high-rise structures. However, these buildings present extra challenges to fire officers. Practices that may prevent future fire fighter fatalities in high-rise structures include:

- Comply with your organization's SOPs
- Consider bringing the big attack line first
- Beware of weather conditions
- Assemble an adequate crew

Fire Marks

Fire Fighter Fatalities in Residential High-Rise Fires

There have been three high-rise residential fires since 1994 in which members of the first-arriving fire crew have died. The first was the April 11, 1994 Regis Towers fire that killed Memphis Fire Department Lieutenant Michael Mathis and Fire Fighter William Bridges. The second was the Dec. 18, 1998 Vandalia Houses fire that killed FDNY Lieutenant Joseph Cavalieri and Fire Fighters Christopher Bopp and James Bohan, members of Ladder 170's inside team. The third fire was the Oct. 13, 2001 Four Leaf Tower fire that killed Houston Fire Department Captain Jay Jahnke.

Specific information, including safety recommendations, is available on each incident from a variety of sources. The National Institute for Occupational Safety and Health completed in-depth reviews of the FDNY and the Houston Fire Department incidents that are available online.

General High-Rise Considerations

A working fire within a high-rise structure requires more fire fighters and an expanded IMS. The incident commander is responsible for the overall management of the incident. Because of the complexity of a high-rise fire, the incident commander often expands the IMS to include planning, operations, and logistics sections. The incident commander also usually appoints staff to fill the positions of safety officer, liaison, and public information officer on the command staff.

High-rise fires require a tremendous amount of personnel deployed in a carefully concerted effort to achieve the strategic objectives. This requires dividing the incident into manageable units (i.e., branches), each with a supervisor. The last operations section position that a fire officer might be called on to fill at a high-rise incident is the staging area supervisor. The staging area supervisor is responsible for managing all activities within the staging area, including layout, check-in functions, and traffic control within the area. All fire officers must understand their relationship within the IMS. The large-scale operation of a high-rise fire necessitates a large IMS structure, which is necessary to maintain a manageable span of control and to maintain a margin of safety and accountability.

The fire officer may also be called on to fill a position within the logistics section. Like the operations section, the logistics section is divided into branches. Typically, there is a support branch and a service branch. The **service branch** is headed by the service branch director and is responsible for communications and fire fighter rehabilitation. The **support branch** is headed by the support branch director and is responsible for ensuring adequate supplies, personnel, and equipment. A high-rise incident requires large amounts of personnel and support equipment, so a support branch director must be established as an early part of the command staff.

The support branch director is often located at a base. A **base** is an area where the primary logistics functions are coordinated and administered. The incident command post may be colocated with the base, and there is only one base per incident. Spare SCBA cylinders are the first priority for the support branch.

It is important to quickly gain control of the lobby. The **lobby control officer** controls the entering and exiting of both civilians and fire fighters in the lobby. In addition, the lobby control officer oversees the use of the elevators, operates the local building communication system, and assists in the control of the heating, ventilating, and air conditioning systems. The lobby control officer reports to the logistics section chief or, if that position is not established, the incident commander.

A labor-intensive task is the **stairwell support** group. These are the folks who move equipment and water supply hoselines up and down the stairwells. The stairwell support unit leader reports to the support branch director or the logistics section chief. One practice is to position fire fighters on every third floor. Equipment moves up the high-rise in an assembly-line fashion, as the third-floor fire fighter moves equipment up to the sixth-floor fire fighter.

You Are the Fire Officer: Conclusion

You must start by making an accurate size-up of the situation. You need to consider all of the information that you have gathered before the incident, such as that the building does not have a sprinkler system and that there is a center hallway on each floor with solid wood doors to each apartment. You also need to consider the information that you can gather on arrival. Your first observation tells you that you have a serious fire problem and a number of occupants in imminent danger. A mnemonic such as WALLACE WAS HOT could help you remember the basic considerations.

Life safety is your highest incident priority. You have to focus on rescuing or protecting the occupants. Your plan should also include steps to confine the fire and provide ventilation to protect the exits for the occupants. You also have to provide medical care for the occupants after they are removed. After the occupants have been removed and the fire has been controlled, you can turn your attention to property conservation.

This incident exceeds the capabilities of the resources responding with you. You should call for a second alarm assignment without delay and request additional EMS units. As the initial incident commander, you are responsible for assigning all of the first-alarm companies as they arrive.

You might assign the first-arriving truck company to set ladders and rescue any occupants that are in imminent danger, while at the same time opening up the structure for ventilation. You will order the first engine company to stretch an attack line to the second floor in order to confine the fire to the room of origin. If the fire is still contained within one room, that line should be able to extinguish it.

The rescue company would be directed to perform a primary interior search of the second and third floors. The second-due engine will establish a supply line to your apparatus and stretch a back-up attack line to the second floor. The second-arriving truck company will check for extension and ventilate, and the third-arriving engine will become the RIC. The entire first-alarm assignment is committed to the life safety strategy in this situation.

Wrap-Up

Ready for Review

- The primary responsibility of a fire officer is to supervise the company while assigned tasks are being performed.
- The initial responsibilities of the incident commander include:
  - Conducting a scene size-up
  - Developing an action plan
  - Assigning resources
  - Developing a command structure
  - Ensuring that the plan is safely completed
- Size-up is a systematic process of gathering and processing information to evaluate the situation and then translating that information into an action plan.
- Lloyd Layman's five steps for analyzing emergency situations are:
  - Facts
  - Probabilities
  - Own situation
  - Decisions
  - Plan of operations
- The NFA's three basic phases of size-up include:
  - Phase one: preincident information
  - Phase two: initial size-up
  - Phase three: on-going size-up
- A risk/benefit analysis must be incorporated into the size-up process.
- Once an accurate size-up has been performed, the incident commander must develop an incident action plan.
- Single-family dwellings are the most frequent source of civilian fire deaths and may have the widest variety of occupant activity and construction features.

Hot Terms

Base The location at which the primary logistics functions are coordinated and administered. The incident command post may be colocated with the base. There is only one base per incident.

Defensive attack When suppression operations are conducted outside the fire structure and feature the use of large-capacity fire streams placed between the fire and the exposures to prevent fire extension.

Incident action plan The objectives reflecting the overall incident strategy, tactics, risk management, and member safety that are developed by the incident commander. Incident action plans are updated throughout the incident.

Iowa State University Rate-of-Flow Formula Determine the gallons per minute flow for fire knockdown by dividing the cubic size of the area (length × width × height) by 100. This is the flow needed to be provided for 30 seconds to achieve fire control.

Lobby control officer The lobby control officer controls the entering and exiting of both civilians and fire fighters in the lobby. Also oversees the use of the elevators, operates the local building communication system, and assists in the control of the heating, ventilating, and air conditioning systems.

National Fire Academy fire flow formula To provide a first-arriving fire officer the ability to quickly determine the needed fire flow. Determined on a per-floor calculation, the length is multiplied by the width, and the result is divided by three.

Offensive attack An advance into the fire building by fire fighters with hoselines or other extinguishing agents to overpower the fire.

Personnel accountability report (PAR) Systematic method of accounting for all personnel at an emergency incident.

Rapid intervention crew (RIC) A minimum of two fully equipped personnel on site, in a ready state, for immediate rescue of disoriented, injured, lost, or trapped rescue personnel.

Service branch A major division within the logistics section of the IMS. Oversees the communications, medical, and food units.

Stairwell support This group moves equipment and water supply hoselines up and down the stairwells. The stairwell support unit leader reports to the support branch director or the logistics section chief.

Support Branch A major division within the logistics section of the IMS. Oversees the supply, facilities, and ground support units.

Transitional attack A situation in which an operation is changing or preparing to change.

You respond to a residential structure fire as the shift battalion chief. Upon arrival, you see a 2-story structure with about 50% involvement of the second floor. You find one of your engine companies is beginning to conduct a quick attack. Additional incoming units include two more engines and three tenders.

1. What should be included in your IAP?

   A. Development of tactics to fulfill strategies

   B. Development of strategies from tactics

   C. Incident time line

   D. Size-up

2. What does the "C" in RECEO VS stand for?

   A. Control

   B. Containment

   C. Confine

   D. Capture

3. What is two in two out rule?

   A. Minimum of two people on the fire ground

   B. If two people go interior, two must be outside

   C. Minimum of 1 person outside in all situations

   D. Minimum of two people must carry a ladder

4. What is a PAR?

   A. A golf term

   B. Relates to two in two out

   C. Roll call taken at an incident

   D. Tactical benchmark

www.FireOfficer.jbpub.com

Chapter Pretests

Interactivities

Hot Term Explorer

Web Links

Review Manual

www.FireOfficer.jbpub.com

Fire Cause Determination

Technology Resources

www.FireOfficer.jbpub.com

- Chapter Pretests
- Interactivities
- Hot Term Explorer
- Web Links
- Review Manual

Chapter Features

- Voices of Experience
- Getting It Done
- Assessment Center Tips
- Fire Marks
- Company Tips
- Safety Zone
- Hot Terms
- Wrap-Up

NFPA 1021 Standard

Fire Officer I

4.5* **Inspection and Investigation.** This duty involves performing a fire investigation to determine preliminary cause, securing the incident scene, and preserving evidence, according to the following job performance requirements.

4.5.1 Evaluate available information, given a fire incident, observations, and interviews of first-arriving members and other individuals involved in the incident, so that a preliminary cause of the fire is determined, reports are completed, and, if required, the scene is secured and all pertinent information is turned over to an investigator.

    (A) Requisite Knowledge. Common causes of fire, fire growth and development, and policies and procedures for calling the investigators.

    (B) Requisite Skills. The ability to determine basic fire cause, conduct interviews, and write reports.

4.5.2 Secure an incident scene, given rope or barrier tape, so that unauthorized persons can recognize the perimeters of the scene and are kept from restricted areas, and all evidence or potential evidence is protected from damage or destruction.

    (A) Requisite Knowledge. Types of evidence, the importance of fire scene security, and evidence preservation.

    (B) Requisite Skills. The ability to establish perimeters at an incident scene.

Fire Officer II

5.5 **Inspection and Investigation.** This duty involves conducting inspections to identify hazards and address violations and conducting fire investigations to determine origin and preliminary cause, according to the following job performance requirements.

5.5.2 Determine the point of origin and preliminary cause of a fire, given a fire scene, photographs, pertinent data, and/or sketches, to determine if arson is suspected.

    (A) Requisite Knowledge. Methods used by arsonists, common causes of fire, basic cause and origin determination,

fire growth and development, and documentation of preliminary fire investigative procedures.

    (B) Requisite Skills. The ability to communicate orally and in writing and to apply knowledge using deductive skills.

Additional NFPA Standards

NFPA 921, *Guide for Fire and Explosion Investigations*

NFPA 1033, *Professional Qualifications of Fire Investigators*

Knowledge Objectives

After studying this chapter, you will be able to:

- Describe the role of the fire officer in determining the cause of a fire.
- List the common causes of fire.
- Explain when to request a fire investigator.
- Describe how to find the point of origin.
- Describe fire patterns.
- Describe how to determine the cause of the fire.
- Describe how to conduct interviews.
- Describe how to determine the cause of a vehicle fire.
- Describe how to determine the cause of a wildland fire.
- Describe the fire cause classifications.
- Describe the indicators of incendiary fires.
- Discuss arson.
- Discuss the legal considerations of fire cause determination.
- Discuss how to write an investigation report.

Skills Objectives

After studying this chapter, you will be able to:

- Determine the point of origin and the preliminary cause of a fire.
- Demonstrate how to secure the scene using rope or barrier tape to prevent unauthorized persons from entering the incident scene.
- Complete a fire incident report.

You Are the Fire Officer

You have successfully managed the extinguishment of your first structure fire as an incident commander. It was a relatively small, but unusual, fire in a shop that builds custom fine furniture from rare woods and metals. When the first attack line hit the fire in the workshop, the fire flashed brighter. You called for a larger back-up line, and the second engine company was able to extinguish the fire within a couple of minutes. Most of the fire was confined to a large industrial metal wastebasket that was under a cutting lathe. There was some fire extension to the cutting machine and the wall behind it and smoke damage throughout the shop.

The fire investigator is tied up on another call and asks you to look it over and determine whether it will be necessary to call in another investigator on overtime. Your sense of satisfaction immediately turns to anxiety. You have filled out dozens of National Fire Incident Reporting System (NFIRS) reports for automobile fires, food-on-the-stove fires, and other routine events, but you have never written up an industrial fire. This will be a new experience.

Walking back to the fire building, you run into the owner of the adjoining business who asks whether it is arson. He indicates that the company is in dire financial straits. You know that you will not need a fire investigator if you determine that the cause was accidental, but you will have to call one if you suspect arson.

You have some recollection of the occupancy from a familiarization visit 6 months earlier. Now you are looking at empty bins, where the rare woods were stored at that time. You also notice aluminum and magnesium stock on another shelf and a buildup of chips and shavings around the cutting lathe.

1. How will you determine the origin of this fire?
2. How will you determine the cause of this fire?
3. How will you secure the scene?

Introduction to Fire Cause Determination

Once the fire is extinguished and before the property is turned back over to the owner, a preliminary investigation must be conducted to determine how the fire started. The investigation serves several purposes, beginning with fire prevention. One of the best starting points for efforts to prevent future fires is to understand the causes of fires that have occurred in the past. In addition, the investigation is also important to establish the responsibility for fires that could have been caused by criminal acts, negligence, or fire code violation. In many cases, an investigation results in legal actions, ranging from prosecution for arson or related crimes to civil litigation over deaths, injuries, or property damage.

The fire officer who was in command of the incident is usually responsible for conducting the preliminary investigation, as well as completing the NFIRS documents or a local equivalent report. Sometimes, the incident commander delegates this responsibility to another fire officer, typically the officer in charge of the first-arriving company.

The fire officer who conducts the preliminary investigation must determine whether the fire incident requires a formal fire investigation. Most jurisdictions provide criteria for initiating a formal fire investigation and for requesting the response of an investigator or investigation team. In order to make this determination, the fire officer has to consider all of the circumstances as well as the cause, or probable cause, of the fire. Any fire that results in a serious injury or fatality meets the criteria for a formal investigation. Any fire that appears to be arson or related to some type of crime also meets the criteria. In most jurisdictions, any fire that causes significant damage, in which the cause cannot be established, would also meet the criteria for calling a fire investigator. Some agencies automatically dispatch a fire investigator to all working structure fires and to other specified incident types.

The legal responsibility for conducting fire investigations is usually defined by state legislation or regulations. The fire officer should determine the laws that apply to fire investigations and which agency is responsible for conducting investigations in different circumstances. This information should be available from the fire marshal's office for the jurisdiction or from an equivalent agency.

In situations where there is no formal investigation and a fire investigator does not respond to the scene to determine the cause and origin, the incident commander is responsible for determining and reporting the fire cause.

Common Causes of Fires

An exhaustive list of fire causes would be very long; however, a relatively small group of causes are responsible for a large number of fires (▼ **Table 17-1**). The largest fire cause category is unknown because the cause is never determined for nearly half of all fires. Among the fires for which a cause determination is made, incendiary and suspicious fires are the largest category, followed by open flames, cooking-related fires, electrical fires, and fires caused by smoking.

Although these data do not help in determining the cause of a particular fire, it is important to understand why fires occur. This information is particularly important in developing strategies to prevent or reduce the number and the severity of fires.

Requesting an Investigator

The fire officer should be able to determine a point of origin and a cause, or probable cause, of most fires. On small or routine incidents, this could be the only investigation that is conducted, and in those cases, the fire officer must carefully document the findings of the preliminary investigation. Typically, this documentation is accomplished with an NFIRS

Table 17-1 Common Causes of Fires	
Unknown	48.8%
Incendiary/suspicious	14.3%
Open flame	6.4%
Cooking	5.9%
Electrical distribution	4.6%
Smoking	3.2%
Heating	3.1%
Other heat	3.0%
Exposure	2.3%
Children playing	2.2%
Appliances	2.2%
Other equipment	2.2%
Natural	1.7%

Source: United States Fire Administration. *Fire in the United States 1989–1998.* Arlington, Virginia, National Fire Data Center, 2001.

incident report or an alternative reporting system that is used by the fire department.

When the officer is unable to determine the origin or cause of a fire, or when additional expertise is required, the fire officer may request a fire investigator to come to the scene. A qualified fire investigator has specialized training in determining the cause and the origin of fires and, in most cases, is certified in accordance with NFPA 1033, *Professional Qualifications of Fire Investigators.* The investigator could be a fire department member, an employee of a county or state fire marshal's office, or a law enforcement officer. In some cases, a fire department investigator is responsible for determining the cause and the origin of fires, and a law enforcement agency is responsible for any subsequent criminal investigation, if the cause is determined to be arson. State and federal law enforcement agencies become involved in some investigations, particularly where arson is suspected.

The agency or organization that is responsible for conducting fire investigations usually has a set of guidelines to help guide the fire officer in determining when to request an investigator. Generally, these situations include any fire that is suspected of being arson or suspected to involve criminal activity. In these cases, the investigator has more expertise in determining and documenting the exact cause and origin, as well as interviewing witnesses or suspects and gathering physical evidence. One of the fire investigator's primary responsibilities is to develop a well-documented case to be forwarded to the prosecutor. If the case goes to trial, the investigator must have credibility and experience in courtroom procedures to ensure that the facts are presented accurately.

A fire investigator should also be called whenever there is a civilian or fire service casualty. Any fire investigation that involves a death has to be particularly thorough and well documented. Whenever a fire death or serious injury occurs, it is in the public interest to determine the cause of the fire. If the cause of the fire was deliberate, a crime has occurred and must be fully investigated. If the cause was accidental, the information gathered in the investigation is important as a means to identify and evaluate fire prevention strategies. Whenever a death occurs, the coroner or medical examiner's office also must be contacted.

Fire investigators often respond to serious fires because of the degree of damage, even if the cause is known. Serious is a relative term, because a room-and-contents fire in a small community would probably be considered a serious event, whereas hundreds of similar fires might occur each year in a large city. A fire investigator might also be called for situations that could have caused great harm, such as a fire in a hospital or college dormitory, even if it was quickly controlled.

The criteria for requesting a fire investigator generally include situations in which the fire officer on the scene is unable to determine the cause of the fire; however, this often depends on the circumstances of the fire and the extent of the damage. Many agencies do not have the resources to dispatch a fire investigator to minor fires unless the situation meets one of the other criteria. The fire officer has to evaluate the circumstances of the situation in relation to local guidelines and policies; however, it is a good practice to request an investigator whenever the facts do not seem to make sense or there is a compelling reason to know the exact cause of the fire. For instance, the fire officer may determine that the fire began in a microwave oven and is concerned that this may be an indication of a defective product design. Often, in this type of situation, the insurance company also conducts an investigation to determine the exact cause.

When a fire does not meet the criteria for calling a fire investigator, local policy usually dictates that the fire officer on the scene conduct the investigation and complete the report. This policy often applies to fires that cause limited damage, with no deaths or injuries, and show no indication of criminal activity. For example, it is unlikely that an investigator would be dispatched to an engine compartment fire in an older-model car. The exact cause might be undetermined; however, the loss is relatively small, no injuries were reported, and there is no indication of a crime.

In many cases, the fire officer is able to determine the exact cause of a fire. Sometimes, the cause cannot be established with certainty; however, a probable cause can be identified. As noted in the statistics, the causes of many fires are never determined or at least are not reported.

Finding the Point of Origin

The first step in most fire investigations is to determine the point of origin. The **point of origin** is the exact physical location where a heat source and a fuel come in contact with each other and a fire begins. The point of origin is usually determined by examination of fire damage and fire pattern evidence at the fire scene. Flames, heat, and smoke leave distinct patterns that can often be traced back in order to identify the area or the specific location where the heat ignited the fuel and the burning first occurred. A fire investigator usually starts in the area where the least amount of damage occurred and follows the patterns back toward the area of greatest fire damage.

In many cases, eyewitnesses can provide valuable information that can assist in determining the point of origin. Sometimes, a witness actually saw what happened and can describe where it occurred. In other cases, witness accounts can identify the area where the fire was first observed or where indications such as visible smoke or a burning odor were noted. Sometimes, the witnesses can describe what was in the suspected area of origin before the fire and what normally occurred in that area.

Determining the point of origin requires the analysis of information from four sources:

1. The physical marks, or **fire patterns**, left by the fire
2. The observations reported by persons who witnessed the fire or were aware of conditions present at the time of the fire

3. Analysis of the physics and chemistry of fire initiation, development, and growth as an instrument to related known or hypothesized fire conditions capable of producing those conditions
4. Noting the location where electrical arcing has caused damage, as well as the electrical circuit involved. An electrical arc is a luminous discharge of electricity from one object to another, typically leaving a blackening of objects in the immediate area

Fire Growth and Development

To determine the point of origin, the fire officer must understand fire growth and development. Fire fighters learn the basic concepts of fire behavior in Fire Fighter I and II courses. The fire officer should also understand the three methods of heat transfer: conduction, convection, and radiation. The fire officer must be able to apply these basic concepts in order to understand fire growth and interpret the spread of a fire.

When a fire occurs in a room, the heat and products of combustion rise until they reach the ceiling. They then spread out laterally until they reach the walls. The heat and smoke then begin to bank down lower and lower into the room until the room is filled or until an open doorway or window is encountered. At that point, the heat and smoke tend to flow outward through the opening into the adjoining room or space, filling that area from the ceiling down. This process continues until the entire structure is filled with heat and smoke or until an opening to the outside is found, allowing the smoke and heat to escape from the structure.

While this process is occurring, the fire is continuing to develop and grow in the area of origin. The fire follows the same growth pattern as the smoke and heat, rising toward the ceiling, spreading out, and banking down. When a room fire reaches the point that flames are spreading across the ceiling, tremendous heat energy is radiated down onto the combustible contents within the room, such as sofas, chairs, and tables. As these items are heated, they begin to release combustible fuels into the atmosphere and eventually reach their ignition temperatures. As each item ignites, the rate of heat release increases dramatically, and the temperature increases even faster. Within a few seconds, the entire room flashes over, or becomes fully involved in fire. The fire then quickly spreads to adjoining rooms. The process often causes windows to break or holes to burn through to the outside of the structure, releasing heat and smoke and making the fire visible on the exterior of the structure.

If the fire does not have a fresh supply of oxygen, the fire slowly dies down to a smoldering phase. This may occur before or after a flashover has occurred. At this point, the fire sometimes completely dies out and self-extinguishes. In other cases, a fresh supply of oxygen is introduced, and the fire resumes the fully-developed phase.

Fire Patterns

The area of origin and, ultimately, the point of origin can often be identified by interpreting fire patterns. The fire pattern provides the fire officer with a history of the fire. A flaming fire produces a plume of smoke, heat, and flame. As a fire burns up against a wall, it spreads up and out, creating a V- or U-shaped pattern. The origin of the fire is typically at the base of the V- or U-shaped pattern (▼ **Figure 17-1**). This type of pattern is also known as a movement pattern because it allows the fire officer to trace the fire and smoke patterns back to the origin.

The second type of pattern indicates the intensity of the fire. The intensity pattern indicates how much heat (energy) was transferred to the surrounding area and objects. The intensity is indicated by the response of various materials to the fire's rate of heat release and heat flux. The intensity may produce a line of demarcation, which can indicate the area closest to the point where the greatest amount of heat was produced.

The analysis of char is closely related to the fire intensity pattern. **Char** is the blackened remains of a carbon-based material after it has been burned. For the fire officer, the depth of char can assist in determining the direction of fire spread; generally, the deeper the char, the longer the fire burned and therefore the closer to the area of origin (▶ **Figure 17-2**).

Depth of char is only one indicator of the apparent duration and intensity of a fire. Depth of char can be influenced by several factors, including different types of wood and the fire's intensity, so deep charring in locations that are remote from the area of origin. For example, there can be deep charring at a location where a fire vents through a window or doorway, because of the intensity of the fire as it combines with the fresh air in that area.

(**Figure 17-2**) Depth of char.

Determining the Cause of the Fire

The specific cause of a fire can be determined only after the point of origin has been located. The cause refers to the particular set of circumstances and factors that were necessary for the fire to have occurred. The cause determination can be approached as a three-step process:

- The first step is to determine of the source of ignition. This generally includes some device or piece of equipment that was involved in the ignition. A competent ignition source must be present to ignite the fuel.
- The second step is to determine the fuel that was first ignited. Both the type of material and the form of the material should be identified.
- The third step is to determine the circumstances or human actions that allowed the ignition source and the fuel to come together, resulting in a fire.

This three-step process only describes the factors that must be determined in order to conclusively establish the cause of a fire. A systematic and scientific process should be used to perform each step. It is not sufficient to identify and focus on a possible cause that fits the circumstances that were noted; in order to be certain that the cause has been properly identified, the fire officer or fire investigator must eliminate any alternative theories or explanations. For example, before a fire investigator can make a determination that a fire was deliberately started, all possible accidental causes must be considered and eliminated. Any explanation of a fire cause is only potential or theoretical until alternative explanations have been eliminated. Potential causes should be ruled out only if there is definite evidence that they could not have caused the fire. For example, an electric heater can be ruled out if it was unplugged and had not been used for several weeks before the fire occurred.

The fire officer does not have to consider every possible fire cause in relation to every fire that occurs. The facts of the situation can be considered in relation to previous knowledge

(**Figure 17-1**) Often, the point of a V-pattern is near or at the point of origin.

to identify a list of potential causes for the fire. The fire officer must then apply deductive reasoning, considering each possible cause one by one and sequentially eliminating each cause that is not supported by the evidence. For example, when investigating a house fire where an outside wall was ignited from the exterior in proximity to the electrical service, the fire officer must consider any potential source of ignition that could start a fire in that location. The fire officer would identify lightning as a potential cause of the fire. The officer would then determine whether there was lightning in the area at the time of the fire. If it can be established that there was no lightning, then lightning can be eliminated from the list of possible causes.

The fire officer might also consider that the fire could have been caused by an electrical short circuit. If careful examination of the electrical equipment indicates that no short circuit occurred, that possibility can be eliminated.

The fire officer would also know that hot coals or ashes can start fires. The presence of a wood-burning fireplace would cause the officer to ask the homeowner whether it had been used recently and, if so, whether the ashes had been placed outside. The owner might state that he had built a fire in the fireplace the night before and had deposited the ashes outside this morning, thinking that they were out. The location where the ashes were placed coincides with the point of origin of the fire. From these facts, the fire officer could deduce that the fire resulted from hot ashes being dumped too close to a combustible exterior wall.

The cause of a fire cannot be established until all potential causes have been identified and considered and only one cannot be eliminated. After all investigative possibilities have been exhausted, if there are still two or more potential causes, the cause of the fire is considered undetermined.

Source and Form of Heat Ignition

The **source of ignition** is the energy source that caused the material to ignite. The source of ignition must have been located at or near the point of origin. The fire officer may be only able to infer a probable ignition source; some sources of ignition remain at the point of origin in recognizable form, whereas others may be altered, destroyed, or removed. For example, the source of ignition could be a short circuit in an electrical device. The form of the heat of ignition would be an electrical arc. Visible evidence of the short circuit would be present, although the arc could have significantly damaged and altered the device itself.

If the equipment that provided the source of ignition was a cigarette lighter, the form of the heat of ignition would be an open flame. The person who started the fire could have removed the source of ignition by placing it back in his pocket and leaving the scene. If the pilot light flame on a gas water heater ignited the fire, the source of ignition is likely to be found unaltered at the point of origin.

The source of ignition must have enough energy and must remain in contact with the fuel long enough to cause it to ignite. A competent ignition source has three components:

1. **Generation.** The ignition source must produce sufficient heat energy to raise the fuel to its ignition temperature.
2. **Transmission.** Sufficient heat energy must be transmitted from the source to the fuel to raise the fuel to its ignition temperature. Heat can be transferred through conduction, convection, or radiation.
3. **Heating.** The heat transfer from the source to the fuel must continue long enough for the fuel to be heated to its ignition temperature.

Material First Ignited

The fire officer also needs to identify the material that was first ignited. The **type of material** first ignited refers to the nature of the material itself. For example, the type of material might be cotton. The **form of material** tells how that material is used. For example, cotton could be in the form of cotton plants in a field, baled cotton fibers, rolls of cotton thread, woven cloth, or clothing.

The physical configuration of the fuel is an important characteristic in ignition. A 12- by 12-inch wooden beam and a pile of wood shavings are both wood; however, the beam is much more difficult to ignite. This information is significant in determining what type of heat source could have ignited the material. A pile of wood shavings could be ignited by a dropped match, whereas the wooden beam would have to be exposed to a more powerful source of heat.

Ignition Factor or Cause

The third important factor in determining the cause of a fire is the sequence of events that brought together the source of ignition and the fuel. The cause of a fire could be a human act that was either accidental or deliberate. In the case of an accidental cause, negligence could be a factor; this is the failure to exercise appropriate care to avoid an accident. The fire cause could also be related to a mechanical failure; a poor design or improper assembly of a device; a worn-out piece of equipment; a natural force, such as lightning; or several other explanations.

Failure analysis is a logical, systematic examination of an item, component, assembly, or structure and its place and function within a system, conducted in order to identify and analyze the probability, causes, and consequences of potential and real failures. If the cause of a fire is determined to be related to some malfunction that occurred within a device or system, failure analysis is performed to identify what happened and why. The failure analysis could identify a design error, malfunctioning component, inadequate maintenance, operator error, or some other factor.

Fire Analysis

Fire analysis is the scientific process of examining a fire occurrence to determine all of the relevant facts, including the origin, cause, and subsequent development of the fire, as well as determining the responsibility for whatever occurred. Fire analysis brings together all of the available information that can be obtained to examine what happened. The fire officer may need to construct a timeline of events to establish the sequence of events that led up to the fire.

Conducting Interviews

Interviews are an important source of information in determining the cause and origin of many fires. The fire officer may have to interview fire victims, witnesses, fire fighters, and possibly suspected perpetrators. Each individual may have relevant information that will help to identify the cause and origin of a fire. For example, a witness might indicate that she saw lightning strike the building a few seconds before seeing smoke and flames. In another case, a witness might report that a can of gasoline was spilled accidentally and a few minutes later the basement erupted in flames. A witness to another fire could have seen a man speeding away from the scene in a green pickup moments before she saw flames in the window of the house.

Fire fighters can also be valuable sources of information. The first-arriving crew might report that the front door was broken open when they arrived. They might also have valuable information about the color of the smoke or flames or the area where the fire was burning when they arrived.

When preparing for an interview, the fire officer should first review the facts that are already known or believed to be known. This allows the interviewer to recognize whether the witness is corroborating that information or providing conflicting information. The fact that the witness has a different version of the story does not necessarily indicate that the person is lying; the person could be misinterpreting the information, could have made a different observation, or could have an inaccurate memory. It could be that the original information was inaccurate and the witness is providing good information. The presence of disputed or conflicting information means that further investigation is required for the actual facts to be determined.

A witness statement should be disregarded only if it can be established with certainty that the information is incorrect. For example, if a witness indicates that he saw a man running from the back of a house before the fire started, yet the investigator has already determined that there are no footprints in the snow in the backyard, the interviewer would have to suspect that the witness is either lying or mistaken.

The interviewer should be careful not to underestimate the information a witness might be able to provide. Open-ended questions allow a witness to tell what they saw or know, whereas questions that seek a "yes" or "no" answer limit the exchange of information. For example, the fire officer should ask the first-arriving crew to describe what they saw when they arrived, not ask whether there were flames showing through the front windows. After the witness provides an overall description, more direct questions can be asked to clarify the facts.

Interviews that are conducted to determine the preliminary cause and origin of a fire are typically conducted at the fire scene. The officer should introduce himself or herself to the witness and note the person's name and contact information in case a follow-up interview is needed. Interviews are normally conducted with only one individual at a time, so that the comments made by one person do not influence the responses of another. The fire officer should not interrupt the witness unless it is essential for clarification. After the interview is complete, the officer should thank the individual for providing assistance and provide his or her contact information in case the witness thinks of anything else.

At times, a witness can be reluctant to make a full disclosure of information. This can occur for various reasons, including the possibility that the person was involved in causing the incident and is trying to conceal information. A special type of interview, called an interrogation, is used when an individual who is attempting to conceal information is being questioned. Interrogations require special knowledge and skills that are beyond the scope and duties of most fire officers. Only a trained fire or police investigator should conduct interrogations. The Fifth Amendment to the United States Constitution protects against self-incrimination, so when suspects who are under arrest are questioned, it is important to ensure that they are advised of their rights. Only trained police officers should conduct interviews when it is determined that a crime has been committed.

Once an interview is complete, the information should be documented. The documentation could be in the form of the interviewer's notes, or if a more formal documentation is needed, a hand-written statement could be requested from the witness. Another alternative is to record the witness statement with a tape recorder or video camera.

Vehicle Fire Cause Determination

When discussing fire cause and origin determination, most fire officers naturally think of investigating a structure fire. In reality, most fire departments respond to more vehicle fires than structure fires, so a company officer must also be prepared to investigate a vehicle fire. Chapter 15 of NFPA 921, *Guide for Fire and Explosion Investigations*, provides a standard method to conduct a vehicle fire investigation.

According to the U.S. Fire Administration, 66% of vehicle fires were caused by mechanical failure or malfunction. The second leading cause, which accounted for nearly 20% of vehicle fires, was arson. Some cars are burned to defraud insurance companies, whereas others are burned for revenge or to conceal other crimes.

The same basic principles apply to any fire investigation. The fire officer should trace the fire's development back to the point of origin, looking for the area where the most fire damage occurred and the areas of low burn. Like structural fires, there is often a distinctive burn pattern, such as a "V," that help identify the area of origin; however, this should be viewed with caution because of the characteristics of the materials that are present within most vehicles. Fires involving the fuel or tires tend to be particularly intense.

Once the point of origin is established, the specific cause must still be determined. If the cause is related to a mechanical malfunction, the exact cause may be difficult to establish without an automotive expert. Additional factors should be considered to determine whether more extensive investigative techniques are required. Most fire departments do not conduct extensive investigations on vehicle fires unless arson is suspected, because of the time and expense involved.

The first objective in most vehicle fire investigations is to look for indications of arson. Because some expensive vehicles are burned when the owners are unable to keep up with the high payments, insurance companies often conduct a more thorough investigation if they suspect that the fire was not accidental. The vehicle should be examined for its general condition. Is it a new, expensive vehicle burning at the end of a dead-end street in an unpopulated area? Have items of value, such as an expensive stereo, been removed? Are there indications of accelerant use inside the vehicle? Fuel lines should be checked to see whether they have been loosened. The condition of the tires might indicate that new tires were removed and replaced with old tires before the fire.

Several potential sources of accidental ignition must be considered for a vehicle fire. The mechanical sources include the electrical system, exhaust system, catalytic converter, and turbocharger. In older-model cars, a carburetor backfire can cause a fire within the engine compartment. Smoking materials cause many passenger compartment fires.

When conducting the investigation, note the make, model, and year, as well as the vehicle identification number (VIN). This information allows for a review of previous fires that have occurred in the same make, model, and year. Some makes and models have a history of fires, and the reporting of this information is useful in identifying product defects. A diagram of the scene should be drawn. When possible, photographs should be taken before the vehicle is removed.

An interview should be conducted with the owner and/or operator to determine when the vehicle was last driven and how far, total vehicle mileage, operating abnormalities, date of last service and when it was last fueled, how was the vehicle equipped (e.g., compact disc players, custom wheels), and what personal items were in the vehicle.

Additionally, if the vehicle was being driven, the officer should determine the speed and any loads that were being pulled, any abnormal indications before the fire, when and where the smoke and/or fire were observed, what actions were taken by the driver, and how much time elapsed before the fire was extinguished.

Wildland Fire Cause Determination

According to the National Interagency Fire Center, there were nearly 86,000 wildland fires in 2003, and nearly 5 million acres were burned. A large proportion of these fires are caused by human factors, which need to be identified in order to develop effective fire prevention programs. In cases of arson, a thorough investigation is required to pursue criminal prosecution.

The characteristics of wildland fires are quite different from those of structural fires. The National Wildfire Coordinating Group has developed professional qualifications standards for wildland fire investigators. Wildland fires are influenced by environmental conditions, including topography, fuel load, wind, and weather. Wildfires tend to spread vertically through convection, from lower vegetation to taller vegetation, and horizontally through radiation. The rate of spread varies, depending on the type of material burning and its density, the wind speed and direction, the humidity and fuel moisture content, the slope of the terrain, and natural features, such as valleys. All of these factors must be considered by the investigator.

When a fire burn pattern on the side of a hill is investigated, the point of origin is most likely on the lower part of the slope, but not necessarily at the lowest point. The fire burns up the hill very rapidly, but it also burns down the hill at a slower rate. The wind also dramatically affects fire progression, on flat ground or on a slope.

Evaluating the degree of burn on the fuels helps establish the direction of travel. Ash residue can be also evaluated to establish the direction it was blown, which indicates the wind direction. When a tree burns and falls, the remaining trunk is usually burned at an angle, creating a point. The point is generally on the side of the stump opposite the direction of fire approach.

Once the area of origin has been determined, the investigator can look for clues to help determine the cause of the fire. The following are examples of evidence that might be found at the point of origin:

- Campfire remains
- Time-delay devices
- Cigarette remains
- Lighters
- Multiple ignition points
- Splintered trees (indicating lightning strikes)
- Fulgurites (glassy, rootlike residue resulting from lightning strikes)
- Piles of fuel that are prone to spontaneous combustion
- Barrels used to burn trash

- Fallen electrical wires
- Trees on power lines
- Railroad tracks

Although none of these observations necessarily determines the exact cause of a fire, they are helpful in identifying potential causes. The area of origin should be carefully preserved and photographed. Witnesses should be interviewed and bystanders noted and interviewed as needed. Once the investigation is complete, a report needs to be completed.

Fire Cause Classifications

Once the facts are known, the circumstances of the fire are generally divided into four classifications according to NFPA 921. They include:

1. **Accidental fire cause.** All of those fires for which the proven cause does not involve a deliberate human act to ignite a fire or spread fire into an area where it should not go.
2. **Natural fire cause.** Fires caused by lightning, earthquakes, wind, and other natural forces without human intervention.
3. **Incendiary fire cause.** Any fire that is deliberately ignited under circumstances in which the person knows that the fire should not be ignited.
4. **Undetermined fire cause.** Any fire in which the cause cannot be proved is classified as undetermined. The fire may still be under investigation.

Accidental Fire Causes

Many accidental fires result from human activities that are not intended to start or spread a fire. For example, careless smoking causes many accidental fires. Finding evidence of smoking, such as a lighter, ashtray, or cigarettes near the point of origin, may suggest that smoking could be involved as an ignition factor.

Cooking fires are the leading cause of home fires in the United States. The most frequent ignition cause is unattended cooking. When the fire originates on the stove, a pan is found on one of the burners, and the burner knob is on, the likely cause of the fire is accidental. Heating is the third leading cause of residential fires, with the two primary ignition causes being improper maintenance and combustibles located too close to the heating device. Other accidental fire causes include refueling of gasoline-powered equipment near an ignition source, placement of fireplace ashes in a combustible container, and sparks from welding.

When wood is continuously subjected to a moderate level of heat, below its normal ignition temperature over a long period, it starts to break down into carbon. This chemical decomposition, called **pyrolysis**, results in a gradual lowering of the ignition temperature of the wood until auto-ignition occurs. Pyrolysis should be considered if the area of origin includes steam pipes, flue pipes for a fireplace, or a wood-burning stove. The use of zero-clearance fireplace flues has caused an increase in this type of ignition. Fluorescent light ballasts can also cause pyrolysis to occur.

Electricity and electrical appliances are involved in many accidental fire ignitions. The most common electrical fire scenario is misuse by the occupant, such as overloading electrical circuits, using lightweight extension cords for major appliances, or operating too many devices for the electrical service. Any of these errors can overload the wiring or the device, causing heat and eventually igniting a fire.

Many accidental fires have resulted from leaving automatic coffeemakers turned on. After a few hours, the remaining liquid boils off and the unit begins to overheat. If the automatic temperature sensor fails or is not present, the unit eventually gets hot enough to ignite the plastic parts. Nearby combustible materials may become involved once the plastic is burning. An altered temperature sensor or missing thermal cutout unit may indicate an attempt to deliberately ignite a fire.

Large appliances, like dishwashers and clothes dryers, may fail if their high-temperature controls or timing mechanisms fail. Excessive lint buildup that impedes air exchange and overheats the dryers is a frequent accidental scenario at multiple-family buildings with community laundry rooms. Dishwashers are susceptible to plastic dishes and utensils falling onto the element. Always check for an object on the heating element when the dishwasher is the point of origin.

Electrical devices and appliances that start fires usually produce evidence of electrical damage on their power supply cord relatively close to the device or appliance. The wire insulation near the appliance often melts because of the heat produced by the burning appliance or other materials ignited by it. The melted insulation usually results in energized conductors' contacting each other, resulting in a short circuit in the cord. An exception would be an appliance or device malfunction that results in the circuit breaker or fuse tripping, so that the power cord is de-energized before the insulation melts.

Natural Fire Causes

Lightning is a powerful force that causes many fires. A lightning strike in the forest can ignite a tree, and a strike on a building can have a tremendous destructive effect, sending several thousand volts through the electrical wiring system. This surge can destroy televisions, computers, telephone systems, and anything else that is plugged into the electrical system. Lightning can also follow antenna wires, telephone wires, metal plumbing, or the steel structure of a building and ignite combustibles that come in contact with them.

When lightning is suspected, the fire officer should look for a contact point, usually near the top of the structure, concentrating on roof peaks with metal edging, antennas, or large metal objects, such as air-handling units. If the building's electrical system is involved, the surge usually fuses or

VOICES OF EXPERIENCE

❝ Fire fighters should never give out any investigation information to anyone. ❞

While serving as an arson investigator, I was assigned to investigate a fire that had occurred at a fast food restaurant. When I arrived on the scene, the incident commander met with me and explained what he had found upon arrival. Apparently, the fire had started in the lower level of the building, adjacent to an entrance. The doorway of the entrance was open to ventilate the area. I thanked the incident commander and began an external inspection of the restaurant.

As I was inspecting the exterior of the building, a man approached me and identified himself as the owner. He asked me what had happened. I explained to him that when I knew more after my inspection, I would find him and speak to him. I asked him to remain on the property to answer any questions. I then informed him that he couldn't enter the building until the investigation was completed.

As I examined the interior of the building, I made my way into the lower level and examined the area where the fire was discovered by the incident commander. As I turned and looked out the door, I saw a fire fighter speaking to the owner of the building. The fire fighter was pointing to the area where the fire was discovered. I asked the incident commander to please ask the fire fighter to come see me.

The fire fighter told me that he was explaining what had happened to the owner of the building. I asked him not to speak to anyone regarding the incident or any fire incident under investigation. The fire fighter went on to explain that the owner was a decent man and a former employer of his. Once again I asked the fire fighter not to speak to anyone regarding the incident or any other fire under investigation.

Always remind your fire fighters that if a member of the public asks a question, they should be polite and refer the citizen to the senior officer on the scene. Fire fighters should never give out any investigation information to anyone. Everyone likes to feel important and be in the know; however, a small amount of information given to the wrong person can have a disastrous effect on a fire investigation.

Timothy A. Hawthorne
Cranston, Rhode Island Fire Department
16 years of service

melts the main fuses or circuit breakers. The surge can travel through television cables, phone lines, power lines, or even natural gas or propane lines, so the point of entry for each utility should be examined for scorching. The joints of aluminum rain gutters may show the same signs.

Be cautious with aluminum-sided houses where the electrical service is mounted on the side of the house. The entire house exterior could be energized if a lightning strike has fused the circuit breaker box.

Various weather services track lightning strikes throughout the United States. If there are indications that lightning could have caused a fire, the fire officer should check with weather officials to determine whether any lighting strikes were recorded in the area. Many fire investigators commonly include a notation, such as, "It was a clear and calm night with no lightning observed," in the introduction to their reports to establish that this potential source of ignition was eliminated.

Natural forces are more powerful than human forces. Earthquakes, tornadoes, floods, and hurricanes can all cause fires. Natural forces often cause power lines to contact trees or fall onto structures or cause sheets of metal to be blown across power lines. Earthquakes may also cause gas mains to break and release flammable gases into structures where they find an ignition source. Extremely hot lava from a volcano can easily start a forest fire or incinerate a community.

Incendiary Fire Causes

Because an incendiary fire is an intentional occurrence, the direct cause is a person; however, the methods and the reasons for starting fires vary tremendously. Because of the complexity in determining an incendiary fire, a section of this chapter is devoted to explaining why people intentionally set fires and what evidence might indicate that a fire was intentional.

An incendiary fire is one that is intentionally started when the person knows it should not be started. An incendiary fire is not necessarily arson. **Arson** is the crime of maliciously and intentionally or recklessly starting a fire or causing an explosion. The legal definition of arson as a criminal act is defined differently within different jurisdictions. The fire officer may be involved in making the determination of a fire's cause and origin and classifying it as incendiary, but ultimately, a prosecuting attorney or grand jury decide whether it will result in arson charges. The fire officer should be aware of the local or state judicial system that governs his or her jurisdiction.

Undetermined Fire Causes

No matter how much training and experience a fire investigator has acquired, how many resources are available, or how much effort is extended, sometimes the cause of a fire cannot be determined. This may be because of the extensive damage caused by the fire or the firefighting activities. This may occur

when there are multiple reasons that the fire could have started that cannot be ruled out. The evidence may not exist to make a definitive determination of the cause. For example, if a gas can is found next to the water heater, it may be impossible to determine whether the vapors were accidentally ignited from the pilot light or intentionally ignited by an arsonist. Without additional evidence, either situation is possible, so the cause cannot be determined with certainty.

The absence of any logical cause is another reason for classifying the cause of a fire as undetermined. Obviously, something caused the fire, but the lack of evidence or knowledge precludes the fire investigator from determining the cause.

Indicators of Incendiary Fires

The primary difference between an incendiary fire and an accidental fire is that these are intentional actions, sometimes with malicious intent. It is essential for the fire officer to eliminate both accidental and natural causes before making a determination of an incendiary cause. Failure to eliminate accidental and natural causes makes it impossible to prove an incendiary cause.

Once the other causes have been eliminated, there are many conditions or factors that may indicate an intentional fire. These typically fall under five general categories:
1. Disabled built-in fire protection
2. Delayed notification and/or difficulty in getting to the fire
3. Accelerants and trailers
4. Multiple points of origin
5. Tampered or altered equipment

Disabled Built-in Fire Protection

A disabled sprinkler system is most likely to be encountered in large industrial or commercial fire losses. When evaluating the incident, the fire officer should identify the status of fire suppression and alarm systems at the time the fire started. If a building was protected by a central station monitoring company, that service should be one of the first entities to call 911. If the call came from another source, particularly witnesses outside the property, the system might have been disabled or impaired.

The fire officer should also look for damaged or vandalized sprinkler hook-ups, hose cabinets, hard-wired smoke detectors, and high-rise communication systems. At a high-rise under construction, a tractor-trailer–sized dumpster was placed in front of the standpipe/sprinkler connection to prevent the fire department from being able to use the built-in fire protection system.

Delayed Notification and/or Difficulty in Getting to the Fire

One of the reasons businesses use a central station monitoring service is to ensure a prompt notification to the fire department when a smoke detector, water-flow, or manual pull station is activated. Without such a service, a fire has additional time to grow. In an urban environment, an interior fire could have been flaming for 20 to 25 minutes before a passerby observed flames coming from a window. If the business is in a rural setting or an isolated industrial park, the first notification may be after the roof collapses and there are flames shooting up into the sky.

The fire officer should be alert for conditions or situations that delay the fire department's ability to get to the fire. Malfunctioning keys and key cards, vandalized doors, and stock or other materials blocking access are conditions to note. In addition, points of origin that are in the attic, basement, or closet should receive special consideration. Professional arsonists prefer these places to start a fire because they are prone to going unnoticed for longer periods of time. Sometimes, an arsonist starts another fire to divert resources to a different location, delaying response to the real target property.

Accelerants and Trailers

<u>Accelerants</u> are agents, often an ignitable liquid, used to initiate a fire or increase the rate of fire growth. Trained canines or detection devices can usually detect ignitable liquid accelerants. In addition, ignitable-liquid–fueled fires often leave distinct burn and char patterns (▶ **Figure 17-3**).

<u>Trailers</u> are materials used to spread a fire from one area of a structure to another, causing a fire to grow more quickly. Materials most often used as trailers are:
- Paper towels
- Black gunpowder
- Film wrapped in paper rags
- Kerosene or other combustible liquids
- Gasoline or other flammable liquids
- Decorative streamers

Figure 17-3 Ignitable-liquid–fueled fires leave distinct burn and char patterns.

- Cotton batting
- Paper
- Sheets or rolls of fabric softener
- Newspapers
- Combinations of these items

Trailers usually leave a distinct fire pattern that resembles the material's shape and often runs from one room to the next. Liquids typically have irregular edges and areas that are burned where the liquids pooled because of low spots. Sometimes, a trailer of ignitable liquid is poured from one electrical outlet to another to simulate an electrical fire in an attempt to mislead the investigator.

Multiple Points of Origin

Professional arsonists often set multiple ignition points inside a building. This is sometimes done in case one burns out prematurely as well as to maximize the amount of fire growth before the fire department can respond. In other cases, the arsonist is attempting to cause confusion or to trap the occupants by blocking access to all of the exits.

Sometimes, an arsonist sets fire to multiple buildings at the same time, presenting the fire department with an even more complicated problem. The responding units may be sufficient to manage one fire, but two or three buildings on fire at the same time is a different situation.

The appearance of multiple points of origin does not always prove that a fire was intentional. In some cases, material falling from the ceiling and burning at the floor level can create a secondary "U" or "V" pattern, resembling an additional point of origin. There may be nothing at that location that could have caused the fire to ignite.

Multiple points of origin can also occur when a major electrical surge causes ignitions at different locations in a building's electrical system or an overpressure in a gas system causes every pilot light to become a torch. Situations have occurred where fires started simultaneously in several kitchens and water heater closets.

Tampered or Altered Equipment

The fire officer should always look for indications at a fire scene that appear to be unusual. Starting from the outside, look for indications of a forcible entry into the structure before the fire department arrived. If the fire origin was electrical, look for electrical devices that could have been altered or appear unusual. A professional arsonist is likely to try to make a fire appear accidental. Arsonists tend to damage irons, heating equipment, and stoves to make structure fires appear accidental. Starting a fire in the engine compartment could make a car fire appear to be accidental.

Arson

Arson is the crime of maliciously and intentionally, or recklessly, starting a fire or causing an explosion. National statistics indicate that one in every four fires is of incendiary origin.

According to the NFPA, the number of structure fires reported as intentional in the United States decreased by 61% from 1978 to 2001. The NFPA statistics also indicate that related deaths fell by 24% during the same period. Unfortunately, this number does not reflect all intentional fires. The figures include only structural fires and only those that are proved to be intentional, omitting any fires in which the cause was undetermined or suspicious.

Arson has consistently had the highest rate of juvenile involvement when compared with all other Federal Bureau of Investigation index crimes (the most serious felonies). The National Institute of Occupational Safety and Health conducts an investigation after every fire fighter line-of-duty death. These investigations have shown that arson and intentionally set fires are a contributing factor in fire fighter deaths in structure fires.

Arson Motives

What motivates people to deliberately start fires? It is important to understand why people set fires in order to apprehend and prosecute arsonists. A motive is not necessary for the crime of arson to be proved, but a jury is more likely to convict an arsonist if they understand why the crime was committed. Special fire investigation units or law enforcement agencies are usually involved in the investigation of an arsonist's motive. There are six basic motives for arson:

1. Profit
2. Crime concealment
3. Excitement
4. Spite/revenge
5. Extremism
6. Vandalism

Profit

Monetary gain is frequently the motive for arson. Most often, the plan is to collect insurance money. An individual or company might feel that collecting insurance proceeds is the easy way out of a financial crisis. Indicators of insurance fraud include inability to meet payments, failure to complete business contracts, poor sales volume, or lack of supplies and inventory. When the circumstances are suspicious, investigators verify the insurance coverage amounts and look for recent changes in the policy. In some cases, the insured makes claims for burned inventory that was actually removed before the fire or never existed.

Elaborate schemes have been conducted to manipulate supply and demand for various products, by destroying inventory or manufacturing facilities in order to cause a shortage and drive up prices. Arson has also been used for extortion or to eliminate competition. Sometimes, a contractor or individual starts a fire and then offers to repair the damaged property. Many vacant buildings have been burned to save the cost of demolition.

In some cases, a fire is started because the individual wants to relocate or wants the insurance money to renovate the property. This type of arson occurs most often in residential properties.

Crime Concealment

Arson is often used to conceal other crimes. A business owner or employee may elect to burn the business in order to destroy records that show embezzlement of cash, supplies, or inventory. Burglars might set fire to a building to destroy evidence of their entry, eliminate fingerprints, or even conceal the fact that a theft occurred before the fire. Arson is also used to destroy evidence of other crimes, including murder, or to create a distraction while a crime is taking place in another location.

Excitement

Sometimes, a fire is started for the excitement of the arsonist, who could be seeking thrills, attention, or recognition. The arsonist could be planning to make a dramatic rescue to gain praise or to obtain recognition for discovering or extinguishing the fire. The arsonist may simply enjoy the excitement and spectacle that are generated by a fire. Unfortunately, fire fighters, security guards, and even police officers are sometimes found to be involved in starting these fires.

Spite/Revenge

Spite and revenge fires are the most deadly forms of arson because of the intense emotions that are involved. The arsonist may intentionally set a fire in the exit stairway of an occupied building and strikes while the occupants are asleep. These fires are usually set in retaliation for an injustice, real or imagined. Frequently, the process is initiated by hatred, jealousy, or other uncontrollable emotions, such as a lover's quarrel, divorce, or bar fight. In many cases, the individual has consumed alcohol or drugs before starting the fire.

Extremism

Extremism for a variety of causes has been the reasoning for setting fires for many centuries. Abortion clinics, religious institutions, businesses that were ecologically damaging, and labor disputes have all been prone to this motive in re-

cent years. With the recent emphasis on terrorism, this factor should be on each investigator's mind in a wide variety of circumstances.

An extremist may want to cause a monetary loss to the person or business or simply to bring attention to a cause. In many cases, the arsonist has both goals in mind. The second motivation can often prove tragic. Radical activists know that a large loss of life focuses attention on their cause and may cause others to withdraw from the targeted activities. Frequently, incendiary devices are used in these crimes.

Vandalism

The motive for vandalism is simply to cause damage for its own sake. Vandalism is most often directed toward schools, abandoned structures, vegetation, and trash containers. In most cases, the fire setter is within walking distance of his or her home and does not return to the scene of the crime.

Legal Considerations

There are many legal considerations that must be considered when investigating fires. Although the law provides that there is a public interest in determining the cause and origin of a fire, there are also the competing interests of a citizen's rights to privacy and due process. If a case goes to court, the fire officer who investigated the fire is usually called to testify and may be challenged on issues of proper procedure.

Searches

It has been widely upheld that when a fire has occurred and the fire department has been called, the fire department has the right to determine the cause and origin of the fire. This process must be accomplished in accordance with the law. In *Michigan v. Tyler* (1978), the US Supreme Court held:

"Fire officials are charged not only with extinguishing fires, but with finding their cause. Prompt determination of the fire's origin may be necessary to prevent its recurrence, as through the detection of continuing dangers such as faulty wiring or a defective furnace. Immediate investigation may also be necessary to preserve evidence from intentional or accidental destruction."

The fire officer must take care to avoid an unlawful search and seizure, which is prohibited by the Fourth Amendment of the US Constitution. Typically, no search warrant is needed to enter a fire scene and collect evidence, when the fire department remains on scene for a reasonable length of time to determine the cause of the fire and as long as the evidence is in plain view of the investigator. This principle has been reaffirmed by the US Supreme Court in *Michigan v. Clifford* (1984).

The aftermath of a fire often prevents exigencies that will not tolerate the delay necessary to obtain a warrant or to secure the owner's consent to inspect fire-damaged premises. Because determining the cause and origin of a fire serves a compelling public interest, the warrant requirement does not apply to such cases.

The plain view doctrine allows for potential evidence to be seized during the processing of a fire scene, if the fire investigator had a legal right to be there and the evidence is in plain view. Under *Michigan v. Clifford*, the court held that as fire fighters remove rubble or search other areas where the cause of fires is likely to be found, an object that comes into view during such process may be preserved. In the same decision, the court also found that once the investigator has determined where the fire started, the scope of the search authority is limited to that area. After the cause and the origin have both been determined, a search warrant or consent is required for any further search.

If re-entry is needed after the fire department leaves the scene or to conduct a search for evidence of a crime after the cause and origin have been determined, the investigator must obtain a search warrant or receive permission from the occupant. The general provisions were clarified in *Michigan v. Tyler*:

- No search warrant is needed when fighting a fire or remaining on scene for a reasonable period of time to determine the cause of a fire and any evidence is admissible under the plain view doctrine.
- Administrative search warrants are needed for re-entry that is not a continuation of a valid search, when the purpose is to determine the cause of the fire.
- A criminal search warrant is needed when re-entry is not a continuation of a valid search and the purpose is to gain evidence for prosecution.

Securing the Scene

A fire officer who conducts a preliminary fire cause investigation and suspects that a crime has occurred should immediately request the response of a fire investigator. When this occurs, the scene must be secured in order to protect any evidence that exists. If the fire department leaves the scene unsecured, any evidence that is collected after that point could be called into question. The fire officer must ensure that fire department personnel maintain custody of the scene until the investigator arrives.

Protecting the scene includes preventing unauthorized personnel from entering into the scene. To create a security perimeter, the fire officer can use fire line tape or police crime scene tape secured to objects, such as trees and fence posts. Natural barriers can also be used to aid in securing the area. Objects such as fences or hedges can be used along with barrier tape to completely surround an area.

All access to and from the area must be controlled. To be certain that no unauthorized personnel enter the area, a fire fighter or law enforcement officer may need to be posted to limit access. Posting a guard preserves the chain of custody over the scene and any evidence that is present until the fire investigator arrives. Failure to do so may require the fire department to get a warrant to return to the fire scene.

Barrier tape may also be placed across a doorway to prevent unauthorized entry into a room or building. To secure and protect even smaller areas of evidence, the fire officer may decide to cover it with a plastic sheet or tarp.

The number of fire personnel that are allowed into the secured area should be strictly limited. Each person who enters the area increases the likelihood that the scene will be contaminated. The area should be treated as a crime scene, and only activities that are essential to control the emergency or protect the scene should be conducted. Fire fighters should never collect artifacts as souvenirs of their firefighting adventure, particularly when a scene is being secured.

Evidence

Evidence includes material objects as well as documentary or oral statements that are admissible as testimony in a court of law. Evidence proves or disproves a fact or issue. The fire officer must consider three types of evidence.

1. **Demonstrative evidence**: Tangible items that can be identified by witnesses, such as incendiary devices and fire scene debris
2. **Documentary evidence**: Evidence in written form, such as reports, records, photographs, sketches, and witness statements
3. **Testimonial evidence**: Witnesses speaking under oath.

If the fire officer has determined that the fire requires a formal investigation, then every effort should be made to protect and preserve the fire scene evidence. The structure, contents, fixtures, and furnishings should remain in their prefire locations, as intact and undisturbed as possible.

Evidence plays a vital role in the successful prosecution of arson cases. In order to prove arson, the fire investigator must rule out all potential accidental and natural causes of the fire. The investigator must consider all possible circumstances, conditions, or agencies that could have brought together a fuel, an ignition source, and an oxidizer, resulting in a fire or combustion explosion.

Artifacts, in the context of fire evidence, could include the remains of the material first ignited, the ignition source, or other items or components that are in some way related to the fire ignition, development, or spread. An artifact could also be an item on which fire patterns are present; in which case preservation of the artifact is not for the item itself, but for the fire pattern that appears on the item.

Protecting Evidence

One part of the fire investigator's job is to dig out the fire scene. Like an archeologist, the fire investigator removes each layer of fire debris. The investigator's ultimate goal is to identify the point of origin and the cause of the fire. Because fire follows the rules of science, an analysis of how the fire spread assists in determining where it originated and whether the cause was accidental or intentional.

Fire scene reconstruction is the process of recreating the physical scene before the fire occurred, either physically or theoretically. As debris is removed, the contents and structural elements are replaced in their prefire positions, as much as possible. Where the damage and destruction are too extensive to physically restore the scene, whatever information is available is used to fill in the blanks. The investigator interprets the fire scene and documents the fire development by examining the damage to objects, devices, and surfaces. Throughout the process, the investigator must concentrate on locating, examining, and preserving evidence.

The challenge for the fire officer is often to determine when to stop fire suppression or overhaul operations in order to preserve evidence for the investigator. The fire officer must ensure that the fire is extinguished without destroying valuable evidence. From the fire investigator's viewpoint, the least amount of firefighting work is the best. The less the fire fighters disturb, the more intact the scene remains.

The worst case occurs when the fire investigator arrives at an apartment fire and discovers that the fire company has removed all of the fire debris, including the fire-damaged ceilings, walls, and doors. This prevents the investigator from being able to evaluate the evidence in the context of the fire. This would be akin to the police trying to reconstruct a motor vehicle collision from which both cars have been removed to the junkyard and no witnesses remain on the scene: it is possible, but very difficult (▼ **Figure 17-4**).

The fire officer is responsible for protecting the fire scene evidence from the public and from excessive overhaul and salvage (▶ **Figure 17-5**). In addition, the fire officer is the first step in the chain of evidence that is vital to successful prose-

cution of arson cases. Once the fire department arrives, it is responsible for preventing evidence contamination; this requires fire fighters to stay until the fire investigator arrives. The chain of evidence requires that evidence remain secured and documented, from the fire scene to the courtroom. Most investigators document all physical evidence before collecting it by taking 35-mm photographs.

Legal Proceedings

As a company officer, you may be called on to testify in courtroom proceedings, as a witness or as an expert witness. Although a witness can provide the court with testimony based only on their personal knowledge and observations, an expert witness has scientific, technical, or other specialized knowledge that can be relied on to interpret the facts. The role of an expert witness is to assist the judge and jurors to understand the evidence or to determine the true facts in an

Figure 17-5 The fire officer must protect the fire scene evidence from the public and from excessive overhaul and salvage.

Figure 17-4 Evidence must be preserved during overhaul activities.

issue. An expert witness is allowed to give an opinion based on the facts or the data of the case.

If you are called on to testify in court, the first step is to fully prepare. This includes reviewing all reports, photographs, and diagrams of the incident. Also, review any previous depositions or testimony that you have given on the incident. The prosecutor typically wants to meet with you before going to court to review your testimony and qualifications.

When testifying:
- Dress appropriately
- Follow the prosecutor's directions
- Sit up with both feet on the floor
- Avoid gesturing
- Keep answers short and to the point
- Use language a jury can understand
- Be courteous and patient
- Be honest
- Do not hesitate or avoid answering questions
- Speak clearly and loudly
- If you do not remember, do not guess

Remember that the defense counsel's job is to try to discredit you and your statements. It is essential to remain calm and cool, even when the lawyer is asking questions or making statements that are designed to make you look bad. Your role is to present factual information that you can support.

Getting It Done

Coordinated Overhaul and Salvage

Although it is important not to destroy evidence during fire suppression operations, the fire officer's first responsibility is to maintain a safe fireground. That means promptly controlling and extinguishing the fire. After the searches are completed and the fire is declared under control, the fire officer has to consider evidence preservation. Overhaul and salvage operations should be coordinated with the fire investigator, if the investigator is on the scene. For example, a fire investigator may wish to photograph or document the conditions of a ceiling before it is removed during overhaul operations. If the investigator's response is delayed, the fire officer should attempt to identify and protect the area of origin, limiting overhaul in that area to the absolute minimum required to ensure that the fire does not rekindle.

NFPA 921 makes the following recommendations:
- Use caution with straight or solid-stream water patterns; they can move, damage, or destroy physical evidence.
- Restrict the use of water for washing down, and try to avoid possible areas of origin.
- Refrain from moving any knobs or switches.
- Use caution with power tools in the fire scene. Refuel away from the fire scene.
- Limit the number of fire fighters performing overhaul and salvage until the fire investigator is finished documenting the scene.

Every fire officer should be familiar with NFPA 921, *Guide for Fire and Explosion Investigations*. Failure to know and follow the procedures outlined in this document may quickly lead to an embarrassing situation, particularly when expert testimony is being given.

Documentation and Reports

All fires must be properly documented and reported according to the fire department's standard procedures. Most fire departments use the NFIRS reporting system or a variation of it. The NFIRS format incorporates all of the data that are needed for a standard report. The basic report includes the incident number, alarm time and date, location of the incident, property ownership, building construction and occupancy type, weather conditions, responding units and personnel, and numerous other factors that are required for full documentation of an incident.

The data collected through NFIRS provides important information for the fire department to identify risk factors and trends and to plan the most efficient utilization of resources to prevent fires and respond to emergencies. This information also flows into the state and national database systems to provide a better understanding of the overall fire problem. These data are used to assist the fire service and elected officials in determining where resources should be directed and when laws should be changed.

Preliminary Investigation Documentation

In addition to the basic incident report, the fire officer often writes up a special narrative report if the cause of the fire is incendiary or if unusual circumstances are involved. The narrative report is particularly valuable if the fire officer is called to court in the future. Chapter 14 discusses communications in more detail, but fire officers could use the following format to document activities and observations for the fire investigator:

A. Receipt of the alarm
   1. Who reported the fire?
   2. What time was the alarm received?
   3. Who discovered the fire?
B. Response to the incident
   1. Did the fire company encounter any suspicious activity while responding to the fire?
   2. How much time elapsed from dispatch to arrival?
   3. What were the observed weather conditions?
C. Accessibility at the scene
   1. What was the general condition of the fire?
      a. What were the extent and the intensity of the fire?
      b. What was the location of the fire or fires?
      c. Did anyone meet the fire fighters on arrival?
      d. Were any familiar spectators at the scene?

2. Circumstances on arrival
   a. Was anything unusual, considering the fire load?
   b. Where was smoke and fire coming from?
   c. What was the rapidity and the spread of the fire?
3. Gaining access to building
   a. How did the fire fighters get into the structure?
   b. If entry was forced, how did this occur and who forced entry?
D. Fire suppression
   1. What were your immediate observations on entry?
      a. Where was the fire centered?
      b. Was there any unusual flame, smoke, or odors?
   2. What were the conditions of fire extinguishment?
      a. Did the fire flash when it was hit by water?
      b. Was the fire difficult to extinguish?
   3. Were there obstructions to fire suppression?
      a. Were fire protection systems tampered with?
   4. Was the alarm system functioning properly?
E. Civilian contacts
   1. Did witnesses make statements to fire fighters?
   2. Was the owner at the scene?
   3. What were the names of persons allowed into the fire scene?
F. Scene integrity
   1. Was any physical evidence or artifacts removed from scene?
   2. Were any photos/videos taken during fire suppression?
   3. Did any fire department make holes in walls or ceilings?

The report must be clear, complete, and factual. The fire officer must make no assumptions or speculations. The best report is a narrative that accurately and completely describes what the fire officer observed and what the fire company did.

Investigation Report

If the fire officer is responsible for conducting the full investigation, the officer has to complete an investigation report, which is generally written in narrative form. The information is usually written in a chronological order, beginning with a description of the structure before the event occurred. This description would include the building height and dimensions, construction type, structural condition, occupancy, and utility services. It would then describe the alarm notification information, including the time of the call, the name of the caller, and what the caller said.

The report should fully describe the results of the fire scene examination, beginning with a description of the exterior damage. This is followed by a description of the interior damage, including the determination of fire origin and cause, as well as the examination and elimination of any other pos-

sible fire causes. This section is closed with the officer's opinion and conclusion as to the cause and origin of the fire.

Attached to this report would be the information obtained from interviews and witnesses as well as statements from responders. Attachments should also include statements of evidence that was collected, warrants, and sketches.

The Investigative Action After the Fire Officials Are Gone

Many fire investigations continue long after the fire department has completed activities at the scene. Even if the fire is determined to be accidental or undetermined, there is often continuing investigation by the insurance company. This is not to undermine or question the fire investigator's conclusions but rather to look at the fire from the insurance company's point of view. The fire officer should not be offended that the insurance company is conducting further investigations.

The role of the fire investigator and the insurance company investigator varies. The fire investigator could be interested in determining the fire cause and origin to help prevent future fires or to help prosecute criminal actions (▼ Figure 17-6). The insurance investigator might also be looking at the factors that contributed to the loss, such as the absence or inadequacy of fixed fire protection systems and whether the applicable codes were followed.

The insurance company has a major financial interest in any fire loss. The insurance company is often interested in determining other parties that could have liability for the loss, even if there is no criminal responsibility. Although a criminal prosecution requires proof beyond a reasonable doubt, a civil action involving the insurance company is decided by a preponder-

Figure 17-6 The fire investigator is interested in determining fire cause and origin to help prevent future fires and prosecute criminal actions.

ance of the evidence. Many insurance cases are not settled for several years after the fire is extinguished.

Summary

Determining the initial origin and cause of fires is a responsibility of the company officer. This determination may lead to requesting a fire investigator, or it may conclude with the officer's report. Either way, it is essential for a fire officer to be able to make this determination.

The origin is the point where the fire began. After it is determined, the fire officer must determine what material was first ignited, how it was ignited, and why this occurred. This allows the fire to be categorized into one of four classifications: intentional, accidental, natural, and undetermined.

If the fire is intentional, the fire officer needs to gather the information that is used by a prosecutor to determine whether it rises to the level of arson. This usually includes writing a detailed narrative with all relevant information about the cause and origin of the fire.

You Are the Fire Officer: Conclusion

Your first step will be to determine the area or point of origin of the fire. You will do this by working from the unburned area toward the burned and examining the fire damage and the intensity of the fire. This allows you to identify the area or point of origin of the fire. Once this is done, you will begin to look for clues as to the cause of the fire. You will need to know what material burned first, what was the source of ignition, and why the two came together. You are then ready to classify the fire as natural, accidental, incendiary, or undetermined.

You may elect to request an investigator, particularly if the cause is incendiary or undetermined. If you call for an investigator, you will need to restrict entry into the area and preserve the evidence. This can be accomplished by posting a fire fighter outside the entrances or by placing barrier tape or ropes at the entrance.

Wrap-Up

Ready for Review

- Know your jurisdiction's criteria on when to call a fire investigator.
- The point of origin is where a heat source and fuel come in contact with each other and start a fire.
- The "U" and "V" fire patterns can lead the fire officer to the point of origin.
- Char can be used to estimate fire intensity and direction.
- The four fire cause classifications are accidental, natural, incendiary, and undetermined.
- When considering intentional fires, look for disabled built-in fire protection, delayed notification/difficulty in getting to the fire, accelerants and trailers, multiple points of origin, and tampered/altered equipment.
- Arson is a crime of maliciously and intentionally, or recklessly, starting a fire or causing an explosion.
- There are six basic motives for arson: profit, crime concealment, excitement, spite/revenge, extremism, and vandalism.
- Evidence is the documentary or oral statements and material objects admissible as testimony in a court of law.
- Artifacts are the remains of the first material ignited, ignition source, or related components; they should be left where they are found.
- Fire investigators systematically remove each layer of fire debris as part of the fire scene reconstruction.
- Delay the overhaul and salvage until the fire investigator arrives and completes the initial scene survey.
- All fires must be properly documented and reported according to the fire department's standard operating procedures.

Hot Terms

Accelerants An agent, often an ignitable liquid, used to initiate a fire or increase the rate of growth or spread of fire.

Arson The crime of maliciously and intentionally, or recklessly, starting a fire or causing an explosion.

Artifacts The remains of the material first ignited, the ignition source, or other items or components in some way related to the fire ignition, development, or spread. An artifact may also be an item on which fire patterns are present, in which case the preservation of the artifact is not for the item itself but for the fire pattern that appears on the item.

Char Carbonaceous material that has been burned and has a blackened appearance.

Demonstrative evidence Tangible items that can be identified by witnesses, such as incendiary devices and fire scene debris.

Documentary evidence Evidence in written form, such as reports, records, photographs, sketches, and witness statements.

Evidence The documentary or oral statements and the material objects admissible as testimony in a court of law.

Failure analysis A logical, systematic examination of an item, component, assembly, or structure and its place and function within a system, conducted in order to identify and analyze the probability, causes, and consequences of potential and real failures.

Fire analysis The process of determining the origin, cause, development, and responsibility, as well as the failure analysis of a fire or explosion.

Fire patterns Physical marks left on an object by the fire.

Fire scene reconstruction The process of recreating the physical scene during fire scene analysis through the removal of debris and the replacement of contents or structural elements in their pre-fire position.

Form of material What the material is being used for. For example, the form of material might be clothing.

Point of origin The exact physical location where a heat source and a fuel come in contact with each other and a fire begins.

Pyrolysis The destructive distillation of organic compounds in an oxygen-free environment that converts the organic matter into gases, liquids, and char.

Source of ignition Devices or equipment that, because of their intended modes of use or operation, are capable of providing sufficient thermal energy to ignite flammable gas-air mixtures.

Testimonial evidence Witnesses speaking under oath.

Trailers Materials used to spread the fire from one area of a structure to another.

Type of material What the material is made of. For example, the type of material might be cotton.

As a fire officer, your engine responds to a reported arson fire that is now out. At your arrival, you find a small outbuilding smoldering down a long driveway. It appears that the fire burned sometime during the night. No one is around, and the homeowner arrived this morning to find the burned structure. You contact your department's fire investigator, and he advises you to begin obtaining preliminary information.

1. What is point of origin?

   A. Fire patterns

   B. Where people observed flames

   C. Physical location where fire begins

   D. Location of fire travel

2. What causes a "V" pattern?

   A. When fire source is against or near a wall

   B. When the fire is free burning in the middle of the room

   C. When there is carpet involved

   D. When the fire burns down from the ceiling

3. What is the source of ignition?

   A. Generation

   B. Heating of equipment

   C. Energy source that caused material to ignite

   D. Location of material that caused ignition

4. What is pyrolysis?

   A. Energy

   B. Power from ignition source

   C. Chemical decomposition

   D. Fission

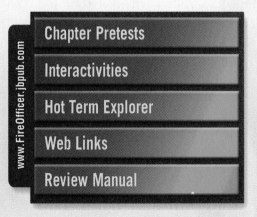

www.FireOfficer.jbpub.com

Chapter Pretests

Interactivities

Hot Term Explorer

Web Links

Review Manual

www.FireOfficer.jbpub.com

Crew Resource Management

NFPA 1021 Standard

Fire Officer I

NFPA 1021 contains no Fire Officer I Job Performance Requirements for this chapter.

Fire Officer II

NFPA 1021 contains no Fire Officer II Job Performance Requirements for this chapter.

Knowledge Objectives

After studying this chapter, you will be able to:

- Discuss the origins of crew resource management (CRM).
- List Dupont's "dirty dozen" human factors that contribute to tragedy.
- Describe the six-point CRM model that can be used in the fire service.
- Describe the five steps in a successful debriefing.

Skills Objectives

There are no skills objectives for this chapter.

You are the incident commander at the scene of a rapidly extending fire in an occupied multi-family dwelling. Crews are operating on multiple floors to fight the fire. Additional alarm companies are arriving and flooding the command post, anxiously seeking assignments. The occupants of the dwelling are nearby, seeking information on the fire's progress. You are beginning to feel overwhelmed.

1. When faced with multiple tasks, how can you best allocate resources to ensure that the emergency situation is mitigated and everyone returns home safely afterward?

Origins of Crew Resource Management

On December 28, 1978, United Airlines Flight #173, a McDonnell Douglas DC8, is making a routine flight from Denver to Portland with 189 passengers and crew onboard. During final approach to Portland International Airport, an unfamiliar "thump" is felt as the landing gear deploys, and the "gear down and locked" light on the cockpit instrument panel does not illuminate. The captain decides to circle the airport while he, the first officer, and the flight engineer attempt to figure out the problem. The flight engineer tells the captain that "15 minutes is really gonna run us low on fuel here." The captain looks at the fuel gauge and decides that the plane can circle while the crew troubleshoots the landing gear issue.

One hour passes as Flight 173 circles the Portland area while the three flight deck crew members seek confirmation that the gear is down and flight attendants prepare the passenger cabin for an emergency evacuation. Aviation traditions of the day hold that the captain is the "infallible head of the ship" and is therefore never questioned. While the crew continues to fuss with the lights and the pilot awaits word on cabin preparations, the plane's engines cough, sputter, and finally go quiet, one at a time. The captain realizes that the plane has run out of fuel moments before it slams into the Oregon pines, narrowly missing a suburban residential community. Ten people are killed, including the flight engineer, and 23 are critically injured. Total disaster is averted because the typical fire that accompanies most air crashes never materializes due to a lack of fuel.

On April 30, 2000, the assistant chief of a small Mid-Atlantic fire department is conducting a live burn exercise at an old farmhouse. He and two other officers ascend a narrow flight of stairs that lead to a low-ceilinged attic. The assistant chief, using a small sprayer, broadcasts diesel fuel on debris in the attic, and then the other officers ignite the debris at several points. The resulting fire increases rapidly in intensity, forcing the two officers to leave. Lacking full protective clothing, they retreat to the second floor, urging the assistant chief to come with them. The assistant chief, in full protective equipment and SCBA, tells them that he is going to remain in the attic to ensure that the fire stays lit.

Brushing their concerns aside, the assistant chief remains in the attic and dies when the fire flashes over. When he does not answer radio calls checking on his welfare, a two-fire fighter rescue team enters the attic but is unable to remove him. The assistant chief's body is removed after the roof has collapsed and the fire is extinguished.

More than 25 years and a seemingly completely unrelated set of circumstances separate the two incidents recounted above. The most significant factor about the first event is that the aviation industry evaluated the facts of the Portland crash, along with several other incidents, and began to look at elements of the event from a unique perspective. Accident reconstruction officials determined that the cause of the plane crash was not a burned out lightbulb on the cockpit console. Rather, the real cause of the crash was a culture that bred inhibited subordinates who would not speak up when something was wrong, as well as leaders who believed that they were infallible.

In the subsequent 25 years, the aviation industry reduced its accident rate by 80%, partly through the development, refinement, and system-wide adoption of a behavioral modification training system known as **crew resource management (CRM)**. Tony Kern, a respected subject matter expert on CRM states: "CRM is designed to train team members how to achieve maximum mission effectiveness in a time-constrained environment under stress. That is a concept with nearly universal utility and timeless applicability." Extensive studies on the application of CRM in the aviation industry by Dr. Robert Helmreich and his staff at the University of Texas reveal that teaching people to work better together and recognizing that errors do occur generate better performance, fewer mistakes, and lower injury rates.

Other industries and fields of practice, most notably the United States military and the medical profession, have adopted and embraced CRM with similar results. Today, a variety of CRM models are used in different types of industries and organizations; however, they are all based on the same basic concepts and principles.

You may ask, "What does this have to do with the fire service?" The short answer is—plenty! Although the number of structure fires has declined significantly over the past 25 years and civilian fire deaths and injuries have been greatly reduced, the number of fire fighter deaths and injuries has remained rather static for the same period. In relation to the number of fires, the rate of fire fighters killed and injured in the line of duty has actually increased.

Many major advances in fire fighter health and safety have been introduced and widely adopted during this period. The thermal properties of protective clothing have been greatly improved, protective trousers have become standard equipment, helmets have been strengthened, and protective hoods have been added to the protective ensemble. Breathing apparatus has been reduced in weight and improved in performance, with positive-pressure regulators, integrated PASS alarms, and heads-up air consumption displays in the face pieces. Portable radios are provided for every fire fighter or at least one member of every team working in a hazardous area. Thermal imagers improve fire fighter visibility in vision-obscured atmospheres. Apparatus cabs are now fully enclosed and reinforced, and seat belt use is mandated. Despite all of these technological advances, fire fighters are still dying at a rate of more than 100 per year, with another 100,000 experiencing lost-time injuries.

National Institute for Occupational Safety and Health reports indicate that human error plays a major role in a large proportion of fire fighter fatalities. In many cases, human error is a contributing factor, rather than the principal cause, because the direct cause of death is closely associated with the fire. More often than not, the human error places the fire fighter in a dangerous situation that could have been recognized and avoided. Breakdowns in communication, fire fighters caught in a flashover or becoming disoriented in a smoke-filled building and running out of air, failure to follow standard operating procedures, or in some cases, failure to establish standard operating procedures all cry out "human error."

The fire service must acknowledge the significant role that human error plays in fire fighter deaths and injuries. History has proven that technology can provide only a finite level of protection in an inherently dangerous environment. History has also illustrated that adopting and embracing the proven principles of CRM are an effective means for reducing the effects of human error and preventing tragedy. The prin-

Safety Zone

The leading cause of fire fighter fatalities continues to be heart attacks and cardiovascular incidents, accounting for almost 50% of all fire fighter fatalities over the past 10 years. Vehicle collisions are the second leading cause of fire fighter fatalities. On average, approximately 25 fire fighters die each year in the United States as a direct result of situations that occur during firefighting activities (structural and wildland) and other emergency operations.

ciples of CRM are particularly applicable to the interaction that occurs within a fire company, involving a company officer and the crew members who work as a team at the scene of an emergency incident.

Human Error

Error is as much a part of life as the sun rising and setting. The noted Roman philosopher Cicero observed more than 2,000 years ago that "to err is human." Some errors are innocuous, such as a "2 + 2 = 5" on a first grader's math test. Others have tragic and far-reaching consequences, like the events described in the opening paragraphs of this chapter. Accepting that human error is an everyday part of life is a basic tenet of CRM training. We have to recognize that human error is the rule and not the exception as the first challenge in the journey to reduce injuries and prevent line-of-duty deaths. The hallmark of CRM is error management through improved communications, decision making, and maintenance of situational awareness.

Accepting that error is a matter of course is not an automatic excuse for every error. The aviation industry and others that use CRM offer this important caution. "Normalizing" error is not a "get out of jail free card" that can be used to avoid taking responsibility for making a mistake. The first priority should always be to prevent errors from occurring. The fundamental concept of CRM is to recognize that some errors will occur, in spite of the best efforts to prevent them, and to take steps to minimize the harmful effects.

The balance to strike is that CRM:

- Acknowledges that errors occur.
- Promotes strategies and concepts that minimize the effects of those errors.
- Creates an environment where all team members remain more alert (situationally aware) of their work setting.

The basic objective of CRM is to cause an appropriate reaction when an error occurs, in order to prevent a catastrophic outcome. One essential element that was recognized through research by Dr. Robert Helmreich and others was the fact that a catastrophic event is usually not the result of a single, standalone action. Posttragedy investigation reports typically identify a chain of error that ends in disaster. The catastrophic outcome results when a critical series of errors or circumstances occur simultaneously or in a particular sequence. The individual contributing factors can often occur without the catastrophic outcome. These observations apply in the aviation environment and in most other human activities.

Human Factors Contributing to Error

Gordon Dupont, a recognized expert on error and human factors, echoed Dr. Helmreich's observation in determining that a "dirty dozen" of human factors contribute to tragedy. Mr. Dupont has an extensive background in aviation maintenance procedures and has studied the similarities between errors that occur in the cockpit and in the maintenance hangar. His dirty dozen are considered a comprehensive list of reasons and ways that humans make mistakes. The 12 factors touch on the gamut of human emotion, action, thought, and action. Mr. Dupont's dirty dozen are:

- Lack of communication
- Complacency
- Lack of knowledge
- Distraction
- Lack of teamwork
- Fatigue
- Lack of resources
- Pressure
- Lack of assertiveness
- Stress
- Lack of awareness
- Norms

Consideration of these 12 factors should lead to the conclusion that they are not unique to aviation mechanics or pilots. Even a casual review of fire fighter injury and line-of-duty death reports would find that at least one of the factors, sometimes several of them, can be identified as causal factors in most situations. The root question for the fire service is, "Are we willing to continue to believe these factors are acceptable causes of injuries and deaths or do we embark on an effort to overcome their impact on firefighting crews? The CRM approach requires institutional change to achieve lifesaving results.

Error Management Model

CRM offers a proven mechanism for dealing with error at all levels. If we recognize that to a certain degree, human error is unavoidable, we must plan to minimize the negative impact of errors that do occur. If we also accept that most disasters and tragedies are not the result of a single event, we recognize that a plan to effectively manage errors will reduce the risk of catastrophic outcomes. The concept of managing errors includes preventing them from occurring, recognizing them when they do occur, reacting appropriately when they are recognized, and taking the necessary steps to minimize their

impact. In this manner, we can interrupt the chain of events that leads to many tragedies. Managing and minimizing error can create a culture that the fire service needs to explore and adopt.

Dr. Robert Helmreich has created an error management model that describes three methods for dealing with error: avoidance, entrapment, and mitigating consequences. The model is shaped like an inverted pyramid to illustrate where the greatest potential for capturing error exists. Error avoidance provides the greatest opportunity for trapping and preventing error from moving to catastrophe. Errors that are not avoided are trapped at the second level. Errors that slip through the first two levels require mitigation (the action taken by emergency responders to minimize the effect of an emergency on a community) (▼ **Figure 18-1**). CRM offers a proven mechanism for dealing with error at all levels.

One important statement regarding CRM and leadership is to establish early on: CRM does not advocate or support an abandonment of legitimate authority. By all accounts, the CRM process enhances authority through an "all for one and one for all" approach. Longtime practitioners of CRM are adamant that:

- Lines of authority are maintained
- Ultimate authority rests with the legitimate ranking officer

CRM Model

Several industries use a variety of CRM models today. Although there are several variations of CRM models, one six-point model is well suited to the fire service. Fire fighters are already aware of most of these points from different aspects of their training and experience. Because CRM involves changing a culture, a model that is easily understood promotes acceptance. The six points of this model are:

- Communication skills
- Teamwork
- Task allocation

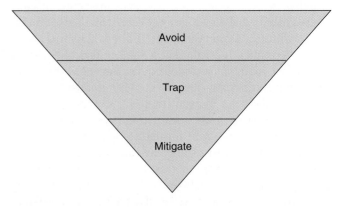

Figure 18-1 Helmreich's error management model.

- Critical decision making
- Situational awareness
- Debriefing

CRM concentrates on minimizing the effects of error by capturing errors before catastrophe strikes, preferably at the first, but certainly at the second, level of Dr. Helmreich's error management model. The synergy that occurs among people, hardware, and information is a pivotal factor. When CRM is applied to an organization, it requires everyone in the chain of command to recognize that:

- *No one* is infallible.
- Humans create technology; therefore, it is fallible.
- Catastrophes are the result of a chain of events, sometimes a long chain and sometimes a short chain.
- Everyone has an obligation to speak up when he or she sees something wrong.
- People who work together effectively are less likely to have accidents.

Adopting CRM within an organization requires both individual and organizational behavioral changes, and in order to be successful, CRM must be practiced by every member of the organization. As noted previously, "CRM is designed to train team members how to achieve maximum mission effectiveness *in a time-constrained environment under stress.*" In order for the team to become more effective, every member of the team must participate.

Communication Skills

The simplest definition of communication is the successful transfer and understanding of a thought from one person to another. Communicating involves two primary forms: verbal and nonverbal. Six simple steps follow a thought or idea from formulation to completion of the process:

1. Formulate the idea.
2. Select the medium for transmission.
3. Transmit the idea.
4. Receive the idea.
5. Interpret the idea.
6. Provide feedback to the sender to ensure understanding.

Communicating becomes complicated when two or more people bring all of their personal traits, biases, and history to the exchange. Overcoming these hurdles is not impossible, but it does require a new look at one of the oldest reasons accidents occur: miscommunication.

Cockpit voice recordings from numerous airline disasters revealed that flight crews engaged in several forms of miscommunication that contributed to catastrophic events. These communication failures included misinterpretations of take-off, altitude, and landing instructions; airlines that fostered and condoned "fighter pilot" mentalities in their captains; a lack of assertiveness on the part of crew members who recognized problems before the captain; and distrac-

tions in the cockpit that inhibited focused communication (e.g., idle chitchat during preflight checks).

CRM suggests that developing a standard language and teaching appropriate assertive behavior are the keys to reducing errors resulting from miscommunication. In the aviation application, CRM advocates maintaining a "sterile cockpit," a cockpit environment where all communication and focus is on flight operations. We expect that the flight crew be focused on getting the plane safely off the ground and to its destination, not engaged in a spirited debate about the greatest golfer in the world as the plane hurtles down the runway.

Let us look at the analogy to a typical firefighting crew. When a fire apparatus is en route to an alarm, the crew members are sitting in an enclosed box, surrounded by switches, gauges, wheels, pedals, and noise and moving down the road. In order for the machines to move efficiently and safely, the people inside need to focus on the mission and communicate effectively. The entire crew should be exchanging only information that is pertinent to responding to and arriving safely at the scene of the alarm.

The CRM-enriched environment generates a climate where the freedom to question is encouraged. Crew members are encouraged to respectfully speak up when they see something that causes concern by using clear, concise questions and observations (**▼ Figure 18-2**). A discrepancy between what is going on and what should be occurring is often the first indication of an error.

(**Figure 18-2**) Crew members are encouraged to respectfully speak up when they see a problem.

The freedom for all crew members to question something that appears unusual or is not understood is a delicate subject for many organizations. This is one of the main reasons why all levels of an organization need to be trained in CRM. CRM advocates speaking directly, yet in a manner that does not challenge the authority of a superior.

The British aviation world initially termed CRM "charm school" because of the philosophy's promotion of specific steps subordinates should take to question authority and authority's specific direction to listen to subordinates. Cockpit voice recordings of a number of air disasters tragically reveal first officers and flight engineers attempting to bring an issue to the captain's attention by speaking indirectly about the subject. CRM promotes subordinates being proactive; using clear, concise questions; and expressing concerns accurately.

Inquiry and advocacy are discrete, learnable skills that promote synergy between the mechanical element and human players. Inquiry is the process of questioning a situation that causes concern. Advocacy is the statement of opinion that recommends what one believes is the proper course of action under a specific set of circumstances. Using the two skills effectively, in concert with each other, requires practice and patience on the part of all crew members. The factor to keep in mind is that communication should not focus on who is right, but on what is right.

One effective tactic is to use specific buzzwords, such as "red light" and "red flag," to signal discomfort with a situation. These terms are cues that open the door to inquiry and advocacy. They should be reserved for situations that involve an immediate risk of injury.

Todd Bishop, a CRM expert, teaches a five-step assertive statement process that encompasses the inquiry and advocacy communications steps. The assertive statement runs like this:

- **Opening/attention getter.** Address the other individual. *"Hey Chief,"* or *"Bill,"* or whatever appropriate moniker is used to get the individual's attention.
- **State your concern.** Use an owned emotion. *"That smoke is really pushing from those windows. I have a bad feeling about it."*
- **State the problem as you see it.** *"It looks like it is going to flash. I think any crews entering that place are going to take a beating."*
- **State a solution.** *"Why don't we vent the roof and give the place a minute or two to vent before we send in the attack team?"*
- **Obtain agreement (aka "buy-in").** *"Does that sound good to you?"*

The inquiry and advocacy process and the assertive statement are essential components of the communication segment of CRM. Of all the components, these two require the most work and attention. They are the toughest to impart and adopt because they often require a wholesale change in interpersonal dynamics. Once mastered, however, inquiry and advocacy can enhance performance, avoid mishaps, and save lives.

Effective listening is also an important communication skill. A common human failing is to start formulating a response before hearing the other person out. One effective listening technique is to purposely refrain from making any response or counterargument until the other individual stops for a breath. This effort requires patience and self-control, but it is essential for CRM to be effective.

Teamwork

CRM promotes the concept of team members working together for the common good. Conceptually, this implies developing effective teams, which requires buy-in by all members, leaders *and* followers, in the effort to be effective, efficient, and safe. Leaders will always be in charge; however, in order to be effective in a CRM environment, they must also be open to suggestion and constructive criticism.

The typical role of a fire officer at the company officer level is to lead a team of fire fighters in the performance of tactical operations. The fire officer's primary responsibility is to develop an effective team environment that supports safe and effective operations. The company must also be prepared to function as a component of a larger team at an incident where multiple companies are working together. The CRM concept of a team applies at both levels.

Leadership

Fire officers formally exercise leadership of the team through a combination of rank and authority. The fire department's structural hierarchy mandates respect for the rank, and in many cases, there is a genuine respect and admiration for the individual. In order for officers to become truly effective leaders, they have to earn the trust and respect of their subordinates and demonstrate the skills of effective leadership. These three components comprise the triangle of leadership (▼ Figure 18-3).

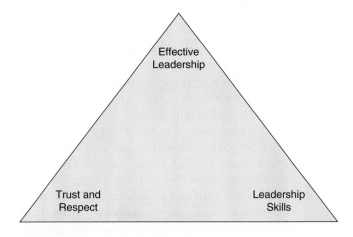

Figure 18-3 The triangle of leadership.

The informal authority to lead is derived through respect. True respect is based on three competencies: personal, technical, and social. Personal competence refers to an individual's own internal strengths, capabilities, and character. Technical competence refers to an individual's ability to perform tasks that require specific knowledge or skills. Social competence refers to the person's ability to interact effectively with other people. All three competencies are essential for leaders to function effectively in a CRM environment.

The fact that a leader must demonstrate effective social skills does not mean that a leader must always be deferential. Social competence requires knowing how to speak to people respectfully, whether praising or chastising them.

Mentoring

Mentors pay special attention to helping others develop their skills. Leaders who are also mentors typically develop the greatest affinity for their people and produce highly effective teams. CRM provides an exceptional vehicle for leaders to impart knowledge and skills to their subordinates.

Leading by example is a highly effective technique. Crews notice a supervisor's personal habits, pick up on his or her thought processes, and constantly evaluate the leader's performance. Traits that are admired and respected by the crew members have a major influence on their future behavior. The crew usually sees more than the supervisor may think. Leading by example requires constant effort on the part of the leader, with no opportunities to turn the positive role model on or off at different times.

One of the basic premises of CRM is that human error is a normal, inevitable occurrence. A leader must be willing to admit to making a mistake. Failing to admit mistakes quickly compromises a leader's authority, and hiding mistakes is a serious character flaw that undermines a leader's credibility. Superior leaders readily admit when they have made mistakes, accept responsibility, and focus their attention on moving forward. This trait creates an environment that fosters open communication, promotes safety, and encourages subordinates' belief in the team.

Mentoring also requires sharing knowledge. In an organization, knowledge is often closely associated with power. Leaders who are insecure about their positions tend to withhold knowledge and fail to share lessons as they are learned. Mistakes are often repeated in such an environment, and errors that could have been avoided are permitted to occur. In the fire service, a lack of knowledge can be a prescription for calamity. A fire officer must maintain technical competence, stay on top of technological advances, and ensure that others have the information that will keep them from being killed or injured.

Technical competence inspires respect; however, technical competency requirements vary from position to position. The street-level fire fighter does not expect the fire chief to pull hose and throw ladders, but the chief must be able to command a fire, mass-casualty incident, or weapons of mass destruction event with the same degree of skill that the fire fighter is expected to use to pull hose and throw ladders.

Handling Conflict

Conflict is an inevitable consequence of human interaction. It is unrealistic to expect that everyone will agree on everything. Whether the issue is a simple disagreement or a full-blown clash of ideologies, conflict will occur. Failing to keep egos and emotions in check often creates conflict and even more often causes conflict to escalate to unresolvable levels.

The fire officer must deal with conflict in a manner that effectively resolves the problem. CRM offers a valuable axiom regarding conflict. The focal point of CRM in conflict resolution is to focus on what is right, not who is right. Keeping that statement in mind allows the leader to focus on the best outcome, which should also be a safe outcome.

Establishing an open climate for error prevention is the paramount goal of CRM. The leader who perceives a subordinate's comment or question as a threat to his or her own authority is part of the problem instead of an essential part of the solution. Leaders are required to keep their own egos in check. A leader who rules by intimidation can often win an argument, but the forces of nature always prevail over words. An officer can order a crew to advance into a dangerous position, but the building will not follow an order to remain standing. An officer who fails to listen or refuses to listen is simply dangerous.

Responsibility

There is a tendency for newly indoctrinated CRM disciples to believe that CRM advocates the creation of "management by committee," in which leadership gives way to consensus. Nothing could be further from the truth! CRM advises leaders to keep their eyes and ears open, especially for input from subordinates. CRM also emphasizes that in order for decision making to be efficient, there has to be someone in charge, someone who has ultimate responsibility for decisions and outcome.

In the cockpit, the captain retains ultimate authority for decision making and ultimate responsibility for getting the plane from portal to portal. Subordinates are encouraged to provide input, but the final decision rests with the recognized authority. Fire service members have to abide by the same rules. Longtime practitioners of CRM are adamant that lines of authority are maintained and that ultimate authority rests with the legitimate ranking officer.

Being a leader requires an individual to exercise authority and accept responsibility, not abdicate authority or shirk responsibility. Firefighting and other emergency operations are conducted in an environment where the risks are real and often immediate. In the end, there are many situations where the leader must evaluate a situation, weigh the alternatives, and quickly make a decision that is based on an intelligent assessment of the options.

Fire fighters have high expectations for their leaders. They expect their officers to manage emergency operations in

VOICES OF EXPERIENCE

❝ Why use one brain when you have access to many? ❞

I have been a fire officer for four years now and have used the concepts of crew resource management (CRM) during almost every incident I've managed. There is really no reason not to. Why use only one brain when you have access to many? Though I have not solicited opinions from my company every time we are in the heat of battle, I have created an atmosphere where my subordinates feel comfortable offering input. Many times the input given comes from a fire fighter who has a different perspective or vantage point than I do and that input needs to be considered.

For example, during a residential structure fire, I was the acting battalion chief. On arrival, Engine 3 reported a fully involved detached garage spreading to the adjacent two-story house. Engine 3 deployed the deck gun onto the garage while a 2½″ line was being set up to protect the house. I realized that although the exterior fire was being suppressed, three bedrooms on the second floor of the house were already fully involved. This presented my challenge for this incident. One engine company and a truck company were on scene while two engine companies were on the way. Engine 3 was committed to the outside operations and the truck company was conducting water supply operations. Meanwhile I had a fightable interior fire that I needed to get a hose-line on. I was going to wait for the next arriving engine to suppress the interior fire. The hand line had already been deployed and was waiting at the front door.

Then the truck company officer suggested that his company go inside and let the next arriving engine take over the water supply responsibilities. The order was given and the truck company entered the house. If the order had not been given at that time, the house would have been a total loss.

Jerrod Vanlandingham
Longmont, Colorado Fire Department
13 years of service

a manner that effectively accomplishes the mission of the organization and sets the stage for safe outcomes. "Mission analysis" requires fire service leaders to look at all situations with a risk-versus-gain mentality. Phoenix Fire Chief Alan Brunacini's mantra of risk/benefit (risk a lot to save a savable life, take a calculated risk to save savable property, and risk nothing to save what is already lost) provides sage words to live by and establishes a concrete foundation for leaders to manage emergency operations.

Followership

Because this book is intended for fire officers, the primary focus on the discussion of CRM is directed toward leadership. A discussion of CRM would not be complete without considering followership. <u>Followership</u> is a term that can be applied to the appropriate actions of those who are led. Just as the leaders must be good leaders, CRM requires the followers to be good followers.

All followers (and we are all followers of something or someone) should perform a self-assessment of their ability to function as part of a team. The self-assessment should consider four critical areas:

1. **Physical condition.** People in good physical condition are more aware, alert, and oriented to their surroundings. Maintaining good physical health is critical to the success of any endeavor; however, it is absolutely critical in the fire service.
2. **Mental condition.** We are constantly pulled in a variety of directions; however, the challenging and dangerous environments where fire fighters have to perform demand our full attention. Fire fighters must ask themselves, "Am I free of distractions that could divert my attention from the task at hand?"
3. **Attitude.** In order to be an effective team member, a fire fighter must be willing to follow orders and be part of a cohesive team.
4. **Understanding human behavior.** The effectiveness of CRM is based on understanding human behavior and interpersonal dynamics in a team environment. In order to be effective using CRM, the team members must understand how their individual behavior relates to each other and to the team as a whole.

In order to be effective team players and maximize the benefits of CRM, each individual must have:

- A healthy appreciation for personal safety.
- A healthy concern for the safety of the crew.
- A respect for authority.
- A willingness to accept orders.
- A knowledge of the limits of authority.
- A desire to help their leader be successful.
- Good communication skills.
- The ability to provide constructive, pertinent feedback.
- The ability to admit errors.
- The ability to keep ego in check.

- The ability to balance assertiveness and authority.
- A learning attitude.
- The ability to perform demanding tasks.
- Adaptability.

These qualities of a good follower are remarkably similar to those required of effective leaders. Followership enhances leadership, which in turn promotes effective teams. Teams that practice CRM make fewer mistakes and are able to recognize and correct errors before they cause tragic outcomes.

Task Allocation

Task allocation refers to dividing responsibilities among individuals and teams in a manner that allows for their effective accomplishment (▼ **Figure 18-4**). Task overload occurs when we exceed our capacity to manage all of the functions and responsibilities that we are trying to manage simultaneously. When task overload arises, safety is quickly compromised.

Think about the individual who is driving down the interstate at rush hour, cell phone to the ear and attempting to write down a telephone number. He looks up and realizes

Figure 18-4 Task allocation is about dividing labor.

that traffic has stopped. Cell phone, paper, and pencil all fall to the floor as he slams on the brakes, too late to avoid a collision. Task overload has occurred.

Knowing one's own limits and the capacity of the team is the first step in the CRM task allocation phase. Each of us has a point at which outside stimuli override our ability to process information and perform effectively. We need to know what we can handle and when we need additional resources to take on some of the required tasks. If the resources are not available, we have to prioritize tasks and focus on what we can do safely and effectively. The alternative is to risk disaster by trying to do too much and losing control over everything. Using all eyes and ears, soliciting input, and recognizing our limits prevents task overload from occurring and safety from failing.

Fire officers generally fall into one of three categories. Some officers are reluctant to admit that they are ever overwhelmed and believe that they become more effective as the situation becomes more hectic. The second group often becomes overwhelmed before the full complexity of the event is even recognized. The third group effectively assesses the incident, calls for additional resources early, and manages to stay ahead of both the incident and the span of control.

Helmreich identified a common trap that is associated with task overload. A NASA/University of Texas study noted that pilots believed they were capably handling multiple tasks, when in reality, after a certain point, mistakes began to appear and performance deteriorated. They were so busy that they failed to recognize when the situation was getting out of control.

A fire officer has to perform a serious self-evaluation to evaluate his or her personal capacity to manage complicated situations and identify weak spots. Training and practice can definitely improve an individual's performance; however, everyone has limitations. When these are known, we can take action early to compensate.

The fire officer must also know the crew's limits. In many fire departments, the same crew members work together as a team for long periods, allowing the fire officer to evaluate strengths and weaknesses and focus on team development. This factor is often much more difficult to determine when the make-up of a crew changes from day to day or the crew is assembled only at the time of alarm. This method leaves little time for sizing up crew members, putting the right people in the right position, and capitalizing on individual strengths for a highly effective team; the fire officer often has to make quick assessments and adjustments.

The emergency scene is the locale where CRM proves its worth, but it is not an appropriate environment to develop the basic skills and build an effective team. The strengths of individuals and crews should be evaluated and built on in multiple nonemergency settings. Training classes, live training exercises, table-top modeling, didactic presentations by experts, mentoring, and exchanging personnel are all effective, proven techniques for enhancing CRM performance. (▶ **Figure 18-5**).

The final advice for task allocation is drawn from the story of the two lions. Both lions are on the hunt when they spot a large elephant in the distance. The first lion says to the second, "I sure am hungry, but how are we going to eat that big thing?" The second lion, his mind already developing strategy and salivating at the potential feast, dryly replies, "One bite at a time."

Critical Decision Making

Although CRM promotes the concept of team involvement in all aspects of an operation, emergency scenes often demand rapid decision making by the crew leader. Input is welcome from all team members; however, the final responsibility resides with the crew leader, and experience and training often play pivotal roles in successful outcomes.

Gary Klein, a leading expert on decision making under pressure, found that fire officers and military combat officers often use a very similar decision-making process. During research conducted in the mid-1980s, urban fire commanders told Klein that they often made fireground decisions based on previous experiences, not on traditional decision-making models. The commanders reported that they often did not

Figure 18-5 Activities such as live training exercises help the officer determine the strengths of individuals.

even consider options; they simply looked at the incident, recalled previous experiences of a similar nature, and applied the processes that had been used to successfully mitigate the previous incidents. Many of the fire officers of that era had tremendous experience fighting numerous fires in an urban environment; however, much of that experience has been lost through retirements and a declining number of fires.

Klein observed that the fireground and combat commanders often have to make decisions in environments that are time compressed and dynamic, involved missing or ambiguous data and multiple players, and lacked real-time feedback. This research gave rise to two decision-making models that have gained acceptance over the years: **recognition primed decision making (RPD)**, which describes how commanders can recognize a plausible plan of action, and **naturalistic decision making**, which describes how commanders make decisions in their natural environment.

Effective application of CRM can enhance both of these decision-making processes. Players who have been trained in enhanced communication skills can provide the missing data and help clarify ambiguous information. Better real-time feedback is also provided through the improved communication skills. Each element of CRM establishes a foundation for better decision making.

A recommended practice for improving decision making includes gaining experience, training constantly, improving communication skills, and preplanning incidents. According to longtime CRM practitioners, including CRM at the training level, practicing enhanced communication skills, and taking advantage of improved crew interaction at all levels provide the following benefits:

- Problem identification is enhanced
- Supervisors at all command levels maintain better incident control
- Situational awareness is improved
- Hazards are more rapidly identified
- Resource capability is rapidly assessed
- Potential solutions are more rapidly developed
- Decision making is improved
- Surprises and unanticipated problems are reduced

Situational Awareness

"What did they say?," "I didn't see that," and "I *think* Communications said turn left on Main Street," are three statements that identify an individual's loss of a human factors element known as situational awareness. An "I didn't copy it either," or an "I don't know," expands the loss of situational awareness to other crew members. The loss of situational awareness is frequently the first link in a chain of errors that leads to calamity. Simply stated, **situational awareness** is the accurate perception of what is going on around you.

The triple fire fighter fatality at One Meridian Plaza in Philadelphia (February 23, 1991) is a case in point. Three members of an engine company were assigned to open a

bulkhead door at the top of a stairway to assist with ventilation at the high-rise fire. The crew became fatigued climbing the stairs and radioed that they had left the stairway and were disoriented in heavy smoke on the 30th floor. Command initiated a search effort, sending additional crews up to the 30th floor. Finding no sign of the missing fire fighters at that level, the crews continued their search by moving to higher floors, eventually becoming victims themselves on the 38th Floor. Additional search crews coming down from the roof found the distressed rescue team and safely removed them to the roof for evacuation.

The missing fire fighters were eventually found when a helicopter scanning the exterior of the building observed a lone, broken window on the 28th floor. The missing fire fighters were located and pronounced dead, just inside the broken window.

Other notable fire fighter fatality incidents have followed a similar tragic chain of events, in which fire fighters do not realize that they are in a perilous situation until they are in imminent danger, disoriented, and unable to find a way out. If they are able to communicate their predicament, they are unable to tell potential rescuers where they are located.

One of the common human behavior factors that often leads to a loss of situational awareness is the tendency to ignore or disregard information that is out of context. If an incident commander is busy thinking about an attack strategy, a brief observation or comment about unusual smoke conditions can be easily brushed aside and forgotten. A critical radio message might be acknowledged without really being understood or placed in context in the situation at hand.

Situational awareness affects performance and decision making. When situational awareness is maintained, operations are completed without a hitch. When situational awareness is not maintained, errors occur, performance suffers, and catastrophic events often result. Situational awareness is a human factors element that is somewhat difficult to explain but easy to recognize after an error occurs. Individuals and members acting as a team need to be aware of what situational awareness is and what the consequences of losing situational awareness are so they can maintain the highest state of attentiveness during all phases of an incident. Situational awareness also carries over into nonincident operations. Failing to note at shift change that the booster tank on an engine company is empty or that the SCBA cylinder in the battalion buggy is down to 2000 psi are situational awareness failures that can have unfortunate consequences.

NASA notes in its definition of situational awareness that, "Situational awareness is the awareness of acknowledging and assessing; and is the basis for choosing courses of action of the situation both now and in the future." Returning to our opening paragraph and question, acknowledging the existence of situational awareness, recognizing the pitfalls of losing situational awareness, and adopting strategies for maintaining situational awareness in all phases of life are crucial elements in the larger error management philosophy of CRM.

Several strategies have been suggested by different experts to help maintain situational awareness. Threaded throughout all of these different versions is a message to constantly check and cross-check the situation and operational performance. This constant review helps ensure that the individual and the entire team are focused on concluding the incident in an efficient and injury-free fashion.

Each approach deals with the same six fundamental components. The six steps for maintaining emergency scene situational awareness are:

- **"Fight the fire!"** All members have to focus their attention on the details of the incident, while keeping the larger picture in mind.

   Numerous aviation disaster case studies revealed that air crews became distracted by events that took their attention away from flying the plane. These distractions resulted in "CFIT" (controlled flight into terrain) events. In nontechnical language, the crews stopped flying the plane and it crashed.

- **Assess problems in the time available.** Emergency scenes do not always permit leisurely, timeless periods for decision making. There has to be a balance between rushing headlong into a burning building and waiting until every possible hazard is evaluated before attacking an incipient fire. Seasoned fire officers and incident commanders have learned that taking an extra 30 to 60 seconds to absorb and process as much information as possible often results in better and more confident decisions.

- **Gather information from <u>all</u> sources.** Information gathering at the emergency scene has to be a rapid process. One person cannot see everything, know everything, hear everything, or smell everything. Using the rest of the "crew," as well as additional arriving command officers and people familiar with the building or terrain, should enhance decision making and help keep situational awareness current.

- **Choose the best option.** Once all of the factors have been weighed, choose the option that maximizes results and minimizes risk.

- **Monitor results and alter the plan as necessary.** Maintaining situational awareness requires continual evaluation of the effectiveness of decisions. Having a "plan B" or "plan C" ready for implementation may be necessary. An extra set of eyes at a strategic location on the incident scene can often provide valuable information.

- **Beware of situational awareness loss factors!** Knowing the factors that can lead to a loss of situational awareness or indicate a situational awareness problem is essential. The situational awareness loss factors include:

   - **Ambiguity:** An event, order, or message that has more than one potential meaning.

   - **Distraction:** Anything that takes attention away from the larger mission. Distraction can come in the form of outside influences (the slamming of a door down the hall while you are watching television) to inside influences (cultural biases, personal).

   - **Fixation:** Tunnel vision.

   - **Overload:** More things are happening than one person can process. The first few minutes after arrival at an exceptionally chaotic scene can lead to overload.

   - **Complacency:** When humans perform the same process repeatedly, they tend to become bored due to the hypnotic effect of repetitive action. Inattentiveness breeds carelessness. The term "routine" fire has opened more than one discussion of a fire fighter fatality. The perception or description of any structure fire as routine is the pinnacle of complacency.

   - **Improper procedure:** Advancing a hand line into a structure fire before ventilating and positioning apparatus downwind at a hazardous materials release are examples of improper procedures. Immediate action is required to correct these mistakes.

   - **Unresolved discrepancy:** You are heading westbound on an interstate and you see a vehicle approaching eastbound in the same lane you are traveling in. There is a brief period where your eyes send the signal to the brain, but you mutter, "I don't believe this." The brain has a tendency to recognize that something is not right and to want to right it. This is an often fatal flaw in the human psyche. Indications and observations that are in conflict with expectations must not be ignored. Fire fighters must be trained to recognize unresolved discrepancies as needing attention immediately.

   - **"Nobody fighting the fire":** An incident commander finds his attention drawn to the actions of one specific company. The company is having some difficulty getting its hoseline in service. While the incident commander steps in to help advance the line, he misses vital radio transmissions and dramatic changes in the fire's progress. Shortly after the crew enters the structure, a flashover occurs, killing the crew. Emergency scene commanders cannot allow themselves to be drawn away from the big picture.

Maintaining situational awareness requires relying heavily on training, experience, decision making, professionalism, and good overall mental and physical condition. A solid strategy for improving situational awareness requires the use of mental joggers for the individual and crew. The crew mental joggers ask:

- "What do we have here?"
- "What's going on here?"
- "How are we doing?"
- "Does this look right?"

The personal mental joggers ask:

- "What do I know that they need to know?"
- "What do they know that I need to know?"
- "What do we all need to know?"

Using crew and personal joggers, remaining alert for the indicators of lost situational awareness, using checklists, and constantly analyzing decisions versus outcomes further reinforce CRM concepts.

Debriefing

The debriefing component is not found in all models of CRM. It is mentioned here for two reasons:

1. Fire service personnel are familiar with the concept of a "postincident analysis" or "critique."
2. Improved performance resulting from well-managed debriefing operations has been well documented.

Debriefing offers personnel operating at an incident the opportunity to "replay" the event, extract lessons learned, and evaluate performance. A formal debriefing should emphasize the elements of communication and teamwork (both leadership and followership), evaluate critical decision making, and reinforce the importance of situational awareness ▶ **Figure 18-6**). The process should be conducted by a strong facilitator, who can keep the participants focused and on task. One recommendation is to assign an individual who was not involved in the incident to facilitate the debriefing. This individual can maintain a higher level of objectivity about the actions that were taken. The individual should be respected and familiar with departmental standard operating procedures.

Keeping the process simple and direct maximizes the learning environment, minimizes blame, and keeps everyone focused. Rules of decorum need to be established and followed. An environment of candid communication must be provided by the facilitator and respected by the participants.

One recommended technique is borrowed from debriefs performed by the United States Navy's precision flying team, the Blue Angels. Each mission is debriefed in an environment known as crossing the "blue line." The debriefing is conducted in an open, honest, candid atmosphere, with a goal of improving performance and ensuring team safety. No member of the team is immune from constructive criticism. Egos and feelings are left outside of the blue line, and whatever is said stays inside the blue line.

Fire service debriefs should review the incident, capitalize on the positives, identify shortcomings, discuss strategies for improvement, and allow participants an opportunity to see

Figure 18-6) Debriefing offers personnel the opportunity to "replay" the event, extract lessons learned, and evaluate performance.

and hear the incident through different sets of eyes and ears. Lubnau and Okray recommend a five-step model that lays out a simple formula for conducting a successful debriefing.

Step 1: Just the facts. Start with the dispatch and walk through the incident from beginning to end. Refer only to what actually took place (e.g., time of dispatch, unit or units that responded, time of arrival, apparatus position, weather, building construction).

Step 2: What did you do? Discuss the actions of the individual companies on the incident. Personnel get an opportunity to describe their actions on the scene. The facilitator needs to maintain order, allowing all participants to describe what they did, without interruption, and maintain a strong instructional tone. Points such as the weather, time of arrival, and building construction are rather concrete. Hoseline placement, fire conditions, apparatus placement, and other observations are more argumentative. Recognize that some facts may be difficult to report with precision, and perceptions can vary.

Step 3: What went wrong? Most fire fighters would prefer to admit their own errors than to have their peers point out their shortcomings. Anyone with more than 5 minutes in the fire service is familiar with this step. Kitchen tables are piled high with coffee mugs that are overflowing with the performance sins of the crews from the next house or "C shift." The key here is for the facilitator to remind the group that a debriefing is not a kangaroo court. This step often requires a reminder of the rules of decorum. The environment to set here is one of encouraging all to provide an honest self-evaluation rather than be put under the microscope and be dissected by their peers. The actions of companies or individuals at the incident but not attending the critique should be avoided.

Step 4: What went right? The facilitator and fire chief may want to switch Step 3 with Step 4. A prevailing thought, however, is that getting the poor performance issues out of the way first, then rebuilding the bruised egos by noting positive actions, rebuilds the individual's and team's desire to excel. No fire fighter that this author has ever met hit the street with an attitude that they were going out to purposely perform poorly. Asking what went right challenges all team members to analyze the incident and identify as many actions as possible that had either favorable outcomes or resulted in positive results.

Step 5: What are you going to do about it? Most errors on the incident scene can be divided into two categories: poor personal or crew performance and mechanical breakdown. A second measure of a debriefing's effectiveness consists of actions taken by the department to correct performance and mechanical issues identified during the debriefing. Debriefings become meaningless if the lessons that are identified are not addressed. Department inaction sets the stage for future debriefings to become unruly character assassinations that further divide crews, stations, and eventually the department itself. The swifter the troops see a response from their officers (including the fire chief), the more value they will attach to the process. Whether the action involves additional training (or retraining), budgeting for (or actually purchasing) new equipment, rewriting an existing operational procedure, or issuing valor awards, decisive responses from leadership instill confidence and restore morale.

CRM: The Time Is Now

Fire departments in the United States are the most technologically advanced in the world. American fire apparatus, protective clothing, tools, appliances, and stations are unparalleled. The number of structure fires and the number of civilians killed in fires has fallen each year since the advent of home smoke alarms, the proliferation of "911" emergency communication centers, and the placement of residential sprinklers. A natural conclusion would draw one to predict a reduction in fire fighter deaths. Sadly, this is not the case. Fire fighter deaths and injuries have decreased, but when compared with the number of deaths and injuries per incident, the death and injury rate has remained comparatively steady. The percentages of fire fighter death and injury in two specific categories (stress and fireground operations) have also remained stable through the past decade. Fire fighters killed in motor vehicle collisions have increased through the same period, in spite of improved cab design, mandatory seat belt rules, traffic light control devices, and improved vehicle suspension systems.

The focus of this chapter was to point out that as much as the fire service touts itself as a "breed set apart," it is made up of people, just like the aviation industry, the medical profession, the military, and all other types of organizations. Many years and millions of dollars have been spent inventing technological "patches" and repairs to interdict injury and death rates. The aviation industry was the first to recognize the role of human error on its operations. Human behavior specialists, pilots, industry executives, and crew members developed and refined CRM into a highly regarded system of error management and disaster reduction for that industry within one generation. Subsequent applications of CRM have met the needs of several other fields of activity.

Since the introduction of CRM in the aviation industry, nearly 25 years ago, almost 3,000 fire fighters have died in the line of duty (this excludes the 343 FDNY fire fighters killed in the World Trade Center attack of September 11, 2001). In the next 10 years, another 1,000 fire fighters can be expected to die in the line of duty. Because we know about the relationship between human error, catastrophe avoidance, and CRM, the time has come to adopt CRM in the fire service; it cannot wait another year.

You Are the Fire Officer: Conclusion

The answer to our beleaguered incident commander's dilemma lies in CRM. His most valuable tools are arriving with the multiple alarm companies. Additional alarms bring resources in the form of people, hardware, and information. The commander's first actions should include identifying who can best assist him with mitigating the emergency and ensuring safety. The department should have existing standard operating procedures that further enhance operational efficiency.

The standard operating procedures rapidly place companies at the incident commander's disposal. Resources must be committed to the fire attack/rescue operations. Appointing a safety officer, division commander or commanders, a radio operator, and an occupant liaison provides the necessary breathing space for the incident commander to maintain, and even improves situational awareness. The incident commander's role is to mitigate the incident by staying focused on strategy and safety. A command advisor (in the form of a chief officer coming in on the extra alarm) should be used as rapidly as possible to provide redundancy in the command post. At least one of the arriving extra alarm companies should be assigned to the command post to take care of radio communications, accountability, and occupant liaison. An existing standard operating procedure that spells out which company will assist the incident commander on arrival provides the necessary structure to bolster the command post immediately, without compromising operational efficiency.

Task allocation rapidly divides responsibilities among resources, allowing the incident commander time to think, make clearer decisions, and communicate effectively. Clear communication means fewer errors. Fewer errors lead to operations that are more efficient, as well as a safer fireground. Everyone goes home a little tired and a little sore, but professionally satisfied and more experienced to handle future emergencies.

Wrap-Up

Ready for Review

- The aviation industry has reduced its accident rate by 80% through the development, refinement, and system-wide delivery of a behavioral modification training system known as crew resource management (CRM).
- National Institute for Occupational Safety and Health reports on fire fighter fatalities routinely indicate that human error plays a role in each fatality.
- The proven principles of CRM are an effective means of reducing the effects of human error and preventing tragedy.
- Gordon Dupont determined that a "dirty dozen" of human factors contribute to tragedy:
  - Lack of communication
  - Complacency
  - Lack of knowledge
  - Distraction
  - Lack of teamwork
  - Fatigue
  - Lack of resources
  - Pressure
  - Lack of assertiveness
  - Stress
  - Lack of awareness
  - Norms
- A six-point model serves the fire service well:
  - Communication skills
  - Teamwork
  - Task allocation
  - Critical decision making
  - Situational awareness
  - Debriefing
- CRM suggests that developing a standard language, maintaining a "sterile cockpit," and teaching appropriate assertive behavior are the keys to reducing errors resulting from miscommunication.
- CRM promotes the concept of working together for the common good.
- Respect, trust, and effectiveness comprise the triangle of leadership.
- Know the crew's limits. Once an officer identifies his or her own limits, the crew needs to be evaluated for the same purpose.
- CRM promotes the concept of team involvement in all aspects of operation. In the area of emergency scene decision making, time, experience, and training play pivotal roles in successful outcomes.
- The loss of situational awareness is frequently the first link in a chain of errors that leads to calamity.
- Situational awareness loss factors include:
  - Ambiguity
  - Distraction
  - Fixation
  - Overload
  - Complacency
  - Improper procedure
  - Unresolved discrepancy
  - "Nobody fighting the fire"
- Debriefing offers personnel operating at an incident the opportunity to "rerun" the event, extract lessons learned, and evaluate performance.

Hot Terms

Crew resource management (CRM) A behavioral modification training system developed by the aviation industry to reduce its accident rate.

Followership Leaders can be effective only to the extent that followers are willing to accept their leadership.

Naturalistic decision making Describes how commanders make decisions in their natural environment.

Recognition primed decision making (RPD) Describes how commanders can recognize a plausible plan of action.

Situational awareness The process of evaluating the severity and consequences of an incident and communicating the results.

During a recent training session for department officers, you were introduced to a new concept called crew resource management. During this training, you are taught new concepts to help crew cohesiveness and to help ensure good, proper communications between shift personnel. You look forward to implementing this on your shift and hope to bring a new level of professionalism and cohesion to the shift.

1. There are several skills necessary to be a good follower to be an effective team member and to maximize CRM. What is one of them?

   A. Leave on time

   B. Be respectful

   C. Respect authority

   D. Insubordination is acceptable

2. What is task allocation?

   A. Knowing one's limits

   B. Admitting errors

   C. Being adaptable

   D. Creating a good work setting

3. What is RPD?

   A. Resource planning division

   B. Recognized personnel decision making

   C. Recognition primed decision making

   D. Recognition planned decisions

4. What is one of the situational awareness loss factors?

   A. Insubordination

   B. Incompleteness

   C. Overload

   D. Absence

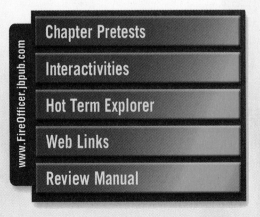

www.FireOfficer.jbpub.com

Chapter Pretests

Interactivities

Hot Term Explorer

Web Links

Review Manual

www.FireOfficer.jbpub.com

Fire Officer Presentation

Perceptions are reality and your demeanor is a reflection of your ethics and values. The image that you portray may or may not project an appropriate level of confidence and professionalism. But what do these things mean, and why are they important? Fire officers should project a professional image for many reasons. One reason is that you want to gain the acceptance and confidence of the public you serve, the fire fighters that you lead, and the supervisors to whom you report. When people project a professional manner, they tend to act in a more professional manner, exuding the characteristics dictated by high ethics and values. A professional image helps you be an outstanding fire officer.

Verbal Presentation

Always project a calm, professional image when you speak, whether on the phone or face-to-face. When speaking face-to-face, whether to a superior, a member of the community, or to a fire fighter, remember to use a calm, even tone of voice. Maintain eye contact with the person to whom you are speaking. This simple act gives an air of confidence, authority, and attentiveness.

In addition, remember to:

- Listen carefully
- Never interrupt
- If you do not know an answer to a question, do not guess

When speaking on the phone, remember to use a clear, pleasant tone of voice. When speaking to a member of the community, be friendly and professional. Avoid using jargon and make certain to answer any questions thoroughly. When speaking with a superior, remember that his or her time is valuable—be concise, honest, and direct.

Physical Presentation

Always present a neat, professional image. Remember that you are representing the entire fire department, so make certain that you put your best face forward. A well-groomed, well-dressed fire officer will exude a calm, professional demeanor. A fire officer with a wrinkled uniform and unkempt hair will exude an air of disorganization. Ask yourself who you would rather interact with—the calm, collected professional, or a frazzled coworker?

Written Presentation

Written communication is critical to the effective operations of any profession. As a fire officer, you will need to communicate with others in the form of written documentation, be it via e-mail or by letter. When you speak to another person, he or she can see your facial expressions, hear what you are saying, and, with a few clarifying questions, can usually understand your message with relative ease. When putting your words in writing, however, these variables are removed; therefore, it is absolutely critical that you articulate your words correctly in order ensure understanding of the message that you are conveying.

Whether you are documenting an incident report for your supervisor, justifying the need for a budget increase to city or county officials, or corresponding with other fire officers, the main point of your message or request should be at the beginning with the explanation or justification following.

Remember, the individual who is reading the correspondence is a busy person; he or she will quickly lose interest in a lengthy letter or e-mail, especially if the main point of the message is found in the last sentence of the last paragraph. If you cloud your message with superfluous text, the recipient will probably end up placing a phone call to you for clarification, thus defeating the purpose of the letter.

If you quote a fact or statistic, you should include a reference where the data can be found. Such references could be a web site, current text or journal, or other notable source, like a textbook. Backing up factual or statistical data is important, especially if you are requesting new equipment for all fire personnel because "studies have shown" that the equipment you are currently using does not provide the best protection as newer equipment on the market does.

Regardless of whom you are corresponding with, the content of your message must remain professional and succinct. To ensure this, consider having a disinterested party read the letter or e-mail before you send it. If he or she can understand what you are stating, then the recipient most likely will as well.

Remember, poorly written communications could undermine your credibility. To avoid this, follow these hints:

- Use a computer whenever possible
- Always use spell check
- Have a pocket guide on grammar available
- Handwriting should be neat and legible
- Write in shorter sentences
- Use paragraphs to switch topics

Voice of the Fire Officer

Every fire officer will be called upon to perform in various roles and knowing how to adapt the Fire Officer Voice (FOV) is critical to appropriately communicate. The confidence behind the FOV is directly impacted by the training, education, and experiences of the individual in the position. The FOV is also one of the most visible methods by which the fire officer is measured and/or perceived. Subordinates and the public react more directly by the FOV than any other action.

The FOV must include the following characteristics and must vary appropriately to the circumstances at hand:

- Confident
- Persuasive
- Passionate
- Honest
- Firm
- Empathetic/Caring
- Diplomatic
- Instructional
- Commanding (when necessary)
- Counseling/Consoling

The voice of a new fire officer is often viewed as rough. Developing the skill of communications (both written and spoken) must be a lifelong endeavor. On the written side, fire officers will be required to correspond on many levels. From the incident narrative to departmental memos to Town Council presentations—the finesse of the FOV is of paramount importance. The measure of the written word is based on proper grammar, spelling, and organization.

The fire officer also requires the confident and persuasive ability to speak in a public forum. A weak and squeaky voice results in disinterest and projects a lack of confidence, especially when situations have escalated to a level of uncertainty and peril. The ability of the fire officer to understand and master the traits discussed will ultimately distinguish whether that officer is average or great and will influence the success of his or her fire service career.

Chapter 1: Administration

1.1* Scope. This standard shall identify the performance requirements necessary to perform the duties of a fire officer and specifically identifies four levels of progression.

1.2 Purpose. The purpose of this standard shall be to specify the minimum job performance requirements for service as a fire officer.

1.2.1 The intent of the standard is to define progressive levels of performance required at the various levels of officer responsibility. The authority having jurisdiction has the option to combine or group the levels to meet its local needs and to use them in the development of job descriptions and specifying promotional standards.

1.2.2 It is not the intent of this standard to restrict any jurisdiction from exceeding these minimum requirements.

1.2.3 This standard shall cover the requirements for the four levels of progression—Fire Officer I, Fire Officer II, Fire Officer III, and Fire Officer IV.

1.3* General.

1.3.1 All of the standards for any level of fire officer shall be performed in accordance with recognized practices and procedures or as defined by an accepted authority.

1.3.2 It is not required for the objectives to be mastered in the order in which they appear. The local or state/provincial training program shall establish both the instructional priority and the program content to prepare individuals to meet the performance objectives of this standard.

1.3.3 The Fire Fighter II shall meet all the objectives for Fire Officer I before being certified at the Fire Officer I level, and the objectives for each succeeding level in the progression shall be met before being certified at the next higher level.

Chapter 2: Referenced Publications

2.1 General. The documents or portions thereof listed in this chapter are referenced within this standard and shall be considered part of the requirements of this document.

2.2 NFPA Publications. National Fire Protection Association, 1 Batterymarch Park, P.O. Box 9101, Quincy, MA 02269-9101.

NFPA 1001, *Standard for Fire Fighter Professional Qualifications,* 2002 Edition.

NFPA 1041, *Standard for Fire Service Instructor Professional Qualifications,* 2002 Edition.

2.3 Other Publications. (Reserved)

Chapter 3: Definitions

3.1* General. The definitions contained in this chapter shall apply to the terms used in this standard. Where terms are not included, common usage of the terms shall apply.

3.2 NFPA Official Definitions.

3.2.1* Approved. Acceptable to the authority having jurisdiction.

3.2.2* Authority Having Jurisdiction (AHJ). An organization, office, or individual responsible for enforcing the requirements of a code or standard, or for approving equipment, materials, an installation, or a procedure.

3.2.3 Labeled. Equipment or materials to which has been attached a label, symbol, or other identifying mark of an organization that is acceptable to the authority having jurisdiction and concerned with product evaluation, that maintains periodic inspection of production of labeled equipment or materials, and by whose labeling the manufacturer indicates compliance with appropriate standards or performance in a specified manner.

3.2.4* Listed. Equipment, materials, or services included in a list published by an organization that is acceptable to the authority having jurisdiction and concerned with evaluation of products or services, that maintains periodic inspection of production of listed equipment or materials or periodic evaluation of services, and whose listing states that either the equipment, material, or service meets appropriate designated standards or has been tested and found suitable for a specified purpose.

3.2.5 Shall. Indicates a mandatory requirement.

3.2.6 Should. Indicates a recommendation or that which is advised but not required.

3.2.7 Standard. A document, the main text of which contains only mandatory provisions using the word "shall" to indicate requirements and which is in a form generally suitable for mandatory reference by another standard or code or for adoption into law. Nonmandatory provisions shall be located in an appendix or annex, footnote, or fine-print note and are not to be considered a part of the requirements of a standard.

3.3 General Definitions.

3.3.1* Comprehensive Emergency Management Plan. Planning document that includes preplan information and resources for the management of catastrophic emergencies within the jurisdiction.

3.3.2 Fire Department. An organization providing rescue, fire suppression, and other related activities. For the purposes of this standard, the term "fire department" shall include any public, private, or military organization engaging in this type of activity. [**1002:**3.3]

3.3.3 Fire Officer I. The fire officer, at the supervisory level, who has met the job performance requirements specified in this standard for Level I.

3.3.4 Fire Officer II. The fire officer, at the supervisory/managerial level, who has met the job performance requirements specified in this standard for Level II.

3.3.5 Fire Officer III. The fire officer, at the managerial/administrative level, who has met the job performance requirements specified in this standard for Level III.

3.3.6 Fire Officer IV. The fire officer, at the administrative level, who has met the job performance requirements specified in this standard for Level IV.

3.3.7 Incident Management System (IMS). A system that defines the roles and responsibilities to be assumed by personnel and the operating procedures to be used in the management and direction of emergency operations; the system is also referred to as an incident command system (ICS).

3.3.8 Job Performance Requirement. A statement that describes a specific job task, lists the items necessary to complete the task, and defines measurable or observable outcomes and evaluation areas for the specific task. [**1000:**2.1]

3.3.9 Member. A person involved in performing the duties and responsibilities of a fire department under the auspices of the organization. A fire department member can be a full-time or part-time employee or a paid or unpaid volunteer, can occupy any position or rank within the fire department, and can engage in emergency operations. [**1500:**3.3]

3.3.10 Promotion. The advancement of a member from one rank to a higher rank by a method such as election, appointment, merit, or examination.

3.3.11 Qualification. Having satisfactorily completed the requirements of the objectives.

3.3.12 Supervisor. An individual responsible for overseeing the performance or activity of other members.

3.3.13 Unit. An engine company, truck company, or other functional or administrative group.

Chapter 4: Fire Officer I

4.1* General. For certification at Fire Officer Level I, the candidate shall meet the requirements of Fire Fighter II as defined in NFPA 1001, Fire Instructor I as defined in NFPA 1041, and the job performance requirements defined in Sections 4.2 through 4.7 of this standard.

4.1.1 General Prerequisite Knowledge. The organizational structure of the department; geographical configuration and characteristics of response districts; departmental operating procedures for administration, emergency operations, incident management systems, and safety; departmental budget process; information management and recordkeeping; the fire prevention and building safety codes and ordinances applicable to the jurisdiction; current trends, technologies, and socioeconomic and political factors that impact the fire service; cultural diversity; methods used by supervisors to obtain cooperation within a group of subordinates; the rights of management and members; agreements in force between the organization and members; generally accepted ethical practices, including a professional code of ethics; and policies and procedures regarding the operation of the department as they involve supervisors and members.

4.1.2 General Prerequisite Skills. The ability to effectively communicate in writing utilizing technology provided by the AHJ; write reports, letters, and memos utilizing word processing and spreadsheet programs; operate in an information management system; and effectively operate at all levels in the incident management system utilized by the AHJ.

4.2 Human Resource Management. This duty involves utilizing human resources to accomplish assignments in accordance with safety plans and in an efficient manner. This duty also involves evaluating member performance and supervising personnel during emergency and nonemergency work periods, according to the following job performance requirements.

4.2.1 Assign tasks or responsibilities to unit members, given an assignment at an emergency operation, so that the instructions are complete, clear, and concise; safety considerations are addressed; and the desired outcomes are conveyed.

(A) Requisite Knowledge. Verbal communications during emergency situations, techniques used to make assignments under stressful situations, and methods of confirming understanding.

(B) Requisite Skills. The ability to condense instructions for frequently assigned unit tasks based on training and standard operating procedures.

4.2.2 Assign tasks or responsibilities to unit members, given an assignment under nonemergency conditions at a station or other work location, so that the instructions are complete, clear, and concise; safety considerations are addressed; and the desired outcomes are conveyed.

(A) Requisite Knowledge. Verbal communications under nonemergency situations, techniques used to make assignments under routine situations, and methods of confirming understanding.

(B) Requisite Skills. The ability to issue instructions for frequently assigned unit tasks based on department policy.

4.2.3 Direct unit members during a training evolution, given a company training evolution and training policies and procedures, so that the evolution is performed in accordance with safety plans, efficiently, and as directed.

(A) Requisite Knowledge. Verbal communication techniques to facilitate learning.

(B) Requisite Skills. The ability to distribute issue-guided directions to unit members during training evolutions.

4.2.4 Recommend action for member-related problems, given a member with a situation requiring assistance and the member assistance policies and procedures, so that the situation is identified and the actions taken are within the established policies and procedures.

(A)* Requisite Knowledge. The signs and symptoms of member-related problems, causes of stress in emergency services personnel, and adverse effects of stress on the performance of emergency service personnel.

(B) Requisite Skills. The ability to recommend a course of action for a member in need of assistance.

4.2.5* Apply human resource policies and procedures, given an administrative situation requiring action, so that policies and procedures are followed.

(A) Requisite Knowledge. Human resource policies and procedures.

(B) Requisite Skills. The ability to communicate orally and in writing and to relate interpersonally.

4.2.6 Coordinate the completion of assigned tasks and projects by members, given a list of projects and tasks and the job requirements of subordinates, so that the assignments are prioritized, a plan for the completion of each assignment is developed, and members are assigned to specific tasks and supervised during the completion of the assignments.

(A) Requisite Knowledge. Principles of supervision and basic human resource management.

(B) Requisite Skills. The ability to plan and to set priorities.

4.3 Community and Government Relations. This duty involves dealing with inquiries of the community and projecting the role of the department to the public and delivering safety, injury, and fire prevention education programs, according to the following job performance requirements.

4.3.1 Initiate action on a community need, given policies and procedures, so that the need is addressed.

(A) Requisite Knowledge. Community demographics and service organizations, as well as verbal and nonverbal communication.

(B) Requisite Skills. Familiarity with public relations and the ability to communicate verbally.

4.3.2 Initiate action to a citizen's concern, given policies and procedures, so that the concern is answered or referred to the correct individual for action and all policies and procedures are complied with.

(A) Requisite Knowledge. Interpersonal relationships and verbal and nonverbal communication.

(B) Requisite Skills. Familiarity with public relations and the ability to communicate verbally.

4.3.3 Respond to a public inquiry, given policies and procedures, so that the inquiry is answered accurately, courteously, and in accordance with applicable policies and procedures.

(A) Requisite Knowledge. Written and oral communication techniques.

(B) Requisite Skills. The ability to relate interpersonally and to respond to public inquiries.

4.3.4 Deliver a public education program, given the target audience and topic, so that the intended message is conveyed clearly.

(A) Requisite Knowledge. Contents of the fire department's public education program as it relates to the target audience.

(B) Requisite Skills. The ability to communicate to the target audience.

4.4 Administration. This duty involves general administrative functions and the implementation of departmental policies and procedures at the unit level, according to the following job performance requirements.

4.4.1 Recommend changes to existing departmental policies and/or implement a new departmental policy at the unit level, given a new departmental policy, so that the policy is communicated to and understood by unit members.

(A) Requisite Knowledge. Written and oral communication.

(B) Requisite Skills. The ability to relate interpersonally.

4.4.2 Execute routine unit-level administrative functions, given forms and record-management systems, so that the reports and logs are complete and files are maintained in accordance with policies and procedures.

(A) Requisite Knowledge. Administrative policies and procedures and records management.

(B) Requisite Skills. The ability to communicate orally and in writing.

4.4.3 Prepare a budget request, given a need and budget forms, so that the request is in the proper format and is supported with data.

(A) Requisite Knowledge. Policies and procedures and the revenue sources and budget process.

(B) Requisite Skill. The ability to communicate in writing.

4.5* Inspection and Investigation. This duty involves performing a fire investigation to determine preliminary cause, securing the incident scene, and preserving evidence, according to the following job performance requirements.

4.5.1 Evaluate available information, given a fire incident, observations, and interviews of first-arriving members and other individuals involved in the incident, so that a preliminary cause of the fire is determined, reports are completed, and, if required, the scene is secured and all pertinent information is turned over to an investigator.

(A) Requisite Knowledge. Common causes of fire, fire growth and development, and policies and procedures for calling for investigators.

(B) Requisite Skills. The ability to determine basic fire cause, conduct interviews, and write reports.

4.5.2 Secure an incident scene, given rope or barrier tape, so that unauthorized persons can recognize the perimeters of the scene and are kept from restricted areas, and all evidence or potential evidence is protected from damage or destruction.

(A) Requisite Knowledge. Types of evidence, the importance of fire scene security, and evidence preservation.

(B) Requisite Skills. The ability to establish perimeters at an incident scene.

4.6* Emergency Service Delivery. This duty involves supervising emergency operations, conducting pre-incident planning, and deploying assigned resources in accordance with the local emergency plan and according to the following job performance requirements.

4.6.1 Develop a pre-incident plan, given an assigned facility and preplanning policies, procedures, and forms, so that all required elements are identified and the approved forms are completed and processed in accordance with policies and procedures.

(A) Requisite Knowledge. Elements of the local emergency plan, a pre-incident plan, basic building construction, basic fire protection systems and features, basic water supply, basic fuel loading, and fire growth and development.

(B) Requisite Skills. The ability to write reports, to communicate orally, and to evaluate skills.

4.6.2 Develop an initial action plan, given size-up information for an incident and assigned emergency response resources, so that resources are deployed to control the emergency.

(A)* Requisite Knowledge. Elements of a size-up, standard operating procedures for emergency operations, and fire behavior.

(B) Requisite Skills. The ability to analyze emergency scene conditions; to activate the local emergency plan, including localized evacuation procedures; to allocate resources; and to communicate orally.

4.6.3* Implement an action plan at an emergency operation, given assigned resources, type of incident, and a preliminary plan, so that resources are deployed to mitigate the situation.

(A) Requisite Knowledge. Standard operating procedures, resources available for the mitigation of fire and other emergency incidents, an incident management system, scene safety, and a personnel accountability system.

(B) Requisite Skills. The ability to implement an incident management system, to communicate orally, to manage scene safety, and to supervise and account for assigned personnel under emergency conditions.

4.6.4 Develop and conduct a post-incident analysis, given a single unit incident and post-incident analysis policies, procedures, and forms, so that all required critical elements are identified and communicated, and the approved forms are completed and processed in accordance with policies and procedures.

(A) Requisite Knowledge. Elements of a post-incident analysis, basic building construction, basic fire protection systems and features, basic water supply, basic fuel loading, fire growth and development, and departmental procedures relating to dispatch response tactics and operations and customer service.

(B) Requisite Skills. The ability to write reports, to communicate orally, and to evaluate skills.

4.7* Health and Safety. This duty involves integrating safety plans, policies, and procedures into the daily activities as well as the emergency scene, including the donning of appropriate levels of personal protective equipment to ensure a work environment, in accordance with health and safety plans, for all assigned members, according to the following job performance requirements.

4.7.1 Apply safety regulations at the unit level, given safety policies and procedures, so that required reports are completed, in-service training is conducted, and member responsibilities are conveyed.

(A) Requisite Knowledge. The most common causes of personal injury and accident to members, safety policies and procedures, basic workplace safety, and the components of an infectious disease control program.

(B) Requisite Skills. The ability to identify safety hazards and to communicate orally and in writing.

4.7.2 Conduct an initial accident investigation, given an incident and investigation forms, so that the incident is documented and reports are processed in accordance with policies and procedures.

(A) Requisite Knowledge. Procedures for conducting an accident investigation and safety policies and procedures.

(B) Requisite Skills. The ability to communicate orally and in writing and to conduct interviews.

Chapter 5: Fire Officer II

5.1 General. For certification at Level II, the Fire Officer I shall meet the requirements of Fire Instructor I as defined in NFPA 1041 and the job performance requirements defined in Sections 5.2 through 5.7 of this standard.

5.1.1 General Prerequisite Knowledge. The organization of local government; enabling and regulatory legislation and the law-making process at the local, state/provincial, and federal levels; and the functions of other bureaus, divisions, agencies, and organizations and their roles and responsibilities that relate to the fire service.

5.1.2 General Prerequisite Skills. Intergovernmental and interagency cooperation.

5.2 Human Resource Management. This duty involves evaluating member performance, according to the following job performance requirements.

5.2.1 Initiate actions to maximize member performance and/or to correct unacceptable performance, given human resource policies and procedures, so that member and/or unit performance improves or the issue is referred to the next level of supervision.

(A) Requisite Knowledge. Human resource policies and procedures, problem identification, organizational behavior, group dynamics, leadership styles, types of power, and interpersonal dynamics.

(B) Requisite Skills. The ability to communicate orally and in writing, to solve problems, to increase team work, and to counsel members.

5.2.2 Evaluate the job performance of assigned members, given personnel records and evaluation forms, so each member's performance is evaluated accurately and reported according to human resource policies and procedures.

(A) Requisite Knowledge. Human resource policies and procedures, job descriptions, objectives of a member evaluation program, and common errors in evaluating.

(B) Requisite Skills. The ability to communicate orally and in writing and to plan and conduct evaluations.

5.3 Community and Government Relations. No additional requirements at this level.

5.4 Administration. This duty involves preparing a project or divisional budget, news releases, and policy changes, according to the following job performance requirements.

5.4.1 Develop a policy or procedure, given an assignment, so that the recommended policy or procedure identifies the problem and proposes a solution.

(A) Requisite Knowledge. Policies and procedures and problem identification.

(B) Requisite Skills. The ability to communicate in writing and to solve problems.

5.4.2 Develop a project or divisional budget, given schedules and guidelines concerning its preparation, so that capital, operating, and personnel costs are determined and justified.

(A) Requisite Knowledge. The supplies and equipment necessary for ongoing or new projects; repairs to existing facilities; new equipment, apparatus maintenance, and personnel costs; and appropriate budgeting system.

(B) Requisite Skill. The ability to allocate finances, to relate interpersonally, and to communicate orally and in writing.

5.4.3 Describe the process of purchasing, including soliciting and awarding bids, given established specifications, in order to ensure competitive bidding.

(A) Requisite Knowledge. Purchasing laws, policies, and procedures.

(B) Requisite Skills. The ability to use evaluative methods and to communicate orally and in writing.

5.4.4 Prepare a news release, given an event or topic, so that the information is accurate and formatted correctly.

(A) Requisite Knowledge. Policies and procedures and the format used for news releases.

(B) Requisite Skills. The ability to communicate orally and in writing.

5.4.5 Prepare a concise report for transmittal to a supervisor, given fire department record(s) and a specific request for details such as trends, variances, or other related topics.

(A) Requisite Knowledge. The data processing system.

(B) Requisite Skills. The ability to communicate in writing and to interpret data.

5.5 Inspection and Investigation. This duty involves conducting inspections to identify hazards and address violations and conducting fire investigations to determine origin and preliminary cause, according to the following job performance requirements.

5.5.1 Describe the procedures for conducting fire inspections, given any of the following occupancies, so that all hazards, including hazardous materials, are identified, approved forms are completed, and approved action is initiated:

    (1) Assembly
    (2) Educational
    (3) Health care
    (4) Detention and correctional
    (5) Residential
    (6) Mercantile
    (7) Business
    (8) Industrial
    (9) Storage
   (10) Unusual structures
   (11) Mixed occupancies

(A) Requisite Knowledge. Inspection procedures; fire detection, alarm, and protection systems; identification of fire and life safety hazards; and marking and identification systems for hazardous materials.

(B) Requisite Skills. The ability to communicate in writing and to apply the appropriate codes.

5.5.2 Determine the point of origin and preliminary cause of a fire, given a fire scene, photographs, diagrams, pertinent data and/or sketches, to determine if arson is suspected.

(A) Requisite Knowledge. Methods used by arsonists, common causes of fire, basic cause and origin determination, fire growth and development, and documentation of preliminary fire investigative procedures.

(B) Requisite Skills. The ability to communicate orally and in writing and to apply knowledge using deductive skills.

5.6 Emergency Service Delivery. This duty involves supervising multi-unit emergency operations, conducting pre-incident planning, and deploying assigned resources, according to the following job requirements.

5.6.1 Produce operational plans, given an emergency incident requiring multi-unit operations, so that required resources and their assignments are obtained and plans are carried out in compliance with approved safety procedures resulting in the mitigation of the incident.

(A) Requisite Knowledge. Standard operating procedures; national, state/provincial, and local information resources available for the mitigation of emergency incidents; an incident management system; and a personnel accountability system.

(B) Requisite Skills. The ability to implement an incident management system, to communicate orally, to supervise and account for assigned personnel under emergency conditions; and to serve in command staff and unit supervision positions within the Incident Management System.

5.6.2 Develop and conduct a post-incident analysis, given multi-unit incident and post-incident analysis policies, procedures, and forms, so that all required critical elements are identified and communicated and the approved forms are completed and processed.

(A) Requisite Knowledge. Elements of a post-incident analysis, basic building construction, basic fire protection systems and features, basic water supply, basic fuel loading, fire growth and development, and departmental procedures relating to dispatch response, strategy tactics and operations, and customer service.

(B) Requisite Skills. The ability to write reports, to communicate orally, and to evaluate skills.

5.7 Health and Safety. This duty involves reviewing injury, accident, and health exposure reports, identifying unsafe work environments or behaviors, and taking approved action to prevent reoccurrence, according to the following job requirements.

5.7.1 Analyze a member's accident, injury, or health exposure history, given a case study, so that a report including action taken and recommendations made is prepared for a supervisor.

(A) Requisite Knowledge. The causes of unsafe acts, health exposures, or conditions that result in accidents, injuries, occupational illnesses, or deaths.

(B) Requisite Skills. The ability to communicate in writing and to interpret accidents, injuries, occupational illnesses, or death reports.

NFPA 1021, *Standard for Fire Officer Professional Qualifications*, 2003 Edition	Corresponding Textbook Chapter
Fire Officer I	
4.1	1
4.1.1	1, 3, 5
4.1.2	14
4.2	4, 8
4.2.1	6, 9, 14
4.2.1 (A)	6, 9, 14
4.2.1 (B)	14
4.2.2	6, 9, 14
4.2.2 (A)	9, 14
4.2.2 (B)	9, 14
4.2.3	6, 7
4.2.3 (A)	7
4.2.3 (B)	7
4.2.4	8
4.2.4 (A)	8
4.2.4 (B)	8
4.2.5	3, 4, 5, 8
4.2.5 (A)	3, 4, 5, 8
4.2.5 (B)	3, 4, 5, 8
4.2.6	4
4.2.6 (A)	4
4.2.6 (B)	4
4.3	10
4.3.1	10
4.3.1 (A)	10
4.3.1 (B)	10
4.3.2	11
4.3.2 (A)	11
4.3.2 (B)	11
4.3.3	10
4.3.3 (A)	10
4.3.3 (B)	10

NFPA 1021, *Standard for Fire Officer Professional Qualifications*, 2003 Edition	Corresponding Textbook Chapter
4.3.4	10
4.3.4 (A)	10
4.3.4 (B)	10
4.4	3
4.4.1	11
4.4.1 (A)	11
4.4.1 (B)	11
4.4.2	3
4.4.2 (A)	3
4.4.2 (B)	3
4.4.3	13
4.4.3 (A)	13
4.4.3 (B)	13
4.5	17
4.5.1	17
4.5.1 (A)	17
4.5.1 (B)	17
4.5.2	17
4.5.2 (A)	17
4.5.2 (B)	17
4.6	12, 16
4.6.1	12
4.6.1 (A)	12
4.6.1 (B)	12
4.6.2	16
4.6.2 (A)	16
4.6.2 (B)	16
4.6.3	15, 16
4.6.3 (A)	15, 16
4.6.3 (B)	15, 16
4.6.4	6, 15
4.6.4 (A)	6, 15

NFPA 1021, *Standard for Fire Officer Professional Qualifications,* **2003 Edition**	Corresponding Textbook Chapter
4.6.4 (B)	.6, 15
4.7	.6
4.7.1	.6
4.7.1 (A)	.6
4.7.1 (B)	.6
4.7.2	.6
4.7.2 (A)	.6
4.7.2 (B)	.6

Fire Officer II

NFPA 1021, *Standard for Fire Officer Professional Qualifications,* **2003 Edition**	Corresponding Textbook Chapter
5.1	1
5.1.1	1, 5
5.1.2	13
5.2	.8
5.2.1	.4, 8, 9
5.2.1 (A)	.4, 8, 9
5.2.1 (B)	.4, 8
5.2.2	.8
5.2.2 (A)	.8
5.2.2 (B)	.8
5.3	10
5.4	10, 11, 13
5.4.1	11
5.4.1 (A)	11
5.4.1 (B)	11
5.4.2	13
5.4.2 (A)	13
5.4.2 (B)	13

NFPA 1021, *Standard for Fire Officer Professional Qualifications,* **2003 Edition**	Corresponding Textbook Chapter
5.4.3	13
5.4.3 (A)	13
5.4.3 (B)	13
5.4.4	10, 14
5.4.4 (A)	10, 14
5.4.4 (B)	10, 14
5.4.5	14
5.4.5 (A)	14
5.4.5 (B)	14
5.5	12, 17
5.5.1	12
5.5.1 (A)	12
5.5.1 (B)	12
5.5.2	17
5.5.2 (A)	17
5.5.2 (B)	17
5.6	12, 16
5.6.1	.6, 15, 16
5.6.1 (A)	.6, 15, 16
5.6.1 (B)	.6, 15, 16
5.6.2	15
5.6.2 (A)	15
5.6.2 (B)	15
5.7	.6
5.7.1	.6
5.7.1 (A)	.6
5.7.1 (B)	.6

Glossary

Accelerants An agent, often an ignitable liquid, used to initiate a fire or increase the rate of growth or spread of fire.

Accident An unplanned event that interrupts an activity and sometimes causes injury or damage. A chance occurrence arising from unknown causes; an unexpected happening due to carelessness, ignorance, and the like.

Accreditation A system whereby a certification organization determines that a school or program meets with the requirements of the fire service.

Actionable items Employee behavior that requires an immediate corrective action by the supervisor because dozens of lawsuits have shown that failing to act will create a liability and a loss for the department.

Adoption by reference Method of code adoption in which the specific edition of a model code is referred to within the adopting ordinance or regulation.

Adoption by transcription Method of code adoption in which the entire text of the code is published within the adopting ordinance or regulation.

Arson The crime of maliciously and intentionally, or recklessly, starting a fire or causing an explosion.

Artifacts The remains of the material first ignited, the ignition source, or other items or components in some way related to the fire ignition, development, or spread. An artifact may also be an item on which fire patterns are present, in which case the preservation of the artifact is not for the item itself but for the fire pattern that appears on the item.

Assessment centers A series of simulation exercises to identify a candidate's competency to perform the job that is offered in the promotion examination.

Assistant or division chiefs Midlevel chief who often has a functional area of responsibility, such as training, and answers directly to the fire chief.

Authority having jurisdiction An organization, office, or individual responsible for enforcing the requirements of a code or standard, or for approving equipment, materials, an installation, or a procedure.

Automatic sprinkler system A system of pipes with water under pressure that allows water to be discharged immediately when a sprinkler head operates.

Base The location at which the primary logistics functions are coordinated and administered. The incident command post may be colocated with the base. There is only one base per incident.

Base budget The level of funding that would be required to maintain all services at the currently authorized levels, including adjustments for inflation, salary increases, and other predictable cost changes.

Battalion chiefs Usually the first level of fire chief; also called district chief. These chiefs are often in charge of running calls and supervising multiple stations or districts within a city. A battalion chief is usually the officer in charge of a single-alarm working fire.

Binding arbitration The resolution of a dispute by a third and neutral party, one who is not personally involved in the dispute and may be expected to reach a fair and objective decision based on an informal hearing, at which the disputants may argue their cases and present all relevant evidence. It is usually agreed in advance that such a decision will be binding and not subject to appeal.

Bond A certificate of debt issued by a government or corporation; a bond guarantees payment of the original investment plus interest by a specified future date.

Brainstorming A method of shared problem solving in which all members of a group spontaneously contribute ideas.

Branch A supervisory level established in either the operations or logistics function to provide a span of control.

Branch director A supervisory position in charge of a number of divisions and/or groups. This position reports to a section chief or the incident commander.

Budget An itemized summary of estimated or intended expenditures for a given period, along with financing proposals for financing them.

Captain The second rank of promotion, between the lieutenant and battalion chief. Captains are responsible for a fire company and for coordinating the activities of that company among the other shifts.

Catastrophic theory of reform When fire prevention codes or firefighting procedures are changed in reaction to a fire disaster.

Central tendency An evaluation error that occurs when a fire fighter is rated in the middle of the range for all dimensions of work performance.

Chain of command The superior-subordinate authority relationship that starts at the top of the organization hierarchy and extends to the lowest levels.

Char Carbonaceous material that has been burned and has a blackened appearance.

Chief's trumpet An obsolete amplification device that enabled a chief officer to give orders to fire fighters during an emergency; precursor to a bullhorn and portable radios.

Chronological statement of events A detailed account of the fire company activities as related to an incident or accident.

Class specification A technical worksheet that quantifies the knowledge, skills, and abilities (KSAs) by frequency and importance for every classified job within the local civil service agency.

Coach To provide training to an individual or a team.

Collective bargaining Method whereby representatives of employees (unions) and employers determine the conditions of employment through direct negotiation, normally resulting in a written contract setting forth the wages, hours, and other conditions to be observed for a stipulated period (e.g., 3 years). Term also applies to union-management dealings during the terms of the agreement.

Command staff Positions that are established to assume responsibility for key activities in the Incident Management System that are not a part of the line organization that include safety officer, public information officer, and liaison officer.

Company journal Log book at the fire station that creates an extemporaneous record of the emergency, routine activities, and special activities that occurred at the fire station. The company journal also records any fire fighter injury, liability-creating event, and special visitors to the fire station.

Complaint Expression of grief, regret, pain, censure, or resentment; lamentation; accusation; or fault finding.

Conflict A state of opposition between two parties. A complaint is a manifestation of a conflict

Consensus document A code or standard developed through agreement between people representing different organizations and interests. NFPA codes and standards are consensus documents.

Construction type The combination of materials used in the construction of a building or structure, based on the varying degrees of fire resistance and combustibility.

Contrast effect An evaluation error in which a fire fighter is rated on the basis of the performance of another fire fighter and not on the classified job standards.

Controlling Restraining, regulating, governing, counteracting, or overpowering.

Crew resource management (CRM) A behavioral modification training system developed by the aviation industry to reduce its accident rate.

"Data dump" question A promotional question that asks the candidate to write or describe all of the factors or issues covering a technical issue, such as suppression of a basement fire in a commercial property.

Decision making The process of identifying problems and opportunities and resolving them.

Defensive attack When suppression operations are conducted outside the fire structure and feature the use of large-capacity fire streams placed between the fire and the exposures to prevent fire extension.

Demographics The characteristics of human populations and population segments, especially when used to identify consumer markets. Generally includes age, race, sex, income, education, and family status.

Demonstrative evidence Tangible items that can be identified by witnesses, such as incendiary devices and fire scene debris.

Demotion A reduction in rank, with a corresponding reduction in pay.

Dimensions Attribute or quality that can be described and measured during a promotional examination. There are between five and 15 dimensions that would be measured on a typical promotional examination. The six most common include Oral Communication, Written Communication, Problem Analysis, Judgment, Organizational Sensitivity, and Planning/Organizing.

Discipline A moral, mental, and physical state in which all ranks respond to the will of the leader. Also the guidelines that a department sets for fire fighters to work within.

Diversity A fire work force that reflects the community it serves.

Division of labor The production process in which each worker repeats one step over and over, achieving greater efficiencies in the use of time and knowledge; also, the formal assignment of authority and responsibility to job holders.

Division supervisor A supervisory position in charge of a geographical operation at the tactical level.

Divisions A supervisory level established to divide an incident into geographical areas of operations.

Documentary evidence Evidence in written form, such as reports, records, photographs, sketches, and witness statements.

Education The process of imparting knowledge or skill through systematic instruction.

Employee assistance program (EAP) An employee benefit that covers all or part of the cost for employees to receive counseling, referrals, and advice in dealing with stressful issues in their lives. These may include substance abuse, bereavement, marital problems, weight issues, or general wellness issues.

Environmental noise A physical or sociological condition that interferes with the message in the communication process.

Ethical behavior The fire officer makes decisions and demonstrates behavior that is consistent with the department's core values, mission statement, and value statements.

Evidence The documentary or oral statements and the material objects admissible as testimony in a court of law.

Expanded incident report narrative A report in which all company members submit a narrative on what they observed and activities they performed during an incident.

Expenditures The act of spending money for goods or services.

Failure analysis A logical, systematic examination of an item, component, assembly, or structure and its place and function within a system, conducted in order to identify and analyze the probability, causes, and consequences of potential and real failures.

Fair Labor Standards Act (FLSA) 1938 act that provides the minimum standards for both wages and overtime entitlement and spells out administrative procedures by which covered work time must be compensated. Public safety workers were added to FLSA coverage in 1986.

Federal Response Plan (FRP) The Federal Response Plan (for Public Law 93-288, as amended) describes the basic mechanisms and structures by which the federal government will mobilize resources and conduct activities to augment state and local disaster and emergency response efforts.

Finance/administration section A section of the Incident Management System that is responsible for the accounting and financial aspects of an incident, as well as any legal issues that may arise.

Fire analysis The process of determining the origin, cause, development, and responsibility, as well as the failure analysis of a fire or explosion.

Fire chief The highest ranking officer in charge of a fire department. The individual assigned the responsibility for management and control of all matters and concerns pertaining to the fire service organization.

Fire mark Historically, an identifying symbol on a building to let fire fighters know that the building was insured by a company that would pay them for extinguishing the fire.

Fire patterns Physical marks left on an object by the fire.

Fire prevention division or hazardous use permit A local government permit renewed annually after the fire prevention division performs a code compliance inspection. A permit is required if the process, storage, or occupancy activity creates a life-safety hazard. Restaurants with more than 50 seats, flammable liquid storage, and printing shops that use ammonia are three examples of occupancies that may require a permit.

Fire scene reconstruction The process of recreating the physical scene during fire scene analysis through the removal of debris and the replacement of contents or structural elements in their prefire position.

Fire tax district Special service district created to finance the fire protection of a designated district.

Fiscal year Twelve-month period for which an organization plans to use its funds. Local governments' fiscal years are from July 1 to June 30.

Followership Leaders can be effective only to the extent that followers are willing to accept their leadership.

Form of material What the material is being used for. For example, the form of material might be clothing.

Formal communication Represents an official fire department communication. The letter or report is presented on stationery with the fire department letterhead and generally is signed by a chief officer or headquarters staffperson.

Formal written reprimand An official negative supervisory action at the lowest level of the progressive disciplinary process.

Frame of reference An evaluation error in which the fire fighter is evaluated on the basis of the fire officer's personal standards instead of the classified job description standards.

General orders Short-term documents signed by the fire chief and lasting for a period of days to 1 year.

Good faith bargaining A legal requirement arising out of Section 8(d) of the National Labor Relations Act on both the union and the employer. Enforced by the National Labor Relations Board, the parties are required: "To bargain collectively . . . to meet at reasonable times and confer in good faith with respect to wages, hours, and other conditions of employment, or the negotiation of an agreement or any question arising thereunder, and the execution of a written contract incorporating any agreement reached if requested by either party, but such obligation not to compel either party to agree to a proposal or require the making of a concession. . . ."

Governmental Accounting Standards Board (GASB) The mission of the Governmental Accounting Standards Board is to establish and improve the standards of state and local governmental accounting and financial reporting that will result in useful information for users of financial reports and to guide and educate the public, including issuers, auditors, and users of those financial reports.

Grievance A dispute, claim, or complaint that any employee or group of employees may have in relation to the interpretation, application, and/or alleged violation of some provision of the labor agreement or personnel regulations.

Grievance procedure A formal structured process that is employed within an organization to resolve a grievance. In most cases, the grievance procedure is incorporated in the personnel rules or the labor agreement and specifies a series of steps that must be followed in order.

Group A supervisory level established to divide the incident into functional areas of operation.

Group supervisor A supervisory position in charge of a functional operation at the tactical level. This position reports to a branch director, the operations section chief, or the incident commander.

Halo and horn effect Evaluation error in which the fire officer takes one aspect of a fire fighter's job task and applies it to all aspects of work performance.

Hazards Any arrangement of materials and heat sources that presents the potential for harm, such as personal injury or ignition of combustibles.

Health and safety officer The member of the fire department assigned and authorized by the fire chief as the manager of the safety and health program.

Immediately dangerous to life and health (IDLH) Any condition that would do one or more of the following: (a) pose an immediate or delayed threat to life, (b) cause irreversible adverse health effects, or (c) interfere with an individual's ability to escape unaided from a hazardous environment.

Impasse Occurs when the parties have reached a deadlock in negotiations. Also described as the demarcation line between bargaining and negotiation. A declaration of an impasse brings in a state or federal negotiator that will start a fact-finding process and will lead to a binding arbitration resolution.

IMS general staff The group of incident managers composed of the operations section chief, planning section chief, logistics section chief, and finance/administration section chief.

In-basket A promotional examination component in which the candidate deals with correspondence and related items accumulated in a fire officer's in-basket.

Incident action plan (IAP) The objectives reflecting the overall incident strategy, tactics, risk management, and member safety that are developed by the incident commander. Incident action plans are updated throughout the incident.

Incident commander The person who is responsible for all decisions relating to the management of the incident and is in charge of the incident site.

Incident management system (IMS) A system that defines the roles and responsibilities to be assumed by personnel and the operating procedures to be used in the management and direction of emergency operations; the system is also referred to as an incident command system (ICS).

Incident safety officer An individual appointed to respond to or assigned at an incident scene by the incident commander to perform the duties and responsibilities specified in NFPA 1521, *Standard for Fire Department Safety Officer*. This individual can be the health and safety officer, or it can be a separate function.

Incident safety plan The strategies and tactics developed by the incident safety officer based upon the incident commander's incident action plan and the type of incident encountered.

Incident scene rehabilitation The tactical level management unit that provides for medical evaluation, treatment, monitoring, fluid and food replenishment, mental rest, and relief from climatic conditions of the incident.

Informal communications Internal memos, e-mails, instant messages, and computer-aided dispatch/mobile data terminal messages. Informal reports have a short life and are not archived as permanent records.

Interrogatory A series of formal written questions sent to the opposing side. The opposition must provide written answers under oath.

Investigation A systematic inquiry or examination.

Involuntary transfer or detail A disciplinary action in which a fire fighter is transferred or assigned to a less desirable or different work location or assignment.

Iowa State University Rate-of-Flow Formula Determine the gallons per minute flow for fire knockdown by dividing the cubic size of the area (length × width × height) by 100. This is the flow needed to be provided for 30 seconds to achieve fire control.

Job description A narrative summary of the scope of a job. Provides examples of the typical tasks.

Job instruction training A systematic four-step approach to training fire fighters in a basic job skill: (1) prepare the fire fighters to learn, (2) demonstrate how the job is done, (3) try them out by letting them do the job, and (4) gradually put them on their own.

Knowledge, skills, and abilities (KSAs) The traits required for every classified position within the municipality. Defined by a narrative job description and a technical class specification.

Leadership A complex process by which a person influences others to accomplish a mission, task, or objective and directs the organization in a way that makes it more cohesive and coherent.

Leading Guiding or directing in a course of action.

Liaison officer The incident commander's representative or a point of contact for representatives from outside agencies.

Lieutenant A fire officer who is usually responsible for a single fire company on a single shift; the first in line of officers.

Line-item budget Budget format where expenditures are identified in a categorized line-by-line format.

Lobby control officer The lobby control officer controls the entering and exiting of both civilians and fire fighters in the lobby. Also oversees the use of the elevators, operates the local building communication system, and assists in the control of the heating, ventilating, and air conditioning systems.

Logistics Section Responsible for providing facilities, services, and materials for the incident. Includes the communications, medical, and food units within the service branch, as well as the supply, facilities, and ground support units within the support branch. The logistics section chief is part of the general staff.

Logistics Section Chief A supervisory position that is responsible for providing supplies, services, facilities, and materials during the incident. The person in this position reports directly to the incident commander.

Loudermill hearing A predisciplinary conference that occurs before a suspension, demotion, or involuntary termination is issued. Term refers to a Supreme Court decision.

Masonry wall May consist of brick, stone, concrete block, terra cotta, tile, adobe, precast, or cast-in-place concrete.

Metropolitan (metro-sized) fire department A department with more than 400 fully paid fire fighters. The Metropolitan Fire Chiefs is a special interest group organized by the International Association of Fire Chiefs in 1965 to address the needs of large fire departments. Metro Chiefs are also a section of the National Fire Protection Association.

Mini/max codes Codes developed and adopted at the state level for either mandatory or optional enforcement by local governments; these codes cannot be amended by local governments.

Mistake An error or fault resulting from defective judgment, deficient knowledge, or carelessness. A misconception or misunderstanding.

Model codes Codes generally developed through the consensus process with the use of technical committees developed by a code-making organization.

National Fire Academy fire flow formula To provide a first-arriving fire officer the ability to quickly determine the needed fire flow. Determined on a per-floor calculation, the length is multiplied by the width, and the result is divided by three.

National Fire Incident Reporting System (NFIRS) A nationwide database at the National Fire Data Center under the U.S. Fire Administration that collects fire-related data in order to provide information on the national fire problem.

National Incident Management System (NIMS) A system that provides a consistent nationwide approach for federal, state, local, and tribal governments to work effectively and efficiently together to prepare for, prevent, respond to, and recover from domestic incidents, regardless of cause, size, or complexity.

Naturalistic decision making Describes how commanders make decisions in their natural environment.

Occupancy type The purpose for which a building or a portion thereof is used or intended to be used.

Offensive attack An advance into the fire building by fire fighters with hoselines or other extinguishing agents to overpower the fire.

Ongoing compliance inspection Inspection of an existing occupancy to observe the housekeeping and confirm that the built-in fire protection features, such as fire exit doors and sprinkler systems, are in working order.

Operations section Responsible for all tactical operations at the incident. In the national model, the operations section can be as large as five branches, 25 divisions/groups or sectors, or 125 single resources, task forces, or strike teams.

Operations section chief A supervisory position that is responsible for the management of all actions that are directly related to controlling the incident. This position reports directly to the incident commander.

Oral reprimand, warning, or admonishment The first level of negative discipline. Considered informal, it remains with the fire officer and is not part of the fire fighter's official record.

Ordinance A law of an authorized subdivision of a state, such as a city, county, or town.

Organizing Putting together into an orderly, functional, structured whole.

Performance log Informal record maintained by the fire officer listing fire fighter activities by date and with a brief description. Used to provide documentation for annual evaluations and special recognitions.

Personal bias An evaluation error that occurs when the evaluator's personal bias skews the evaluation such that the classified job knowledge, skills, and abilities are not appropriately evaluated.

Personal study journal A personal notebook to aid in scheduling and tracking a candidate's promotional preparation progress.

Personnel accountability report (PAR) Systematic method of accounting for all personnel at an emergency incident.

Personnel accountability system A method of tracking the identity, assignment, and location of fire fighters operating at an incident scene.

Planning Developing a scheme, program, or method that is worked out beforehand to accomplish an objective.

Planning Section Responsible for the collection, evaluation, dissemination, and use of information about the development of the incident and the status of resources. Includes the situation status, resource status, and documentation units as well as technical specialists. The planning section chief is part of the general staff.

Planning section chief A supervisory position that is responsible for the collection, evaluation, dissemination, and use of information relevant to the incident. This position reports directly to the incident commander.

Point of origin The exact physical location where a heat source and a fuel come in contact with each other and a fire begins.

Policies Formal statements that provide guidelines for present and future actions; policies often require personnel to make judgments.

Political action committee (PAC) Organizations formed by corporations, unions, and other interest groups who solicit campaign contributions from private individuals and who distribute these funds to political candidates.

Preincident plan A written document resulting from the gathering of general and detailed data to be used by responding personnel for determining the resources and actions necessary to mitigate anticipated emergencies at a specific facility.

Pretermination hearing An initial check to determine if there are reasonable grounds to believe that the charges against the employee are true and support the proposed termination.

Progressive negative discipline A process for dealing with job-related behavior that does not meet expected and communicated performance standards. Discipline increases from mild to more severe punishments if the problem is not corrected.

Public information officer A command staff position that is responsible for gathering and releasing incident information to the news media and other appropriate agencies. This position reports directly to the incident commander.

Pyrolysis The destructive distillation of organic compounds in an oxygen-free environment that converts the organic matter into gases, liquids, and char.

Rapid intervention crew (RIC) A minimum of two fully equipped personnel on site, in a ready state, for immediate rescue of disoriented, injured, lost, or trapped rescue personnel.

Recency An evaluation error in which the fire fighter is evaluated only on recent incidents rather than on the entire evaluation period.

Recognition primed decision making (RPD) Describes how commanders can recognize a plausible plan of action.

Recommendation report A decision document prepared by the fire officer for the senior staff. The goal is support for a decision or action.

Regulations Governmental orders written by a governmental agency in accordance with the statute or ordinance authorizing the agency to create the regulation. Regulations are not laws but have the force of law.

Rehabilitation The process of providing rest, rehydration, nourishment, and medical evaluation to members who are involved in extended or extreme incident scene operations.

Restrictive duty A temporary work assignment during an administrative investigation that isolates the fire fighter from the public and usually is an administrative assignment away from the fire station.

Revenues The income of a government from all sources appropriated for the payment of the public expenses.

Right-to-work A worker cannot be compelled, as a condition of employment, to join or not to join, or to pay dues to a labor union.

Risk management Identification and analysis of exposure to hazards, selection of appropriate risk management techniques to handle exposures, implementation of chosen techniques, and monitoring of results, with respect to the health and safety of members.

Risk/benefit analysis A decision made by a responder based on a hazard and situation assessment that weighs the risks likely to be taken against the benefits to be gained for taking those risks.

Rules and regulations Developed by various government or government-authorized organizations to implement a law that has been passed by a government body.

Safety officer The person who is responsible for all decisions relating to the management of the incident and is in charge of the incident site.

Safety unit A member or members assigned to assist the incident safety officer. The tactical level management unit that can be composed of the incident safety officer alone or with additional assistant safety officers assigned to assist in providing the level of safety supervision appropriate for the magnitude of the incident and the associated hazards.

Scientific management The breakdown of work tasks into constituent elements; the timing of each element based on repeated stopwatch studies; the fixing of piece rate compensation based on those studies; standardization of work tasks on detailed instruction cards; and generally, the systematic consolidation of the shop floor's brain work.

Sector Either a geographical or a functional assignment.

Service branch A major division within the logistics section of the IMS. Oversees the communications, medical, and food units.

Show and tell Activities in which the fire and rescue department is called to present a program at a school or public gathering on short notice, resulting in the closest fire company taking their apparatus to a location and presenting a program.

Situational awareness The process of evaluating the severity and consequences of an incident and communicating the results.

Source of ignition Devices or equipment that, because of their intended modes of use or operation, are capable of providing sufficient thermal energy to ignite flammable gas-air mixtures.

Span of control The maximum number of personnel or activities that can be effectively controlled by one individual (usually three to seven).

Special evaluation period A designated period of time when an employee is provided additional training to resolve a work performance/behavioral issue. The supervisor issues an evaluation at the end of the special evaluation period.

Spoils system Also known as the patronage system. The practice of making appointments to public office based on a personal relationship or affiliation rather than because of merit. A problem since 1850, the spoils system scandals of the New York City "Tweed Ring" and the Tammany Hall political machine (1865–1871) resulted in Congress passing the Pendleton Civil Service Reform Act of 1883.

Staging A specific function in which resources are assembled in an area at or near the incident scene to await instructions or assignments.

Stairwell support This group moves equipment and water supply hoselines up and down the stairwells. The stairwell support unit leader reports to the support branch director or the logistics section chief.

Standard operating procedures (SOPs) A written organizational directive that establishes or prescribes specific operational or administrative methods to be followed routinely for the performance of designated operations or actions.

Standpipe system An arrangement of piping, valves, hose connections, and allied equipment installed in a building or structure, with the hose connections located in such a manner that water can be discharged in streams or spray patterns through attached hose and nozzles, for the purpose of extinguishing a fire, thereby protecting a building or structure and its contents in addition to protecting the occupants. This is accomplished by means of connections to water supply systems or by means of pumps, tanks, and other equipment necessary to provide an adequate supply of water to the hose connections.

Strategic level Command level that entails the overall direction and goals of the incident.

Strike A concerted act by a group of employees, withholding their labor for the purposes of effecting a change in wages, hours, or working conditions.

Strike team Specified combinations of the same kind and type of resources, with common communications and a leader.

Strike team leader A supervisory position that is in charge of a group of similar resources.

Supervisor's report A form that is required by most state worker's compensation agencies that is completed by the immediate supervisor after an injury or property damage accident.

Supplemental budget Proposed increases in spending to provide additional services.

Support branch A major division within the logistics section of the IMS. Oversees the supply, facilities, and ground support units.

Suspension A negative disciplinary action that removes a fire fighter from the work location; he or she is generally not allowed to perform any fire department duties.

T-account A documentation system, similar to an accounting balance sheet, listing credits and debits, in which a single sheet form is used to list the assets on the left side and liabilities on the right side, appearing like a letter "T."

Tactical-level Command level in which objectives must be achieved to meet the strategic goals. The tactical-level supervisor or officer is responsible for completing assigned objectives.

Tactical worksheet A form that allows the incident commander to ensure all tactical issues are addressed and to diagram an incident with the location of resources on the diagram.

Task force Any combination of single resources assembled for a particular tactical need, with common communications and a leader.

Task force leader A supervisory position that is in charge of a group of dissimilar resources.

Task-level Command level in which specific tasks are assigned to companies; these tasks are geared toward meeting tactical-level requirements.

Termination The organization has determined that the member is unsuitable for continued employment.

Testimonial evidence Witnesses speaking under oath.

Trailers Materials used to spread the fire from one area of a structure to another.

Training The process of achieving proficiency through instruction and hands-on practice in the operation of equipment and systems that are expected to be used in the performance of assigned duties.

Transitional attack A situation in which an operation is changing or preparing to change.

Two-in-two-out rule OSHA Respiratory Regulation (29 CFR 1910.134) that requires a two-person team to operate within an environment that is immediately dangerous to life and health (IDLH) and a minimum of a two-person team to be available outside the IDLH atmosphere to remain capable of rapid rescue of the interior team.

Type of material What the material is made of. For example, the type of material might be cotton.

Unfair labor practices An employer or union practice forbidden by the National Labor Relations Board or state/local laws, subject to court appeal. It often involves the employer's efforts to avoid bargaining in good faith.

Unity of command The management concept that a subordinate should have only one direct supervisor, and a decision can be traced back through subordinates to the manager who originated it.

Use group Found in the building code classification system whereby buildings and structures are grouped together by their use and by the characteristics of their occupants.

Yellow dog contracts Pledges that employers required workers to sign indicating that they would not join a union as long as the company employed them. Declared unenforceable by the Norris-LaGuardia Act of 1932.

Abrashoff, D. Michael. *It's Your Ship. Management Techniques from the Best Damn Ship in the Navy.* New York, NY: Warner Books, Inc., 2002.

Aitchison, W. *The Rights of Fire fighters.* 3rd ed. Portland, OR: The Labor Relations Information System, 2001.

Barry, T. F. and Waterous, L. "Fire Loss Prevention and Emergency Organizations." *Fire Protection Handbook.* 19th ed. Quincy, MA: National Fire Protection Association, 2003.

Benoit, J. and Perkins, K. B. *Leading Career and Volunteer Firefighters: Searching for Buried Treasure.* Halifax, N. S., Canada: Henson College, Dalhousie University, 2001.

Bittel, L. R. and Newstrom, J. W. *What Every Supervisor Should Know.* 6th ed. New York, NY: McGraw-Hill, 1990.

Blake, R. R. and Mouton, J. S. *The Managerial Grid III: The Key to Leadership Excellence.* Revised ed. Houston, TX: Gulf Publishing, 1985.

Bowers, M. H. *Labor Relations in the Public Safety Services.* Chicago, IL: International Personnel Management Association, 1974.

Brannigan, F. L. *Building Construction for the Fire Service.* 3rd ed. Quincy, MA: National Fire Protection Association, 1992.

Brannigan, F. L. "Effect of Building Construction and Fire Protection Systems on Fire Fighter Safety." *Fire Protection Handbook.* 19th ed. Quincy, MA: National Fire Protection Association, 2003.

Brunacini, A. V. *Essentials of Fire Department Customer Service.* Stillwater, OK: International Fire Service Training Association, 1996.

Brunacini, A. V. *Fire Command: The Essentials of Local IMS.* 2nd ed. Phoenix, AZ: Heritage Publishers Inc, 2002.

Bryan, J. L. *Automatic Sprinkler and Standpipe Systems.* Quincy, MA: National Fire Protection Association, 1997.

Bugbee, P. *Man Against Fire: The Story of the National Fire Protection Association, 1896-1971.* Boston, MA: National Fire Protection Association, 1971.

Cantor, D. and Almond, L. *The Burning Issue: Research and Strategies for Reducing Arson* (No. 02FPD00404.). London, UK: Office of the Deputy Prime Minister, 2002.

Carson, W. G. and Klinker, R. L. *Fire Protection Systems: Inspection, Test and Maintenance Manual.* 3rd ed. Quincy, MA: National Fire Protection Association, 2000.

Cole, D. *The Incident Command System: A 25-Year Evaluation by California Practitioners.* Emmitsburg: MD, National Fire Academy Executive Fire Officer Research Report, 2002.

Coleman, R. J. "Are You Ready for the Spotlight, or Are You a Deer in the Headlights?" *Going for Gold: Pursuing and Assuming the Job of Fire Chief.* Albany, NY: Thompson Delmar Learning, 1999.

Corey, M. (Ed.). *Motive, Means and Opportunity: A Guide to Fire Investigation.* Princeton, NJ: American Re-Insurance Company, 1996.

Davis, L. and Colletti, D. *The Rural Firefighting Handbook.* Royersford, PA: Lyon's Publishing, 2002.

DeHaan, J. D. *Kirk's Fire Investigation.* 5th ed. New York, NY: Prentice-Hall, 2002.

Diamantes, D. *Fire Prevention: Inspection and Code Enforcement.* 2nd ed. Albany, NY: Thomson Delmar Learning, 2003.

Edwards, S. T. *Fire Service Personnel Management.* Upper Saddle River, NJ: Brady/Prentice-Hall Health, 2000.

Farr, R. R. and Sawyer, S. F. "Fire Prevention and Code Enforcement." *Fire Protection Handbook.* 19th ed. Quincy, MA: National Fire Protection Association, 2003.

Federal Emergency Management Agency. *Topical Fire Research Series: Highway Vehicle Fires.* Vol. 2, Issue 4. Washington, DC: Federal Emergency Management Agency, 2002.

Federal Emergency Response Agency. *Federal Response Plan.* Washington, D.C.: Federal Emergency Response Agency, April 1999.

FIRESCOPE. *Fire Service Field Operations Guide: ICS 420-1.* 30th ed. Rancho Cordova, CA: Governor's Office of Emergency Services, 2001.

Fried, E. *Fireground Tactics.* Chicago, IL: H. Marvin Ginn, 1972.

Gess, D. and Lutz, W. *Firestorm at Peshtigo: A Town, Its People, and the Deadliest Fire in American History.* New York, NY: Henry Holtz and Company, 2002.

Graham, D. H. *The Missing Protocol: A Legally Defensible Report.* Ashton, MD: Clemens Publishing, 1999.

Griffins, J. S. *Fire Department of New York: An Operational Reference.* 4th ed. Los Alamos, NM: James S. Griffin, 2002.

Hall Jr, J. R. *Children Playing With Fire* (Report). Quincy, MA: National Fire Protection Association, 2001.

Hall Jr, J. R. *Intentional Fires and Arson* (Report). Quincy, MA: National Fire Protection Association, 2003.

Hashagen, P. "New York City Fire Department History." *Fire Department City of New York.* Paducah, KY: Turner Publishing Company, 2000.

Heil, G., Bennis, W. and Stephens, D. C. *Douglas McGregor, Revisited: Managing the Human Side of the Enterprise.* New York, NY: John Wiley & Sons, 2000.

Helmreich, R. and A. Merritt. Culture at Work in Aviation and Medicine: National Organizations and Professional Influences. Aldershot: Ashgate, 1998.

Helmreich, R. "On Error Management: Lessons from Aviation." *British Medical Journal*, 320, 2000.

Helmreich, R. L., Merritt, A. C. and Wilhelm, J. A. "The Evolution of Crew Resource Management in Commercial Aviation." *Aerospace Crew Research Project.* Austin, TX: Department of Psychology, The University of Texas at Austin, August 2001.

Hershey, P., Blanchard, K. H. and Johnson, D. E. *Management of Organizational Behavior: Leading Human Resources.* 8th ed. New York, NY: Prentice Hall, 2000.

IFSTA. *Introduction to Fire Origin and Cause.* 2nd ed. Stillwater, OK: Fire Protection Publications, 1997.

Ingrassia, P. and White, J. B. *Comeback: The Fall and Rise of the American Automobile Industry.* New York, NY: Simon and Schuster, 1994.

Jones, J. C. "Investigating Fires." *The Fire Chief's Handbook.* 6th ed. Tulsa, OK: PennWell, 2003.

Kern, T. "Culture, Environment, and CRM." *Controlling Pilot Error Series 10.* New York, NY: McGraw-Hill, 2001.

Kimball, W. Y. *Fire Attack 1: Command Decisions and Company Operations.* Boston, MA: National Fire Protection Association, 1966.

Kimball, W. Y. *Fire Attack 2: Planning, Assigning, Operating.* Boston, MA: National Fire Protection Association, 1968.

Klaene, B. and Sanders, R. E. *Structural Fire Fighting.* Quincy, MA: National Fire Protection Association, 2000.

Klein, G., Calderwood, R. and Clinton-Cirocco, A. "Rapid Decision Making on the Fireground." *Proceedings of the Human Factors Society 30th Annual Meeting.* San Diego, CA: HFS, 1986.

Kouzes, J. M. and B. Z. Posner. *The Leadership Challenge.* San Francisco, CA: Jossey-Bass, 2003.

Krebs, D. R. and American Academy of Orthopaedic Surgeons. *When Violence Erupts: A Survival Guide for Emergency Responders.* Boston, MA: Jones and Bartlett Publishers, 2003.

Layman, L. *Fire Fighting Tactics.* Boston, MA: National Fire Protection Association, 1953.

Lewis, B. *Managing in a Cutback Environment.* Stillwater, OK: Fire Protection Publications, 2002.

Lopes III, B. F. "Office Management and Workflow." *The Fire Chief's Handbook.* 6th ed. Tulsa, OK: PennWell, 2003.

Lubnau, T. and R. Okray. *Crew Resource Management for the Fire Service.* Tulsa, OK: Penn Well, 2004.

Maher, P.T. and R.S. Michelson. *Preparing for Fire Service Assessment Centers.* Bellflower, CA: Fire Publications, Inc, 1992.

Maniscalco, P. M. and Christen, H. T. *Understanding Terrorism and Managing the Consequences.* Upper Saddle River, NJ: Pearson Education, Inc, 2002.

Maslow, A. H. *Maslow on Management.* New York, NY: John Wiley and Sons, 1998.

Matejka, M. *Fiery Struggle: Illinois Fire Fighters Build a Union, 1901–1985.* Chicago, IL: Illinois Labor History Society, 2002.

McGregor, D. *The Human Side of Enterprise.* New York, NY: McGraw-Hill Book Company, Inc, 1960.

Murtaugh, M. "Fire Department Promotional Tests." *A New Direction: New Testing Components, New Testing Formats.* Pearl River, NY: Fire Tech Promotional Courses, 1993.

National Transportation Safety Board. United Airlines, Inc., McDonnell-Douglas, DC-8-61, N8082U, Portland, Oregon, December 28, 1978. NTSB Report Number: AAR-79-07, adopted 6/17/1979.

NFSIMS. *Model Procedures Guide for Structural Firefighting.* 2nd ed. Stillwater, OK: Fire Protection Publications, 2000.

NIOSH Fire Fighter Fatality Investigation and Prevention Program. (2003). Report #2000-27. Penton Media Inc., Occupational Hazards. October 2002; Gale Group.

NIOSH (1999). ALERT: Request for assistance in preventing injuries and deaths of fire fighters due to structural collapse. Cincinnati, OH, US Department of Health and Human Services, Public Health Service, Centers for Disease Control and Prevention, National Institute for Occupational Safety and Health.

Norman, J. *Fire Officer's Handbook of Tactics.* 2nd ed. Saddle Brook, NJ: Fire Engineering Penn Well, 1998.

Northrup, H. R. and Morgan, J. D. "The Memphis Police and Firefighters Strikes of 1978: A Case Study." *Labor Law Journal* (January), 40–54, 1981.

Page, J. O. *Effective Company Command.* Alhambra, CA: Borden Publishing, 1973.

Peters, T. and Waterman, R. *In Search of Excellence: Lessons from America's Best-Run Companies.* New York, NY: HarperCollins, 1982.

Phoenix Fire Department. *The Phoenix Fire Department Way.* Phoenix, AZ: Phoenix Fire Department, 2002.

Poynter, D. *The Expert Witness Handbook: Tips and Techniques for the Litigation Consultant.* Santa Barbara, CA: Para Publishing, 1997.

Redsicker, D. R. and O'Connor, J. J. *Practical Fire and Arson Investigation.* 2nd ed. Boca Raton, FL: CRC Press, 1996.

Rhyne, C. *Police and Firefighters: The Law of Municipal Personnel Relations.* Washington, D. C.: The Law of Local Government Operations Project, 1982.

Schrag, Z. M. *Nineteen Nineteen: The Boston Police Strike in the Context of American Labor.* Social Studies Boston; Harvard: 122, 1992.

Smith, Dennis. *Report from Engine Co. 82.* New York, NY: Warner Books, 1972.

Taigman, M. and Dean, S. "Complaints" [Recorded by M. Taigman]. *On Secrets of Successful EMS Leaders: How to Get Results, Advance Your Career, and Improve Your Service* [audio-cassette]. Midlothian, VA: Sempai-Do, 1999.

Taylor, F. W. *The Principles of Scientific Management.* New York, NY: Harper, 1911.

TriData Corporation. *Firefighter Fatality Retrospective Study.* Report FA-220. Emmitsburg, MD: Federal Emergency Management Agency, April 2002.

United States Department of Homeland Security, Federal Emergency Management Agency, and United States Fire Administration. *Firefighter Fatalities in the United States in 2001.* FA-237. Washington, DC: 2002.

United States Department of Homeland Security, Federal Emergency Management Agency, and United States Fire Administration. *Firefighter Fatalities in the United States in 2002.* FA-260. Washington, DC: 2003.

United States Fire Administration. *America Burning Revisited.* Emmitsburg, MD: Federal Emergency Management Agency, 1990.

United States Fire Administration. *Fire in the United States, 1989-1998.* 12th ed. Emmitsburg, MD: Federal Emergency Management Agency, 2001.

United States Fire Administration. *Guide to Developing Effective Standard Operating Procedures for Fire and EMS Departments.* Emmitsburg, MD: United States Fire Administration, 1998.

United States Fire Administration. *Managing Company Tactical Operations.* Emmitsburg, MD: Federal Emergency Management Agency, 1991.

US Airways. *US Airways Crew Resource Management Lesson Plan, Phase I: Initial Training.* Arlington, VA: US Airways.

Watson, L. *Safer Communities: Towards Effective Arson Control* (The Report of the Arson Scoping Study). London, UK: Home Office Research, Development and Statistics Directorate, 1998.

Weiss, M. J. *The Clustered World: How We Live, What We Buy, and What it All Means about Who We Are.* Boston, MA: Little, Brown and Company, 2000.

Wiseman Jr, J. D. *The Iowa State Story: The Iowa Rate-of-Flow Formula and Other Contributions of Floyd W. (Bill) Nelson and Keith Royer to the Fire Service—1951 to 1988.* Stillwater, OK: Fire Protection Publications, 1998.

Ziskind, D. *One Thousand Strikes of Government Employees.* Morningside Heights, NY: Columbia University Press, 1940.

Index

Notes

Notes

Notes

Notes

Notes

Photo Credits

You are the Fire Officer Courtesy of Captain David Jackson, Saginaw Township Fire Department; Voices of Experience © Dennis Wetherhold, Jr.

Chapter 1
Opener © Keith Cullom; 1-1 © Chicago Historical Society #1CHi02808

Chapter 2
2-4 Courtesy of New Hanover County Fire Rescue, Wilmington, NC

Chapter 3
Opener © Glen E. Ellman; 3-5 Courtesy of Captain David Jackson, Saginaw Township Fire Department

Chapter 5
Opener Courtesy of Michael Connolly/FEMA; 5-1 Courtesy of IAFF; 5-2 © Tom Kableka, The Republican-American/AP Photo; 5-3 © George Nikitin/AP Photo

Chapter 6
6-1 Courtesy of NFPA; 6-2 © Tom Carter/911 Pictures; 6-4 Courtesy of Philips Medical Systems. All rights reserved; 6-5 Courtesy of NFPA

Chapter 7
Opener © Chris Mickal/911 Pictures; 7-6 Courtesy of David Hall; 7-9B Courtesy of Captain David Jackson, Saginaw Township Fire Department

Chapter 9
Opener Courtesy of District Chief Chris E. Mickal/New Orleans Fire Department, Photo Unit; 9-1 © Glen E. Ellman; 9-3 Courtesy of Michael Connolly/FEMA; 9-4 Courtesy of Captain David Jackson, Saginaw Township Fire Department

Chapter 10
Opener Courtesy of Captain David Jackson, Saginaw Township Fire Department; 10-1 Courtesy of the City of Hazelwood, Missouri; 10-2 © NFPA; 10-3A Courtesy of Captain David Jackson, Saginaw Township Fire Department; 10-4 © NFPA; 10-5 © Wilfredo Lee/AP Photo; 10-6 Courtesy of Captain David Jackson, Saginaw Township Fire Department; 10-8 Courtesy of Las Vegas Fire and Rescue

Chapter 12
12-1 Courtesy of Dennis Wetherhold, Jr.; 12-4 Courtesy of NFPA; 12-8 © John Foxx/Alamy Images; 12-9 © Joe Sohm/Alamy Images; 12-10 © Ken Hammon/USDA; 12-11 Courtesy of APA-The Engineered Wood Association; 12-14 © Photodisc; 12-15A © Tony Freeman/PhotoEdit

Chapter 14
14-3 © Glen E. Ellman; 14-5 Courtesy of Loudon County Fire-Rescue Services; 14-6 Courtesy of Claims Management Inc.

Chapter 15
Opener © Glen E. Ellman; 15-1 Courtesy of the USDA Forest Service; 15-4 Courtesy of the Fire and Rescue Department of Northern Virginia; 15-7 © Keith D. Cullom

Chapter 16
Opener © Glen E. Ellman; 16-2 © Steven Townsend/Code 3 Images

Chapter 17
Opener Courtesy of Captain David Jackson, Saginaw Township Fire Department; 17-1, 17-2, 17-3 Courtesy of Charles B. Hughes/Unified Investigations & Sciences, Inc.; 17-4 © Joe Giblin/AP Photo; 17-5 © Steve Hamblin/Alamy Images; 17-6 Courtesy of Captain David Jackson, Saginaw Township Fire Department

Chapter 18
Opener © Steven Townsend/Code 3 Images; 18-1 Courtesy of Doctor Robert Helmreich; 18-4 © Glen E. Ellman; 18-5 © 2003, Berta A. Daniels